An Introduction to Electronics

William G. Oldham   University of California, Berkeley

Steven E. Schwarz   University of California, Berkeley

Holt, Rinehart and Winston, Inc.

New York   Chicago   San Francisco   Atlanta
Dallas   Montreal   Toronto   London   Sydney

Preface

In any field of study the student's first course is an important, significant experience. Here he must acquire the fundamentals that he will use throughout his further studies. But also – and perhaps this is even more important – it is here that his attitudes toward the subject first are formed. The first course should teach the student basic principles; but it should also leave him exhilarated by what he has learned, and eager to know more.

In our belief, students feel the greatest desire to learn when a course seems "relevant," or "real" – when the subject is of obvious importance, and its applications are easily seen. What subject fits this description better than present-day electronics? Exciting developments are taking place, of which students are not unaware, and the excitement is contagious. Surely the first course should not waste this opportunity. We propose a first course that deals with the important aspects of electronics as it is practiced today. To be sure, this first course must also teach basic principles. But we believe that it is best to teach fundamentals while teaching topics of current interest. In brief, that is the basic idea of this book.

The Changing Scene

If the first course is to be well-suited to students' needs, it must be responsive to the important changes taking place in electronics. Some trends we feel of particular importance are the following.

(1) Computer science has burst upon the electrical engineering scene. This is a trend of great significance. Growing from almost nothing in the last decade, the significance of computer science, in terms of jobs and man-hours, is approaching parity with all the rest of electronics combined. This change in the direction of electronics progress requires a corresponding shift in educational emphasis. Some subjects are not quite as important as they were; analog circuits used to compose nearly all of electronics, whereas now they amount perhaps to half. In contrast, digital signals and systems are now of great importance, and deserve more than parenthetic mention in an introductory course.

(2) Integrated circuits continue to grow in importance. With entire circuits available in economical, subminiature form, there is a trend in engineering work away from discrete-component design and toward a building-block approach. The introductory course should explain the principles of building-block design, as well as the power of this approach and the wide range of possibilities that it offers.

(3) The electronics curriculum, as always, continues to become more crowded. There is constantly more and more to learn. Thus traditionally "junior" and "senior" subjects must be introduced earlier, to make room for the advanced courses, as the latter grow steadily more specialized and more numerous.

Today it makes a good deal of sense to introduce electronics in the sophomore year. For electrical engineering majors, a sophomore electronics course leaves them ready for intensive advanced work in their junior year, where, with so many courses to choose from, they can already begin to build up strength in areas of particular interest.

For computer sciences majors, of which we are due to see an increasing number, there may be little or no room in the upper-division curriculum for classical advanced electronics courses. It is desirable that they gain some early exposure to electronics, both to place their software studies in perspective and to provide background for specialized computer hardware courses later on. For students in other branches of engineering and science, electronic techniques are of great practical importance. A modern upper-division curriculum in mechanical engineering or physics, for example, may be expected to include several laboratories, where students make use of electronic measurements, simulation, and computation. To these students, electronics is a basic tool, which, like mathematics or basic physics, is best acquired early. Thus a flexible course that relies on minimum prerequisites and that may be taken as early as the sophomore year is ideal.

"Building Blocks" in Electronics

What is meant by the term "electronics" must be constantly redefined. In the 1960s, electronics dealt primarily with circuits that had a fairly small number of elements. To learn electronics was to learn to design and analyze circuits with up to a dozen or so transistors. More than likely, the circuits were analog circuits. But now that we have integrated circuits, the "circuits man" becomes a "systems man." He is called upon to deal with circuits containing perhaps thousands of transistors, such as, for example, those in a computer. Hence it is vital for students to learn a "building-block" approach to electronics. They must also learn about the functions of digital building blocks (almost unknown a decade ago) and the analysis and synthesis of digital systems. For analog work, they must learn to build up systems from operational amplifiers and other analog blocks. This is not to say that design of circuits with individual transistors is no longer important, or can be neglected. But if one is to give attention to topics in proportion to current importance, older topics must somehow make room for the new.

We have tried to solve this problem with a twofold approach. Part of the time the student's attention is directed to application of building blocks in systems; at other times attention is directed to the internal design of blocks. Since the internal design of blocks is a subject very much like discrete-circuit design, an acquaintance with the latter is thereby gained. The relative weights placed on the two areas — applications of blocks and their internal design — can be adjusted to suit the students' needs and interests. Many of the chapters of this book fall clearly into one or the other of the categories, and can be included or not as needed. Thus applications of blocks and internal circuit design can be dealt with in whatever depth may be desired. Broken down in these terms, the chapters sort out as follows:

General background: Chapters 1, 2, 4
Special background for internal circuit design: Chapters 5, 6, 7
Internal circuit design: Chapters 8, 10, 11, 13
Applications of building blocks: Chapters 3, 9, 12

Ways to Use This Book

We have designed this book to be used in an introductory electronics course which may be from one quarter to one year in length. It is organized so that when used in a one-quarter or one-semester course, the instructor has a good deal of flexibility in his choice of topics. A number of suggested courses are discussed in the *teaching manual,* available on request. Assuming that the introductory chapters 1, 2, and 3 have been covered, the following prerequisite relationships among the later chapters exist:

Chapter	Prerequisite
4	–
5	–
6	5
7	–
8	6
9	–
10	6
11	10
12	–
13	5

In teaching a general one-quarter course, we have on occasion taken the some-what radical step of omitting Chapter 4, on phasors. (Digital is here. Phasors are not as vital as they once were!) The book is specifically designed with this possibility in mind. If Chapter 4 is omitted there is time, even in a one-quarter course, for some work in digital and analog systems. (The phasor techniques of Chapter 4 are used in Chapters 10 and 12; however, phasor analysis is always contained in separate sections at the end of these chapters, so that a knowledge of phasors is optional.)

With regard to previous courses needed as prerequisites, the student should be acquainted with differential calculus and (if Chapter 4 is to be covered) with complex numbers. A college-level course in electrical physics is desirable but not absolutely essential; such a course may be taken concurrently.

Some Apologies

We know, of course, that some compromises have been required. An introductory course at the sophomore level cannot take the place of a one-year junior series. The student who wants a deeper knowledge must go further — we hope to help him want to. On the other hand, there are many subjects that we have not attempted to cover. Our feeling here is that the more subjects are covered, the less is said about each. There is probably some minimum amount that is worth saying about a given subject; rather than say less, we have in many cases left a subject for another time. Our efforts have been directed toward trying to do a complete job of treating the subjects selected, eliminating others so that as thorough a job as possible could be done on those remaining.

There are no doubt a few exceptions. Once or twice we could not resist a few words on an extraneous topic. We did this because we thought the topic was interesting, and hoped that students would think so too.

Acknowledgments

In writing this book we have been influenced and helped by many of our colleagues. Professors D. O. Pederson and J. R. Whinnery have given us the benefit of their advice on numerous occasions, and through their wide-ranging knowledge of electronics as an evolving discipline, have shaped our way of thinking. Professors A. C. English, T. E. Everhart, A. J. Lichtenberg, and Charles Turner, among others, have taught from these notes, and made valuable suggestions for their improvement. Dr. William Duxler read over the entire manuscript, and corrected numerous errors. Assistance and hospitality were provided us (WGO and SES, respectively) by the Institute of Electronics of the Technical University, Munich, and by the IBM Corporation Research Laboratory in Zurich, where parts of this book were written. Our discussions with the faculty of nearby community colleges have helped us, particularly in our efforts to coordinate our course with theirs.

We have two of the world's really great wives, Nancy and Janet, who typed, numbered, pasted, and put up with more than anyone could reasonably expect.

And lastly, we wish to thank our students here at Berkeley, who have always let us know when we needed to do better.

William G. Oldham
Steven E. Schwarz

Berkeley, California
January 1, 1972

Contents

Preface   iii

Foreword   What is Electronics?   x

1. Principles of Electric Circuits   1

1.1 Information and Signals   1
1.2 Electrical Definitions and Principles   4
1.3 Circuit Elements   17
1.4 Node and Loop Analysis   35
1.5 Voltage and Current Dividers   43
    Summary   45
    References   46
    Problems   46

2. Techniques of Circuit Analysis   53

2.1 Equivalent Circuits   54
2.2 Nonlinear Circuit Elements   68
2.3 Natural Response and Forced Response   74
2.4 Superposition   90
    Summary   92
    References   93
    Problems   94

3. Active Circuits and Building Blocks   99

3.1 Dependent Voltage and Current Sources   101
3.2 Linear Models for Active Devices and Blocks   105
3.3 Building Blocks for Analog Systems   109
3.4 Building Blocks for Digital Systems   114
    Summary   124
    References   125
    Problems   125

4. Sinusoidal Analysis and Phasors   131

4.1 Properties of Sinusoids   132
4.2 Phasors   135
4.3 Impedance   148
4.4 Applications of Sinusoidal Analysis   159
    Summary   171
    References   172
    Problems   173

5. Properties and Applications of Diodes   179

5.1 Electrons and Holes   180
5.2 *p n* Junctions   187
5.3 The Diode as a Circuit Element   196
5.4 Rectifier Circuits   207
5.5 Diode Logic Circuits   214
   Summary   219
   References   220
   Problems   221

6. Junction Transistors   227

6.1 Transistor Principles   228
6.2 Transistor Operation in the Active Mode   233
6.3 Large Signal Operation   243
6.4 Transistor Characteristics: Comparison with Theory   257
   Summary   263
   References   263
   Problems   264

7. Integrated Circuit Technology   268

7.1 Modern Semiconductor Technology   275
7.2 Integrated Circuits   292
   Summary   313
   References   313
   Problems   314

8. Introduction to Digital Circuits   317

8.1 The Transistor Inverter   318
8.2 Diode-Transistor Logic   329
8.3 The Flip-Flop   350
   Summary   354
   References   355
   Problems   355

9. Analysis and Synthesis of Digital Systems   359

9.1 Switching Algebra   361
9.2 Analysis and Synthesis of Combinational Circuits   379
9.3 Characteristics of Basic Sequential Circuits   391
9.4 Analysis and Synthesis of Flip-Flop Circuits   395
   Summary   417
   References   418
   Problems   418

10. Introduction to Amplifiers   424

10.1 Biasing Circuits   428
10.2 The Small-Signal Transistor Model   431
10.3 Common-Emitter Amplifier Circuits   438
10.4 Multistage Common-Emitter Amplifiers   446
10.5 Improved Transistor Models   448

10.6 Frequency Response of Amplifier Circuits   451
      Summary   462
      References   463
      Problems   464

11. **Design Techniques for Linear Circuits   469**

11.1 Differential Amplifiers   473
11.2 Output Circuits   483
11.3 Interstage Coupling   488
11.4 The Complete Amplifier   490
      Summary   493
      References   494
      Problems   495

12. **Operational Amplifiers and Their Applications   498**

12.1 Characteristics of Operational Amplifiers   502
12.2 Operational Amplifier Circuits   508
12.3 Design Considerations in Op-Amp Circuits   525
      Summary   541
      References   542
      Problems   542

13. **MOS Transistors and Integrated Circuits   547**

13.1 Characteristics of MOS Transistors   554
13.2 MOS Digital Circuits   563
13.3 MOS Integrated Circuits   584
      Summary   588
      References   590
      Problems   590

Appendix I The Binary Number System   594

      Summary   599
      References   599
      Problems   600

Appendix II An Elementary Theory of *pn* Junctions   602

II.1 Current Flow in Semiconductors   602
II.2 *pn* Junction Theory   611
      Summary   624
      References   624
      Problems   624

Foreword

What is Electronics?

At one time it might have been said that electrical engineering dealt with electrical lighting and power. Today, when the field includes such subjects as transistor circuits, systems theory, and computers, one might well ask, just what *is* the unifying thread of the entire field? A reasonably good reply might be that most of today's electrical engineering deals with the handling, communication, and storage of *information.* The field known as *electronics* is a major subdivision of electrical engineering, dealing with the design and application of electrical apparatus for the processing of information.*

Electronics is a large field, with relationships to both applied mathematics and applied physics. Its boundaries are not clearly defined. The most physics-oriented work in electronics is that dealing with the principles of electronic devices, and resembles solid state, quantum, or plasma physics. The most mathematics-oriented work in electronics deals with optimal design of circuits and logical design of computers; it is closely allied to the mathematical fields of topology, algebra, and logic. At the center of the field of electronics is the *electronic circuit,* a connection of circuit elements such as resistors, capacitors, and transistors, designed to do a certain task. In this book such outlying subjects as device physics (Chapter 5) and computer mathematics (Chapter 9) are touched upon, in order to put the field in perspective. The principal emphasis, however, is on electronic circuits. Chapters 1 through 4 deal with the basic principles of circuits and with techniques for their analysis. Chapters 5 through 7 introduce the semiconductor devices – diodes, transistors, and integrated circuits – that are at the heart of electronics as we know it today. Chapters 8 through 13 then deal with the analysis, design and applications of complete electronic circuits.

For many years electronics has been one of the most exciting and challenging fields a student could choose to study. This is, if anything, even more true today than in the past. We are in a period of tremendous progress, thanks largely to the development of integrated circuits and of the digital computer. The techniques available to the electronics engineer are now vastly more powerful than they have ever been before. Yet the possibilities opened up by the new advances have only begun to be explored. In the next decade we may expect to see safe electrically-controlled automobiles, video telephones, and handy computers supplementing man's intelligence for every conceivable purpose. The advances in the decade after that are probably beyond our ability to imagine. But whatever those advances may be, you, who are today beginning a study of electronics, will have a chance to make them come about.

---

* Historically, electronics received its name from the fact that vacuum tube operation depends on controlled motion of electrons in a vacuum. Now vacuum tubes are no longer of central importance, and the scope of the field has greatly expanded. Thus the term "electronics" has undergone a considerable change of meaning.

Chapter 1

Principles of Electric Circuits

To begin our study of electronics, we shall consider its most characteristic structure, the electric circuit. In this chapter we shall be concerned with basic definitions, circuit principles, and the most fundamental techniques of mathematical circuit analysis.

It is reasonable to begin the discussion by explaining just why electrical circuits are of interest. What, for example, is the relationship between the handling of information and the flow of currents in a circuit? This question is addressed in Section 1.1; here we introduce the electrical representations of information, which are known as *signals*. Then with this as motivation, we go on in Section 1.2 to review the physics of electricity and electric circuits. Section 1.3 deals with circuit *elements,* the component parts of which circuits are made. Then we are ready to proceed, in Sections 1.4 and 1.5, with the subject of *circuit analysis*. This term refers to a family of mathematical techniques, by which the operation of circuits can be predicted and described.

1.1 Information and Signals

A concept of fundamental importance is that of *electrical representation of information*. A moment of reflection reveals that the word "information" refers to thoughts that are represented in some physical form. For example, the words on a printed page are marks which represent thoughts, and amount to a form of information storage. An electromechanical form of information storage is a tape recording. Even a simple electrical capacitor could serve as a simple information-storage element: one could say that when charged it means "yes" and when uncharged it means "no," or that when charged it means "one" and when uncharged, "zero."

As another example, let us imagine a temperature detector, which develops a voltage that is proportional to its temperature. Suppose two long wires are attached to this detector. A distant observer can measure the voltage between his ends of the wires, and learn the temperature where the detector is located. The time-varying voltage containing the information is called a *signal*.

Two particularly important types of signals may be distinguished. The simple case just mentioned, involving a voltage proportional to temperature, gives rise to what is called an *analog signal*. In the case of analog signals an electrical quantity, such as voltage, is a continuous function of some physical quantity, such as temperature, the value of which is to be transmitted. A common example of an analog signal is the voltage generated by a microphone; this voltage is proportional to the acoustic pressure to which the microphone is subjected.

Another important kind of signal is the *digital signal*. Here the information to be transmitted consists of *numbers* (or is converted to numerical form) and the *numbers* are then transmitted. A familiar example of the transmission of digital information is found in a telephone dialing system. Here a sequence of pulses ("clicks" or, in modern systems, frequency-coded "beeps") is used to transmit a multidigit number to an electronic switchboard. Digital signals are widely used in modern computers, and therefore are rapidly gaining in importance.

The same information can always be represented in either digital or analog form. For instance, let us imagine that temperature information is to be transmitted in analog form. Our detector supplies us with a voltage proportional to the temperature. Suppose that a temperature of 3.2°K gives 0.32 V, 3.5°K gives 0.35 V, and so forth. The time-varying voltage is then the analog signal, as shown in Figure 1.1(*a*). On the other hand, suppose the information is to be transmitted in digital form. Then the analog signal from the temperature detector would be sent to an analog-to-digital converter. One way this device could work would be to respond to an analog signal of 0.32 V by emitting a pulse of voltage of magnitude 3 V, followed by a pulse of magnitude 2 V [Figure 1.1(*b*)]. The sequence of pulses may be repeated as often as new temperature information is needed.

The electronic circuits that are used in processing information tend to be of one type, when the information is represented in analog form, and of quite a different type, when the information is represented in digital form. For

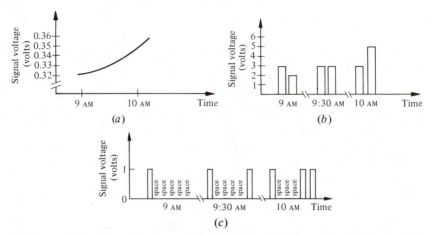

Figure 1.1   Comparison of analog and digital signals. The information being transmitted is that the temperature changed from 3.2° at 9 AM to 3.3° at 9:30 AM to 3.5° at 10 AM. Figure (*a*) shows an analog signal, (*b*) shows a digital signal, and (*c*) shows a more common type of digital signal, in which the numbers transmitted are binary rather than decimal. A "space" is an interval of time where the absence of voltage indicates a "zero" is being sent. The digital signals could be sent oftener if more frequent readings of temperature were needed.

analog signals, continuous ranges of voltages are used; hence circuits for handling analog signals are sometimes called *continuous-state circuits,* or simply analog circuits. On the other hand, the voltages which represent the digits of a digital signal must fall into discrete ranges. In the case of Figure 1.1(*b*), the height of each pulse must be within a specific range in order to signify a certain number. For instance, any pulse in the range of 1.8 to 2.2 V could be regarded as signifying 2; any pulse between 2.8 and 3.2 V could signify 3; and so forth. There would have to be 10 different voltage ranges to correspond with numbers from 1 to 10, and ten different kinds of pulses would be possible. Circuits intended to handle signals of this kind are sometimes called *discrete-state circuits,* or simply *digital circuits.*

The simplest kind of digital circuit is one with only *two* states. Such a circuit can be built of switches, or from circuits which act like switches. The two states of a switch are "off" and "on"; each state corresponds to a digit, which is 0 or 1. The digital signal illustrated by Figure 1.1(*b*) is essentially *decimal,* because there are ten states; each pulse belongs to one of ten different kinds. However, in practice *two-state* systems are almost always used; each pulse then belongs to one of only two different kinds, which may be called "off" and "on," or "zero" and "one."

The logical way to handle digital information in a two-state circuit is to use the binary number system, which is reviewed in Appendix I. In the binary number system all digits are either 0 or 1: The number 32 is represented by

100000, 33 is 100001, and 35 is 100011. A digital signal containing, in binary form, the same information as in Figures 1.1(*a*) and 1.1(*b*) is shown in Figure 1.1(*c*).

The subject known as electronics deals primarily with the generation, processing, and transmission of electrical signals. For example, a radio receiver is an electronic device which operates on the weak electrical signal generated in an antenna by a distant radio station. The receiver processes the signal from the antenna, producing another signal (containing the same information) which the loudspeaker can convert to audible sound. Similarly, a computer transforms the digital signals from a keyboard or card reader into new electrical signals, which represent the answer to the problem.

A radio receiver is a complex piece of apparatus, and a computer is orders of magnitude more complex. Fortunately, even a very complex apparatus is usually composed of simpler subunits. One may call an entire signal processing apparatus a *system*. The system is composed of *circuits,* and for ease of understanding these may be regarded as composed of *subcircuits*. Generally, the best way to understand the operation of a system is to analyze it in terms of circuits or subcircuits of manageable size.

Electronics, then, deals with the processing of electrical signals. We must now focus our attention on our basic tool, the electric circuit, by means of which signal processing is brought about.

1.2   Electrical Definitions and Principles

An understanding of electronic circuits must be built upon a basic knowledge of electrical physics. One must be familiar with electrical quantities, such as *charge, current,* and *voltage,* which are used in describing the operation of circuits. Moreover, the concept of *electric circuit* itself must be understood, as well as the physical laws by which circuits are governed. The purpose of this section is to summarize the basic electrical quantities and fundamentals of electric circuits. It is expected that to most readers, the material of this section will not be entirely new. Nonetheless, perusal of this section will serve as review, and will also familiarize the reader with our definitions, notation, and point of view.

Electrical Quantities   The branch of physics known as *electricity* seeks to describe electrical phenomena in terms of mathematical equations. The quantities in these equations must describe specifically electrical conditions, and hence are called electrical quantities. To understand the role of these quantities in describing electricity, it is only necessary to consider the comparable quantities in other branches of physics. Imagine trying to describe the motion of a particle without using the words "position" or "velocity"; imagine trying to describe, without using the word "temperature," conditions in an oven!

As with other physical quantities, electrical quantities must be expressed in a consistent system of *units*. The system of units to be used throughout this book is known as the rationalized MKS system. In this system the units of length, mass, and time are, respectively, *m*eter, *k*ilogram, and *s*econd; hence the abbreviation MKS. (The term "rationalized" refers to a choice of certain numerical factors, so that some equations are simplified; however, this need not concern us further here.) The rationalized MKS system is used almost universally in electrical engineering. A summary of this system of units is given in Table 1.1.

Table 1.1 The MKS System of Units

Quantity	Symbol	Unit	Abbreviation of Unit
Length	l	meter	m
Mass	m	kilogram	kg
Time	t	second	sec
Energy	ϵ	joule	J
Force	F	newton	N
Power	P	watt	W
Charge	Q	coulomb	C
Current	I or i	ampere	A
Potential (or voltage)	V or v	volt	V
Electric field	E	volt per meter	V/m
Resistance	R	ohm	Ω
Capacitance	C	farad	F
Inductance	L	henry	H

NOTE: Other units may be derived from the units in the table by means of decimal-multiplier prefixes, as follows:

Prefix	Abbreviation	Multiplies Unit by
pico-	p	10^{-12}
nano-	n	10^{-9}
micro-	μ	10^{-6}
milli-	m	10^{-3}
kilo-	k	10^{3}
mega-	M	10^{6}
giga-	G	10^{9}

For example, 1 nanosecond (nsec) is 10^{-9} second; 1 kilovolt (kV) is 1000 V. These units are often more convenient than the basic MKS units; for instance, it is handier to write "1.6 μA" than "0.0000016 A." However, it should be remembered that almost all equations in this book are true statements only when all quantities are expressed in the basic MKS units. Numerical errors can occur if the prefixed units are carelessly used instead.

The three specifically electrical quantities that are most often encountered are charge, current, and voltage. We shall now review the meaning of these terms in some detail.

Charge   Each kind of atomic particle carries a certain amount of *charge*. In the rationalized MKS system of units, charge is measured in *coulombs* (C). Experimentally it is found that the neutron has a charge of zero. The proton has a charge of $+1.6 \times 10^{-19}$ coulombs; this amount of charge is given the symbol q. The electron has a charge which is exactly equal to that of the proton but opposite in sign; that is, the charge of an electron is -1.6×10^{-19} C, or $-q$. It is also found experimentally that the charge q is the smallest unit of charge that exists in nature. Larger charges occur when many charged particles are collected together. Two particles with charges of the same sign (that is, both positive or both negative) repel each other, but particles with opposite charge attract.

An atom consists of a heavy nucleus surrounded by one or more orbiting electrons. Atoms of different elements have different numbers of protons in their nuclei. The number of protons in the nucleus is characteristic of the element; this number is known as the atomic number (Z). For example, atoms of hydrogen (for which $Z = 1$) have one proton in each of their nuclei; silicon nuclei have 14 protons. (Except in the case of hydrogen, atomic nuclei also contain neutrons. However, since neutrons are uncharged, they need not concern us further here.) Around the nucleus move Z electrons, bound to the vicinity of the nucleus by the attraction of the protons with their opposite charge. Thus an atom contains a positive charge of value Zq in its nucleus and a negative charge of value $-Zq$ in its electron cloud; hence the charge of a complete atom is 0.

It is possible to remove an electron from its atom. The separation requires a certain amount of energy (how much depends on the kind of atom) known as the *ionization energy*. The electron which has been removed is then known as a free electron, and of course it still bears a charge $-q$. The remainder of the atom which is left behind is then known as an *ionized atom,* or *ion.* When an atom loses *one* electron, it is said to be *singly ionized*. A singly ionized atom has one more proton than it has electrons, and therefore it has a charge $+q$. It is possible for free electrons and ions to be present simultaneously in a material. If in a given volume there are N_1 singly charged ions and N_2 free electrons, the total charge in that volume is given by $Q = N_1 q + N_2(-q) = q(N_1 - N_2)$.

In most cases of interest large numbers of free electrons or ions are present, and those are distributed throughout a volume. Thus it is meaningful to speak of *charge density* (ρ), which is defined as being the charge per unit volume. In MKS units charge density is expressed in coulombs per cubic meter (C/m³).

EXAMPLE 1.1

An impure silicon crystal contains equal densities of singly charged positive ions and free electrons throughout its volume, except for a small region whose dimensions are 5×10^{-4} cm by 1 mm by 1 mm, where only the ions are present. The density of ions everywhere in the crystal is $10^{22}/m^3$. Find the total charge of the crystal.

SOLUTION

In the region where both ions and free electrons are present, their opposite charges cancel, and the charge density is 0. Hence the total charge of the crystal arises from the region containing ions only. The charge density here is given by the number of ions per unit volume times the charge of each ion; the latter quantity is $+q$, or 1.6×10^{-19} C. Thus

$$\rho = (10^{22}) \cdot (1.6 \times 10^{-19}) \ C/m^3 = 1.6 \times 10^3 \ C/m^3$$

The total charge Q is equal to the charge density (charge per unit volume) ρ times the volume of the charged region. In computing the latter we must convert the given dimensions to MKS units:

$$V = (5 \times 10^{-6} \ m) \cdot (10^{-3} \ m) \cdot (10^{-3} \ m) = 5 \times 10^{-12} \ m^3$$

Thus

$$Q = \rho V = (1.6 \times 10^3 \ C/m^3) \cdot (5 \times 10^{-12} \ m^3) = 8 \times 10^{-9} \ C \qquad \blacksquare^1$$

Current A flow of electrical charge from one region to another is called a *current*. According to convention, a current from region A to region B is said to have a value which is positive in sign, if the flow of charge is such as to make the total charge of region A more negative and that of region B more positive. Figure 1.2(a) illustrates a case in which positive charges are moving from A to B. Because such a flow tends to increase the positive charge in region B, the current from A to B is positive. In Figure 1.2(b) negative charges are moving from A to B. This flow tends to reduce the net positive charge in region B, because the negative charge cancels whatever positive charge is there. Thus in Figure 1.2(b), the value of the current flowing from A to B is negative.

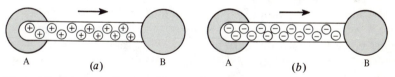

A   (a)   B   A   (b)   B

Figure 1.2   Schematic diagrams illustrating current flow. In (a) the current flowing from region A to region B is positive. In (b) the current flowing from A to B is negative.

The mathematical value of the current flowing from A to B is defined as the amount of positive charge that moves from A to B per unit time. In the MKS system of units, the unit of current is the *ampere;* a current of 1 A is, by definition, a flow of charge at the rate of 1 C/sec. In Figure 1.2(a) let us say that $+Q$ C move through the "pipeline" from A to B over a time of T sec. Then the current, in the direction A to B, is given by $I = Q/T$.

---

[1] The symbol $\blacksquare$ denotes the end of the example.

Since a current that tends to increase the positive charge in region B is called a positive current directed from A to B, a current that tends to reduce the positive charge in B is a negative current directed from A to B. In the case of the previous paragraph, a charge $+Q$ flowed from A to B in time T. Such a flow tends to reduce the positive charge in A, and hence the flow can also be called a negative current flowing from B to A. In general, it is equivalent to say either that a current flows in a given direction and has the value I_1 or that a current flows in the opposite direction and has the value $-I_1$. Note that one cannot describe a current simply by stating its value, any more than one can describe the motion of an auto by simply stating its speed in miles per hour; the direction of the current must be specified as well.

EXAMPLE 1.2

In Figure 1.2(b) 10^{13} electrons move from region A to region B each second. What is the current in the direction from B to A?

SOLUTION

The charge that moves from A to B each second is equal to 10^{13} times the electronic charge. Recalling that the charge of an electron is -1.6×10^{-19} C, we find that the current flowing from A to B is given by

$$I_{A \to B} = 10^{13} \cdot (-1.6 \times 10^{-19}) \text{ A} = -1.6 \times 10^{-6} \text{ A}$$
$$= -1.6 \ \mu\text{A}$$

where the symbol μA stands for microamperes (one microampere $= 10^{-6}$ A). However, we are asked to find the current in the direction B to A. This is the negative of the current in the direction A to B; thus

$$I_{B \to A} = +1.6 \ \mu\text{A} \qquad \blacksquare$$

In Figure 1.2 the current is shown flowing through a tubelike structure. Let the cross-sectional area of this tube be S. Then we can define the *current density J* as being the current which flows per unit of cross-sectional area. If the total current which flows from A to B is I, then the current density is $J = I/S$ A/m². The direction of the current density is the same as that of the current.

EXAMPLE 1.3

Inside a silicon crystal, particles with charge $+q$ are present with a density of 10^{22}/m³. All the charged particles are moving through the crystal in the $+z$ direction at a speed of 10 m/sec. Find the current density.

SOLUTION

Since the motion carries positive charge in the $+z$ direction, that is the direction of the current density. We must calculate how much charge passes through the plane $z = $ constant, per unit area, per unit time. To do this, consider the diagram on page 9. The particles are all moving parallel to one another, toward the plane. Consider those which pass through the portion of the plane indicated in the diagram. The area of this surface element is S. All particles destined to pass through S are located in a "tube" perpendicular to the plane, whose cross-sectional area is S. We may think of the contents of this tube as moving steadily toward the plane at speed v, where v is the particle velocity. The volume which passes through the tube and through the surface element per unit time is equal to Sv. The charge which moves through the element of surface is equal to the charge per unit volume ρ times the volume that moves through the tube, past the surface element. Therefore the charge moving through the surface element each second is $Sv\rho$. We are asked to find the current density, or current per unit area. This is obtained by dividing the current through the surface element by its area S. This gives

$$J = \frac{Sv\rho}{S} = v\rho$$

In the present case the charge density is

$$10^{22}\ q/\text{m}^3 = 10^{22}\ (1.6 \times 10^{-19}\ \text{C})/\text{m}^3 = 1.6 \times 10^3\ \text{C}/\text{m}^3$$

Thus $J = v\rho = (10\ \text{m/sec})(1.6 \times 10^3\ \text{C}/\text{m}^3) = 1.6 \times 10^4\ \text{C}/\text{m}^2\text{sec} = 1.6 \times 10^4\ \text{A}/\text{m}^2$.

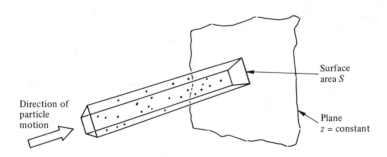

Surface area S

Direction of particle motion

Plane z = constant

Voltage   To each point in space there can be assigned a value of *electrical potential,* or *voltage.* The physical significance of the electrical potential at a given point is that it is proportional to the *energy* of a charge when the charge is located at that point. In fact, if a charge of magnitude Q is located at a place where the potential is V, the energy ϵ of the charge is given by

$$\boxed{\epsilon = QV}^2 \tag{1.1}$$

In the rationalized MKS system, the unit of potential is the *volt* (V). The unit of energy is the *joule* (J).

If no forces other than electrical forces act on a charge, it will tend to move from a place where its energy is higher to a place where its energy is lower. According to Equation (1.1), a positive charge tends to move to a place

---

[2] A box drawn around an equation indicates that it is important or will be used frequently.

of lower voltage. However, the energy of a negative charge decreases when it moves from a lower voltage to a higher voltage. Thus if only electrical forces act, negative charges tend to move from places where the potential is smaller toward places where it is larger.

Force, Energy, and Power   In order to describe the interaction of the electrical potential with charged particles, the notion of *electric field* is useful. The electric field, like current and current density, possesses direction as well as magnitude. The direction of the electric field is always from the region of higher potential toward the region of lower potential. Its magnitude is obtained from the potential by differentiation. For example, if the potential is a function of the coordinate x only, then the electric field E is given by

$$E = -\frac{dV(x)}{dx} \tag{1.2}$$

assuming that E is regarded as positive when oriented in the $+x$ direction. The units of electric field in the MKS system are volts per meter.

The force acting on a charge Q, when the latter is placed in an electric field E, is given by

$$F = QE \tag{1.3}$$

The unit of force in the MKS system is called the *newton* (N). The direction of the force exerted on a positive charge is the same as the direction of the field, and therefore is directed from regions of higher potential toward regions of lower potential. A negative charge experiences a force in the opposite direction from that of the field.

EXAMPLE 1.4

The voltage between $x = 0$ and $x = 1$ m is given by $V(x) = 10x$ V. Find the magnitude and direction of the force acting on a charge $+q$ placed between $x = 0$ and $x = 1$.

SOLUTION

From Equation (1.2) the electric field in the $+x$ direction is given by

$$E = -\frac{dV(x)}{dx} = -\frac{d}{dx}(10x) = -10 \text{ V/m}$$

From Equation (1.3) we have

$$F = qE = (1.6 \times 10^{-19} \text{ C})(-10 \text{ V/m}) = -1.6 \times 10^{-18} \text{ N}$$

The minus sign signifies that the force is actually in the $-x$ direction. This is as we expect, since we know that the electric field is directed from regions of higher to regions of lower potential, and that a positive charge experiences a force in the same direction as the field. ∎

It has already been mentioned that the energy of a particle with charge Q, located at a point where the potential is V, is given by $\epsilon = QV$. However, energy is a relative quantity; only differences in energy actually have physical significance. One can designate any condition as being that of zero energy, and measure other energies relative to the one designated as zero. The physical significance of the potential at a point is that it equals the energy of a unit charge placed at that point. Therefore the same reasoning applies to voltage as to energy: *The choice of which place is at zero potential is an arbitrary one.* Only differences in potential have physical significance. For example, a statement with physical significance is that point A is 10 V higher in potential than point B. We may arbitrarily choose the potential of point B to be 0 V; the potential at A is then 10 V. Or, we may just as well designate the potential at A to be 0 V; the potential at B is then -10 V.

When a charge $+Q$ moves from region 1, which is at potential V_1, to region 2, which has a lower potential V_2, the energy of the charge QV is reduced. The energy which is lost by the charge must be converted into some other form, such as heat. Suppose the process occurs steadily: each second some charge moves from V_1 to V_2. Then there is a steady conversion of energy from electrical form to some other form. The rate at which energy is converted from one form to another is called *power*. The unit of power in the MKS system is the *watt* (**W**). By definition a power of 1 **W** is a rate of energy transfer of 1 J/sec.

Each time a charge $+Q$ Coulombs moves from region 1 to region 2, a quantity of energy $Q(V_1 - V_2)$ joules is converted. But the rate at which charge moves from region 1 to region 2 is called the current, $I_{1 \to 2}$. Thus the rate of energy conversion, that is, the power, is given by

$$P = (V_1 - V_2) \cdot I_{1 \to 2} \tag{1.4}$$

For brevity this is usually abbreviated to the shorter statement $P = VI$.

EXAMPLE 1.5

Electrons are released at electrode A, where the potential is V_A, and fall to electrode B, where the potential is V_B. Assume that $V_A < V_B$ and that the absolute value of the current is I_1. When the electrons hit electrode B their energy is converted to heat. Find the power being converted from electrical form to heat.

SOLUTION

Since electrons, which carry negative charge, are moving from A to B, the direction of the current is from B to A. Thus $I_{B \to A}$ is positive, and therefore $I_{B \to A} = I_1$. From Equation (1.4) we have

$$P = (V_B - V_A) \cdot I_{B \to A} = (V_B - V_A)I_1 \qquad \blacksquare$$

The Electric Circuit   An *electric circuit* is a closed path or combination of paths through which current can flow. In most cases the greater part of the circuit is composed of good electrical conductors, which we shall call *wires*. (The conductors need not actually be in the form of conventional wires; the term is only used here for convenience.) Good electrical conductors tend to be at nearly the same potential everywhere. In electronics, wires are usually "idealized"; that is, it is usually assumed that wires are *perfect* conductors, and therefore are at *exactly* the same potential everywhere. The reader should remember that by convention, in this book and almost everywhere in electronics, *all points along a wire are assumed to be at exactly the same voltage.*

Table 1.2   Symbols Used in Circuit Diagrams

Symbols used in this book	Meaning	Alternate Symbol
	Wires connected	
	Wires not connected	
I	The current flowing in the wire has the value I, taken as positive in the direction indicated	
V_A	The voltage at the indicated node is V_A	
V_{CC}	Terminal; the voltage at the terminal is V_{CC}	
$+$ v $-$	The terminal marked "+" is higher in potential than the terminal marked "−" by the voltage v	$+$ v $-$
R	Resistance, value R	
C	Capacitance, value C	
L	Inductance, value L	
M	Mutual inductance, value M. The dots (often omitted) pertain to the sign of M; see text	
$-$ $+$ V_0	Battery. The fat end of the symbol (here designated "−") represents the negative terminal. The "+" and "−" signs are often omitted. The nominal voltage is V_0.	$-$ $+$

It is also generally assumed (although this is also an idealization) that wires cannot store any charge within themselves. Therefore all current which enters one end of a wire must exit at the other end.

In addition to wires, a circuit contains *circuit elements,* such as resistors, capacitors, and transistors. A circuit element has two or more terminals where wires may be connected. As with wires, we assume that all current entering one terminal of a two-terminal element must exit at the other terminal. Thus the current through a two-terminal element is a well-defined quantity. For each kind of circuit element there is some relationship between the voltages at the terminals and the currents which flow through the terminals. Ohm's law (to be discussed in Section 1.3) is an example of such a relationship.

Electric circuits are shown by *circuit diagrams.* The symbols used in this book's circuit diagrams are listed in Table 1.2. Regrettably, symbols used for the same thing by different writers vary, particularly those in different areas of electrical engineering practice. The symbols used in this book are those most often used in electronics at the present time. Some alternative symbols, which are not used in this book but may be seen elsewhere, are also listed in Table 1.2. It will be noted that the symbol most often used in electronics for "wires not connected" means just the opposite, "wires connected," in the alternative system. Thus in reading circuit diagrams one must be aware of what symbology is being used.

An example of an electronic circuit is shown in Figure 1.3(a). This circuit contains four circuit elements; three of these have two terminals and one is a three-terminal element. Since each wire may be at a different potential, there are four different potentials in this circuit. The names of these potentials are indicated on the diagram; they are designated as V_1, V_2, V_3, and V_4. Note, however, that all points marked V_1 are at the same voltage, because the corresponding wires are connected together.

Figure 1.3(b) shows the same circuit as Figure 1.3(a); in this figure, however, the currents rather than the voltages are labeled. Each circuit element is assumed to have a total charge of zero at all times. Therefore, the current entering one terminal of a two-terminal element is the same as the current going out at the other terminal. From Figure 1.3(b) we see that there are three different currents in this circuit. The names of these three currents are indicated in the diagram, and arrows show the direction assigned as the positive direction for each current. Notice that all the currents marked I_2 must be equal.

When two points are not connected together by any circuit element at all, an *open circuit* is said to exist between them. Putting it another way, the two points are said to be open-circuited. If a wire is broken, an open circuit is created; no current can flow through the open circuit, although there may be a potential difference across it. Conversely, when two points are connected together by an ideal wire, a *short circuit* is said to exist between them; in other words, the two points are short-circuited. By the nature of an ideal wire, no potential difference can exist between points that are short-circuited, although current may flow from one to the other.

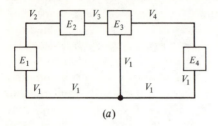

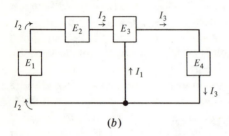

Figure 1.3 A circuit containing four circuit elements. In (*a*) voltages are labeled; we see that there are four different voltages in the circuit. In (*b*) the currents are labeled; we see that there are three different currents in the circuit.

In this book we shall follow general usage by referring to any quantity which is constant in time as a *dc quantity* (for example, "dc voltage," "dc current"). A quantity which varies in time (either sinusoidally or any other way) will be called *ac quantity*. Moreover, *dc voltage and currents will be represented by upper-case symbols* (I_b, V_2); *ac voltages and currents will be represented by lower-case symbols* (v_e, i_A).

The procedure whereby the voltages and currents of a given circuit are calculated is known as *circuit analysis*. In general, a circuit is completely analyzed when all voltages and currents have been found.

Branches and Nodes   We shall now introduce some additional terms which are useful and important in circuit analysis. To do this, let us consider a circuit containing a number of two-terminal elements, such as the one shown in Figure 1.4. We select the points in the circuit which are connected by wires to two or more circuit elements. These points are called *nodes*. There are three nodes in Figure 1.4, designated as nodes *A*, *B*, and *C*.

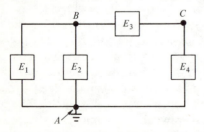

Figure 1.4 A circuit containing three nodes, two of which are principal nodes, and three branches. Node *A* is the reference node, as indicated by the ground symbol.

It is useful to identify those nodes which are connected to *three* or more circuit elements. Such a node is called a *principal node*. In Figure 1.4 nodes *A* and *B* are principal nodes; node *C* is not.

We shall define any path which connects two principal nodes as a *branch*. There are three branches in Figure 1.4. One is the path through E_1, connecting principal nodes B and A; one is the path between B and A through E_2; and the third is the path connecting B and A by way of E_3 and E_4. These terms will be useful in Section 1.4 when we develop techniques of circuit analysis.[3]

Ground   As we pointed out earlier, it is possible to define the potential at any one node in a circuit to be equal to zero. Such a node is then known as the *reference node,* or equivalently, is said to be *grounded.* The chosen reference node is indicated on the circuit diagram by the "ground" symbol ($\pm$). In Figure 1.4, for example, we see that node A is designated as the reference node. Defining one node as being at zero voltage affords a simplification in stating the other voltages. For instance, with node A grounded the statement "node B is higher in potential than node A by 5.4 V" is simplified to "the potential at node B is 5.4 V."

Often more than one point on a circuit diagram is designated by a ground symbol. This is just a kind of shorthand. What is really meant is that all such points are to be regarded as connected together, and the voltage at their common connection point is defined to be zero. This trick makes the circuit diagram look simpler because the wires which actually connect all the grounded points do not have to be shown on the diagram.

Kirchhoff's Laws   The two statements known as Kirchhoff's laws are fundamental to all problems of circuit analysis. In general, these laws, together with a knowledge of the properties of all circuit elements, are sufficient for calculation of the voltage and current everywhere in a circuit.

The first of Kirchhoff's laws, known as *Kirchhoff's current law,* arises from the physical assumption that no charge can accumulate in a wire: what current flows in one end, must flow out the other. More generally, if a certain amount of current flows into a node through one wire, an equal amount must leave the node through other wires. Kirchhoff's current law is a formal statement of this idea: it states that *the sum of all currents entering a node is zero.* A positive current *leaving* a node is the same as a *negative* current *entering* a node. Thus unless all currents are zero, at least one of the currents entering a node must be negative.

EXAMPLE 1.6

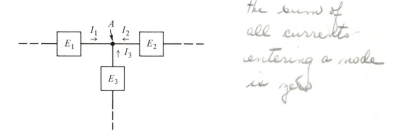

<hr />

[3] Note that according to our definitions, a circuit with no principal nodes has no branches. Other definitions of the term "branch" are sometimes used.

It is known that the current I_1 entering node A through element E_1 equals $+15$ mA. The current I_2 is known to equal $+32$ mA. What is I_3?

SOLUTION

According to Kirchhoff's current law, the sum of the currents entering the node is zero. We therefore write

$$I_1 + I_2 + I_3 = 0$$
$$I_3 = -I_1 - I_2 = -15 \text{ mA} - 32 \text{ mA} = -47 \text{ mA}$$

The magnitude of the current through E_3 is 47 mA, and the fact that I_3 is negative tells us that the actual direction of current through E_3 is outward from node A. ∎

Note that Kirchhoff's current law may also be stated as the sum of all currents leaving a node is zero. To show this, let us assume that I_1, I_2, and I_3 are the only three currents entering a node. Then by Kirchhoff's current law, $(I_1 + I_2 + I_3) = 0$. Now if the current entering a node through a certain wire is I_1, the current going outward from the node through that wire is $-I_1$. Therefore the sum of all currents flowing outward is $-I_1 - I_2 - I_3 = -(I_1 + I_2 + I_3) = -(0) = 0$.

The second of Kirchhoff's laws, known as *Kirchhoff's voltage law*, can be deduced by the following argument. The voltage at any point of a circuit at any instant of time has a certain value. Suppose we begin at one point of the circuit, pass through a circuit element, and note the decrease, or "drop," in potential. We then pass through another element, note this second drop in potential, and add it to the first drop. Suppose now that after passing through several elements we arrive back at the original point. The sum of all the voltage drops that were experienced must now add up to zero, since we have returned to the original potential. The closed itinerary just described is known as a "loop," and the foregoing argument can be formalized into what is known as Kirchhoff's voltage law: *The sum of all the voltage drops around a complete loop is zero.* We could also count increases, or "rises" in potential while going around the loop; thus an alternative statement of Kirchhoff's voltage law is that the sum of voltage rises around a closed loop is zero. Again we note that some of these voltage drops or voltage rises must have negative values.

EXAMPLE 1.7

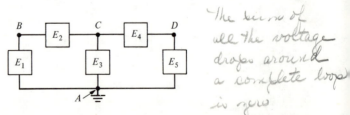

Let us call the voltage at node B, V_B, at node C, V_C, and at node D, V_D. In terms of these three voltages, write an equation expressing Kirchhoff's voltage law for the loop $A \rightarrow B \rightarrow C \rightarrow D \rightarrow A$.

SOLUTION

We are free to add either voltage drops or voltage rises as we go around the loop. Let us add rises. From the presence of the ground symbol we know that node A is at zero potential. The rise as we go from A to B is $(V_B - V_A) = (V_B - 0) = V_B$. The rise as we go from B to C is $(V_C - V_B)$. Adding the voltage rises around the entire loop and setting the sum equal to zero gives

$$V_B + (V_C - V_B) + (V_D - V_C) + (0 - V_D) = 0$$

The correctness of this equation is of course obvious by inspection. ∎

1.3   Circuit Elements

Idealized Circuit Elements   As has already been seen, an electric circuit consists of an interconnection of circuit elements. The elements found in an actual circuit may include resistances, capacitances, diodes, transistors, and various other electrical or electromechanical elements. The general behavior of an element may be quite complex. A complete description of the behavior of each circuit element would have to take into account its operation under conditions of extremely high voltage and current, unusual temperatures, and so on. Such complex specifications for an element would be unwieldy to use, and would make the understanding of circuit operation a difficult matter. Fortunately, most electronic circuits do not involve extreme conditions. Therefore we can substitute, for the more general specifications, an idealized behavior. This is a simplified description of an element's properties, which is valid within a specific range of operating conditions. Idealized circuit elements are valuable tools for understanding circuit operation. One must be aware, however, that there are limits to the range of operating conditions over which the idealization is a good model of the actual physical element.

As an example of idealization, let us discuss electrical wires. All wires at room temperature have a finite, non-zero electrical resistance. Yet throughout this book our assumption will be that wires are ideal, that is, they have zero resistance, and that all points along a wire are at exactly the same potential. This is a useful idealization because the resistance of a copper wire a few centimeters long is usually negligible compared with other resistances in the circuit, and thus has no significant effect on the circuit's operation. It should be noted, however, that the idealization is useful only within a limited range of situations. If the same kind of wire were used to send electrical power over a thousand miles, we would soon find its resistance to be very important.

I-V Relationships   Each ideal circuit element is characterized by a relationship between the current through the element and the potential difference between the terminals of the element, or vice versa. This relationship, which we shall call the current-voltage (or *I-V*) relationship of the ideal element, is a good approximation to the behavior of the corresponding real circuit element, providing the current, potential difference, and other quantities (such as temperature) are kept within certain bounds. Each *I-V* relationship contains a parameter, such as resistance (R), inductance (L), or capacitance (C), the value of which is assumed to be a constant, not dependent on current or voltage and not changing with time.

It should be noted that the current through any circuit element depends on the *difference* between the potentials at its terminals. This is a reflection of the general rule that only *differences* in potential are of physical significance.

A composite circuit element can be constructed out of two or more individual circuit elements. Two cases of this which are very often encountered are illustrated in Figure 1.5. A combination of two circuit elements connected head-to-tail, as shown in Figure 1.5(*a*), is called a *series* combination; the side-by-side connection of Figure 1.5(*b*) is called a *parallel* combination. The two elements E_1 and E_2 may be of the same type (for example, two resistors) or of two different types (for example, a resistor and a capacitor). Both the series and the parallel connections give rise to new composite circuit elements, again having two terminals. If E_1 and E_2 are ideal circuit elements with known properties, the *I-V* relationships of their series or parallel combinations can be calculated. Some of these combinations occur very often, and it is useful to recognize them when they occur.

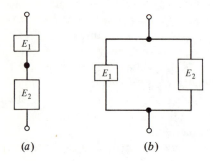

(*a*)   (*b*)

Figure 1.5 Composite circuit elements made from combinations of two simpler circuit elements. (*a*) Series combination; (*b*) parallel combination.

The properties of several important ideal circuit elements will now be individually described.

Resistance   The property of a material which causes it to obstruct, or resist, the flow of current when a potential difference is applied is known as its electrical *resistance*. All real materials at room temperature possess this property. We have taken the point of view that wires are ideal, and have no resistance; this approximation is a good one provided that the actual resistance of the wires is negligibly small compared with the other resistances in the circuit. A resistor therefore may be regarded as an element of comparatively resistive material, to which two wires having negligibly low resistance are connected. The physical construction of a resistor is shown in Figure 1.6(*a*). The symbol for this circuit element is given in Figure 1.6(*b*).

The *I-V* relationship for the ideal resistance, which is known as Ohm's law, will now be stated. In Figure 1.6(*b*) the two terminals of the resistor have been designated as A and B. (The letters are not part of the symbol and are included in the figure for purposes of explanation.) Let the potential at terminals A and B be V_A and V_B, respectively. Then the statement of Ohm's law is

$$I_{A \to B} = (V_A - V_B)/R \qquad \text{(Ohm's law)} \tag{1.5}$$

where the symbol $I_{A \to B}$ stands for the current, taken as positive in the direction from A to B. The constant R is a parameter known as the resistance; in the MKS system R is stated in ohms. Ohm's law is an idealization of the behavior of an actual resistance. The approximation that is generally used, that the physical resistor obeys Ohm's law, is based on two assumptions: that the resistor is subjected only to a limited range of operating conditions (for example, the manufacturer's limit on power dissipation is not exceeded); and that the circuit as a whole will not be much affected by the inevitable small deviations of the real resistor from ideal behavior.

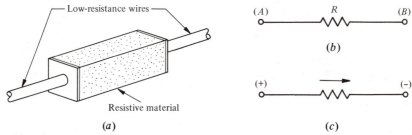

Figure 1.6 Resistance. A typical structure for a resistor is shown in (a). The corresponding circuit symbol is shown in (b). Figure (c) illustrates the direction of current flow: from the terminal that is more positive toward that which is less positive.

It is important to be familiar with the direction in which current flows through the resistor. Figure 1.6(c) presents schematically the definition of the previous paragraph. The meaning of Figure 1.6(c) may be stated in several ways which are equivalent to that of the previous paragraph, but may add to the reader's familiarity with Ohm's law: (a) *the direction of current is from the terminal which is more positive to the one which is less positive;* (b) *if one moves from one terminal of the resistor to the other in the direction in which current flows, a decrease in potential is experienced.*

EXAMPLE 1.8
    Terminal A of Figure 1.6(b) is at a potential of -3 V and terminal B is at a potential of -2 V. In which direction does current flow?

SOLUTION
    In a resistance, the direction of current is from the more positive to the less positive. Terminal B is less negative at terminal A, which is the same as being more positive. Thus the direction of current is from B to A. ∎

EXAMPLE 1.9
    If current is known to flow through a resistor in the direction from terminal A to terminal B, is a voltage rise or a voltage drop experienced as one goes from B to A?

SOLUTION

Since the direction of flow in a resistance is from the more positive potential to the less positive, terminal A must be higher in potential than B. Thus a rise is experienced as one goes from B to A.   ■

Often in circuit calculations an arbitrary assignment of a direction of current flow is made. For example, the current from terminal A to terminal B of some resistor may be chosen as an unknown to be solved for. When the solution is obtained, it is quite possible that the value of this unknown current may turn out to be negative. This simply means that the actual direction of current flow is opposite to what was arbitrarily chosen as positive.

EXAMPLE 1.10

In Figure 1.6(b) let the direction of the current I be defined as being from B to A. The potential at A is 3 V, the potential at B is -2 V, and the resistance is 2.5 Ω. What is the value of I?

SOLUTION

According to Ohm's law, Equation (1.5), we find that the current in the direction from A to B is $(V_A - V_B)/R = (3 + 2)/2.5 = 2$ A. The current I, however, is the current from B to A, which is the negative of the current from A to B. Thus $I = -2$ A. The actual direction of current flow is from A to B.   ■

EXAMPLE 1.11

Let the potential at A be called V_A and the potential at B be V_B. The current I_1 is defined as flowing from A to B and the current I_2 from B to A. The resistance between A and B is R. What are I_1 and I_2? What is the actual direction of current flow?

SOLUTION

According to Equation (1.5), I_1, the current from A to B, is equal to $(V_A - V_B)/R$. The current I_2, in the opposite direction, is the negative of I_1, and therefore $I_2 = (V_B - V_A)/R$. However, one cannot say what the actual direction of current flow is until the values of V_A and V_B are known.   ■

The question of direction of current flow often causes a great deal of difficulty for students new to electric circuits, especially when negative currents or voltages are involved. A second reading of this discussion and Examples 1.8 through 1.11 is recommended.

An important property of the resistance is its ability to convert energy from electrical form into heat. In the last section it was pointed out that when a current I flows from a place at a higher voltage to one where the voltage is lower by an amount V, then the power converted is given by $P = VI$. This is the case in a resistor; the voltage drop V is, according to Ohm's law, equal to IR. Thus when a current I flows through a resistance R, the power converted into heat is

$$\boxed{P = I^2R}$$ (1.6)

Alternatively, since $I = V/R$, we may also state that when a potential difference V exists across a resistance R, the power converted into heat is

$$P = V^2/R \qquad (1.7)$$

In practice, resistors must be designed so that they can dissipate the necessary amount of heat. Generally, the manufacturer states the maximum power dissipation of a resistor in watts. If more power than this is converted to heat by the resistor, the resistor will overheat and be damaged.

EXAMPLE 1.12
The power dissipation of a 47,000-Ω resistor is stated by the manufacturer to be 1/4 W. What is the maximum dc voltage that may be applied? What is the largest dc current that can be made to flow through the resistor without damaging it?

SOLUTION
The maximum voltage is obtained from Equation (1.7), where $P = 1/4$ W. We have

$$V^2_{\max} = RP$$
$$V_{\max} = \sqrt{(4.7 \times 10^4)(0.25)} = 114 \text{ V}$$

Similarly, to find the maximum current we use Equation (1.6):

$$I^2_{\max} = P/R$$
$$I_{\max} = \sqrt{(0.25)/(4.7 \times 10^4)} = 2.3 \times 10^{-3} \text{ A} = 2.3 \text{ mA} \qquad \blacksquare$$

Series and parallel combinations of two resistances occur very often; therefore the *I-V* relationships of these combinations will now be obtained.

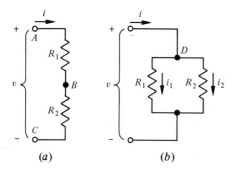

Figure 1.7   Composite circuit elements composed of two resistances (a) in series; (b) in parallel.

(a)   (b)

A composite circuit element made of two resistors in series is shown in Figure 1.7(a). We see that the current flowing through both resistors is the same and is equal to i. Using Ohm's law, Equation (1.5), we find that the voltage at point A, v_A, is higher than that at point B, v_B, by the amount iR_1; in other words, $v_A - v_B = iR_1$. Similarly, $v_B - v_C = iR_2$. The total voltage across the element v equals $v_A - v_C$, which is the same as $(v_A - v_B) + (v_B - v_C)$. Therefore $v = iR_1 + iR_2 = i(R_1 + R_2)$. We see by comparing with Equation (1.5) that the *I-V* relationship of the series combination is the same as that of a single resistor with value $(R_1 + R_2)$. Because of the identical nature of its I-V relationships, we say that the series combination is *equivalent* to a single resistance of value

$$R = R_1 + R_2 \quad \text{(resistors in series)} \tag{1.8}$$

Further use will be made of the concept of equivalence in Chapter 2 and in later chapters.

A parallel combination of two resistances is shown in Figure 1.7(*b*). We observe that the voltages appearing across resistances R_1 and R_2 are the same, and are equal to v. According to Equation (1.5) $i_1 = v/R_1$ and $i_2 = v/R_2$. Applying Kirchhoff's current law to node D (note that the current entering node D through R_1 is $-i_1$), we have $i - i_1 - i_2 = 0$. Therefore

$$i = i_1 + i_2 = v\left(\frac{1}{R_1} + \frac{1}{R_2}\right)$$
$$= v\left(\frac{R_2 + R_1}{R_1 R_2}\right) \tag{1.9}$$

This may be rewritten as

$$i = v \bigg/ \left(\frac{R_1 R_2}{R_1 + R_2}\right) \tag{1.10}$$

Again comparing with Equation (1.5), we see that the parallel combination is equivalent to a single resistance with value:

$$R = \frac{R_1 R_2}{R_1 + R_2} \quad \text{(resistances in parallel)} \tag{1.11}$$

As a shorthand, the resistance given by Equation (1.11) is sometimes abbreviated as $R_1 \| R_2$; that is, by definition $R_1 \| R_2 = R_1 R_2/(R_1 + R_2)$. More complex combinations of resistances can be shown to be equivalent to single resistances by repeated use of Equations (1.9) and (1.11).

EXAMPLE 1.13
The combination of resistors shown below has two terminals designated A and B. It is desired to replace it with a single resistance connected between terminals A and B. What should the value of this resistance be, so that the resistance between the terminals is unchanged?

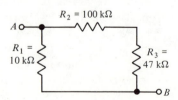

SOLUTION

On inspection it is seen that R_2 and R_3 are in series and that their series combination is in parallel with R_1. Thus the resistance between A and B is given by

$$R = (R_2 + R_3)\|R_1 = \frac{R_1(R_2 + R_3)}{R_1 + R_2 + R_3}$$

$$= \frac{(10,000)(147,000)}{157,000} = 9360 \ \Omega$$

The three resistors may be replaced by a single resistor of this value without altering the resistance between A and B. ∎

Capacitance   The phenomenon of electrical *capacitance* is illustrated in Figure 1.8(a). When a voltage is applied across the capacitor, positive charge accumulates on the capacitor plate at higher potential and negative charge accumulates on the plate at lower potential. Suppose in Figure 1.8(a) that side A is higher in potential than side B, and the potentials on the two sides are v_A and v_B, respectively. Then the amount of positive charge appearing on plate A is given by

$$Q = C(v_A - v_B) \tag{1.12}$$

where the constant C is a parameter known as the capacitance. In the MKS system the capacitance is stated in farads. The circuit symbol for the capacitor is shown in Figure 1.8(b).

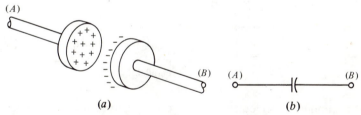

Figure 1.8 Capacitance. A possible physical structure is shown in (a). Figure (b) shows the corresponding circuit symbol. Which side of the symbol is curved has no significance.

The *I-V* relationship for the capacitor can be obtained by differentiating Equation (1.12) with respect to time. If in Figure 1.8 current flows into terminal A from the left, additional positive charge will be deposited on side A; that is, Q increases. In fact, $i = dQ/dt$. Since the negative charge on plate B must be equal in magnitude to the positive charge on plate A, an equal current i must flow to the right out of plate B. Thus a current i seems to flow straight through the device in the direction A to B. From Equation (1.12) we see that increasing Q causes v to increase. Upon differentiation, (1.12) gives

$$\frac{dQ}{dt} = C\frac{d}{dt}(v_A - v_B) \tag{1.13}$$

But since $dQ/dt = i_{A \to B}$, the current in the direction of A to B, we have

$$i_{A \to B} = C \frac{d}{dt} (v_A - v_B)$$

(1.14)

This is the *I-V* relationship for the capacitance.

Several interesting features may be seen from Equation (1.14). First, the current depends not on the voltage across the element, but on its rate of change. Furthermore, if the voltage across a capacitor does not change with time, no current flows. Conversely, if the current is zero, the voltage across the capacitor remains constant. One can, for example, place a certain charge on the plates of a capacitor. According to Equation (1.12), there will then be a potential difference across it. We then break the connections. With the connections open, i must be zero, and hence the potential difference across the capacitor must remain constant. We thus observe that an ideal capacitor, once charged and disconnected, will retain a potential difference for an indefinite length of time.

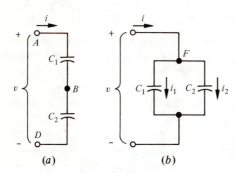

(a)   (b)

Figure 1.9 Composite circuit elements composed of two capacitances in (a) series; (b) parallel.

Series and parallel combinations of capacitors are often encountered. A series combination is shown in Figure 1.9(a). We observe that the current through both capacitors is the same and equals i. Therefore $(d/dt)(v_A - v_B) = i/C_1$ and $(d/dt)(v_B - v_D) = i/C_2$. The total voltage across the composite element is $v = v_A - v_D = (v_A - v_B) + (v_B - v_D)$. Thus dv/dt is given by

$$\frac{dv}{dt} = \frac{d}{dt}[(v_A - v_B) + (v_B - v_D)] = \frac{i}{C_1} + \frac{i}{C_2}$$

(1.15)

$$= i\left(\frac{C_1 + C_2}{C_1 C_2}\right)$$

This may be written in a form comparable with Equation (1.14):

$$i = \left(\frac{C_1 C_2}{C_1 + C_2}\right)\frac{dv}{dt}$$

(1.16)

Thus the series combination of capacitances has the same *I-V* relationship as a capacitor with the value

$$C = \frac{C_1 C_2}{C_1 + C_2} \quad \text{(capacitances in series)} \tag{1.17}$$

For the case of capacitances in parallel, we refer to Figure 1.9(b). It is seen that the voltages across both capacitors are the same, and equal v. Thus $i_1 = C_1(dv/dt)$ and $i_2 = C_2(dv/dt)$. By using Kirchhoff's current law at node F, we find that $i = i_1 + i_2$. Thus $i = (C_1 + C_2)(dv/dt)$, and the parallel combination of capacitors is equivalent to a single capacitor with the value

$$C = C_1 + C_2 \quad \text{(capacitances in parallel)} \tag{1.18}$$

It should be noted that not all capacitors have the physical structure shown in Figure 1.8(a). Many different types of construction are used, in order to obtain values of capacitance which are small, large, precise, or adjustable. In practical cases an important parameter of a capacitor is its *working voltage*. This value, which is generally specified by the manufacturer, is the maximum voltage which can safely be applied between the capacitor terminals. Exceeding this limit may result in breakdown, for instance by formation of an electric arc between the capacitor plates. It should also be noted that all circuits contain unintentional capacitances which arise whenever two wires or circuit elements happen to be near to one another. These unintentional (or "parasitic") capacitances can have serious effects on the operation of a circuit.

Inductance   The phenomenon of *inductance,* or more properly, *self-inductance,* is magnetic in origin. When a wire carries current, a magnetic field is established in its neighborhood. If the current, and hence the magnetic field are time-varying, the field can, in turn, act back on the wire and give rise to a voltage across it. In order to make the magnetic field denser and thus enhance the self-inductance, the wire is usually wound into a coil, as shown in Figure 1.10(a). Sometimes the coil is wound around a core of iron or other material to increase the inductance further. The circuit symbol for the element is shown in Figure 1.10(b). For an ideal inductance, the *I-V* relationship is given by

$$(v_A - v_B) = L \frac{d}{dt} i_{A \to B} \tag{1.19}$$

The constant L is a parameter called the *inductance;* in the MKS system the unit of inductance is the *henry.* From Equation (1.19) we observe that there

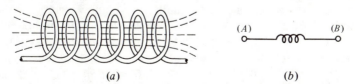

(A)   (B)

(a)   (b)

Figure 1.10   Inductance. Inductors are usually made in the form of a coil of wire, as shown in (a). The dashed lines represent the lines of magnetic field. Figure (b) shows the circuit symbol for the element.

is no voltage across the inductance when the current through it is constant. Under dc conditions, it acts like an ideal wire, or short circuit.

The *I-V* relationships for combinations of two inductances in series or parallel may be found by the same techniques as were used above for resistance and capacitance. In the light of our previous findings, it is not surprising that series and parallel combinations of inductances have *I-V* relationships like those of single inductances. If the values of the individual inductors are L_1 and L_2, the combinations are equivalent to individual inductances with the values

$$L = L_1 + L_2 \qquad \text{(inductances in series)}$$
(1.20)

and

$$L = \frac{L_1 L_2}{L_1 + L_2} \qquad \text{(inductances in parallel)}$$
(1.21)

In practice there often is considerable resistance in the wire of a coil, and there are also sizable capacitances between the various windings. These features of a real, non-ideal inductor may be represented in a circuit diagram by modifying the inductor symbol to the form shown in Figure 1.11. Here a combination of ideal elements has been chosen which together represent the behavior of a real, non-ideal inductor more closely than an ideal inductance by itself could do. Such a combination of ideal elements is called a *model* for the practical inductor. Techniques for modeling real circuit elements will be used extensively in later chapters.

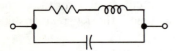

Figure 1.11   A possible model for a real inductor.

Mutual Inductance   The magnetic field produced by a current flowing in a coil can also interact with a second coil, if the latter is wound on the same core or is simply placed nearby. If the magnetic field is time-varying, a voltage is induced across the second coil. Similarly, a time-varying current in the second coil will give rise to a voltage across the first. This mutual interaction between two coils is described by the term *mutual inductance*. A structure of two or more coils designed to exhibit mutual inductance is known as a *transformer*. One possible structure for a transformer is shown in Figure 1.12(*a*). The circuit symbol for a transformer is shown in Figure 1.12(*b*). The two windings are known as the *primary* and *secondary* windings.

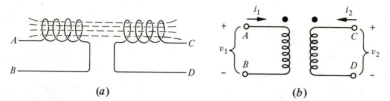

Figure 1.12   Mutual inductance. One possible physical arrangement for a transformer is shown in (a). Figure (b) shows the symbol for a transformer. For
Equation (1.22) to describe the element correctly, the sign convention indicated
by the two dots must be used. Voltages v_1 and v_2 are measured with the positive
sides toward the dots; currents i_1 and i_2 are taken as positive when flowing into
the terminals marked by the dots.

In order to state equations describing the operation of an ideal transformer, sign conventions must be observed. Two dots are shown in the symbol of Figure 1.12(b). Let v_1 and v_2 be the voltages across the two windings,
taken with their positive sense toward the dots. The currents i_1 and i_2 through
the primary and secondary windings are taken as positive, when directed into
the terminals marked by the dots. (Sometimes the dots are not shown in a circuit diagram. This usually means that signs happen to be unimportant in that
particular circuit.) Then the equations of the ideal transformer are

$$v_1 = L_1 \frac{di_1}{dt} + M \frac{di_2}{dt} \qquad (1.22)$$

and

$$v_2 = L_2 \frac{di_2}{dt} + M \frac{di_1}{dt} \qquad (1.23)$$

The constant M is called the *mutual inductance,* and like self-inductance is
measured in henrys. We see that v_2, the voltage across the secondary, is influenced both by the current in the primary, through the mutual inductance,
and by the current in the secondary, through L_2, the self-inductance of the
secondary. For v_1 the case is similar. From Equation (1.23) we observe that
constant currents in the primary have no effect on the voltage in the secondary.
If the primary current is composed of a dc current plus an ac current, only the
ac part affects the secondary. Therefore transformers can be used to couple
circuits together with respect to ac, while keeping them separate with respect
to dc. The most important application of transformers, however, is probably
that of converting a sinusoidal voltage to another of different amplitude. When
the voltage is increased, the device is known as a step-up transformer; in the
opposite case, as a step-down transformer. Step-up and step-down transformers are particularly important in work with 60-Hz ac currents such as
are found in power lines, where large iron-core transformers are used. In
electronics their most frequent application is in high-frequency radio com-

munication circuits, where much smaller units, often without cores, can be used.

Ideal Voltage Source   An idealization very useful in electric circuit theory is that of a component which possesses, between its two terminals, a specified difference of potential. Circuit symbols for such ideal elements are shown in Figure 1.13. The defining property of this ideal circuit element is that *the potential at the terminal marked* (+) *is higher than that at the terminal marked* (−) *by the indicated number of volts.* In other words, the potential difference between the terminals is *controlled* to have a certain value. This circuit element is commonly known as an *ideal voltage source* or *ideal voltage generator.*

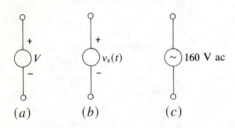

(a)   (b)   (c)

Figure 1.13   Circuit symbols for ideal voltage sources. (*a*) The voltage between the terminals is controlled so that the (+) terminal is higher in potential than the (−) terminal by the value *V*. (*b*) The voltage between the terminals has the value $v_s(t)$, which is time-varying. (*c*) The voltage between the terminals varies sinusoidally in time. The amplitude of the sinusoid is 160 volts.

Although the potential difference across the ideal voltage source is controlled to have a specific value, this value need not be constant in time. The potential difference across a voltage source can be specified to have a constant value (for example, 1.5 V), or the value of the potential difference may change with time (for example, it might be 160 sin 377*t*, where *t* is time in seconds). One may think of a constant-voltage source as being an idealization of a battery. On the other hand, a voltage source designated 160 sin 377*t* could be an idealization of the conventional 60-cycle voltage obtainable in house wiring. It should be noted that the *current* through an ideal voltage source takes on whatever value is dictated by the circuit as a whole. The sign of the current through a voltage source is *not* determined by the polarity of the source; the direction of the current depends on the circuit as a whole. (This is a fundamental difference from the resistance. The sign and magnitude of the current through an ideal resistor do depend exclusively on the potential across it.)

EXAMPLE 1.14

Find the magnitude and sign of the current *I* indicated by the vertical arrow in the sketch.

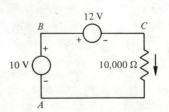

SOLUTION

Point B of the circuit is 10 V higher in potential than point A. Point C is 12 V lower in potential than point B. Therefore point C is 2 V lower in potential than point A. Using Ohm's law we find that

$$I = \frac{(V_C - V_A)}{10^4}$$

$$= \frac{(-2)}{10^4} = -2 \times 10^{-4} \text{ A}$$ ∎

Ideal Current Source   Another very useful idealized circuit element is the *ideal current source,* symbols for which are shown in Figure 1.14. The ideal current source has the property that the current through it, in the direction shown by the arrow, is controlled to have the designated value. The potential difference across the current control has whatever value and sign is dictated by the entire circuit. We see that the current and voltage sources are quite similar in principle. However, while a dc voltage source may be regarded as an idealized battery, there is no everyday example of a real element of which the ideal current source is an idealization.

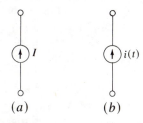

(a)   (b)

Figure 1.14   The circuit symbol for an ideal current source. (*a*) The current in the direction of the arrow is controlled to the value I, independent of the voltage across the element. (*b*) The current is controlled to have the value $i(t)$, which varies in time.

EXAMPLE 1.15

Consider the following circuit:

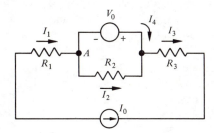

Find $I_1, I_2, I_3,$ and I_4.

SOLUTION

The current source determines the value of I_1. This is clear because the current through R_1 is the current controlled by the current source. Taking the direction indicated for positive I_1 into consideration, we see that

$$I_1 = -I_0$$

Similarly

$$I_3 = -I_0$$

The current through R_2 is determined, according to Ohm's law, by the voltage across it, and the latter is controlled to have the value V_0. Taking into account the direction in which I_2 is regarded as positive, as indicated on the diagram, we have

$$I_2 = -\frac{V_0}{R_2}$$

To find I_4, the current through the voltage source, we use Kirchhoff's current law, setting the sum of the currents leaving node A equal to zero. Thus

$$I_2 + I_4 - I_1 = 0$$

$$I_4 = I_1 - I_2 = (-I_0) - \left(-\frac{V_0}{R_2}\right) = \frac{V_0}{R_2} - I_0$$

The value of I_4 could be either positive or negative, depending on the relative magnitudes of V_0 and I_0. ∎

Ideal Voltmeter and Ammeter   Idealizations of two common measuring devices, the voltmeter and the ammeter, are sometimes useful. In practical circuits meters are used to measure voltage and current at various places in a circuit. The ideal voltmeter and ammeter symbols may also be used in a circuit diagram to identify a particular current or voltage of interest in a circuit.

Various symbols for the ideal voltmeter and ideal ammeter are in use. Because its appearance suggests a property of the element, we shall represent the ideal voltmeter by the symbol shown in Figure 1.15. When used in a circuit diagram, this symbol designates two terminals (the points where the symbol is connected) between which the potential difference is to be found. The ideal nature of this element lies in the property that *no current flows through an ideal voltmeter*. Because of this property, an ideal voltmeter can be connected between any two points of a circuit without producing any change in voltages or currents anywhere in the circuit. All the voltmeter does is to call attention to the potential difference between two points. The sign convention for the measured voltage is the following: The measured voltage is equal to the potential at the $(+)$ terminal minus that at the $(-)$ terminal. The measured voltage may be either positive or negative. Its sign is usually important.

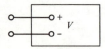

Figure 1.15 Circuit symbol for the ideal voltmeter.

The ideal ammeter will be represented by the symbol shown in Figure 1.16. It registers the current which flows through it, taken as positive when flowing in the direction of the arrow. The ideal ammeter is idealized in that *the potential difference between the two terminals of an ideal ammeter is zero.*

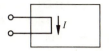

Figure 1.16 The circuit symbol for the ideal ammeter. The meter reads the current in the direction designated by the arrow.

For this reason an ideal ammeter may be inserted into any wire of a circuit without producing any change in voltage or current anywhere in the circuit.

The symbols and *I-V* relationships of all the above simple circuit elements are shown in Table 1.3.

Table 1.3 Symbols and Properties of Simple Circuit Elements

Name	Symbol	I-V Relationship
Resistance	A —⟋⟍⟋⟍— R —⟋⟍— B	$I_{A \to B} = \dfrac{V_A - V_B}{R}$
Capacitance	A ——∣⊢—— C —— B	$i_{A \to B} = C \dfrac{d}{dt}(v_A - v_B)$
Inductance	A —ɷɷɷ— L —— B	$v_A - v_B = L \dfrac{d}{dt}(i_{A \to B})$
Mutual inductance	M	(see text)
Ideal voltage source	A + V_0 − B	$V_A - V_B = V_0$
Ideal current source	A ——⊖—— B I_0	$I_{A \to B} = I_0$
Ideal voltmeter	A, B + V −	$I_{A \to B} = 0$
Ideal ammeter	A, B ↓ I	$V_A - V_B = 0$

Current-Voltage Characteristics The *I-V* relationships of many circuit elements and combinations of circuit elements can be described graphically by graphs known as current-voltage (*I-V*) characteristics. The *I-V* characteristic of an element is nothing more than a graph of the current through the element, as a function of the voltage across its terminals. Let the sign conventions for current and voltage be as shown in Figure 1.17. Then as a first example we may consider an ideal resistance R. According to Ohm's law, we have (with the sign conventions as in Figure 1.17) $I = V/R$. This dependence of I upon V is graphed in Figure 1.18(a); the graph has the form of a straight line through the origin, with slope $1/R$.

Figure 1.17 Sign conventions for voltage and current to be used in graphing the I-V characteristic of a circuit element. Positive current is defined as into the (+) terminal.

The I-V characteristics of a voltage source, current source, voltmeter, and ammeter are shown in Figures 1.18(b), 1.18(c), 1.18(d), and 1.18(e), respectively. Note, for example, how Figure 1.18(b) illustrates the simple relationship (or perhaps we should say non-relationship) between I and V in a voltage source. We see here that the potential difference between the terminals has its specified value, regardless of the value of I. The properties of the ideal current source, voltmeter, and ammeter are similarly evident in Figures 1.18(c), 1.18(d), and 1.18(e).

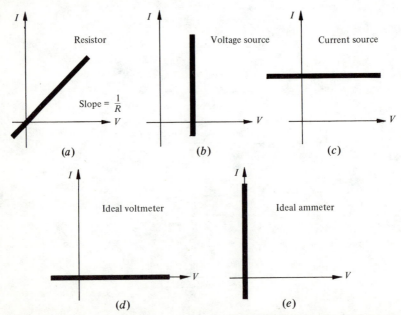

Figure 1.18   The current-voltage relationships for a number of common elements. (a) resistor, (b) voltage source, (c) current source, (d) ideal voltmeter, (e) ideal ammeter.

I-V characteristics are very useful for describing complex circuit elements, such as diodes or transistors. Unlike the resistor, whose *I-V* characteristic is prescribed by a very simple formula, more complex circuit elements may have *I-V* relationships which are mathematically complicated, or not given by any mathematical prescription at all. The properties of such elements may be conveniently described graphically by their *I-V* characteristics. Another instance when *I-V* characteristics are useful is in describing the properties of composite circuit elements, that is, two-terminal circuit elements built of more than one simple element.

EXAMPLE 1.16

Find the *I-V* characteristic of the circuit in sketch (*a*). The positive senses of *I* and *V* are as indicated.

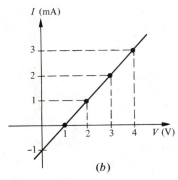

(*a*)

SOLUTION

We can proceed by plotting a graph. When the current is zero, there is, according to Ohm's law, no potential difference across the resistor. Hence terminal *A* is 1 V higher in potential than terminal *B*, and $V = +1$. When $I = 1$ mA, terminal *A* is 1 V higher in potential than node *C*; consequently $V = 2$. (Notice the importance of the sign of the voltage drop across the resistance.) When $I = 2$ mA, $V = 3$, and so on. The *I-V* plot is shown in sketch (*b*).

(*b*)

It should be noted that the capacitor and the inductor cannot be described graphically by *I-V* characteristics. The reason for this is that the current through a capacitor at a given time is not determined by the *voltage* across it at that time, but rather by the *rate of change* (time derivative) of the voltage.

Consequently the current through a capacitor depends not only on what the present voltage across it is, but also on what the voltage was at an earlier time.

It is interesting to observe that the sign of the power being transferred into a two-terminal circuit element or circuit can be determined, if the point on the I-V graph where operation occurs is known. Referring again to Figure 1.17, we see that if both I and V are positive, current flows through the element from the terminal at higher voltage to the terminal at lower voltage. This means that power is being transmitted *into* the element. The amount of power being transmitted is $P = VI$. If the values of V and I are positive, the point on the I-V graph describing the operation of the circuit is in the first quadrant of the I-V plot. Thus operation in the first quadrant of the I-V plot always means power is transferred into the circuit element. Operation in the third quadrant implies the same thing; V and I are then both negative and the product $P = VI$ is again positive. Operation in the second or fourth quadrants, however, implies that the power being transmitted into the element is negative; in other words, that the element is sending power out instead of taking it in.

From Figure 1.18(a) we see that all parts of a resistor's I-V characteristic lie in the first and third quadrants. Therefore a resistor can only receive power and can never send power out. On the other hand, the I-V characteristic of a voltage source with positive value lies in both the first and fourth quadrants. Therefore, depending on the rest of the circuit, the voltage source can either send power out or take power in. As an illustration of the latter case, one might think of a battery being charged.

EXAMPLE 1.17

Let us give the composite circuit element of Example 1.16 the name E_1. A voltage source V_2 is connected to terminals A and B of E_1 as shown. Using the I-V characteristic obtained in Example 1.16, find the power received or emitted by E_1, (1) when $V_2 = 3$ V and (2) when $V_2 = 0.5$ V.

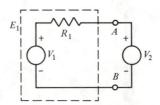

SOLUTION

(1) Referring to the I-V characteristic obtained in Example 1.16, we see that the point on the I-V characteristic corresponding to $V = 3$ volts is in the first quadrant. Therefore E_1 is receiving power from the external voltage source. The current, according to the graph, is 2 mA; therefore the power transmitted into E_1 is (3 V) • (2 mA) = 6.10^{-3} W. Some of this power is being taken in by source V_1, and the remainder is converted to heat by R_1.

(2) When $V = 0.5$ V, $I = -0.5$ mA. Operation is now in the fourth quadrant, so E_1 is sending power out. The power leaving E_1 is (0.5 V) • (0.5 mA) = 0.25 milliwatt. This power must originate in source V_1, as R_1 is only capable of consuming power. ∎

1.4   Node and Loop Analysis

In circuit analysis the usual goal is to calculate voltages or currents at various places in a circuit. In general this can be done, provided that the *I-V* relationships of all the elements in the circuit are known. In this section two general techniques for circuit analysis will be described: the *node method* and the *loop method*. In general either method can be used with any circuit, although at times one may require less effort than the other.

The Node Method   The state of a circuit is completely known if the voltage at all nodes and the current through each of the branches is known. One of the node voltages can arbitrarily be set equal to zero. Thus the number of unknowns to be solved for is equal to the number of branches plus the number of nodes, minus one. In the node method of analysis, one solves first for all unknown node voltages. This is usually done by writing an equation expressing Kirchhoff's current law for each node where the voltage is unknown. These equations state that the sums of the currents flowing into each node are zero. The currents, in turn, are expressed in terms of the unknown node voltages by means of the *I-V* relationships for the branches. The resulting set of simultaneous equations is then solved for the unknown node voltages. Finally, branch currents can be then obtained using the calculated node voltages and the *I-V* relationships of the branches.

As a demonstration of how the node method is used, let us calculate the branch currents and node voltages for the circuit of Figure 1.19. Before we can proceed, we must locate the node voltages which are to be found. The four nodes have been designated as A, B, C, and D. One node can arbitrarily be chosen as the reference node, which by definition has potential zero. In this case node D has been selected as the reference node and marked with the ground symbol. The potentials of other nodes are then defined by comparison with the potential at node D. For example, a node 2 V lower in potential than node D is said to have a potential of -2 V.

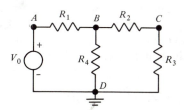

Figure 1.19   A circuit which has four nodes and three branches. Node D has been chosen arbitrarily as the reference node.

We may now proceed with solution by the node method. There are four node voltages in the problem, but they are not all unknown. Because of our choice of the ground point we know that $V_D = 0$, and because of the voltage source we know also that $V_A = V_0$. This leaves V_B and V_C as unknown node voltages. They may be found by simultaneous solution of the two equations expressing Kirchhoff's current law for nodes B and C.

Let us use the form of Kirchhoff's law which states that the sum of all currents entering node B is zero. The current entering node B through R_1 has, by Ohm's law, the value $(V_A - V_B)/R_1$. Since $V_A = V_0$, we may write this current as $(V_0 - V_B)/R_1$. Similarly the current entering through R_2 is $(V_C - V_B)/R_2$. The current entering node B through R_4 is $(V_D - V_B)/R_4$, but since $V_D = 0$, this current is simply $(-V_B)/R_4$. The statement of Kirchhoff's current law for node B is then

$$\frac{(V_0 - V_B)}{R_1} + \frac{(V_C - V_B)}{R_2} + \frac{(-V_B)}{R_4} = 0 \tag{1.24}$$

In similar fashion we may write the node equation for node C:

$$\frac{(V_B - V_C)}{R_2} + \frac{(-V_C)}{R_3} = 0 \tag{1.25}$$

Solving Equations (1.24) and (1.25) simultaneously for V_B and V_C, we have

$$V_C = V_0 \bullet \left[\frac{R_3 R_4}{R_1 R_2 + R_1 R_3 + R_1 R_4 + R_2 R_4 + R_3 R_4} \right]$$
$$V_B = V_0 \bullet \left[\frac{R_4 (R_2 + R_3)}{R_1 R_2 + R_1 R_3 + R_1 R_4 + R_2 R_4 + R_3 R_4} \right] \tag{1.26}$$

The node voltages are now all known, and the branch currents may be found by simple applications of Ohm's law; thus $I_1 = (V_0 - V_B)/R_1$, $I_2 = (V_B - V_C)/R_2$, $I_3 = (V_C)/R_3$, and $I_4 = (V_B)/R_4$. Recapitulating, in the node method equations representing Kirchhoff's current law are written for the nodes whose voltages are unknown, and solved simultaneously for the node voltages. Branch currents are then found from the node voltages by applications of Ohm's law.

A case requiring special treatment occurs when two nodes where the voltages are unknown are connected together by a voltage source. Consider the circuit shown in Figure 1.20(a). Here the voltage source V_2 forms the branch between nodes A and B, where the voltages are unknown. In this case one cannot write an expression for the current flowing from B to A in terms of V_B and V_A. The current through the voltage source is not a function of the voltages [see Figure 1.18(b)]. Thus the sum of currents entering node A or B cannot be written and set equal to zero in the usual way.

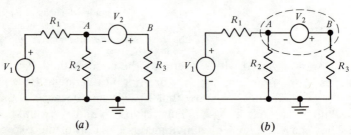

(a)   (b)

Figure 1.20   (a) A circuit in which a modified technique must be used for nodal analysis. This technique involves setting equal to zero the sum of all currents entering the region enclosed by the dashed line shown in (b).

To circumvent this difficulty, we mentally draw a dashed line around the two nodes giving difficulty, as shown in Figure 1.20(*b*). Using reasoning similar to that of Kirchhoff's current law, we can see that the sum of all currents entering the region bounded by the dashed line must be zero. This must be true since charge cannot accumulate at the nodes, and all currents flowing into one end of a wire or circuit element must flow out the other end. But it is easy to write the currents entering the dashed region and set this sum equal to zero:

$$\frac{V_1 - V_A}{R_1} + \frac{(0) - V_A}{R_2} + \frac{(0) - V_B}{R_3} = 0 \qquad (1.27)$$

This provides one equation in the two unknowns V_A and V_B. A second equation is obtained from the known property of the voltage source V_2:

$$V_A + V_2 = V_B \qquad (1.28)$$

These two unknowns V_A and V_B may now be found by simultaneous solution of Equations (1.27) and (1.28).

Often before proceeding with nodal analysis, one can simplify a circuit by reducing the number of nodes. This can be done by combining series combinations of elements into composite elements, and using the *I-V* relationship for the composite element. This eliminates the node between the two series elements that are combined.

EXAMPLE 1.18

Find the voltage V_0 in the circuit in sketch (*a*).

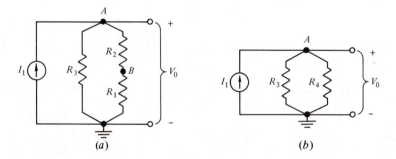

(*a*)   (*b*)

SOLUTION

The voltage V_0 is identical with the node voltage V_A. The voltage at node B is also unknown. Node B can be eliminated from the circuit, however, by replacing the series combination of R_1 and R_2 by the single resistor $R_4 = R_1 + R_2$, which has been shown to have the same *I-V* characteristic. The circuit then becomes that shown in sketch (*b*).

There is now only one unknown node voltage remaining, $V_A(\equiv V_0)$. Writing Kirchhoff's current law at node A, we have

$$I_1 - \frac{V_A}{R_3} - \frac{V_A}{R_4} = 0$$

The solution is

$$V_A = I_1 \cdot \frac{R_3 R_4}{R_3 + R_4} = I_1 \cdot \frac{R_3(R_1 + R_2)}{R_3 + R_1 + R_2}$$ ∎

EXAMPLE 1.19

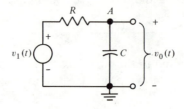

The value of the voltage source is $v_1(t)$, which varies with time. Obtain a differential equation for $v_0(t)$.

SOLUTION

We proceed by writing a statement of Kirchhoff's current law for node A, where the voltage is $v_0(t)$. To do this we use the I-V relationship for a capacitor, Equation (1.14). The node equation is

$$\frac{v_1(t) - v_0(t)}{R} + C\frac{d}{dt}[(0) - v_0(t)] = 0$$

Rearranging slightly, we have the differential equation

$$\frac{d}{dt}v_0(t) + \frac{v_0(t)}{RC} = \frac{v_1(t)}{RC}$$ ∎

The Loop Method   In the loop method of analysis, one defines special currents known as *loop currents* or *mesh currents*. These are related to the branch currents in a simple way, so that in the loop method, unlike the node method, the branch currents are obtained first. Finally, the node voltages can be found from the branch currents using the I-V relationships of the branches.

Let us again consider the circuit of Figure 1.19, which was earlier analyzed by the node method. In Figure 1.21(a) this circuit is redrawn; two mesh currents, I_1 and I_2, have been selected as shown. The mesh currents are circulating currents assumed to flow around the loops of the circuit. The branch currents can each be expressed in terms of the loop currents. For instance, in Figure 1.21(a), the current flowing from A to B through R_1 is equal to I_1. The current flowing from C to B through R_2 is equal to $-I_2$. When two mesh currents both flow through a branch, the branch current is the algebraic sum of the mesh currents. Thus the current flowing from B to D through R_4 is equal to $(I_1 - I_2)$.

There is considerable freedom in the possible choice of mesh currents.

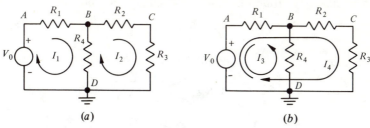

Figure 1.21   For loop analysis of this circuit, two mesh currents must be defined. One possible choice is shown in (*a*); an alternate choice is shown in (*b*).

All that is required is that every branch of the circuit has at least one independent mesh current flowing through it. The directions of the mesh currents (that is, clockwise or counterclockwise) may be freely chosen, and different directions may be used for different mesh currents in the same circuit. In planar circuits like that of Figure 1.21(*a*), one chooses as many mesh currents as there are loops in the circuit—two in that case. In more complex or three-dimensional circuits, the following formula may be used to find the number of mesh currents needed for solution:

(number of mesh currents) = (number of branches) − (number of principal nodes) + 1

For instance, in Figure 1.21(*a*) there are two principal nodes (*B* and *D*) and three branches, or paths, between principal nodes (via V_0, R_1, via R_4, and via R_2, R_3). Thus the requisite number of mesh currents is $3 - 2 + 1 = 2$. A possible alternate choice of mesh currents for this same circuit is shown in Figure 1.21(*b*). This choice is no better or worse than that of Figure 1.21(*a*). With the two mesh currents defined as in Figure 1.21(*b*), the current from *C* to *B* through R_2 is equal to $-I_4$. The current from *A* to *B* through R_1 is equal to $(I_4 - I_3)$.

The values of the loop currents are now the unknowns to be solved for. A number of equations equal to the number of unknown loop currents is needed. These equations are obtained by writing equations expressing Kirchhoff's voltage law: that the sum of the voltage drops (or voltage rises) around any closed path is zero. Any non-identical closed paths through the circuit may be used; often for convenience one chooses the paths to be the routes of the mesh currents, but this is not necessary.

As a demonstration of the procedures, let us consider the circuit of Figure 1.21(*a*), with the mesh currents I_1 and I_2 as there defined. We shall first write an equation stating that the sum of the voltage drops experienced, as one follows the closed path *A–B–D–A*, is equal to zero. The voltage drop in going from *A* to *B* through R_1 is $I_1 R_1$. The drop experienced in going from *B* to *D* through R_4 is $(I_1 - I_2)R_4$. In going from *D* to *A* through the voltage source, a drop of $-V_0$ is experienced. Setting the sum of the voltage drops around the closed path equal to zero, we have

$$I_1 R_1 + (I_1 - I_2)R_4 - V_0 = 0 \tag{1.29}$$

Proceeding similarly around the loop B–C–D–B, we obtain

$$I_2R_2 + I_2R_3 + (I_2 - I_1)R_4 = 0 \tag{1.30}$$

Solving simultaneously, we find

$$I_1 = V_0 \cdot \frac{R_2 + R_3 + R_4}{R_1R_2 + R_1R_3 + R_1R_4 + R_2R_4 + R_3R_4} \tag{1.31}$$

and

$$I_2 = V_0 \cdot \frac{R_4}{R_1R_2 + R_1R_3 + R_1R_4 + R_2R_4 + R_3R_4} \tag{1.32}$$

The individual branch currents can now simply be obtained from I_1 and I_2, as stated above. Finally, node voltages may be found by using the calculated values of I_1 and I_2 and Ohm's law. For instance

$$V_C = I_2R_3 = V_0 \cdot \frac{R_3R_4}{R_1R_2 + R_1R_3 + R_1R_4 + R_2R_4 + R_3R_4} \tag{1.33}$$

The result is in agreement with Equation (1.26), obtained via the node method.

EXAMPLE 1.20

In the circuit of Figure 1.21, calculate the current flowing from C to D through R_3, using the mesh currents of Figure 1.21(b).

SOLUTION

Let us first add the voltage drops around the closed path A–B–D–A. The resulting loop equation is

$$R_1(I_4 - I_3) - R_4I_3 - V_0 = 0$$

Next (since the choice of closed paths is arbitrary) let us use the path C–B–D–C. The resulting loop equation is

$$-R_2I_4 - R_4I_3 - R_3R_4 = 0$$

Solving simultaneously for I_3 and I_4, we have

$$I_3 = -V_0 \cdot \frac{R_2 + R_3}{R_1R_2 + R_1R_3 + R_1R_4 + R_2R_4 + R_3R_4}$$

$$I_4 = V_0 \cdot \frac{R_4}{R_1R_2 + R_1R_3 + R_1R_4 + R_2R_4 + R_3R_4}$$

The current from C to D through R_3 is equal to I_4. We note that this agrees with the value of I_2 found in Equation (1.32). ∎

It is only necessary to write as many loop equations as there are unknown mesh currents in the problem. Sometimes the value of a mesh current can be seen by inspection. For example, in Figure 1.22 I_1 is the current which goes through the current source; therefore $I_1 = 20$ mA. There is only one unknown mesh current, I_2, remaining; therefore writing a single statement of Kirchhoff's voltage law around one closed path will be sufficient. Sometimes, too, it is possible to reduce the number of unknown mesh currents, by replacing a parallel combination of elements by an equivalent single element. The circuit of Figure 1.21(a), for example, can be analyzed with only a single mesh current if R_2, R_3, and R_4 are replaced by a single resistor, of value $R_5 = R_4 \| (R_2 + R_3)$, connected between B and D.

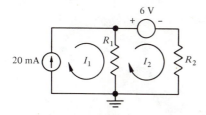

Figure 1.22 The value of the mesh current I_1 must be 20 mA, as the current passing through the current source is I_1. There is only one unknown mesh current in this circuit.

A case requiring special treatment occurs when two unknown mesh currents both pass through a current source. Consider the circuit of Figure 1.23(a). Here it is impossible to write Kirchhoff's voltage law for the closed path A–B–D–A, because one cannot write the voltage drop across a current source as a function of the current through it. One way to handle the problem is to write Kirchhoff's voltage law for the closed path A–B–C–D–A, thus obtaining an equation in the unknowns I_1 and I_2. A second equation is then obtained from the basic property of the current source: $I_0 = I_2 - I_1$. A more elegant way of handling the difficulty is to alter the choice of mesh currents to that shown in Figure 1.23(b). Now only one of the mesh currents, I_3, passes through the current source and therefore $I_3 = I_0$. There is then only one unknown mesh current, I_4, remaining in the problem. It may be found by writing a single loop equation for the closed path A–B–C–D–A.

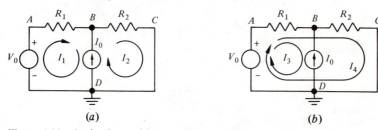

(a)   (b)

Figure 1.23 A circuit requiring special treatment when analyzed by the loop method. If two unknown mesh currents pass through the current source, as in (a), one cannot write the loop equation for the closed path A–B–D–A. The difficulty can be circumvented by a different choice of mesh currents, as in (b).

EXAMPLE 1.21
   Find the current from B to C through R_2 in the circuit of Figure 1.23.

SOLUTION
   Let us choose mesh currents as in Figure 1.23(b). Then the current asked for equals I_4, and we see by inspection that $I_3 = -I_0$. Writing an equation expressing Kirchhoff's voltage law for the closed path A–B–C–D–A we have

$$R_1(I_4 + I_3) + R_2 I_4 - V_0 = 0$$

Replacing I_3 by $-I_0$ and rearranging, we have

$$I_4(R_1 + R_2) = I_0 R_1 + V_0$$
$$I_4 = \frac{I_0 R_1 + V_0}{R_1 + R_2}$$ ∎

EXAMPLE 1.22

Obtain a differential equation for the current $i(t)$ in the circuit shown. The voltage $v(t)$ is a function of time.

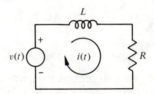

SOLUTION

The voltage drop across the inductor is given by Equation (1.19). Adding voltage drops clockwise around the circuit, we have

$$L \frac{d}{dt} i(t) + R i(t) - v(t) = 0$$

Rearranging slightly, this becomes

$$\frac{d}{dt} i(t) + \frac{R}{L} i(t) = \frac{v(t)}{L}$$ ∎

Table 1.4   The Loop and Node Methods of Circuit Analysis

I   Node Method
  1. Select datum node and locate nodes where voltage is unknown.
  2. (a) Express currents into each node as functions of known and unknown node voltages and (b) write equations stating that the sum of the currents into each node is equal to zero.
  3. Solve equations obtained in step 2 simultaneously for unknown node voltages.
  4. Obtain desired branch currents from node voltages found in step 3 and the *I-V* relationships of the branches.

II   Loop Method
  1. Select mesh currents such that at least one independent mesh current passes through each branch.
  2. (a) Express voltage drops across each element as functions of known and unknown mesh currents and (b) write equations stating that sums of voltage drops around closed paths are zero.
  3. Solve equations obtained in step 2 simultaneously for unknown mesh currents.
  4. Obtain branch currents in terms of the mesh currents found in step 3 and obtain desired node voltages from the branch currents and the *I-V* relationships of the branches.

Comparison of Node and Loop Methods Whether analysis is more easily performed by the node method or the loop method depends on the particular circuit under consideration. When one specifically needs to know a node voltage somewhere in the circuit, the node method may take fewer steps; when one needs to know a current, the loop method is favored, all other things being equal. On the other hand, if there are more loops in the circuit than there are nodes, the node method will probably require the solution of fewer simultaneous equations, and conversely. In most cases the method requiring fewer simultaneous equations is the best one to use.

The steps of the node and loop methods are recapitulated in Table 1.4.

1.5 Voltage and Current Dividers

The Voltage Divider A subcircuit known as the *voltage divider* occurs so often that it is worthy of special attention. Figure 1.24 shows a typical voltage-divider circuit. The part of the circuit which is properly the voltage divider consists of resistors R_1 and R_2 and is enclosed by the dotted line. The input terminals of the voltage divider are connected, in this case, to an ideal voltage source, and the output terminals are connected to an ideal voltmeter.

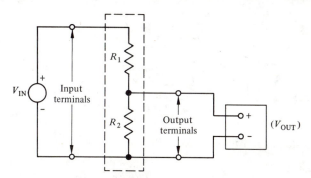

Figure 1.24 The voltage-divider circuit. The voltage indicated by the ideal voltmeter is a simple fraction of the input voltage.

We ask, for a given voltage at the input terminals, what voltage is measured by the ideal voltmeter at the output? Let us answer this question using the node method of solution. Designating the output voltage as V_{OUT}, we use Kirchhoff's current law for the node at the upper output terminal:

$$\frac{V_{IN} - V_{OUT}}{R_1} - \frac{V_{OUT}}{R_2} = 0 \tag{1.34}$$

Solving, we have

$$\boxed{V_{OUT} = V_{IN} \cdot \frac{R_2}{R_1 + R_2} \qquad \text{(voltage divider)}} \tag{1.35}$$

This is the general voltage-divider formula; the authors recommend that it be memorized. Very often formal circuit analysis can be avoided when a circuit can be recognized as a simple voltage divider.

The voltage appearing across R_1 in Figure 1.24 is equal to V_{IN} minus the voltage across R_2. Thus the voltage across R_1 is $V_{IN} (1 - R_2/[R_1 + R_2])$ $= V_{IN} (R_1/[R_1 + R_2])$. The symmetry between this expression and Equation (1.35) is readily apparent: the expression for the voltage across R_2 has R_2 in its numerator; that for the voltage across R_1 has R_1 in its numerator.

In using the voltage-divider formula, one must be careful to include all resistances which are present in the circuit. For instance, in Figure 1.25 we have a circuit very similar to that of Figure 1.24, but the voltmeter is not ideal. On the contrary, it is more like an actual physical voltmeter which, as a rule, consists of a large internal resistance in parallel with an ideal voltmeter. The circuit of Figure 1.25 is nonetheless a voltage divider, but R_3 must be regarded as being in parallel with R_2 so that the voltage-divider formula may be used.

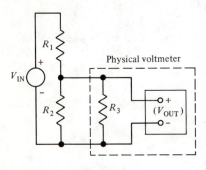

Figure 1.25 A voltage-divider circuit with a load. The resistance R_3 of the non-ideal voltmeter must be taken into account in calculating the output voltage.

The output voltage for Figure 1.25 is therefore

$$V_{OUT} = V_{IN} \frac{(R_2 \| R_3)}{(R_2 \| R_3) + R_1} \tag{1.36}$$

Expanding $R_2 \| R_3$, we have

$$V_{OUT} = V_{IN} \cdot \frac{R_2 R_3}{R_2 R_3 + R_1 R_2 + R_1 R_3} \tag{1.37}$$

EXAMPLE 1.23
Find the voltage indicated by the ideal voltmeter in the circuit shown.

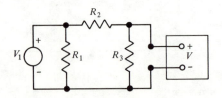

SOLUTION

We observe that the voltage across the terminals of R_1 is V_1, regardless of the value of R_1 (provided only that $R_1 \neq 0$). Using the voltage-divider formula, we have

$$V = V_1 \cdot \frac{R_3}{R_2 + R_3}$$ ∎

The Current Divider A similar subcircuit that can sometimes be recognized in larger circuits is the *current divider,* shown in Figure 1.26. In this figure the parallel combination of R_1 and R_2 gives a net resistance of $R_1 R_2 / (R_1 + R_2)$ across the current source. Therefore by Ohm's law, $(V_A - V_B) = I_{IN} \cdot R_1 R_2 / (R_1 + R_2)$. The values of I_1 and I_2 can now be found from $I_1 = (V_A - V_B)/R_1$, $I_2 = (V_A - V_B)/R_2$:

$$I_1 = I_{IN} \cdot \frac{R_2}{R_1 + R_2} \tag{1.38}$$

and

$$I_2 = I_{IN} \cdot \frac{R_1}{R_1 + R_2} \tag{1.39}$$

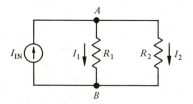

Figure 1.26 A current-divider circuit. The values of I_1 and I_2 are given by Equations (1.38) and (1.39).

Note that although there is a similarity between these equations and Equation (1.35) for the voltage divider, there is also a subtle difference. In (1.35) the fraction appearing in the expression for the voltage across R_2 has R_2 in its numerator. In (1.39) the fraction in the expression for the current through R_2 has R_1, not R_2, in the numerator. This is as one would expect intuitively; as R_1 is increased, more current is diverted to the path through R_2.

Summary

- A time-varying voltage or current which represents information is called a signal. Two common types of signal are the analog signal and the digital signal. An analog signal consists of a voltage (or current) proportional to the value of a physical quantity. In the case of digital signals, the information is first converted to numbers, and the numbers are transmitted. The electronic circuits used in connection with the two kinds of signals tend to be quite different. We refer to the two basic kinds as analog circuits and digital circuits.

- Electrical circuits are subject to basic physical laws. Among the most important of these are the two known as Kirchhoff's laws. Kirchhoff's current law states that the sum of all currents entering a node is zero. Kirchhoff's voltage law states that the sum of all voltage drops that occur around any complete closed path in the circuit equals zero.

- Circuits are interconnections of circuit elements. Each circuit element is characterized by a relationship between current and voltage, known as its *I-V* relationship. Several circuit elements can be connected together to make a composite circuit element which also has an *I-V* relationship. Except when capacitors or inductors are present, the *I-V* relationship of an element (either simple or composite) can be displayed as a graph, known as the *I-V* characteristic.

- The calculation of the voltages and currents that exist in a circuit is known as circuit analysis. Two basic methods of circuit analysis are the node method and the loop method. The process of analysis can be made easier if familiar subcircuits are recognized. Two common subcircuits are the voltage divider and the current divider.

References

Treatments at approximately the same level as this book:

Smith, R. J. *Circuits, Devices, and Systems.* New York: John Wiley & Sons, 1970.

Pederson, D. O., J. J. Studer, and J. R. Whinnery. *Introduction to Electronic Systems, Circuits, and Devices.* New York: McGraw-Hill, 1966.

Durling, A. E. *An Introduction to Electrical Engineering.* New York: The Macmillan Company, 1969.

More advanced treatments:

Desoer, C. A., and E. S. Kuh. *Basic Circuit Theory.* New York: McGraw-Hill, 1969.

Wing, Omar. *Circuit Analysis.* New York: Holt, Rinehart and Winston, 1970.

Skilling, H. H. *Electrical Engineering Circuits.* 2nd ed. New York: John Wiley & Sons, 1965.

An extensive collection of drill problems and worked examples will be found in:

Edminster, J. A. *Electric Circuits.* In Schaum's *Outline Series.* New York: McGraw-Hill, 1965.

Problems

1.1   A current flows through a wire from point A to point B. If 10^{16} electrons pass through the wire each second, what is the current in the direction from A to B?

1.2   There are 10^{20} free electrons per cubic meter in a certain crystal. If these electrons move in the $+z$ direction at an average velocity of 10^3 m/sec, what is the
magnitude of the current density? What is its direction? ANSWER: 1.6×10^4
A/m^2 in the $-z$ direction.

1.3   Between $x = 0$ and $x = L$, the voltage is given by $V(x) = V_0 \sin (\pi x/L)$ V. Find
the magnitude and direction of the force acting on an electron that is located at
$x = x_1$, if $0 < x_1 < L$.

1.4   In a TV picture tube, a thin beam of electrons is emitted by an electron gun at
the back of the tube. This beam falls through a potential increase of V_0 volts
and then strikes the screen. Let us assume that when the beam of electrons hits
the screen, some of their kinetic energy is converted to visible light and the rest
into heat. If the beam current is 10 μA and $V_0 = 15$ kV, what is the total power
being converted to heat and light at the screen?

1.5   An electron that is initially stationary falls through a potential increase of 100 V.
Find its final velocity. Express this velocity as a fraction of the velocity of light.
The mass of an electron is 9.1×10^{-31} kg.

1.6   Use Kirchhoff's voltage law to find the voltage V indicated in Figure 1.27.

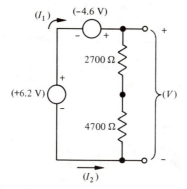

Figure 1.27

1.7   What is the current I_1 indicated in Figure 1.27? The current I_2?

1.8   A current of 2 mA flows through an 8.2-kΩ resistor in the direction from A to B.
Is A or B at higher potential? What is the magnitude of the potential difference?

1.9   Let I_1 be defined as the current through a 510-Ω resistance in the direction from
A to B. The potential at A is 6 V and that at B is -5 V. What is I_1? Note that the
sign as well as the magnitude of I_1 must be determined. ANSWER: $I_1 = +22$ mA.

1.10   A current of 0.5 mA flows through a resistance R from A to B. V_A (the potential
at A) $= 9$ V and $V_B = 3$ V. What is R?

1.11   "If one swims upstream against the current in a resistance, one experiences a
voltage rise." Verify this rule and explain its meaning.

1.12   A manufacturer supplies resistors with the following power dissipation ratings:
1/4, 1/2, 1, 2, 5, and 10 W. The resistors with higher power dissipation ratings
are larger and more expensive. In a particular application you intend to connect
a 75-Ω resistor between points A and B, where the potentials, respectively, are
-6.3 and 2.1 V. Which size resistor should you choose?

1.13   Find the resistance between terminals A and B in Figure 1.28. Assume $R_1 = 2.2$ kΩ, $R_2 = 4.7$ kΩ, $R_3 = 3.3$ kΩ, $R_4 = 10$ kΩ.

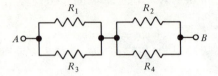

Figure 1.28

1.14   Calculate the resistance of a combination of three resistances, with values R_1, R_2, and R_3, connected in parallel.

1.15   In Figure 1.29 what is the current I_1 through the 100-Ω resistance in the direction indicated? What is the current I_2 through the voltage source?

1.16   In Figure 1.29 how much power is being produced by the voltage source? By the current source? By the resistance? Is energy conserved?

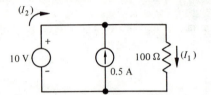

Figure 1.29

1.17   Obtain and graph the I-V characteristic of the circuit shown in Figure 1.30. For what range of voltages V does power *enter* the circuit?

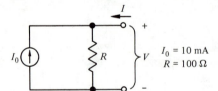

$I_0 = 10$ mA
$R = 100$ Ω

Figure 1.30

1.18   In the circuit of Figure 1.31 calculate the voltage V_1 by the node method. ANSWER: $V_1 = [R_1R_3I_0/(R_1 + R_2 + R_3)]$.

1.19   In the circuit of Figure 1.31, calculate the voltage V_1 by the loop method.

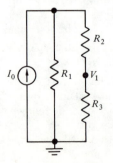

Figure 1.31

1.20   Find the current I_1, as indicated in Figure 1.32, by the loop method.

1.21   Find the current I_1 of Figure 1.32 using the node method.

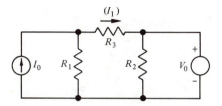

Figure 1.32

1.22   In the circuit of Figure 1.33, (1) How many principal nodes are present? (2) How many branches? (3) How many mesh currents are required to analyze this circuit?

1.23   Write a set of loop equations sufficient for a solution for I_5, as indicated in Figure 1.33. It is not asked that you solve the equations.

1.24   Write a set of node equations for the circuit of Figure 1.33, plus any additional equations necessary for calculation of I_5. It is not asked that you solve the equations.

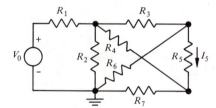

Figure 1.33

1.25   Consider the circuit of Figure 1.34.
(1) Using the node method, write equations sufficient for a solution for V_{OUT}.
(2) Solve the equations obtained in (1).

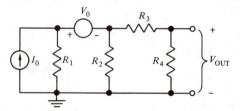

Figure 1.34

1.26   Using the loop method, calculate the current I_1 in Figure 1.35.

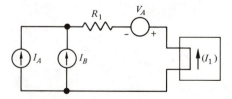

Figure 1.35

1.27   Calculate the current I_1 in Figure 1.36.

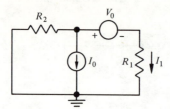

Figure 1.36

1.28   The circuit of Figure 1.37 is known as a *bridge circuit*. When the voltage V_1 equals zero, the bridge is said to be *balanced*. Suppose $R_1 = 27$ kΩ, $R_2 = 47$ kΩ, $R_3 = 10$ kΩ. What must R_4 be in order to balance the bridge?

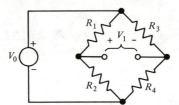

Figure 1.37

1.29   Referring to the voltage-divider circuit of Figure 1.24, what is the limit of V_{OUT} (a) as R_1 approaches infinity? (b) As R_1 approaches zero? (c) As R_2 approaches infinity? (d) As R_2 approaches zero? In each case give a physical explanation for your result.

1.30   Use the voltage-divider formula to find the voltage V_1 in Figure 1.38(a) and in Figure 1.38(b).

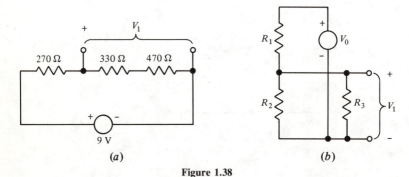

(a)   (b)

Figure 1.38

1.31   Check the answer to Problem 1.18 using the current divider formula.

1.32   Derive Equation (1.20), which states the inductance of two inductances in series.

1.33   Derive Equation (1.21), which states the inductance of two inductances in parallel.

1.34 In the circuit of Figure 1.39, the switch is opened at time $t = 0$. Find the voltage $v(t)$ across the capacitor. The value of the current source I_0 is constant.

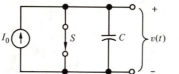

Figure 1.39

1.35 In the circuit of Figure 1.40, the value of the voltage source $v_0(t)$ varies with time. However, its value is known to have been zero at least until time $t = -12$ sec, at which time the capacitor was uncharged. Write a loop equation for the circuit. It is not asked that you solve the equation.

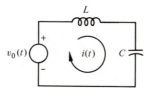

Figure 1.40

1.36 Write a node equation for the voltage $v_1(t)$ in the circuit of Figure 1.41. The value of the source voltage $v(t)$ is a function of time. It is not asked that you solve the equation.

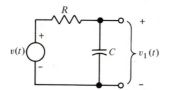

Figure 1.41

1.37 Prove that voltage $v_1(t)$ is the same in the circuits of Figure 1.42(a) and 1.42(b).

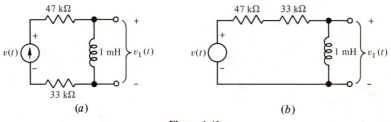

(a)   (b)

Figure 1.42

1.38   Figure 1.43(*a*) shows an infinitely long chain of resistors, all of the same value *R*. Find the resistance one would measure between its input terminals, *A* and *B*. *Suggestion:* Note that if two additional resistors are connected as shown in Figure 1.43(*b*), the resistance between the terminals *C* and *D* will be the same as that formerly measured between *A* and *B*.

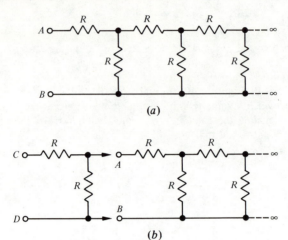

(*a*)

(*b*)

Figure 1.43

Chapter 2

Techniques of Circuit Analysis

In Chapter 1 the principles of electric circuits were discussed. Here we consider techniques by which circuit operation can be analyzed. Every engineer must be conversant with basic principles, but nonetheless every engineer is a specialist. Like medical doctors, engineers learn what kinds of problems are likely to arise. They must have not only basic knowledge, but practical and efficient techniques for treating the problems.

One simplifying technique, which is almost always used in complex circuit problems, is that of breaking the circuit up into pieces of manageable size, and analyzing the pieces individually. When this is done, some of the circuit's subunits may be recognized as already familiar. The voltage divider, introduced in Chapter 1, is an example of a circuit subunit which occurs very frequently. Experienced circuit engineers can recognize dozens of common circuits at sight, and are thus able to understand complex circuits very quickly. On the other hand, if a subunit is not familiar, it often can be reduced to a simpler form, thus making its function easier to see. Methods for doing this are the subject of Section 2.1.

Circuits containing *nonlinear* circuit elements are very common in electronics. Diodes and transistors, for example, are nonlinear circuit elements. The analysis of circuits containing nonlinear elements requires special methods. Section 2.2 is devoted to presenting two such methods. One is a simple graphical approach, handy for getting a quick idea of how a circuit will operate. The second is a numerical technique, well-suited as a basis for computer analysis of nonlinear circuits.

When energy-storage circuit elements — capacitors or inductors — are present in a circuit, analysis presents difficulties not encountered in purely resistive circuits. In general, the node and loop equations governing the circuit then become differential equations. It is customary to divide the kinds of behavior of such a circuit into two types, known, respectively, as the circuit's *natural response* and its *forced response*. In Section 2.3 these two kinds of circuit behavior are described and examples of their analysis are given. This section is particularly intended to give the reader a physical idea of the action of capacitors or inductors in a circuit. In the case of forced response, the discussion here may be regarded as an introduction to the powerful phasor techniques of Chapter 4.

In Section 2.4 a technique known as *superposition* is described. This, too, is a tool which may be invoked to simplify circuit analysis. Superposition methods are useful when more than one current or voltage source is present in a circuit.

2.1   Equivalent Circuits

Often one is interested in a subunit of a circuit only to the extent that it affects the rest of the circuit. The internal workings of the subunit do not matter in such a case; only the relationships between voltage and current at its terminals, where it is connected to the circuit of interest. One is then said to be concerned only with the *terminal properties* of the subunit. As a gross example, consider a radio receiver plugged into a standard ac power line. The radio and the power line together form a complete circuit; however, let us suppose that we are mainly interested in the operation of the radio. The power line, with its connecting wires, transformers, and generator, may then be regarded as an uninteresting subunit of the complete circuit. Fortunately, we do not need to concern ourselves with the detailed structure of the power line. All we must know are its terminal properties: what voltage is available at the power plug and what voltage variations (if any) will occur as current is drawn. A convenient way to handle such a situation is to imagine that the power line has been replaced by an *equivalent circuit,* which is a simpler entity that has nearly the same terminal properties as the original subunit. Once the power line has been replaced on the circuit diagram by a much simpler equivalent circuit, analysis of the remainder of the circuit (the radio) can proceed. Of course, when such a substitution is made, nothing further can be learned about the internal operation of the replaced subunit; by means of the equivalent circuit, the features of the original subunit have been eliminated from the problem.

We shall now state (without formal proof) a principle that will allow equivalent circuits to be found. Provided that the original subunit contains only resistances, voltage sources, and current sources, *an adequate equivalent circuit will be one which has an* I-V *relationship identical to that of the original subunit.*[1] As a simple example, consider the subunit of Figure 2.1(*a*). Using the formula for parallel resistances, the *I-V* characteristic of this subunit is seen to be given by the equation $V = (1.1 \times 10^4)I$. From the parallel resistance formula [Equation (1.11)] it is clear that the circuit of Figure 2.1(*b*) has an identical *I-V* characteristic, and therefore is an equivalent circuit for that of Figure 2.1(*a*). If subunit (*a*) appears anywhere within a larger circuit, it clearly may be replaced by (*b*) without changing the operation of the rest of the circuit. Nor can (*b*) be distinguished from (*a*) by external measurements made at their respective terminals. We note, however, that in internal behavior the two circuits are different; in (*b*) the current flows through a single element, while in (*a*) it divides and a part flows through each of three elements.

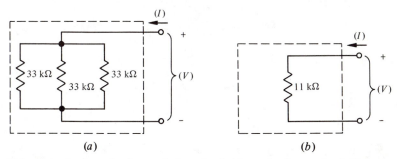

(*a*)   (*b*)

Figure 2.1   An example of an equivalent circuit. The replacement subunit (*b*) has an *I-V* characteristic identical to that of subunit (*a*). The two cannot be distinguished by measurements made at their respective terminals.

In the example just given the simplest equivalent was obvious, and could be found by inspection. However, for a large class of circuit subunits, there exist two simplest equivalent circuits. These are known as the *Thévenin* and *Norton equivalents*. They are found through application of the Thévenin and Norton theorems. The technique is based on identity of *I-V* characteristics for the original subcircuit and its equivalent.

Let us consider a circuit consisting of an arbitrary number of resistances and voltage and current sources having constant values. Let the circuit have two terminals for connection to an external circuit. Then it can be shown that its *I-V* relationship has the form $I = aV + b$. That is, when graphed on the *I-V* plane the *I-V* relationship is a straight line.

---

[1] The present discussion can be extended to circuits containing capacitances and inductances. This will be done in Chapter 4.

EXAMPLE 2.1

Consider the following composite circuit element, which has two terminals for external connection. Let I and V be defined at those terminals, as shown in sketch (a). Find the I-V relationship.

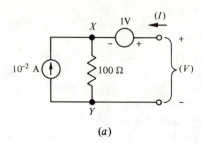

(a)

SOLUTION

Designating node voltages V_x and V_y as shown, we can write a node equation for node x:

$$10^{-2} + \frac{V_y - V_x}{100} + I = 0$$

which becomes

$$V_x = V_y + 100\, I + 1$$

The potential difference V is given by

$$V = (V_x - V_y) + 1$$

so we find that

$$V = (V_y + 100\, I + 1 + 1) - V_y = 100\, I + 2$$

By algebraic rearrangement we have

$$I = 10^{-2} V - 2 \times 10^{-2}$$

which is of the linear form $I = aV + b$, as stated above. The I-V characteristic, plotted on the I-V graph in sketch (b), is a straight line.

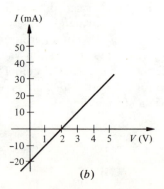

(b)

EXAMPLE 2.2

Calculate the *I-V* relationship of the following circuit and show that it is the same as that of the circuit in Example 2.1.

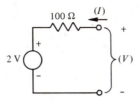

SOLUTION

Writing a loop equation we have

$$V - 100\,I - 2 = 0$$

Manipulating algebraically, we have

$$I = 10^{-2}\,V - 2 \times 10^{-2}$$

which is identical to the *I-V* relationship of Example 2.1. This circuit is therefore an equivalent for the previous circuit.   ■

EXAMPLE 2.3

The two circuits in sketches (*a*) and (*b*) have been shown (in the two previous examples) to be equivalents. Let each of these be connected to an external circuit, as shown in sketch (*c*), where the value of *R* may be freely chosen. Show that the current (*I*) indicated by the ideal ammeter is the same whether subcircuit (*a*) or its equivalent (*b*) is used, and that this is true regardless of the value of *R*.

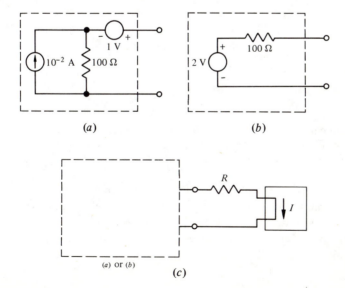

(*a*)   (*b*)

(*a*) or (*b*)

(*c*)

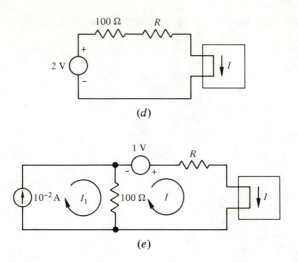

(d)

(e)

SOLUTION

When (b) is used, the entire circuit is as shown in sketch (d). From Ohm's law we immediately find that

$$I = \frac{2}{100 + R} \quad [\text{subcircuit } (b)]$$

When (a) is used, the entire circuit is as shown in sketch (e). For the sake of variety, let us use the loop method to calculate I. Two equations containing the two unknowns (I_1 and I) are obtained. The first is trivial since I_1 is controlled by the current source:

$$I_1 = 10^{-2} \text{ A}$$

The other loop equation is

$$100 (I - I_1) - 1 + IR = 0$$

Eliminating I_1, we have

$$I = \frac{2}{100 + R} \quad [\text{subcircuit } (a)]$$

which is identical to what was obtained using subcircuit (b). This example illustrates the statement that the operation of the remainder of a circuit is unaffected when a sub-circuit is replaced by its equivalent. ■

In this section we shall discuss the methods whereby two types of equivalent circuits, known, respectively, as Thévenin equivalents and Norton equivalents, may be found. Beginning first with the Thévenin type, we shall find equivalents for subcircuits composed of ideal resistances and sources with constant values. The technique is then extended to subcircuits in which time-varying sources are present. In addition to finding equivalents for ensembles of ideal elements, one can also construct an approximate equivalent for a physical element, such as a phonograph pickup or a battery; this technique is the next to be discussed. Finally, the second type of equivalent subcircuit, the Norton equivalent, is introduced.

The Thévenin Equivalent   The general form of the Thévenin equivalent circuit is shown in Figure 2.2. The values V_T and R_T may be called the *Thévenin voltage* and *Thévenin resistance*, respectively.

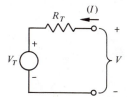

Figure 2.2 The general form of a Thévenin equivalent circuit. V_T is the Thévenin voltage and R_T is the Thévenin resistance.

In order to discuss the *I-V* characteristic, the positive directions for *I* and *V* are arbitrarily chosen to be as indicated on the figure. (These conventions are the same as previously used in connection with *I-V* characteristics. They must be adhered to when making use of equations derived in this section.[2]) Using Ohm's law, it is easily seen that the *I-V* characteristic of the Thévenin circuit is

$$I = \frac{V}{R_T} - \frac{V_T}{R_T} \tag{2.1}$$

The original circuit had an *I-V* relationship of the form $I = aV + b$. In order that the Thévenin circuit have the same *I-V* characteristic as the original circuit, it is only necessary to choose the quantities V_T and R_T so that $1/R_T = a$, $-(V_T/R_T) = b$. The problem of finding the Thévenin equivalent circuit thus amounts to evaluating V_T and R_T.

Clearly, one way to obtain the values of V_T and R_T is simply to obtain the algebraic form of the *I-V* characteristic of the original circuit, thus finding *b* and *a*. However, since only two points are needed to determine completely a straight line, finding the value of *I* for two different values of *V* amounts to finding all the information present in the *I-V* characteristic, and thus should be sufficient for finding *b* and *a*. Suppose we obtain, by study of the original

---

[2] We have throughout this book used the convention that positive current through a set of terminals is directed *into* the (+) terminal. This is consistent with the convention normally used for *I-V* relationships. Furthermore, in more complex circuits it avoids the confusion of having different conventions for different terminal pairs. However, many books do use the opposite sign convention for *I*, and regard it as positive when flowing *out* of the (+) terminal. One is of course free to use this opposite convention. In that case, however, Equation (2.5) becomes $R_T = +V_{oc}/I_{sc}$, and Equation (2.8) becomes $I_N = +I_{sc}$.

circuit, the value of V at the terminals when I is zero. (Note that this is the voltage which would be measured by an ideal voltmeter connected to the terminals.) Let us call this value V_{OC} (for "open-circuit" V). From Equation (2.1) we then have

$$0 = \frac{V_{OC}}{R_T} - \frac{V_T}{R_T} \qquad (2.2)$$

or

$$\boxed{V_T = V_{OC}} \qquad (2.3)$$

Thus if we can calculate V_{OC}, the value of V_T has been obtained. One may then proceed by obtaining, by study of the original circuit, the value of I when V is zero. (Note that this is the current which would be measured by an ideal ammeter connected between the terminals.) Let us call this value I_{SC} (short-circuit current). From Equation (2.1) we then have

$$I_{SC} = (0) - \frac{V_T}{R_T} \qquad (2.4)$$

Inserting the value of V_T obtained in Equation (2.3), we have

$$\boxed{R_T = -\frac{V_{OC}}{I_{SC}}} \qquad \text{(See footnote}^3\text{)} \qquad (2.5)$$

Thus when V_{OC} and I_{SC} have been computed, both parameters of the Thévenin equivalent can be found.

EXAMPLE 2.4
   Find the Thévenin equivalent of the circuit given in sketch (a).

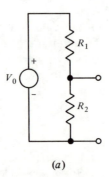

(a)

___

[3] In using this formula it should be remembered that I_{SC} is assumed positive when directed *into* the (+) terminal. Except in very unusual cases, the value of R_T is always positive.

SOLUTION

We proceed in two steps, first finding the Thévenin voltage from Equation (2.3) and then the Thévenin resistance from Equation (2.5). To obtain the former, it is necessary to calculate the open-circuit voltage, which is the voltage which would be measured by an ideal voltmeter at the terminals. In this case V_{OC} can be found by inspection, since the circuit is a voltage divider (Section 1.5). From Equations (2.3) and (1.35) we have

$$V_T = V_{OC} = V_0 \cdot \frac{R_2}{R_1 + R_2}$$

Next we calculate I_{SC}, which is the current measured by an ideal ammeter connected between the terminals, shown in sketch (b). (Note that the convention for the positive direction of I must be that used in Figure 2.1.) To simplify the calculation we choose an arbitrary point of zero potential, as shown. We then write the node equation for the node between R_1 and R_2, noting that the potential of this node is zero:

$$\frac{V_0 - (0)}{R_1} + \frac{(0) - (0)}{R_2} + I_{SC} = 0$$

from which we have

$$I_{SC} = -\frac{V_0}{R_1}$$

From Equation (2.5) we then obtain the Thévenin resistance:

$$R_T = \frac{V_{OC}\,R_1}{V_0} = \frac{R_1 R_2}{R_1 + R_2}$$

The complete Thévenin equivalent circuit is shown in sketch (c).

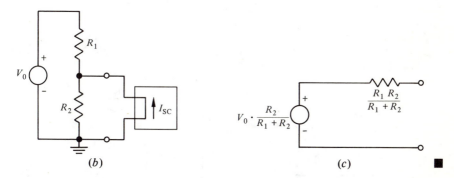

(b)   (c)   ∎

There is an alternative method for finding the Thévenin resistance which sometimes is easier to use than the one given above. The steps in the alternate method are as follows: (1) Locate all voltage and current sources inside the subcircuit whose equivalent is to be found. (2) Replace all voltage sources by short circuits. (3) Replace all current sources by open circuits. (4) The remaining circuit now contains only resistances, in series and parallel combinations. Determine what resistance now exists between the subcircuit's two terminals. This resistance is equal to the Thévenin resistance.

EXAMPLE 2.5
    Find the Thévenin resistance for the subcircuit given in sketch (a) (which is the same as that of Example 2.4) by means of the alternative method.

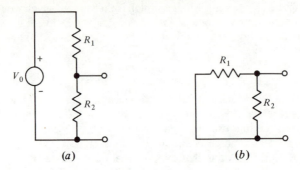

(a)   (b)

SOLUTION
    The first step is to replace the voltage source by a short circuit. The connection obtained is that shown in sketch (b). We observe that in this modified subcircuit, the parallel combination of R_1 and R_2, which is called $R_1 \| R_2$, appears across the terminals. Thus $R_T = R_1 \| R_2 = R_1 R_2/(R_1 + R_2)$, in agreement with Example 2.4. ∎

    In some cases it may not be possible to see the value of resistance between the terminals simply by inspection. In such a case the following procedure should be used. Connect a voltage source, supplying a constant test voltage V_{TEST}, across the terminals of the subcircuit. Then a current flows through the subcircuit; the value of the current is determined by the effective resistance of the subcircuit. Let us call this current I_{TEST}. The Thévenin resistance is found from Ohm's law: $R_T = V_{\text{TEST}}/I_{\text{TEST}}$.

EXAMPLE 2.6
    Find the Thévenin resistance for the subcircuit of Example 2.5 by means of a test voltage V_{TEST} applied between the terminals.

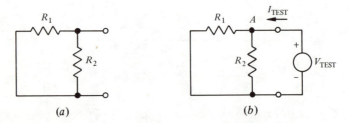

(a)   (b)

SOLUTION
    After the subcircuit's internal voltage source has been replaced by a short circuit, the modified subcircuit is as shown in sketch (a). We connect a voltage source V_{TEST} across the terminals. Now we have the circuit shown in sketch (b). Next we calculate the value of I_{TEST}, the location and direction of which are shown on the diagram. We can write a node equation for the node designated as A. Setting to zero the sum of the currents leaving the node, we have

$$\frac{V_{\text{TEST}}}{R_1} + \frac{V_{\text{TEST}}}{R_2} - I_{\text{TEST}} = 0$$

Solving

$$I_{\text{TEST}} = V_{\text{TEST}} \left(\frac{R_1 + R_2}{R_1 R_2} \right)$$

The Thévenin resistance is therefore

$$R_T = \frac{V_{\text{TEST}}}{I_{\text{TEST}}} = \frac{R_1 R_2}{R_1 + R_2}$$

as we found before.

It may seem that the method used in this example is unnecessarily elaborate compared to that of Example 2.5. However, cases where the use of V_{TEST} is necessary will later be encountered. ■

Thévenin Equivalent of a Circuit with Time-Varying Sources   The method of the previous section may be used unaltered when the original circuit contains time-varying sources. The value of the Thévenin voltage will then vary with time.

EXAMPLE 2.7
Find the Thévenin equivalent of the following circuit.

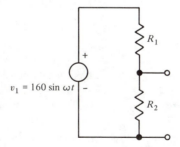

SOLUTION
The circuit is identical to that of Example 2.4 except that the constant voltage V_0 is replaced by the time-varying voltage $160 \sin \omega t$. Except for this replacement the steps of the solution are identical. The Thévenin parameters are

$$v_T = v_1 \cdot \frac{R_2}{R_1 + R_2} = 160 \sin \omega t \, \frac{R_2}{R_1 + R_2}$$

$$R_T = \frac{R_1 R_2}{R_1 + R_2}$$ ■

Finding the Thévenin equivalent of a circuit containing several dc sources presents no special difficulties. However, if more than one time-varying source is present, some care must be employed in the algebra. For instance, the sum of two sinusoidal voltages with amplitude 1 V is always a sinusoidal voltage, but the amplitude of the sum may not be 2 V. Thus caution must be used.

EXAMPLE 2.8
Find the Thévenin equivalent of the following circuit.

$$R_1 = 1000 \qquad v_1 = 10 \sin(100\,t) \text{ V}$$
$$R_2 = 2000 \qquad v_2 = 10 \cos(100\,t) \text{ V}$$

SOLUTION
We proceed by finding v_{OC} and i_{SC}:

$$v_{OC} = v_1 + v_2 = 10[\sin(100\ t) + \cos(100\ t)] \text{ V}$$
$$i_{SC} = -\frac{(v_1 + v_2)}{R_1 + R_2}$$

Now we find the Thévenin parameters

$$v_T = v_{OC}$$

and

$$R_T = \frac{-v_{OC}}{i_{SC}} = R_1 + R_2 = 3000 \ \Omega$$

Note that v_T is a sinusoid, but its amplitude is not 20. There is a trigonometric identity: $\sin x + \cos x = \sqrt{2} \sin(x + \pi/4)$, which is easily verified by expanding the right-hand side. Thus the amplitude of the sinusoid v_T in this case is equal to $10\sqrt{2}$.   ∎

Thévenin Equivalent of a Physical Element   In the previous sections attention has been directed to the replacement of a composite subcircuit containing several ideal elements by a Thévenin equivalent. The same principles may be used, however, to obtain an equivalent for a two-terminal *physical* element. Such an element is not a combination of ideal elements, but rather is a physical "black box" whose innards may not even be known. However, for a Thévenin equivalent to exist, the physical element must have a linear *I-V* characteristic. In point of fact, many physical elements do not have linear *I-V* characteristics. Even then, because it is so convenient to have a Thévenin equivalent available, one may wish to evade the restriction by assuming that the *I-V* characteristic of the physical element is at least approximately linear. In many cases the results obtained by such an approximation are accurate enough.

Replacement of the original element by a Thévenin equivalent depends on identity of the *I-V* characteristics. Two points defining the linear *I-V* characteristic of a physical element can be obtained by actual measurement. That is, a voltmeter which has a very high internal resistance (and therefore is a good approximation to an ideal voltmeter) is connected to the terminals of the physical element to measure V_{OC}, and a low-resistance ammeter is used to measure I_{SC}. Equations (2.3) and (2.5) then give the Thévenin parameters. V_T may be time-varying.

One example of a physical element to which this procedure might be applied is a battery. Another is a phonograph pickup.

EXAMPLE 2.9

It is desired to find the Thévenin equivalent of a flashlight cell. Measurement of V_{OC} with a high-resistance voltmeter gives 1.50 V, and a (very brief)[4] measurement of I_{SC} with a low-resistance ammeter gives -0.5 A.

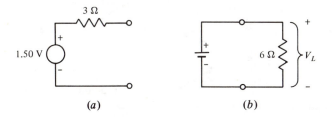

(a)   (b)

SOLUTION

Using Equations (2.3) and (2.5) we find that the equivalent is as shown in sketch (a). This example illustrates the fact that the equivalent of a battery is not simply an ideal voltage source, but rather a voltage source in series with a resistance. Practically, this makes an important difference. Even though the battery is nominally a 1.5-V cell, the voltage across its load will not, in general, be 1.5 V, due to voltage drop in the 3-Ω resistance. For instance, let a 6-Ω load be connected to the cell, as shown in sketch (b). Replacing the cell with its Thévenin equivalent and using the voltage-divider formula, Equation (1.35), we find that the voltage across the load V_L is given by

$$V_L = \frac{6}{6+3}\,1.5 = 1.0 \text{ V} \qquad \blacksquare$$

EXAMPLE 2.10

It is desired to find the Thévenin equivalent of a battery, while avoiding the necessity of a short-circuit measurement, which may damage it. Show that this may be done by measuring the open-circuit voltage of the battery, and then measuring the voltage when a known external resistance R is connected across the terminals.

---

[4] Drawing such a large current as I_{SC} for even as short a time as one second would probably cause damage to the battery. A better method is described in Example 2.10.

SOLUTION

The Thévenin voltage is obtained in the usual way, by measuring V_{OC}. Replacing the battery with its assumed Thévenin equivalent, the situation when R is connected across its terminals is as given in the sketch shown.

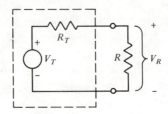

The voltage V_R across the known resistance is now measured. By use of the voltage-divider formula we see that the Thévenin voltage is related to the measured V_R by

$$V_R = V_T \cdot \frac{R}{R + R_T}$$

Rearranging, we have

$$R_T = R\left(\frac{V_T}{V_R} - 1\right)$$

Since $V_T = V_{\mathrm{OC}}$, which, like V_R is obtained by measurement, all quantities on the right are known and R_T can be found.   ∎

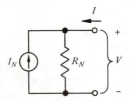

Figure 2.3 The general form of a Norton equivalent circuit. I_N is the Norton current and R_N is the Norton resistance.

Norton Equivalent   The principle of the Norton equivalent circuit is similar to that of the Thévenin equivalent. However, the equivalent circuit takes a different form, shown in Figure 2.3. As in the case of the Thévenin equivalent, the values of the source (now a current source) and resistance are to be chosen so as to give the Norton equivalent the same *I-V* characteristic as the circuit to be replaced. The *I-V* characteristic of the Norton circuit (Figure 2.3) is found (by writing a node equation) to be

$$I = \frac{V}{R_N} - I_N \tag{2.6}$$

Proceeding as in the Thévenin case, we calculate I_{SC} for the original circuit (I_{SC} being what would be measured by an ideal ammeter connected to the terminals). Inserting this value and $V = 0$ into Equation (2.6)

$$I_{\mathrm{SC}} = \frac{(0)}{R_N} - I_N \tag{2.7}$$

or

$$\boxed{I_N = -I_{\text{SC}}} \qquad \text{(See footnote}^5) \qquad (2.8)$$

Next we calculate V_{OC} (which is what would be measured by an ideal volt-meter at the terminals of the original circuit). Inserting this value and $I = 0$ into Equation (2.6)

$$(0) = \frac{V_{\text{OC}}}{R_N} - I_N \qquad (2.9)$$

or, using (2.8)

$$R_N = \frac{V_{\text{OC}}}{I_N} = -\frac{V_{\text{OC}}}{I_{\text{SC}}} \qquad (2.10)$$

By comparing Equation (2.10) with Equation (2.5) we see that the Norton resistance has the same value as the Thévenin resistance.

EXAMPLE 2.11
    Find the Norton equivalent of the subcircuit in sketch (a), whose two terminals are A and B.

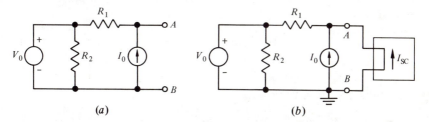

(a)   (b)

SOLUTION
    First we find I_{SC}, which is the current that is indicated by an ideal ammeter con-nected to the terminals, shown in sketch (b). To obtain the current through the ammeter, we write the node equation for node A. For convenience, let us define the potential of B as the zero of potential. Remembering that since there is no voltage drop across an ideal ammeter, the potential at A is zero, we have

$$\frac{V_0 - (0)}{R_1} + I_0 + I_{\text{SC}} = 0$$

or

$$I_{\text{SC}} = -\frac{V_0}{R_1} - I_0$$

Next, to obtain V_{OC} we find the reading of an ideal voltmeter connected to the terminals, shown in sketch (c).
    Writing a node equation for node A (at which point the voltage is V_{OC}) we have

---

[5] Remember that I_{SC} is taken as positive directed *into* the (+) terminal.

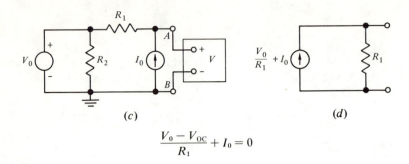

$$(c) \qquad\qquad\qquad\qquad (d)$$

$$\frac{V_0 - V_{\text{OC}}}{R_1} + I_0 = 0$$

or

$$V_{\text{OC}} = I_0 R_1 + V_0$$

Inserting into Equations (2.8) and (2.10) the values we have found for V_{OC} and I_{SC} gives

$$I_N = \frac{V_0}{R_1} + I_0$$

and

$$R_N = R_1$$

The Norton equivalent circuit is as shown in sketch (d).

It is interesting to note that the value of R_2 in the original circuit has no effect on the Norton parameters. This reflects the fact that R_2 plays no part in determining the I-V characteristics of the original circuit. ∎

2.2   Nonlinear Circuit Elements

All the circuit elements discussed up to now belong to the class known as linear circuit elements. The I-V relationships of linear elements are represented mathematically by linear algebraic equations or linear differential equations. However, other circuit elements exist which do not exhibit linear properties. In this section we shall be concerned with the important class of elements for which a graphical I-V characteristic exists, but does not have linear form. In general, the I-V characteristic of such an element is a curve of any form, which can be determined only by measurements on the nonlinear element. Our concern here is with methods for analyzing circuits containing such nonlinear elements.

Two techniques for analysis will be considered. The first, a graphical method, is useful for making estimates and for gaining physical insight into circuit operation. The second, a numerical method, is readily incorporated into a computer program, so that computerized circuit analyses can be performed.

The Graphical Method   Let us consider a circuit consisting of voltage and current sources, resistances, and a single nonlinear element which we shall call N_1. Such a circuit is shown in Figure 2.4(a). We are interested in calcu-

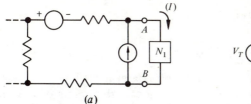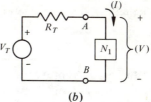

(a)   (b)

Figure 2.4 (*a*) An arbitrary circuit containing a single nonlinear element N_1, as well as voltage and current sources and resistances. (*b*) Simplified circuit obtained by replacing all the circuit except N_1 by its Thévenin equivalent.

lating the voltage across the nonlinear element $(V_A - V_B)$ and the current through it, designated in the figure as I. As a first step the problem may be simplified by breaking the circuit at the points A,B and replacing everything to the left of A,B by its Thévenin equivalent. Thus to find I and $(V_A - V_B)$, it will be sufficient to analyze the circuit of Figure 2.4(*b*). To simplify the notation, we shall define $(V_A - V_B) \equiv V$.

In order to demonstrate the graphical method, let us assume that $V_T = 5$ V and $R_T = 1000 \ \Omega$. We must also specify the I-V characteristic of N_1. Adopting the sign conventions given in Figure 2.5(*a*), let us assume that the I-V characteristic of N_1 is as shown in Figure 2.5(*b*). We shall give the mathematical function depicted in Figure 2.5(*b*) the name $I_N(V)$.

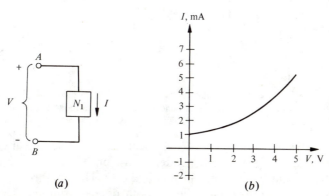

(a)   (b)

Figure 2.5 The I-V characteristic of the nonlinear element N_1. (*a*) Sign conventions; (*b*) I-V characteristic.

Our next step is to obtain the I-V characteristic for the remainder of the circuit of Figure 2.4(*b*), exclusive of N_1. This subcircuit is shown in Figure 2.6(*a*). We shall need an I-V characteristic expressed in terms of the same variables as were used in the I-V characteristic of the other part of the circuit. Therefore our variables must be chosen as shown in Figure 2.6(*a*). [Note the direction of positive I, chosen to agree with Figure 2.5(*a*).] We shall obtain I as a function of V for this subcircuit. Referring to Figure 2.6(*a*), we add voltage drops around the loop $A \rightarrow B \rightarrow V_T \rightarrow R_T \rightarrow A$. This gives

$$V - V_T + IR_T = 0 \qquad (2.11)$$

Rearranging, we find the relationship

$$I = \frac{V_T}{R_T} - \frac{V}{R_T} \qquad (2.12)$$

Clearly the graph representing this equation is a straight line. To locate the
line, two points are sufficient. It is easiest to locate the points where the line
crosses the $I = 0$ axis and where it crosses the $V = 0$ axis. We see that when
$I = 0$, $V = V_T$ when $V = 0$, $I = V_T/R_T$. Using the given values $V_T = 5$ V and
$R_T = 1000$ Ω, we see that the I-V characteristic passes through the points $I = 0$,
$V = 5$ V, and $V = 0$, $I = 5$ mA. The two points and the line through them, which
is the I-V characteristic, are shown in Figure 2.6(b). This line, the I-V characteristic of all the circuit except the nonlinear element, is known as the *load line*.
Let us refer to the mathematical function shown in Figure 2.6(b) as $I_L(V)$.

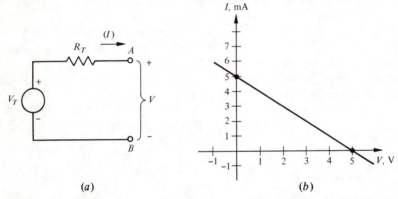

(a) (b)

Figure 2.6 I-V characteristic of part of the circuit of Figure 2.4(b) to the left of A,B.
The sign conventions, shown in (a), are chosen to agree with those of Figure 2.5(a).
The I-V characteristic is shown in (b). This curve is for $V_T = 5$ V, $R_T = 1000$ Ω.

Now the dependence of I on V, as specified by the nonlinear element, is
as shown in Figure 2.5(b). The dependence of I on V, as specified by the remainder of the circuit, is as shown in Figure 2.6(b). The point representing the
circuit's operation therefore must lie on both of these curves. Thus to find
the solution all we need to do is superimpose the two curves; the point of crossing is the solution. The two curves are shown superimposed in Figure 2.7. We
see that the values which satisfy both parts of the circuit are $I \cong 2.2$ mA,
$V \cong 2.6$ V. This point on the I-V graph is known as the *operating point* or
bias point.

Numerical Methods   Now let us reconsider the circuit of Figure 2.4(b),
using non-graphical methods which are suitable for solution by computer. We
have already chosen symbols for the two mathematical functions of V shown
in Figures 2.5(b) and 2.6(b); that of Figure 2.5(b), representing the nonlinear

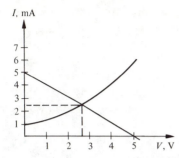

Figure 2.7 Graphical solution of the circuit of Figure 2.4(*b*). The crossing of the load line and the *I-V* characteristic of the nonlinear element is the operating point.

element, is called $I_N(V)$, and that of Figure 2.6(*b*), the load line, is called $I_L(V)$. The equation which must be satisfied is $I_L(V) = I_N(V)$, or, equivalently

$$I_L(V) - I_N(V) = 0 \qquad (2.13)$$

The quantity $I_L(V) - I_N(V)$ is itself a function of V, which we may call $f(V)$. Then we are looking for the value of V for which $f(V) = 0$. The function $f(V)$ is shown in Figure 2.8. For computer solution, this function must somehow be entered into the computer as a special function. The most usual way of doing this is to construct a mathematical function which approximates it. In our present example $I_N(V)$, the *I-V* characteristic of N_1 shown in Figure 2.5(*b*), is closely approximated by the function $10^{-3} \exp\{V/3\}$ A, where V is in volts. From Equation (2.12) we have that $I_L(V) = 10^{-3}(5 - V)$ A. Thus $f(V) = I_L(V) - I_N(V)$ is represented approximately by the function $g(V) = [5 - V - \exp\{V/3\}]$ mA. The task of the computer is to find the zero of this function.

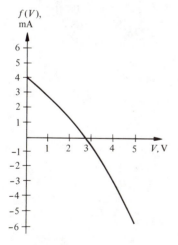

Figure 2.8 The function $f(V) = I_L(V) - I_N(V)$. The operating voltage across the nonlinear element is the value of V which makes $f(V) = 0$.

A simple method of finding the zero is available if, as is usually the case, one initially has a rough idea of the region where the zero lies. In that case one can instruct the computer to divide the region into small intervals, and calculate the value of $g(V)$ at each interval. In the example of Figure 2.8, one might instruct the computer to calculate the value of $g(V) = [5 - V - \exp\{V/3\}]$ for $V = 0$, $V = 0.1$, $V = 0.2$ ... on up to $V = 5$. The computer would then select the value of V for which $g(V)$ has the smallest absolute value. For greater accuracy, the divisions of the region could be made smaller: $g(V)$ could be calculated for $V = 0$, $V = 0.01$, $V = 0.02$,.... This would, of course, require more computer time.

A more interesting and efficient technique is that known as the *Newton–Raphson algorithm*. Here an iterative technique is used. An initial guess, which is probably quite incorrect, is made at the solution for V. The steps of the algorithm show how one can find a second guess at V which is closer to the correct answer. One proceeds by a sequence of gradually improving guesses until an answer within the desired accuracy is reached.

As an example of the Newton–Raphson technique, let us again find the zero of the function $g(V) = 5 - V - \exp\{V/3\}$. The principle of the technique is illustrated by Figure 2.9, in which the function $g(V)$ again appears. Giving to our first arbitrary guess the symbol $V^{(1)}$, let us choose $V^{(1)} = 1.0$ V. Here the superscript (1) indicates that this is the value of V obtained in the first step of the procedure. We then compute, or instruct the computer to compute, the value $g(V^{(1)})$, which we shall call $g^{(1)}$. We also compute the value of the derivative dg/dV, evaluated at $V = V^{(1)}$; we shall call this quantity $(dg/dV)^{(1)}$. Now suppose the straight line tangent to $g(V)$ at the point $V = V^{(1)}$ is constructed. This straight line intersects the $g = 0$ axis at a point which, in general, is nearer to the solution than $V^{(1)}$, as shown in the figure. The point of intersection of the tangent line with the $g = 0$ axis will be our second value for V, called $V^{(2)}$. But the slope of the tangent line is $(dg/dV)^{(1)}$. In other words, the tangent of the angle Θ shown in Figure 2.9 is equal to $-(dg/dV)^{(1)}$. Thus from geometry it is seen that

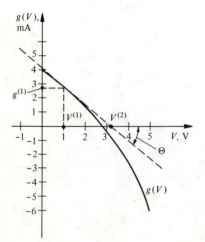

Figure 2.9 Graphical depiction of a typical iteration step in the Newton–Raphson technique. The solid curve is $g(V)$; the dashed line is tangent to $g(V)$ at $V = V^{(1)}$.

$$V^{(2)} = V^{(1)} - g^{(1)} \Big/ \left(\frac{dg}{dV}\right)^{(1)} \qquad \text{(Newton–Raphson algorithm)} \qquad (2.14)$$

Equation (2.14) states the procedure by which each successive approximation to the solution is obtained from the one before. The procedure can be terminated at the Nth approximation, when $V^{(N+1)} - V^{(N)}$ is found to be less than the required maximum error.

EXAMPLE 2.12

Find the zero of the function $g(V) = 5 - V - \exp\{V/3\}$ by the Newton–Raphson algorithm.

SOLUTION

It is convenient to set up the work in tabular form. We make the initial guess $V^{(1)} = 1.0$, then calculate successively $g^{(1)}$, $(dg/dV)^{(1)}$, $V^{(2)} = V^{(1)} - g^{(1)}/(dg/dV)^{(1)}$, and so forth. In general, one can always calculate (or instruct the computer to calculate) the derivative by means of the approximate formula

$$\frac{dg}{dV}\bigg|_{V = V^{(1)}} \cong \frac{g(V^{(1)} + \delta V) - g(V^{(1)})}{\delta V}$$

where δV is an arbitrarily chosen small number. In this case, however, an analytic expression for $g(V)$ exists, and it is easier to use the formula for dg/dV obtained by conventional differentiation:

$$g(V) = 5 - V - \exp\{V/3\}$$
$$\frac{dg}{dV}\bigg|_{V=V^{(1)}} = -1 - \frac{1}{3}\exp\{V^{(1)}/3\}$$

The calculation, in tabular form, is

N	$V^{(N)}$	$g^{(N)}$	$\left(\dfrac{dg}{dV}\right)^{(N)}$	$V^{(N+1)}$
1	1.000	2.605	−1.465	2.780
2	2.780	−0.320	−1.847	2.607
3	2.607	+0.003	−1.799	2.609
4	2.609			

We see that after three iterations the values of $V^{(3)}$ and $V^{(4)}$ differ by only 0.002, or 0.1%. The answer $V = 2.609$ is probably correct within that accuracy. ∎

It may at first seem that the Newton–Raphson technique is more laborious than the graphical method. The Newton–Raphson technique really comes into its own, however, when *several* nonlinear elements are present in a circuit. This is a common occurrence, since transistors are nonlinear elements, and most circuits contain more than one transistor. In such cases, several nonlinear equations in several variables must be solved simultaneously. This is almost impossible by graphical methods, but an appropriate form of the

Newton–Raphson technique works well even when many simultaneous non-linear equations are to be solved.

2.3 Natural Response and Forced Response

We shall now turn our attention to analysis of circuits containing *inductance* or *capacitance*. Alone among the common circuit elements, these two elements exhibit the phenomenon of *energy storage*. For example, when a capacitor is charged, energy is stored in it. Later, the capacitor can discharge through a circuit, releasing the energy. The energy-storage action of capacitors and inductors gives them a kind of "memory." For example, suppose last week a current flowed into a capacitor and charged it. If the charge has not been allowed to leak away in the meantime, the voltage on the capacitor today is determined by the current that flowed a week ago.

The operation of a circuit containing capacitors or inductors takes on, as one might expect, some of the properties of its energy-storage elements, most notably the characteristic of "memory." This characteristic manifests itself in various ways. Let us consider two different examples.

As a first example, suppose that the input to a circuit containing energy-storage elements is suddenly changed. At first the circuit, with its sense of memory, may not be much influenced by the new value of the input, so that the output may remain at the same value that it had before the input was changed. Gradually, the memory of the old input fades, while as time passes the effect of the new input becomes stronger; the output voltage of the circuit now begins to change. Finally, the circuit completely "forgets" about the old input, and the output comes to rest at the value that it would always have had, if the input had always had the new value. This kind of response to a change of input is called *transient response*. One encounters it when equipment is turned off or on; for example, the "click" heard from a radio loudspeaker, at the moment when the radio is turned on, results from the circuit's transient response. The click sound comes from transient currents which flow for a short time, until the circuit "forgets" that it has just been roused from an "off" condition.

On the other hand, suppose that the input of the circuit *continues* to change. In that case, the circuit in general is continually responding to previous changes of input, while fresh changes of input are continually occurring. For our second example, we shall consider what happens if the input voltage to the circuit is a sinusoidal function of time.[6] After such an input has gone on for a long time, the transient, caused by turning the sinusoidal input on, will have had time to die out. The output will then stabilize into a form which, like the input, is sinusoidal. However, the output at any instant is affected by the inputs

---

[6] It is assumed here that the reader has had some acquaintance with sinusoidal functions. If a review of the subject is desired, Section 4.1 on properties of sinusoids may be read at this time.

at previous times, because of the circuit's memory. The output sinusoid will, in general, be different if the maxima of the input sinusoid are far apart in time than if they are close together, and thus easier to remember from one to the next. In other words, the circuit responds differently to inputs of different *frequency*. This effect is familiar in "hi-fi" amplifiers. A manufacturer may specify that his amplifier is "flat" over 20 to 20,000 Hz. By this statement he is at once boasting that his amplifier gives equal amplification for signals of all frequencies in the range 20 to 20,000 Hz, and confessing that a signal does not receive equal treatment when its frequency lies outside this range. The response to a continuing sinusoidal input, after the starting transient has died away, is known as the *steady-state response* of the circuit to the sinusoidal input. Calculation of sinusoidal steady-state response is a common problem in circuit analysis.

The two examples given above correspond to two basic kinds of circuit behavior. In the first example, the input was suddenly changed to a new value and there held constant. The output changed to a new value in a manner characteristic of the circuit. Even if the input were suddenly removed altogether, the output would, in general, not become zero immediately, but would change to zero in a characteristic way. This behavior, which arises from the nature of the circuit rather than the details of the input, is known as the *natural response* of the circuit. In the second example, on the other hand, the output of the circuit was not free to takes its natural form because of the constantly changing input. The resulting behavior is known as the *forced response* of the circuit. In general, both types of response are simultaneously present. However, in many cases, one is primarily concerned with only one or the other of these two kinds of response. Hence it is convenient to study the two kinds of behavior separately.

Natural Response   In this section we shall discuss the natural response of circuits when excited by inputs which undergo sudden changes from one value to another. Such inputs are known as *step inputs;* one example is shown in Figure 2.10. Response to step inputs is important in connection with digital circuits, where pulses composed of voltage and current steps are commonly encountered.

Figure 2.10   An example of a time-varying voltage with a single step. At time $t = 0$ the voltage suddenly changes from V_1 to V_2.

We shall begin by establishing some rules which will be of great utility. To obtain the first rule, let us recall the *I-V* relationship of the capacitor:

$$i = C \frac{dv}{dt} \tag{2.15}$$

This equation states that the rate of change of voltage on a capacitor is proportional to the current supplied to the capacitor. In particular, the voltage across the capacitor cannot change instantaneously, as this would make dv/dt infinitely large, and by Equation (2.15), an infinite current would be required. Therefore we may conclude that *in the instant following a sudden event in a circuit (such as a step input), the voltages across the capacitors of the circuit may be considered to be the same as they were just before the event.*

Now let us shift our attention to times long after the event. Suppose for example that the event was a change of input voltage from V_1 to V_2. After a very long time, it is reasonable to expect that the currents and voltages of the circuit will have stabilized to constant values, and that these values will be just the same as if the input voltage was always V_2. But in this ultimate steady-state condition, the current through the capacitor is zero. This must be so because, from Equation (2.15), no current flows through the capacitor unless the voltage across it is changing. After the steady state has been reached, the voltages are no longer changing; *therefore at times long after a sudden event in a circuit, the currents through the capacitors must become zero and the capacitors act as open circuits.* This is our second rule. As has already been mentioned, the condition of constant voltages and currents reached after a long time is known as the steady-state response of the circuit to the input. The time-varying voltages and currents which occur between the time the input changes and the time when the steady state is reached are known as the transient response of the circuit.

Two similar (but not identical) rules govern the action of inductors. The *I-V* relationship for the inductor is

$$v = L\frac{di}{dt} \tag{2.16}$$

We see from Equation (2.16) that no instantaneous change in the current through an inductor can occur, since this would require an infinite voltage to be present. Therefore in the instant following a sudden event in a circuit, the currents through the inductors of a circuit have the same value that they had just before the event. Furthermore, we also see from (2.16) that after the transient has had time to die out, and the currents in the circuit have become constant, the voltages across the inductors must approach zero. Therefore at times long after a sudden event in a circuit, the voltages across the inductors become zero and they act as short circuits.

As an example of the application of these rules, let us consider the circuit of Figure 2.11(*a*). This particular circuit contains one resistance and one capacitance. Let us assume that the input voltage is provided by a special source which applies the voltage zero for $t < 0$ and the voltage V_1 (a constant) for $t > 0$. Such a source could consist of an ordinary dc voltage source and a switch, as illustrated in Figure 2.11(*b*). We wish to find the output voltage $v_{OUT}(t)$ for times (1) very soon after and (2) very much later than the input step.

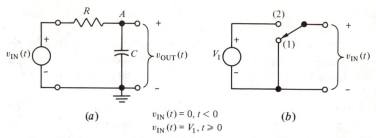

Figure 2.11   (*a*) A simple *RC* circuit with step function input. The input voltage equals zero for all times $t < 0$ and equals V_1 (a constant) for all times $t \geq 0$. (*b*) A source which would produce $v_{IN}(t)$. At time $t = 0$ the switch is moved from position (1) to position (2).

Prior to the input step, $v_{IN} = 0$ and (assuming v_{IN} has been zero for a long time) $v_{OUT} = 0$ also. Immediately after the step, the voltage across the capacitor must, according to the first rule above, still be what it was just before the step. Therefore, immediately after the time $t = 0$, the value of v_{OUT} still has the value zero.

In order to find the limit of v_{OUT} as $t \to \infty$, we take note of the second rule for capacitors, which tells us that in the steady state the capacitor current will be zero. In that case the current through the resistor, and therefore the voltage drop across the resistor, will also be zero. The potential being the same at both ends of the resistor, we see that in the steady state $v_{OUT} = V_1$.

EXAMPLE 2.13

At time $t = 0$ the value of i_{IN} is suddenly changed from I_1 to I_2 in the circuit shown. (1) What is the value of i_L, the current through the inductor, immediately after $t = 0$? (2) What is the limit of i_L as $t \to \infty$?

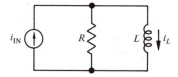

SOLUTION

(1) At times before $t = 0$, the circuit was in a steady state. Since no currents were changing then, the voltage across the inductor was zero. The voltage across the resistor was therefore also zero. Hence for $t < 0$, no current was flowing through the resistor, and i_L must have been equal to I_1. Since the current through L cannot change suddenly, immediately after $t = 0$, i_L must still equal I_1.

(2) As the steady state is reached, currents again become constant. The voltage across L approaches zero. Therefore the voltage across R approaches zero, and consequently the current through R approaches zero. Thus all of i_{IN} must flow through L and the final value of i_L is I_2. ∎

We have now seen how to calculate the value of a voltage or current at the start of the transient and at its end. An important piece of information, however, remains to be obtained. We still do not know the *duration* of the transient, or to put it another way, how long a time is required by the circuit to make its natural response to the step input and reach a new steady-state condition. In order to obtain this information, let us consider the circuit of Figure 2.11(a) in greater detail.

Regardless of what kind of elements may be connected to a node, Kirchhoff's current law must always hold. Thus we can write a node equation for node A in Figure 2.11(a). The current through C is given by Equation (2.15) one must carefully observe the sign conventions for this current, which are as stated in Equation (1.14). Using the fact that $v_A = v_{OUT}$, the node equation becomes

$$\frac{v_{OUT} - v_{IN}}{R} + C \frac{d}{dt} (v_{OUT}) = 0 \qquad (2.17)$$

We observe that the node equation has the form of a first-order differential equation; that is, no derivatives of the unknown higher than the first appear. It can be shown that any circuit containing *one* capacitance or *one* inductance is described by a first-order differential equation; consequently circuits containing one capacitor or one inductor are called *first-order circuits*. First-order circuits are important because in practice it often happens that only one capacitance or inductance in a circuit has a strong effect on the circuit's behavior. Real circuits actually have many small, accidental capacitances and inductances in their wiring, but in many cases only a single large capacitance has much effect on operation. In this chapter we shall consider only first-order circuits.

In the present case, we are interested in evaluation v_{OUT} at times greater than zero. For such times $v_{IN} = V_1$. The equation obeyed by v_{OUT} for $t > 0$ can be obtained from Equation (2.17) with this substitution. To simplify the calculations, it is convenient at this point to add a term dV_1/dt to the equation. Since V_1 is a constant, its derivative equals zero; thus no error is made by introducing this term. The equation for v_{OUT} now has the form

$$\frac{d}{dt}(v_{OUT} - V_1) + \frac{(v_{OUT} - V_1)}{RC} = 0 \qquad (2.18)$$

It now remains to find a solution $v_{OUT}(t)$ which satisfies this equation. The solution must also have the property that $v_{OUT} = 0$ for values of t only slightly greater than zero, as we have already shown. The function $v_{OUT}(t)$ which satisfies these requirements is given in Equation (2.23); the reader may wish to substitute this function back into Equation (2.18) and test its behavior near $t = 0$, thus affirming that it is the solution. However, since it is of interest to see how the solution is found, we shall show how it is done.

We proceed by rearranging Equation (2.18) into the form

$$\frac{d(v_{\text{OUT}} - V_1)}{(v_{\text{OUT}} - V_1)} = -\frac{1}{RC} \, dt \tag{2.19}$$

We may now integrate both sides of the equation, obtaining

$$\ln(v_{\text{OUT}} - V_1) = -\frac{t}{RC} + K \tag{2.20}$$

where K is a constant of integration, the value of which is as yet undetermined. Exponentiating both sides of Equation (2.20), we have

$$v_{\text{OUT}} - V_1 = e^K \, e^{-t/RC} \tag{2.21}$$

The value of e^K can now be found from the "initial condition," that is, from the fact that at $t = 0$, v_{OUT} must equal zero. When $v_{\text{OUT}} = 0$ and $t = 0$ are substituted into Equation (2.21), we find that

$$0 - V_1 = e^K \tag{2.22}$$

Actually it is only the value of e^K that we need, and this is given directly by Equation (2.22). (The value of K itself is a complex number, but that need not concern us here.) Substituting $e^K = -V_1$ into Equation (2.21), and rearranging, we have

$$v_{\text{OUT}} = V_1 (1 - e^{-t/RC}) \tag{2.23}$$

which is the desired solution. The behavior of v_{OUT} as a function of time is graphed in Figure 2.12, with time shown in terms of the ratio t/RC. We see that for large values of t, v_{OUT} approaches as a limit the value V_1, in accordance with our earlier prediction. The output shown in Figure 2.12 arises from an input which, after the initial step, is held constant. Therefore the duration of the transient is not determined by the properties of the input, but rather is characteristic of the circuit itself. In other words, the output we have found illustrates the circuit's natural response.

Referring again to Figure 2.12, we observe that for times earlier than $t = RC$, the output is still near its initial value; for times later than $t = RC$, the

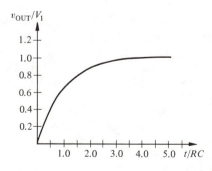

Figure 2.12 Output voltage of the circuit shown in Figure 2.11(a) as a function of time. This output is the response to a step function input voltage which changes from zero to the value V_1 at time $t = 0$.

output nears its final value. Thus we can adopt, somewhat arbitrarily, the time RC as an order of magnitude for the duration of the transient. Since we are dealing with the circuit's natural response, this time is a property of the circuit and not of the details of the input. Therefore we expect that the transient resulting from any sudden event at time $t = 0$ will have a duration on the order of RC. We shall refer to this characteristic time RC as the *time constant* of the circuit.

Having found the duration of the transient for the circuit of Figure 2.11(*a*), we can now state a simple method for finding it, for the general class of RC circuits containing one capacitor. Figure 2.11(*a*) has the form of a Thévenin equivalent circuit connected in parallel with a capacitor; therefore any circuit connected in parallel with a capacitor may be reduced to Figure 2.11(*a*). One must only find the Thévenin resistance R_T of the subcircuit connected across the capacitor terminals. The circuit time constant is then equal to $R_T C$. The natural response of the circuit will always exhibit transients of this duration.

EXAMPLE 2.14
 In the circuit in sketch (*a*), the capacitor is initially uncharged, before switch S_1 is closed at time $t = 0$.
 (1) What is the value of the output voltage v_3 before the switch S_1 is closed?
 (2) What is the value of v_3 at the instant just after the switch S_1 is closed?
 (3) What is the value of the output long after the switch is closed?
 (4) Approximately how long must we wait for the output to reach this value?

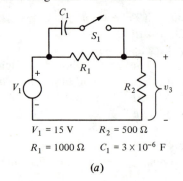

$V_1 = 15$ V $R_2 = 500\ \Omega$
$R_1 = 1000\ \Omega$ $C_1 = 3 \times 10^{-6}$ F

(*a*)

SOLUTION
 (1) Before the switch is closed, the circuit consists of a simple voltage divider. Hence before the switch is closed, $v_3 = R_2 V_1 / (R_1 + R_2) = 5$ V.
 (2) Since the capacitor is initially uncharged, the voltage across it must be zero prior to and therefore immediately after the switch closes. Hence immediately after the switch closes the voltage across R_2 (the output voltage) must equal V_1. Thus $v_3 = 15$ V just after $t = 0$.
 (3) At a time long after the closing of the switch, no current flows through the capacitor, which is then equivalent to an open circuit. Hence for large t the circuit (with C_1 replaced by an open circuit) is a voltage divider composed of R_1 and R_2. Referring to Equation (1.35) we see that the steady state value of v_3 is $R_2 V_1 / (R_1 + R_2) = 5$ V.

(4) To determine the time for the output voltage to make the change from 15 down to 5 V, we compute the time constant. We are interested in the time constant for the case $t > 0$, that is, with the switch closed. The subcircuit connected across the capacitor is thus as shown in sketch (b).

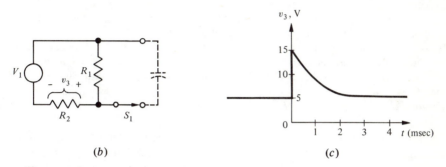

(b) (c)

The Thévenin equivalent of a subcircuit of this form was obtained in Example 2.4. There it was found that $R_T = R_1 \| R_2 = R_1 R_2 / (R_1 + R_2)$. Thus the circuit time constant is given by

$$R_T C = \frac{R_1 R_2}{R_1 + R_2} \cdot C$$

Using the given values for the circuit elements, we find that the time constant has the value 10^{-3} sec. The duration of the transient is of this order of magnitude. A sketch of $v_3(t)$ is shown in sketch (c).   ∎

Similar general rules apply to first-order circuits containing one inductor. To find the time constant of such a circuit, one obtains the Thévenin equivalent of all the circuit except the inductor, as seen from the two points where the inductor is connected. The time constant is then equal to L/R_T. (The proof of this statement is deferred to the problems.)

EXAMPLE 2.15
    In the circuit in sketch (a), the switch is opened at time $t = 0$. Find the current i_L through the inductor: (1) just before the switch is opened; (2) just after the switch is opened; (3) a long time after the switch is opened. (4) What is the approximate duration of the transient?

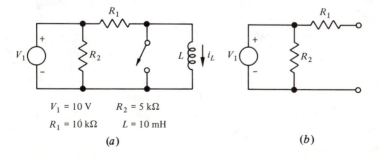

$V_1 = 10$ V $R_2 = 5$ kΩ
$R_1 = 10$ kΩ $L = 10$ mH

(a) (b)

SOLUTION

(1) Before $t = 0$, the switch was closed, short-circuiting the inductor. The current i_L was then zero.

(2) Since the current through the inductor cannot change suddenly, it must also be zero just after the switch is opened.

(3) A long time after the opening of the switch, the voltage across L must have dropped to zero. The current i_L is then equal to V_1/R_1.

(4) To find the time constant it is necessary to find the Thévenin equivalent of the subcircuit in parallel with the terminals of L. Since the switch is open during the period of interest, this subcircuit is as shown in sketch (b). For this subcircuit $V_{OC} = V_1$ and $I_{SC} = -V_1/R_1$; therefore $R_T = -V_{OC}/I_{SC} = R_1$. The time constant, and hence the duration of the transient, is $L/R_T = L/R_1 = 10^{-6}$ sec $= 1$ μsec. ■

To summarize, in the above we have described a simple technique predicting the natural response of first-order circuits. This simple technique, based on the idea of a circuit time constant, does not give the precise mathematical form of the transient, but it does provide its initial and final values and its duration, which are enough for most purposes. If the precise mathematical form of the transient should be needed, a differential equation method like that of Equations (2.17) through (2.23) can be used.

Before leaving the subject of natural response, it is interesting to consider its relevance to digital circuits. Digital signals often have the form of nearly rectangular pulses, as shown in Figure 2.13(a). Let us suppose that the circuit of Figure 2.11(a) exists somewhere in a digital system, and that its input voltage v_{IN} has the form shown in Figure 2.13(a). Let us further suppose that the circuit time constant RC is much smaller than the pulse duration T_1. Then the sequence of events is this: When the initial rise of v_{IN} occurs, a transient response takes place. We have already found the shape of the resulting v_{OUT}, as shown in Figure 2.12. By assumption, the duration of the transient is much less than the pulse duration T_1. Therefore the system is in a new steady state for most of T_1, until the downward fall of v_{IN} causes a second transient of the same duration as the first. The approximate form of the circuit's output is thus shown in Figure 2.13(b). We see that the output pulse shape is only slightly distorted, compared with v_{IN}.

Now let us suppose that the circuit time constant RC is larger, so that it is comparable with T_1. In that case, the initial transient will not have time to approach its limiting value before the second, downward transient begins. The expected form of the output pulse is then as shown in Figure 2.13(c). We see that the output pulse is now quite badly distorted. Finally, if the time constant is made much larger than T_1, the output shown in Figure 2.13(d) occurs. The distortion of the pulse is now extreme; and moreover, because the upward transient hardly gets started before the downward one begins, the maximum value attained by the output pulse is small compared with the value of the input. This would be highly undesirable in a digital system, where the diminished pulse might be interpreted as a nonpulse, or space; that is, as a "zero" instead of a "one."

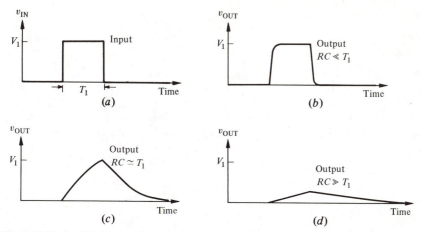

Figure 2.13 Let the input voltage to the circuit of Figure 2.11(a), v_{IN}, be as shown in (a). If the circuit time constant RC is much less than the input pulse duration T_1, an output of the form shown in (b) is obtained. If $RC \simeq T_1$ the output has the form shown in (c). For the case $RC \gg T_1$, the output shown in (d) is obtained.

The moral of this discussion is that for correct handling of pulsed signals, sufficiently short circuit time constants are required. Stray wiring capacitances and the internal capacitances of circuit elements like transistors are always present, and in general it is impossible to reduce the value of the circuit time constant to zero. Its value should be estimated by the circuit designer, however, and care should be taken that it is not so large as to interfere with the intended operation of the circuit.

Forced Response   When the input voltage of a circuit is constantly being changed by an external signal source, the circuit's behavior is known as its *forced response*. Although in general the changing input could be any function of time, the case of by far the greatest interest is that of a *sinusoidal* input. It is only this case that we shall consider here.

Let us again consider the circuit of Figure 2.11(a). However, let us choose a different form for the input voltage $v_{IN}(t)$. In the discussion of natural response we used an input voltage which was a step function; now to study forced response we shall let $v_{IN}(t) = V_1 \cos \omega t$. The circuit then appears as shown in Figure 2.14.

Once again we can proceed by writing a node equation for node A, where the voltage is v_{OUT}. Setting the sum of the currents flowing out of this node equal to zero, we have

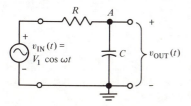

Figure 2.14   To study forced response we use the same circuit previously considered in the discussion of natural response. The input voltage, however, is now a sinusoid.

$$\frac{v_{\text{OUT}} - V_1 \cos \omega t}{R} + C \frac{d}{dt} v_{\text{OUT}} = 0 \tag{2.24}$$

Upon rearrangement, this becomes

$$\frac{d}{dt} v_{\text{OUT}} + \frac{1}{RC} v_{\text{OUT}} = \frac{1}{RC} V_1 \cos \omega t \tag{2.25}$$

The solution for v_{OUT} can be obtained by the standard techniques of differential equations. Here it is sufficient to state the solution, which is

$$v_{\text{OUT}}(t) = V_2 \cos(\omega t + \phi) \tag{2.26}$$

where

$$V_2 = \frac{V_1}{\sqrt{1 + \omega^2 (RC)^2}} \tag{2.27}$$

$$\phi = \tan^{-1}(-\omega RC) \equiv \cos^{-1} \frac{1}{\sqrt{1 + (\omega RC)^2}} \tag{2.28}$$

[The last statement in (2.28) follows from the trigonometric identity $\tan^2 x + 1 = \sec^2 x$.] The output voltage is a sinusoid with the same frequency as the input voltage. Its amplitude has the value V_2, given by Equation (2.27). The quantity ϕ is a constant known as the *phase,* or *phase angle*; its value is given by Equation (2.28). The phase angle is often an important quantity; however, our principal concern for now will be the output amplitude V_2. That the function given in Equation (2.26) does indeed satisfy the differential equation (2.25) can be demonstrated by substitution. The check is left as an exercise.

A graph of Equation (2.27), which states the amplitude of the output sinusoid V_2 as a function of frequency, is shown in Figure 2.15(*a*). Because it is difficult to see what happens over a large range of frequencies on a linear plot like Figure 2.15(*a*), it is common to graph functions of this kind on a log-log plot. A log-log plot of the same function is shown in Figure 2.15(*b*). We see

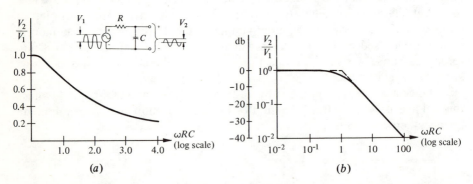

Figure 2.15 Plots of the ratio of output amplitude to input amplitude for the circuit of Figure 2.14, as given by Equation (2.26). Figure (*a*) is a plot on linear coordinates; (*b*) is a log-log plot of the same function, showing its behavior over a much greater frequency range. A decibel scale is also provided in (*b*).

that on the log-log plot, the function $V_2(\omega)$ is asymptotic to a pair of straight lines. These lines, if projected to the point where they meet, intersect at $\omega RC = 1$, that is, at $\omega = 1/RC$. This frequency is known as the *break frequency* or *corner frequency* of the circuit. Interestingly, the corner frequency is found to be equal to the reciprocal of the time constant found in the discussion of the circuit's natural response.

Two vertical scales are shown in Figure 2.15(*b*). In addition to the logarithmic scale of V_2/V_1, there is another scale calibrated in *decibels* (db). A decibel scale is often used as a means of comparing the value of some quantity (in this case V_2/V_1) to some arbitrary reference value (in this case chosen to be unity). In general, if the quantity being described is V and the reference value is V_0, then V is said to be above the reference value by a number of decibels given by[7]

$$\boxed{\text{decibels} = 20 \log_{10} (V/V_0)} \qquad (2.29)$$

EXAMPLE 2.16

By how many decibels is V_2/V_1 above the reference level of unity when $V_2/V_1 = 0.5$?

SOLUTION

The number of decibels is given by Equation (2.29):

$$\begin{aligned} \text{db} &= 20 \log_{10} (0.5/1.0) = 20 \log_{10} (0.5) \\ &= 20 \, (-0.301) = -6.02 \end{aligned}$$

The result is negative because V_2/V_1 is less than the reference level; one can also say V_2/V_1 is 6.02 db *below* the reference level. In practice one rounds the number off; it is customary, when the numbers differ by a factor of 2, to say that one is 6 db below the other.   ∎

We observe from Figure 2.15(*b*) that above the break frequency, the slope of the function V_2/V_1 is -20 db per decade; that is, every time the frequency increases by a factor of 10, V_2/V_1 drops by another 20 db.

The log-log graph of amplitude versus frequency is known as a *Bode plot*. (A full-fledged Bode plot also has a second curve showing the phase as a function of frequency.) Bode plots are almost always used instead of linear plots such as Figure 2.15(*a*). This particular kind of circuit is known as a *low-pass filter*. As is seen from the Bode plot, low-frequency sinusoids are passed from input to output, but high-frequency sinusoids are "filtered out."

From Figure 2.15(*b*) we observe that at frequencies well below $1/RC$, $V_2/V_1 \cong 1$. In fact, referring to Equations (2.26) and (2.28), we see that

---

[7] The unit being defined here is known more specifically as the *voltage decibel*. There is another unit also in use called the *power decibel*, defined by $\text{db} = 10 \log_{10} (P/P_0)$. Power decibels are used when power levels, rather than voltages or currents, are being compared. In this book the word "decibel" always means "voltage decibel."

$v_{OUT}(t) \cong v_{IN}(t)$ in this frequency range. On the other hand, for frequencies above $1/RC$, V_2/V_1 decreases to values much less than 1. These observations can be understood through the following argument. Let a sinusoidal voltage $v(t) = V_C \sin \omega t$ be applied to a capacitor. The current which flows is $i(t) = C(dv/dt) = C\omega V_C \cos \omega t$; thus the current is a sinusoid with amplitude $C\omega V_C$. We observe that when ω approaches zero, the current approaches zero; that is, the capacitor acts like an open circuit. On the other hand, when ω becomes very large, very large currents flow through the capacitor even for small voltage amplitudes; thus in this limit the capacitor acts more like a short circuit. From this point of view the form of Figure 2.15(b) can be understood. Looking again at the circuit, Figure 2.14, we think of the capacitor as being replaced by an open circuit at low frequencies. Then at low frequencies no current flows through R, and there is no voltage drop across R; consequently $v_{OUT} = v_{IN}$. On the other hand, when the frequency is high, the capacitor acts more like a short circuit, and one expects the voltage measured across a short circuit to be zero. These features are precisely the ones seen in Figure 2.15(b).

It is possible to obtain output voltages in other circuits containing one capacitor by using the results just obtained for Figure 2.14. For example, consider the circuit of Figure 2.16. The loop equation for this circuit is identical with that for Figure 2.14; therefore the loop currents are the same. The voltage across the capacitor is therefore the same in the two circuits, since it is determined by the current flowing through the capacitor. Thus we can find the voltage v_{OUT} in Figure 2.16 by subtraction:

$$v_{OUT} = V_1 \cos \omega t - v_C \tag{2.30}$$

where v_C is the voltage across the capacitor, already given in Equations (2.26) to (2.28). Thus the output in Figure 2.16 is given by

$$
\begin{aligned}
v_{OUT} &= V_1 \cos \omega t - V_2 \cos(\omega t + \phi) \\
&= V_1 \cos \omega t - V_1 \cos \phi \cos(\omega t + \phi) \\
&= V_1 [\cos \omega t - \cos^2 \phi \cos \omega t + \sin \phi \cos \phi \sin \omega t] \\
&= V_1 \sin \phi \; [\sin \phi \cos \omega t + \cos \phi \sin \omega t] \\
&= V_1 \sin \phi \sin(\omega t + \phi) \\
&= -V_1 \bullet \frac{\omega RC}{\sqrt{1 + (\omega RC)^2}} \bullet \sin(\omega t + \phi)
\end{aligned}
\tag{2.31}
$$

[The last step above is by Equation (2.28), using the trigonometric identity $\sin x = \cos x \bullet \tan x$.] Thus the output in Figure 2.16 is a sinusoid with amplitude V_3 given by

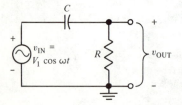

Figure 2.16   A simple RC circuit. The value of v_{OUT} may be found from v_{OUT} for the circuit of Figure 2.14.

$$\frac{V_3}{V_1} = \frac{\omega RC}{\sqrt{1 + (\omega RC)^2}} \tag{2.32}$$

This function of ω is graphed on log-log coordinates in Figure 2.17. As before, the break frequency is $1/RC$, and the form of the curve is consistent with the idea that a capacitor acts like an open circuit at low frequencies and like a short circuit at high frequencies. The results of Equations (2.31) and (2.32) could also, of course, be obtained by solving a differential loop or node equation for the circuit. This circuit acts in the opposite way from that of Figure 2.15; it is a *high-pass filter*.

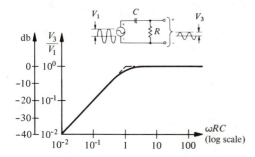

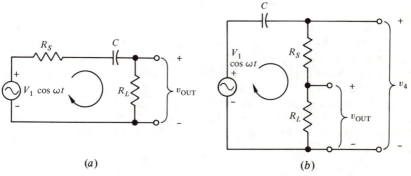

Figure 2.17   Amplitude of output sinusoid for the circuit of Figure 2.16, as a function of frequency.

Voltages in other single-capacitor circuits with sinusoidal excitation can be found by using Figures 2.15 and 2.17. Often the more simple circuits can either be recognized as being of the form of Figure 2.14 or 2.16, or else can be converted into one of those forms by replacing all the subcircuit connected in parallel with C by its Thévenin equivalent.

EXAMPLE 2.17
   From sketch (*a*) deduce the amplitude of the output sinusoid V_0 as a function of frequency.

(*a*)   (*b*)

SOLUTION
   The circuit has the same loop equation and therefore the same loop current as the one given in sketch (*b*). Hence the voltage across R_L (which equals the loop current times R_L) is the same in both circuits, and if we can find the voltage across R_L in the second circuit, we have the desired answer.

The second circuit, however, has the same form as Figure 2.16. Thus V_4, the amplitude of the sinusoid v_4, can be obtained from Equation (2.32) with $(R_S + R_L)$ substituted for R. The desired voltage v_{OUT} is related to v_4 by the voltage-divider formula: $v_{OUT} = v_4 R_L / (R_L + R_S)$. Thus

$$V_0 = \frac{R_L}{R_L + R_S} V_4 = \frac{R_L}{R_L + R_S} \cdot \frac{\omega(R_S + R_L)C}{\sqrt{1 + \omega^2 C^2 (R_S + R_L)^2}} \qquad \blacksquare$$

EXAMPLE 2.18

Sketch the amplitude of the sinusoid v_{OUT}, shown in sketch (a), as a function of frequency.

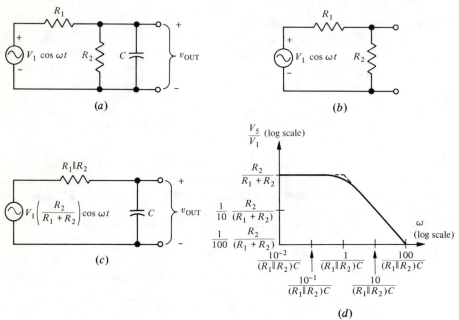

(a)

(b)

(c)

(d)

SOLUTION

This circuit can be reduced to the form of Figure 2.14 by replacing everything in parallel with the capacitor terminals by its Thévenin equivalent. The subcircuit to be replaced is shown in sketch (b). Its Thévenin equivalent was found earlier, in Example 2.7. The result was $v_T = V_1 \cos \omega t \cdot R_2/(R_1 + R_2)$, $R_T = R_1 \| R_2$. Thus the output of the original circuit is the same as the output of the circuit in sketch (c) with R_T and v_T as stated above. This circuit in turn is identical with Figure 2.14 with $R = R_T$, $v_{IN} = v_T$. The output amplitude, which we may call V_5, is therefore given, according to Equation (2.27), by

$$\frac{V_5}{V_1} = \frac{\dfrac{R_2}{R_1 + R_2}}{\sqrt{1 + \omega^2 C^2 (R_1 \| R_2)^2}}$$

and the break frequency, comparing with Figure 2.15, is $[(R_1 \| R_2)C]^{-1}$. Graphed against ω on a log-log plot, the function V_5/V_1 appears as shown in sketch (d). $\blacksquare$

Circuits with One Inductor   Principles similar to those described above for RC circuits apply to circuits containing a single inductor. Let us consider the circuit of Figure 2.18(a). Writing a loop equation for the loop current i_L, we have

$$V_1 \cos \omega t - i_L R - L \frac{di_L}{dt} = 0 \tag{2.33}$$

The solution of this equation, as may be verified by substitution, is

$$i_L = \frac{V_1}{\sqrt{R^2 + (\omega L)^2}} \cos(\omega t + \phi) \tag{2.34}$$

where

$$\tan \phi = -\frac{\omega L}{R} \tag{2.35}$$

The value of v_{OUT} is given by $v_{\text{OUT}} = L \, di_L/dt$:

$$v_{\text{OUT}} = \frac{-V_1 L\omega}{\sqrt{R^2 + (\omega L)^2}} \sin(\omega t + \phi) \tag{2.36}$$

Thus the output voltage is a sinusoid with amplitude V_L given by

$$\frac{V_L}{V_1} = \frac{\omega(L/R)}{\sqrt{1 + (\omega L/R)^2}} \tag{2.37}$$

The dependence of V_L on ω is shown in Figure 2.18(b). We observe that the break frequency is located at $\omega L/R = 1$; that is, at $\omega = R/L$.

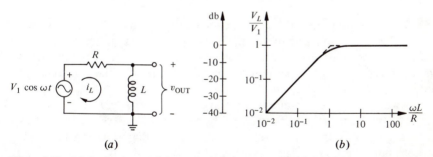

(a)   (b)

Figure 2.18   (a) A circuit containing a single inductance with sinusoidal input voltage. (b) Log-log plot of the amplitude of the output sinusoid versus frequency.

It is seen that at frequencies greater than L/R, $V_L/V_1 \simeq 1$. For frequencies less than L/R, V_2/V_1 decreases to values much less than 1. As in the case of RC circuits, a simple line of reasoning can be used to explain these properties. The voltage across an inductor is given by $v = L(di/dt)$. If $i = I_0 \sin \omega t$, $v = L\omega I_0 \cos \omega t$. Thus the ratio (voltage amplitude)/(current amplitude) for an inductor equals $L\omega$. This ratio is small if ω is given near zero, and we may think of the behavior of an inductor as approaching that of a short circuit at low frequencies. At high frequencies the ratio (voltage amplitude)/(current amplitude) becomes large; hence at high frequencies an inductor acts more like an open circuit. It is easy to see that the curve of Figure 2.18(*b*) is consistent with this line of reasoning.

Other circuits containing a single inductor can be analyzed by applying Equation (2.36), using methods similar to those described for *RC* circuits. Often the simple line of physical reasoning used in the preceding paragraph is adequate for an estimate of circuit behavior.

EXAMPLE 2.19
    Find the amplitude of the output sinusoid at very low frequencies and at very high frequencies. What is the break frequency?

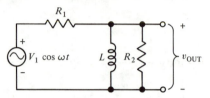

SOLUTION
    At low frequencies the inductor acts like a short circuit, so the output amplitude will approach zero. At high frequencies the inductor acts like an open circuit; in that case the voltage-divider formula applies, and the output amplitude is approximately $V_1 \cdot [R_2/(R_1 + R_2)]$. The circuit can be reduced to that of Figure 2.18(*a*) by making a Thévenin equivalent of the subcircuit connected across the terminals of the inductor. The Thévenin resistance of this subcircuit is $(R_1 \| R_2)$; therefore the break frequency is $(R_1 \| R_2)/L$.   ∎

The foregoing discussion is intended to introduce the behavior of some of the more common circuits in the sinusoidal steady state, and to provide some intuitive feeling for their operation. However, much more powerful techniques than these are available for analysis of circuits in the sinusoidal steady state. Chapter 4 is entirely concerned with a technique known as *phasor analysis,* by which circuits containing arbitrary numbers of *R*'s, *L*'s, and *C*'s can be analyzed, without the use of differential equations. Readers who wish to know more about sinusoidal steady-state response will thus obtain a much deeper understanding of this subject upon reading Chapter 4.

2.4   Superposition

All circuits containing only ideal resistances, capacitances, inductances, and sources are "linear" *circuits.* That is, they are described by linear differential equations. For linear circuits, one can make the following statement: If more than one voltage source or current source is present in any given circuit, the voltage or current at any point in the circuit is equal to the sum of the voltages or currents which would arise from each voltage source or current source acting individually when all other sources are set to zero. This statement, which is known as the principle of superposition, is helpful in analyzing circuits containing more than one source.

The procedure for making use of the principle of superposition is the following. One regards all the voltage and current sources in the circuit, except one, as being set to zero. The voltage (or current) at the desired place in the circuit is then calculated. The process is then repeated with each of the other sources individually turned on. The sum of the individual contributions is the desired voltage or current.

It should be noted that a voltage source, when set to zero, has the effect of making the potential difference between its terminals equal to zero. Therefore a voltage source with value zero is identical with a short circuit. A current source with value zero has zero current flowing through it, and hence is identical to an open circuit. (One sometimes refers to a source whose value has been set to zero as being "turned off." One should remember that this does not mean that its controlling action ceases entirely. Rather, the current or voltage is still controlled, but is constrained to have the value zero.)

EXAMPLE 2.20

Find the current indicated by the ideal ammeter in the circuit in sketch (a).

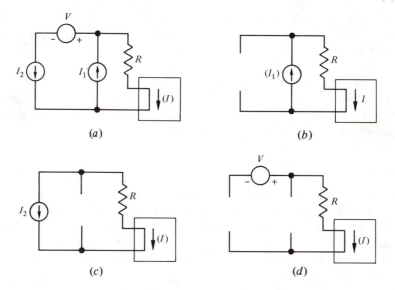

(a)   (b)

(c)   (d)

SOLUTION

To use superposition, we first consider I_1 to be turned on and the other two sources turned off. The situation is then as shown in sketch (b). Let us give the contribution to the ammeter current arising from I_1 the name I_{11}. From sketch (b), it is clear that $I_{11} = I_1$.

When only I_2 is on, as shown in sketch (c), evidently $I_{12} = -I_2$. When only the voltage source is on, the situation is as shown in sketch (d). No current can flow in this circuit, so the contribution from V, I_V, must equal zero. According to the principle of superposition, the actual value of I, the current through the ammeter, is given by

$$I = I_{11} + I_{12} + I_V$$

Substituting the values we have found for I_{11}, I_{12}, and I_V, we have

$$I = I_1 - I_2 \qquad \blacksquare$$

The principle of superposition can be used when both ac and dc sources are present in the circuit. In such cases, it is usually convenient to analyze the circuit separately for ac and for dc. When capacitors are present and the frequency of the ac signal is high enough for the capacitors to be regarded as short circuits, the effective circuits for ac and dc may differ.

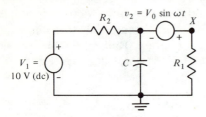

Figure 2.19   Circuit containing both ac and dc voltage sources. It is convenient to use the principle of superposition.

Consider the circuit of Figure 2.19, which contains dc and ac voltage sources. Let it be assumed that ω is large enough for the capacitor to be regarded as a short circuit for ac, although it is of course an open circuit for dc. Let us find the voltage at node x, which we shall call v_x. Making use of the principle of superposition, we first find the contribution of the dc voltage source. In this part of the problem the capacitor is an open circuit. Source v_2, which is set to zero, is a short circuit. Therefore we can use the voltage-divider formula to obtain v_{x1}, the contribution to v_x from source V_1:

$$v_{x1} = 10 \cdot \frac{R_1}{R_1 + R_2}$$

To find the ac part of v_x, which we shall call v_{x2}, we reverse the procedure and turn on only the ac source. Since the frequency is high, the capacitor must now be regarded as a short circuit. Therefore source v_2 is connected directly across R_1, and the ac part of the output voltage v_{x2} is given simply by $v_{x2} = V_0 \sin \omega t$. The total value of v_x is now simply the sum of v_{x1} and v_{x2}:

$$v_x = v_{x1} + v_{x2} = \frac{10\,R_1}{R_1 + R_2} + V_0 \sin \omega t$$

Summary

- Equivalent circuits are circuits which cannot be distinguished from each other by measurements at their terminals. Often circuit analysis can be simplified if a portion of the circuit is replaced by a simpler equivalent. Two general families of equivalents exist for linear circuits: Thévenin equivalents and Norton equivalents.

- Nonlinear circuit elements are those whose *I-V* relationships cannot be expressed as linear equations. Special methods must be used to analyze circuits containing nonlinear elements. Graphical methods are convenient for quickly obtaining insight into circuit operation. However, when several nonlinear elements are present in a circuit, it may be better to use numerical methods such as the Newton–Raphson method. The latter is well-suited for computerized circuit analysis.

- Circuits containing capacitance or inductance are characterized by two types of behavior: natural response and forced response. The natural response of such a circuit is seen when the input voltage is suddenly changed. The circuit may then take a finite length of time to respond to the change. The changes which take place in the circuit after the input has ceased to change are known as the circuit's transient response. The transient response is "natural" because it is a characteristic of the circuit, not of its input. Forced response is seen when the circuit is constantly driven by a continually changing input. The most important case is that where the changing input is sinusoidal in form. In this case the behavior of the circuit is found to be a function of the frequency of the input sinusoid.

- The principle of superposition states that when more than one voltage source or current source is present in a linear circuit, the voltages and currents throughout the circuit are the sum of those which would exist if each source were separately turned on, one by one. Superposition can be used as an aid in circuit analysis.

References

Treatments at approximately the same level as this book:

Smith, R. J. *Circuits, Devices, and Systems.* New York: John Wiley & Sons, 1970.

Pederson, D. O., J. J. Studer, and J. R. Whinnery. *Introduction to Electronic Systems, Circuits, and Devices.* New York: McGraw-Hill, 1966.

Brophy, J. J. *Basic Electronics for Scientists.* New York: McGraw-Hill, 1966.

At a more advanced level:

Desoer, C. A., and E. S. Kuh. *Basic Circuit Theory.* New York: McGraw-Hill, 1969.

Skilling, H. H. *Electrical Engineering Circuits,* 2d ed. New York: John Wiley & Sons, 1965.

Numerical methods for nonlinear circuits are discussed in:

McCracken, D. D., and W. S. Dorn. *Numerical Methods and FORTRAN Programming.* New York: John Wiley & Sons, 1964.

Problems

2.1   Show that the subcircuit of Figure 2.20(*a*) is approximately equivalent to that of Figure 2.20(*b*).

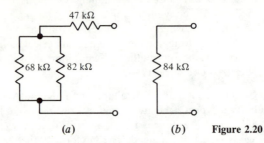

(*a*)   (*b*)   **Figure 2.20**

2.2   Obtain and graph the *I-V* characteristic of the subcircuit shown in Figure 2.21(*a*).

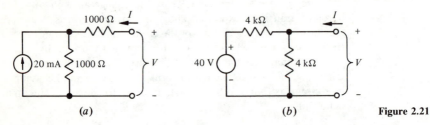

(*a*)   (*b*)   **Figure 2.21**

2.3   Obtain and graph the *I-V* characteristic of the subcircuit shown in Figure 2.21(*b*). Is this subcircuit equivalent to that of Figure 2.21(*a*)?

2.4   Obtain the Thévenin equivalent of the subcircuit shown in Figure 2.21(*a*).

2.5   Obtain the Thévenin equivalent of the subcircuit shown in Figure 2.21(*b*).
ANSWER: $V_T = 20$ V, $R_T = 2$ kΩ.

2.6   Find the Norton equivalents of the subcircuits shown in Figure 2.21(*a*) and 2.21(*b*).

2.7   Consider the subcircuit shown in Figure 2.22. Suppose *V* and *R* are adjustable, but let the ratio *V/R* always be kept equal to I_1, a constant. By constructing a Norton equivalent, show that this subcircuit approaches equivalence to an ideal current source when $V \to \infty$.

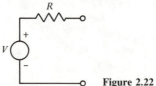

Figure 2.22

2.8   Find the Thévenin resistance of the subcircuit of Figure 2.21(*a*) using the alternate method in which sources are set to zero.

2.9   Find the Thévenin equivalent of the subcircuit shown in Figure 2.23.

2.10   Find the Thévenin equivalent of the subcircuit shown in Figure 2.23, if the 2-A current source is changed to 1 A. (Can the formula $R_T = -V_{OC}/I_{SC}$ be used in this case?)

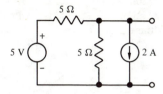

Figure 2.23

2.11   A 10-Ω resistance is connected across the terminals of Figure 2.23. Calculate the current through this resistance in the following ways:
(1) Directly, by means of the node method.
(2) By first constructing the Thévenin equivalent of Figure 2.23.
(3) By means of superposition.

2.12   A battery, whose nominal voltage is 6 V, is found by experiment to have the I-V characteristic shown in Figure 2.24. Suppose that we attempt to construct a Thévenin equivalent for it by measuring V_{OC} and I_{SC}. Graph the I-V characteristic of the equivalent circuit that is obtained, and compare with the actual I-V characteristic of the battery. Why do they not agree?

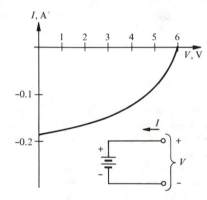

Figure 2.24

2.13   The battery whose I-V characteristic is shown in Figure 2.24 is connected across a 40-Ω resistor.
(1) Use a graphical method to find the voltage across the resistor and the current flowing through it. ANSWER: Using the sign conventions of Figure 2.24, $V = 4.5$ V, $I = -0.115$ A.
(2) Let the battery be replaced by the Thévenin equivalent found in Problem 2.12. Find the voltage across the resistor and current through it in this case.

2.14   Consider the circuit of Figure 2.4(b). Let $V_T = 10$ V, $R_T = 100$ Ω. Furthermore, let the I-V characteristic of the nonlinear element N_1 be given by $I = 0.002 \ V^2$, where I is in amperes and V is in volts.
(1) Write a node equation and solve for the voltage across N_1.
(2) Find the voltage across N_1 graphically.

2.15   Use the Newton–Raphson algorithm to find the value of x which satisfies $e^x = 1$.

2.16   Use the Newton–Raphson algorithm to find the solution of $e^x + x = 0$.

2.17   For the circuit of Figure 2.25:
   (1) Find the voltage across C before the switch is closed.
   (2) Find the voltage across C immediately after the switch is closed.
   (3) Find the voltage across C a long time after the switch is closed.
   (4) Estimate the duration of the transient.
   (5) Sketch the voltage across C as a function of time.

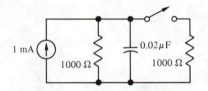

1 mA   1000 Ω   0.02 μF   1000 Ω

Figure 2.25

2.18   In the circuit of Figure 2.26, the switch is closed at time $t = 0$.
   (1) What is the current I through the inductor just after the switch is closed?
   (2) What is I long after the switch is closed?
   (3) What is the approximate duration of the transient?
   (4) Sketch I as a function of time.

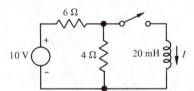

6 Ω   10 V   4 Ω   20 mH   I

Figure 2.26

2.19   Suppose that the coiled filament of a certain light bulb is equivalent, approximately, to a 30-Ω resistance and a 15-mH inductance connected in series. Suppose further that a certain battery is approximately represented by a Thévenin equivalent with $V_T = 1.5$ V, $R_T = 20$ Ω. A switch is suddenly closed, connecting the battery across the bulb. Estimate the time required for the current to increase to a sizable fraction of its steady-state value.

2.20   Refer to Figure 2.13, which illustrates distortion of a rectangular pulse by an RC circuit. If the incoming pulse has magnitude V_1, what is the largest value the circuit time constant RC can have so that the maximum value of v_{OUT} will reach (1) 0.75 V_1? (2) 0.9 V_1? (3) 0.99 V_1?

2.21   Referring to Figure 2.12
   (1) What is the value of v_{OUT}/V_1 when $t = RC$?
   (2) How many db is v_{OUT} below V_1 when $t = RC$?
   (3) At what value of t does $v_{\text{OUT}} = 1/2\ V_1$?
   (4) Discuss the real meaning of the statement that the duration of the transient approximately equals RC.

2.22   We have seen in Figure 2.15(b) that a first-order RC circuit with natural response time RC had a break frequency equal to $(RC)^{-1}$. Give a physical explanation for the appearance of the same factor RC in both expressions.

2.23 In Figure 2.27, the input voltage is a sinusoid with amplitude 10 V. Estimate the amplitudes of the voltages $v_1(t)$ and $v_2(t)$, (1) when the frequency of the input sinusoid is low, and (2) when it is high. (3) Estimate the break frequency which marks the approximate boundary between low-frequency and high-frequency behavior of the circuit. (4) Sketch the appearance of a log-log plot of the amplitude of v_1 versus ω. (5) Sketch a log-log plot of the amplitude of v_2 versus ω.

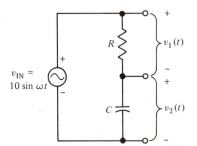

Figure 2.27

2.24 Repeat Problem 2.23, if the capacitor in Figure 2.27 is replaced by an inductor L.

2.25 Sketch a log-log plot of the amplitude of the sinusoidal current $i_1(t)$ which flows through R_1 in Figure 2.28, as a function of frequency. Let $C = 1\ \mu\text{F}$, $R_1 = 5\ \Omega$, $R_2 = 10\ \Omega$.

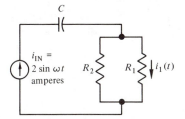

Figure 2.28

2.26 A certain dc voltage increases from 6 to 16 V. Express the change in decibels. What is the change, in decibels, if a voltage increases from 60 to 160 V? If it changes from 160 to 60 V?

2.27 An octave is a change in frequency by a factor of 2. (Thus high C on a piano keyboard, which is one musical octave above middle C, has a frequency two times that of middle C). In Figure 2.17 the increase of V_3/V_1 as a function of ω, at frequencies below the break frequency, can be expressed as "X db per octave." Find the value of X. Do the same for Figures 2.15 and 2.18. Can you state a generalization for first-order circuits?

2.28 Use superposition to calculate the voltage V in the circuit of Figure 2.29.

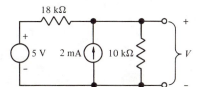

Figure 2.29

2.29 Calculate $v(t)$ in the circuit of Figure 2.30. Assume that all transients have died
      away, so only the forced response is present.

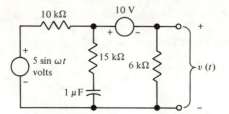

Figure 2.30

2.30 Sketch a graph of the function $y = x^3 - 3x + 4$. Suppose an attempt is made to
      find the value of x for which $y = 0$, by means of the Newton–Raphson method,
      but an unfavorable initial guess is made (for example, $x = 0.8$ or $x = 1.5$). By
      drawing tangents on your graph, study the question of whether the solution will
      eventually be found. Can you suggest some criteria for the initial guess which
      would help guarantee success of the method?

2.31 Verify that Equations (2.26) to (2.28) give a solution for Equation (2.25). (The
      following trigonometric identities are useful: $\tan^2 x + 1 = \sec^2 x$; $\sin x = \tan x \cdot
      \cos x$.)

2.32 In the circuit of Figure 2.31, the switch is closed at time $t = 0$.
      (1) Write a loop equation for the circuit.
      (2) Find the current I immediately after the switch is closed.
      (3) Find a solution for the equation obtained in (1) consistent with the initial
          condition found in (2).

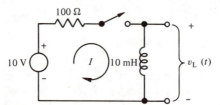

Figure 2.31

2.33 In the circuit of Figure 2.31 write and solve a node equation for $v_L(t)$. Assume
      the switch is closed at time $t = 0$.

2.34 Write a FORTRAN program to perform the Newton–Raphson algorithm.

Chapter 3

Active Circuits and Building Blocks

Resistances, capacitances, and inductances belong to the class known as *passive* circuit elements. A circuit composed entirely of these elements is called a passive circuit. A property of a passive circuit is that the output signal power is less than or equal to the input signal power.

Circuits which are not bound by this restriction are known as *active* circuits. A "hi-fi" amplifier is an example of such a circuit. Watts of power are delivered to the speaker when only milliwatts enter the input. Of course, an active circuit cannot create power out of nothing. Typically, an active circuit has input connections, output connections, and an additional connection through which raw power is supplied. The signal power output, plus the losses of the circuit, must equal the signal power input plus the raw power input. The difference between passive and active circuits is illustrated by Figure 3.1.[1] The graphs display the input power $v_{IN} \times i_{IN}$ and output power $v_{OUT} \times i_{OUT}$ as functions of time. It is clear that the circuit of Figure 3.1(*b*) must be active, because the output signal power exceeds the input signal power. The practical importance of active circuits is extremely great. In fact, "electronic circuit" is almost synonymous with "active circuit."

---

[1] In addition to removing energy, the passive circuit (or the active circuit) can change the shape of a signal. No change of shape is indicated in Figure 3.1.

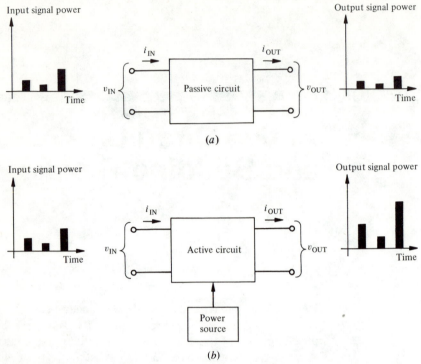

Figure 3.1   The output signal of a passive circuit (*a*) has power less than or equal to that of the input signal. The active circuit (*b*) can produce an output signal power greater than the input signal power.

An active circuit must contain at least one *active circuit element*. The function of an active element is to manufacture a large output signal from raw input power, following instructions derived from the small input signal. At present, the most important active element is the bipolar junction transistor (also called the "BJT" or simply, for short, the "transistor"). Transistors will be introduced when we consider the internal design of active circuits, beginning in Chapter 6.

Recently, there has been a trend away from the use of single active elements in circuits, and toward use of larger active units, which we may call *building blocks*. This development, which is one of great importance, results largely from the advent of integrated circuits. It is now possible to obtain building blocks, in integrated circuit form, which contain numerous transistors, and can by themselves perform entire circuit functions. These devices are often no larger than individual transistors, and may even be no more costly.

As a result the engineer can design systems that are much more sophisticated and complex than would be feasible with individual transistors; yet when it comes to assembly-line production, assembly cost may be reduced, since now only a few interconnections between building blocks need be made. Thus it is not surprising that there is a strong trend in electronic design away from the use of discrete circuit elements, and toward the use of building blocks.

To use modern building blocks, one must have two kinds of knowledge. The first kind has to do with the blocks' external properties: what they do; what kinds exist; how their properties are specified. This, however, is not enough. To attempt to design circuits with building blocks without understanding their insides, is like trying to compose music for the piano without ever having seen one, and not knowing whether it is played with the hands or the feet. Thus a twofold approach must be taken to the subject. In this chapter some of the principal electronic blocks are introduced, and their functions and properties are described. In later chapters we shall go on to consider the internal workings of the blocks.

Clearly, if an electronic block is to be useful, one must have some way of describing what it does. In many cases, particularly in analog circuits, the best way to do this is to construct a *model* of the block. The model is a collection of simple circuit elements which mimics the operation of the actual block, and thus can be used to represent or describe it. A family of special circuit elements, known as *dependent sources,* are used in such models; dependent sources are introduced in Section 3.1. Then in Section 3.2, the circuit models themselves are introduced. With the technique of modeling available, we shall then be ready to describe the most common building block of analog circuits, the *amplifier,* in Section 3.3.

As mentioned in Chapter 1, digital circuits are generally quite different from analog circuits, so it is no surprise that digital building blocks are quite different from analog blocks. The building-block technique is probably even more important in digital work, however, because of the large numbers of circuit blocks needed in a system as large as a computer. The functions and properties of some of the more common digital blocks are discussed in Section 3.4.

3.1 Dependent Voltage and Current Sources

It is very useful, in developing the subject of active devices and circuits, to introduce the concept of dependent voltage and current sources. A dependent source is one whose value cannot be directly changed or adjusted independently of the rest of the circuit. Rather, its value is directly dependent on the value of voltage or current at some other place in the circuit, which we shall call its *reference*. For example, consider Figure 3.1A (*a*), which shows a dependent current source. That the source is dependent is seen from the fact that its value is not specified independently of the rest of the circuit. On the contrary, the diagram indicates that the dependent source has the value βi_1, where β is a constant and i_1 is the reference current. The reference current may be specified as being any current in a circuit, and the dependent source may appear anywhere in the circuit. An example is given in Figure 3.1A (*b*). Except for the fact that the value of the dependent source is specified in this special way, voltages and currents may be calculated using the usual techniques.

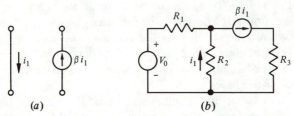

(*a*)   (*b*)

Figure 3.1A   An example of a dependent current source. In (*a*) the value of the current source is β times the current i_1, where β is some specified constant. In (*b*) the dependent source is shown in a circuit. The current βi_1 is controlled by the current i_1, as shown.

EXAMPLE 3.1

Find the voltage across the resistor R_3 in the circuit of Figure 3.1A (*b*). The values of the parameters are $\beta = 50$, $R_2 = R_3 = 10 \text{ k}\Omega$, $R_1 = 100 \ \Omega$, $V_0 = 0.05$ V.

SOLUTION

The circuit is redrawn below. An arbitrary reference node (or ground point) has been chosen, as indicated.

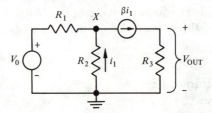

Writing a nodal equation at node *x*, where the potential is V_x, we have

$$\frac{V_0 - V_x}{R_1} - \frac{V_x}{R_2} - \beta i_1 = 0$$

The unknown V_x may be expressed in terms of i_1 by applying Ohm's law to R_2:

$$V_x = -i_1 R_2$$

Solving the two equations above for i_1

$$i_1 = \frac{V_0}{R_1\left(\beta - 1 - \dfrac{R_2}{R_1}\right)}$$

The output V_{OUT} is equal to $\beta i_1 R_3$. Thus

$$V_{OUT} = \frac{\beta R_3 V_0}{R_1\left(\beta - 1 - \dfrac{R_2}{R_1}\right)}$$

$$V_{OUT} = -98 \ V_0 = -4.9 \ \text{V}$$

Note that throughout this example, lower-case symbols have been used for i_1. This is because i_1 in general can be time-varying, even though in this case, with constant input voltage, i_1 happens to be constant. ∎

Both voltage and current sources may be dependent, and their values can depend on either a reference current or a reference voltage. Several examples are shown in Figure 3.2.

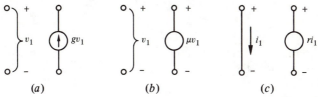

(a)   (b)   (c)

Figure 3.2 Examples of dependent sources. In (a) the current gv_1 is controlled by the voltage v_1. In (b) the voltage μv_1 is controlled by the voltage v_1. In (c) the source voltage ri_1 is controlled by the current i_1. The parameters g, μ, and r are specified constants.

EXAMPLE 3.2
Find the voltage v_{OUT} as a function of v_{IN}.

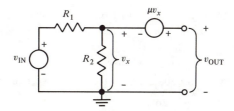

SOLUTION
We need solve for only one unknown node voltage, that called v_x. When this has been found, it will be possible to obtain the value of v_{OUT} from

$$v_{OUT} = v_x + \mu v_x = v_x(1 + \mu)$$

The node equation for v_x is

$$\frac{v_{\text{IN}} - v_x}{R_1} + \frac{(0) - v_x}{R_2} = 0$$

Thus

$$v_x = v_{\text{IN}} \frac{R_2}{R_1 + R_2}$$

$$v_{\text{OUT}} = v_{\text{IN}} \frac{R_2}{R_1 + R_2} (1 + \mu) \qquad\blacksquare$$

An important point to remember is that, although *independent* voltage and current sources can be turned off, for instance in the use of the principle of superposition, one is not at liberty to arbitrarily set *dependent* sources to zero. In general this would produce a contradiction, since the value of the dependent source must always be controlled by the value of current or voltage at its reference point, and the latter might not be zero. The principle of superposition can still be used in active circuits, but only the independent sources are turned on, one at a time; the dependent sources are always left free to do their thing.

EXAMPLE 3.3

Use the principle of superposition to find the current (I) indicated by the ideal ammeter in the circuit in the sketch.

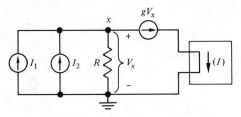

SOLUTION

Let us first find the contribution to I from source I_1 alone; we shall call this contribution I_{11}. This can be found from the value of V_x with only I_1 turned on; we call this value V_{x11}.

$$I_{11} = g\,V_{x,11}$$

To find $V_{x,11}$ we write a node equation for node x, with only source I_1 turned on:

$$I_1 + \frac{(0) - V_{x,11}}{R} - g\,V_{x,11} = 0$$

from which we have

$$V_{x,11} = \frac{I_1 R}{1 + gR}$$

The contribution to I from source I_1 is therefore

$$I_{11} = g \ V_{x,11} = \frac{g \ I_1 R}{1 + gR}$$

When only source I_2 is turned on, identical results are obtained, except that I_1 is replaced by I_2. Hence

$$I_{12} = \frac{g \ I_2 R}{1 + gR}$$

The final result is

$$I = I_{11} + I_{12} = \frac{g \ R (I_1 + I_2)}{1 + gR} \qquad\blacksquare$$

3.2 Linear Models for Active Devices and Blocks

As was pointed out in the introduction to this chapter, it is often useful to think of entire circuits as active building blocks. However, before one can make use of such a block, its properties must somehow be known. A way of specifying or describing its properties must be made available.

To illustrate this problem, let us consider the block shown in Figure 3.3(a). Suppose that a series of measurements is made of v_{OUT} as a function of v_{IN}. These data describe, at least partially, the action of the block; they are presented in tabular form in Figure 3.3(b). The data points can be graphed, giving an alternative, graphical description of the block, as shown in Figure 3.3(c). Clearly the block is nonlinear; v_{OUT} is not linearly proportional to v_{IN}.

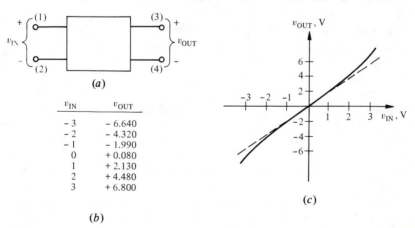

v_{IN}	v_{OUT}
− 3	− 6.640
− 2	− 4.320
− 1	− 1.990
0	+ 0.080
1	+ 2.130
2	+ 4.480
3	+ 6.800

(a)

(b)

(c)

Figure 3.3 Describing the properties of an active block. (a) Circuit symbol for the block; (b) measured values of v_{OUT} as a function of v_{IN}; (c) graph of v_{OUT} versus v_{IN}. The dashed curve is a straight line which fits the $v_{OUT} - v_{IN}$ curve in the range $-1 \ V < v_{IN} < + 1 \ V$.

In Chapter 2 graphical and numerical methods for analyzing circuits containing nonlinear elements were described. Such techniques could be used to analyze circuits containing the block under discussion. But this kind of analysis is slow and unwieldy. Fortunately there is a more efficient method, based on the use of a model for the block.

A model for a block is a collection of simple ideal circuit elements which simulates the behavior of the original block. The model has the same number of terminals as the original block. To analyze a circuit containing the block, one simply replaces the block by its model. This creates a new circuit containing only ideal elements, which now can be analyzed straightforwardly, using the techniques of Chapter 1.

Returning to the block of Figure 3.3, we observe that its I-V characteristic is almost a straight line, so long as v_{IN} is between -1 V and $+1$ V. Thus in this range of input voltages, the block is described approximately by $v_{OUT} = 2\,v_{IN}$. A model which simulates this behavior is shown in Figure 3.4. Thus, to the extent that the block is described by the equation $v_{OUT} = 2\,v_{IN}$, the model is an adequate substitute for the block.

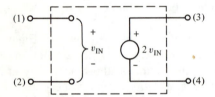

Figure 3.4 A possible model for the block of Figure 3.3. The model is most accurate in the range -1 V $< v_{IN} < +1$ V.

The block may, however, have other properties which we have not considered. If, for example, v_{OUT} depends not only on v_{IN} but also on another circuit condition—say, the output current—then the model could be made more elaborate, so as to simulate this additional behavior as well. In general, one uses the simplest model that is adequate to describe all phenomena that are of interest.

The technique of modeling is useful not only for electronic blocks, but also for individual active devices, such as transistors. It is even useful for passive elements, for example the "real" inductor of Figure 1.11. All that is necessary is that the block or device being modeled have properties that are approximately linear over the range of operating conditions in which one is interested, so that a collection of ideal elements that simulates its action can be made.

One way in which a model can be found for an element is by constructing an approximate Thévenin "equivalent," using the methods of Chapter 2. In Example 2.9, for instance, the Thévenin "equivalent" of a flashlight cell was found. A flashlight cell, like any real physical element, is a nonlinear circuit element; that is, its I-V characteristic is not a straight line over the entire range of voltages and currents. However, so long as the voltage across its terminals

is not too large, the battery and the Thévenin equivalent have *I-V* characteristics that are nearly identical. The battery symbol is shown in Figure 3.5(*a*),
and the Thévenin circuit found in Example 2.9, which we now recognize as a
model for the cell, is shown in Figure 3.5(*b*).

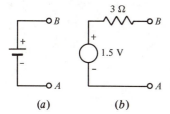

Figure 3.5   A model for a flashlight cell, obtained by constructing its Thévenin equivalent
as was done in Example 2.9. The circuit symbol for the cell is shown in (*a*); the model is
shown in (*b*).

Let us now see how the model is used. Suppose that the flashlight cell is
used in the circuit of Figure 3.6(*a*) and it is desired to know the current *I* flowing through the 10-Ω resistor in the direction shown. We may proceed by substituting the device model for the cell, as shown in Figure 3.6(*b*). By a simple
application of Ohm's law, it is now clear that $I = +1.5 \text{ V}/13 \ \Omega = +0.115\text{A}$. We
see that the necessity for graphical or numerical solution is eliminated. Once
the model is introduced, one need only analyze a circuit containing ideal linear
elements.

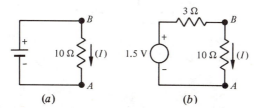

Figure 3.6   (*a*) A circuit containing a flashlight cell.
(*b*) The same circuit, with the model of Figure 3.5(*b*)
substituted for the cell.

It should be noted that in substituting the model of Figure 3.5(*b*) into the
circuit, it is essential that terminal *A* of the model be placed where terminal *A*
of the cell was connected, and the same with terminals *B*. Clearly, if the terminals of the model were reversed, an incorrect answer would have resulted.
We would then conclude that current flowed in the wrong direction. We also
observe that even though the nominal voltage of the battery (also called its
"unloaded" or "open-circuit" voltage) is 1.5 V, this does not mean that there
is 1.5 V across the 10-Ω resistance in Figure 3.6(*a*). On the contrary, there is
a voltage drop in the battery's internal 3-Ω resistance of $0.115 \times 3 = 0.345$ V,
so that the actual voltage across the load is only 1.155 V. This type of effect,
in which the voltage across a load is lowered due to internal resistance of a
source, is known as an *output loading* effect.

EXAMPLE 3.4
Consider two batteries, both of which are described by sketch (*a*), below. One battery consists of four 1.5-V flashlight cells in series. The Thévenin parameters of this battery are $V_T = 6$ V, $R_T = 12 \ \Omega$. The other is an automobile storage battery with $V_T = 6$ V, $R_T = 0.005 \ \Omega$. Each battery is connected in turn to the terminals of a 6-V automobile starter motor, which may be modeled as a 0.075-Ω resistance. Find the actual voltage across the motor V_S in the two cases.

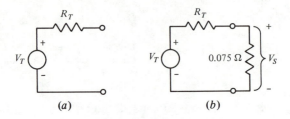

(*a*)   (*b*)

SOLUTION
In both cases the circuit appears as shown in sketch (*b*). Using the voltage-divider formula, we have

$$V_S = V_T \cdot \frac{0.075}{R_T + 0.075}$$

In the case of the flashlight cells, the result is

$$V_S = 6 \cdot \frac{0.075}{12.075} = 0.037 \text{ V}$$

This value is much too low to operate the motor. On the other hand, in the case of the storage battery, the voltage across the motor is

$$V_S = 6 \cdot \frac{0.075}{0.080} = 5.62 \text{ V}$$

which is probably sufficient to run the motor.
In addition to showing why one cannot start a car with flashlight batteries, this example illustrates output loading. When large output currents are required (that is, when the load resistance is small), a source with low internal resistance must be used.   ∎

It should again be noted that a model can represent the device only when the latter is operating under suitably restricted conditions. The model of Figure 3.5(*b*) represents the flashlight cell quite well, so long as only moderate voltages and currents are present. But suppose the cell were connected across a 300-kV power line. Then, if we still believed the model, we would think that a constant current of 100,000 A would flow through the cell. The actual result of such an experiment, however, would more likely be a loud noise, a bright flash, and only small remnants of a flashlight cell. The point is that the model is not adequate for such extreme operating conditions.

EXAMPLE 3.5

The symbol for an *npn* junction transistor is shown in sketch (*a*), below. One possible model for the transistor is the collection of elements shown in sketch (*b*). Here the three terminals of the transistor are labeled E, B, and C, and the corresponding terminals of the model are indicated. (Transistor models will be discussed in later chapters.) The transistor is to be used in the circuit in sketch (*c*). Find the current I_C.

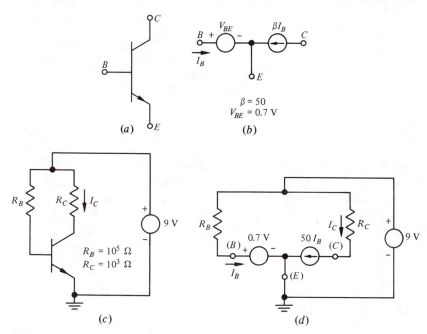

SOLUTION

The circuit is redrawn, using caution that each terminal of the model is connected where the corresponding terminal of the transistor had been. The resulting circuit is as given in sketch (*d*). (Note that the three terminal points of the transistor have been labeled, as a help in avoiding errors.) Using Ohm's law, it is clear that

$$I_B = \frac{9 - 0.7}{R_B} = \frac{8.3}{10^5} = 8.3 \times 10^{-5} \text{ A} = 83 \ \mu\text{A}$$

The current I_C, for which we want to solve, is controlled by the dependent source to have the value βI_B. Therefore

$$I_C = \beta I_B = (50)(8.3 \times 10^{-5}) = 4.15 \times 10^{-3} \text{ A}$$
$$= 4.15 \text{ mA}$$

3.3 Building Blocks for Analog Systems

Using the ideas developed in the preceding section, it is possible to devise models for entire analog blocks. In the present section, we shall consider examples of models for *linear* active circuits, that is, for circuits where output is linearly proportional to input.[2] By far the most important such circuit is the amplifier. Building blocks for digital systems are quite different, and will be considered in the following section.

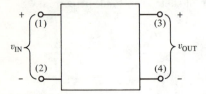

Figure 3.7   One possible circuit symbol for an amplifier block.

One possible circuit symbol for an amplifier block is shown in Figure 3.7. Details of the amplifier's internal construction are not shown. The connection for power input is also omitted from the diagram, although of course one must exist (as the output signal power will usually exceed the input signal power). In order for such a diagram as Figure 3.7 to have any meaning, the properties of the block must be specified. For circuits which have nearly linear properties, a convenient means is a circuit model. We shall now develop a simple model for an amplifier block. This cannot be a model for all amplifiers; depending on their internal construction, different models may be needed. However, the model is suitable for amplifiers of the type most often encountered.

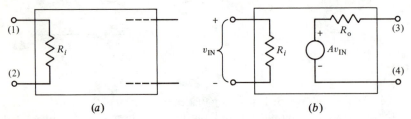

(a)   (b)

Figure 3.8   Development of a model for the amplifier of Figure 3.7. In (a) only the input resistance is shown; in (b) the entire circuit model is given.

The form of the model must, of course, reflect the properties of the amplifier. Let us assume the following property: when a voltage is applied to the input terminals, a current proportional to the voltage flows between them. The constant of proportionality is unchanging throughout the allowed range of operating conditions. Then the circuit model can represent this property by incorporating a resistance of the correct value between the two input terminals. This resistance, which is designated as R_i in Figure 3.8(a), is known as the *input resistance* of the amplifier.

Now transferring our attention to the output portion of the circuit, let

---

[2] More generally, a linear circuit is one whose output is related to input by a linear differential equation.

us assume the following two additional properties: (1) As seen from the two output terminals, the amplifier may be represented by a Thévenin equivalent; (2) the Thévenin voltage is proportional to the voltage existing between the two input terminals, with constant of proportionality A. These additional assumed properties of the amplifier can be represented by the model, if the output part of the model is completed as shown in Figure 3.8(b). The quantity A is known as the *open-circuit voltage amplification*, because when there is an open circuit between the output terminal (3) and (4), there is no voltage drop in R_0, and hence $v_{OUT}/v_{IN} = A$. The quantity R_0 is known as the *output resistance* of the amplifier.

The use of the amplifier model is demonstrated in the following example.

EXAMPLE 3.6

Find the output voltage v_L across the load resistor R_L in the amplifier circuit in sketch (a). Assume that the amplifier is represented by the model of Figure 3.8(b), with $R_S = 10^6 \ \Omega$, $R_L = 250 \ \Omega$, and $v_S = 1$ V.

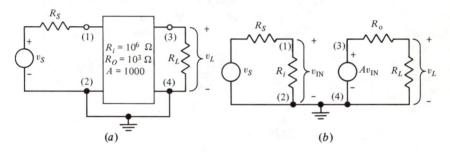

(a)   (b)

SOLUTION

We replace the amplifier by its circuit model and redraw the circuit. The resulting circuit model is as given in sketch (b).

We may find v_{IN} by the voltage-divider formula:

$$v_{IN} = v_S \bullet \frac{R_i}{R_S + R_i}$$

Similarly, we may find v_L in terms of Av_{IN} by the voltage-divider formula:

$$v_L = Av_{IN} \bullet \frac{R_L}{R_L + R_0}$$

The output is therefore given by

$$v_L = A \frac{R_L}{R_L + R_0} \bullet \frac{R_i v_s}{R_s + R_i} = 1000 \bullet \frac{250}{250 + 1000} \bullet \frac{10^6}{10^6 + 10^6} \bullet v_s$$
$$= 100 \ v_s$$

We see that the voltage across the load is affected by the input and output circuitry. Only if $R_S = 0$ and $R_0 = 0$ would the circuit yield a gain equal to the open-circuit gain A. Otherwise both input loading effects, due to R_S, and output loading effects, due to R_0, act to reduce the voltage gain. It is now possible to see why the model of Figure 3.8 was called inadequate. That model represents a special case in which $R_i \cong \infty$ and $R_0 \cong 0$. ∎

The Operational Amplifier   The name *operational amplifier* refers to a particular type of amplifier building block. Operational amplifiers (or "op-amps," as they are known) are multipurpose amplifiers designed for use as elements in analog systems. In the form of integrated circuits, they are small, inexpensive, and extremely versatile; consequently they are in very wide-spread use.

Since op-amps are quite similar, in general, to the simple amplifier of Figures 3.7 and 3.8, they are modeled in similar fashion. However, the op-amp is usually represented in circuit diagrams by a special symbol, shown in Figure 3.9(*a*). This symbol has two input terminals, designated (−) and (+), and one output terminal. With op-amps, it is always understood that the output voltage is measured with respect to ground; hence there is no need to show a second output terminal. Power supply and ground connections to the amplifier are also understood to be present, but are not shown as part of the circuit symbol.

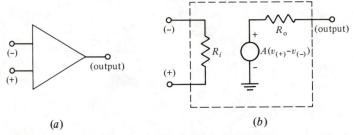

(*a*)   (*b*)

Figure 3.9   The operational amplifier. (*a*) Circuit symbol for an operational amplifier. The power supply and ground connections are understood; therefore they are not shown on the circuit symbol. (*b*) A model for the operational amplifier. The input resistance R_i is typically 50 kΩ to 5 MΩ, the output resistance R_0 is typically 10 Ω to 500 Ω, and the open-circuit voltage amplification is typically 10^4 to 10^5. The symbol $v_{(+)}$ stands for the voltage at the (+) input terminal, and so forth.

A model for the operational amplifier block is shown in Figure 3.9(*b*). We see that the value of the model's dependent voltage source is controlled by the difference in potential between the two input terminals. A positive voltage applied to the (−) input terminal, with the (+) terminal grounded, causes the output to become negative; thus the (−) terminal is called the *inverting input*. A positive voltage applied to the (+) input terminal, with (−) grounded, gives a positive output; hence the (+) input terminal is called the *non-inverting input*. In typical op-amps R_i, the input resistance, is in the range 50 kΩ to 5 MΩ; R_0, the output resistance, is in the range 10 to 500 Ω; and A, the open-circuit voltage amplification is in the range 10^4 to 10^5.

Because of their great practical importance, we shall return to the subject of op-amps later. Chapter 12, in fact, is devoted entirely to the applications of operational amplifiers.

EXAMPLE 3.7

Find the output voltage of the circuit in sketch (a). The operational amplifier parameters are as follows: $R_i = 10^5$ Ω, $R_0 = 0$, $A = 10^5$. Let $R_1 = 10^5$ Ω; $R_2 = 10^7$ Ω.

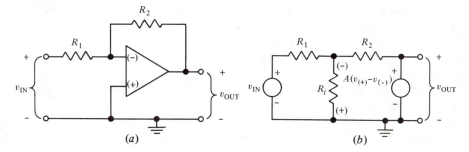

(a)   (b)

SOLUTION

We begin by replacing the op-amp with the model of Figure 3.9(b). The resulting circuit is shown in sketch (b). In order to keep track of the ($-$) and ($+$) input terminals, their locations in the new circuit have been shown.

Let us find $v_{(-)}$, the voltage at the ($-$) input terminal, by means of a node equation. Summing the currents leaving the node marked ($-$), we have

$$\frac{v_{(-)} - v_{\text{IN}}}{R_1} + \frac{v_{(-)}}{R_i} + \frac{v_{(-)} - A(v_{(+)} - v_{(-)})}{R_2} = 0$$

We note that $v_{(+)} = 0$, so this term drops out of the equation. Rearranging, we have

$$v_{(-)} \left[\frac{1}{R_1} + \frac{1}{R_i} + \frac{(1+A)}{R_2} \right] = \frac{v_{\text{IN}}}{R_1}$$

Referring to the numerical values of R_1, R_i, R_2, and A, we see that the last term in the square brackets is by far the largest, so that the first two terms in the brackets may be neglected. Moreover $(1 + A)$ is nearly equal to A. Making these simplifications, we have

$$v_{(-)} \left[\frac{A}{R_2} \right] \cong \frac{v_{\text{IN}}}{R_1}$$

$$v_{(-)} \cong \frac{R_2}{R_1} \frac{v_{\text{IN}}}{A}$$

The output voltage is equal to the voltage of the dependent voltage source, which is $-Av_{(-)}$. Thus

$$v_{\text{OUT}} \cong -\frac{R_2}{R_1} v_{\text{IN}}$$

This particular op-amp circuit is known as an *inverting amplifier*. We note that the voltage amplification is nearly independent of the value of A. It depends only on the values of the resistances R_1 and R_2, which can be controlled to a high degree of accuracy.   ∎

3.4   Building Blocks for Digital Systems

The building-block approach is perhaps even more convenient and important in connection with digital systems than it is with analog systems. However, because the form of the digital signal is quite different from that of the analog signal, the operations performed by the blocks must also be quite different. Before taking up models for digital blocks, therefore, we must digress briefly to discuss the nature of digital signals.

Digital Signals   The most common type of digital signal is the *binary signal*. A binary signal voltage must always have a value that lies inside one or the other of two ranges. For example, a system designer may decide that the two ranges are the range 0 to 1 V and the range 4 to 5 V. All signal voltages must then lie inside one or the other of the two ranges; no other voltages are permitted.

All voltages within the same range have the same meaning. Any signal in the more positive of the two ranges is said to be at the *high level;* any signal in the less positive of the two ranges is at the *low level.* If the ranges are defined as in the preceding paragraph the situation is as illustrated in Figure 3.10. Signal voltages of (for example) 4.1 or 4.8 V would be interpreted as high; signal voltages of 0.2 or 0.7 V would be regarded as low. Any voltage outside the two ranges has no meaningful interpretation, and should not occur in normal operation of the system.

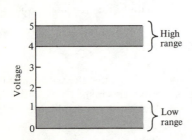

Figure 3.10   Sketch showing ranges of voltage to be interpreted as "high" or "low." Voltages not in the shaded regions do not occur in normal operation of the circuit.

Binary signals are used to represent the digits of binary numbers. (A review of the binary number system is given in Appendix I.). In binary arithmetic there are only two digits possible: 0 and 1. Thus we can let any voltage in the high range represent 1 and any voltage in the low range represent 0. When this assignment is made, one is using *positive logic*. On the other hand, if low-level voltages stand for 1 and high-level voltages stand for 0, one is using *negative logic*.

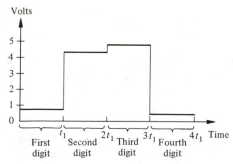

Figure 3.11 Example of a digital signal. It is assumed that the high range is 4 to 5 V and the low range is 0 to 1 V.

In a typical case, a digital signal is a voltage which varies between values in the low range and values in the high range. A digit is represented by a voltage which keeps an appropriate value during a certain unit of time. The units of time are coordinated by a central clock somewhere in the system. When the next unit of time starts, the voltage changes to a value which represents the next digit. An example of a digital signal is shown in Figure 3.11. In this example the first digit is being transmitted from time $t = 0$ to time $t = t_1$; the second from t_1 to $2\,t_1$, and so forth. The central clock causes all signals in the system to change from one digit to the next simultaneously. Thus all signals in the system would stay constant from $t = 0$ to $t = t_1$, at which time they would all change to their next values.

EXAMPLE 3.8

Consider the signal of Figure 3.11. What digits does it represent if positive logic is used? If negative logic is used?

SOLUTION

The sequence of voltages reads low-high-high-low. In positive logic low stands for 0 and high for 1, so the signal stands for 0110. In negative logic low stands for 1 and high for 0, so the signal stands for 1001. ∎

The length of time that is expended in transmitting each digit of course determines the number of digits per second that can be sent. Conventionally, a single binary digit is called a *bit* (a contraction for "binary digit"). The information-carrying capacity of a digital signal is thus measured in terms of bits per second.

EXAMPLE 3.9

Consider again the signal shown in Figure 3.11. If for technical reasons the time occupied in transmitting each digit (t_1) is 2 μsec, what is the rate of information transmission?

SOLUTION

Since the time per digit is 2×10^{-6} sec, one can transmit 5×10^5 digits per second. The information rate is 5×10^5 bits per second.   ■

For convenience, we shall refer to any voltage in the 1 range as being in the **1** state, or as having the significance **1**. Similarly, we shall speak of voltages in the 0 range as being in the **0** state. **Boldface type** will be used to distinguish the logical states from the numbers "1" and "0." The statement **A** = **0** means that the voltage at point A is in the 0 range.

Logic Blocks   We have stated that the information contained in a binary digital signal is a sequence of 0's and 1's. Next let us consider what sort of operations digital building blocks may be expected to perform. For now we shall consider only one particular family of digital blocks, known as logic blocks. Systems of logic blocks are widely used to perform arithmetic functions in computers. Some other types of digital blocks will be considered in Chapters 8 and 9.

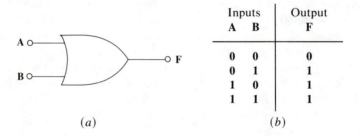

| Inputs | | Output |
A	B	F
0	0	0
0	1	1
1	0	1
1	1	1

(a)   (b)

Figure 3.12   A two-input **OR** gate. The symbol is given in (a), and the truth table in (b).

One important logic block is the **OR** gate, the symbol for which is shown in Figure 3.12(a). The **OR** gate can have any number of inputs; in this example three are shown. The function of the **OR** gate is to execute the **OR** function, which is defined as follows: if and only if the signal at one or more of the inputs has the significance **1**, the output is **1**. This function may be expressed in tabular form, as shown in Figure 3.12(b). A table of this kind is known as the *table of combinations* or *truth table* for the gate. The truth table may, of course, take on different forms, if the number of inputs is different. It must have as many entries as there are possible combinations of inputs. We have seen that a truth table for a two-input **OR** gate has four entries. A three-input **OR** gate would have a truth table with eight entries, as illustrated in Figure 3.13.

A second important digital block is the **AND** gate. The function of this block is to execute the **AND** operation: if and only if *all* of the input voltages have the significance **1**, the output is **1**. The circuit symbol for the **AND** gate is shown in Figure 3.14(a). Again the number of inputs is arbitrary; in this example three are shown. The truth table for the **AND** gate is shown in Figure 3.14(b).

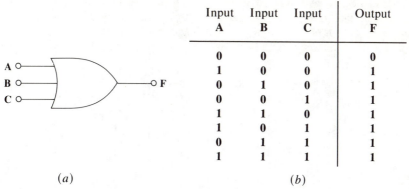

Input A	Input B	Input C	Output F
0	0	0	0
1	0	0	1
0	1	0	1
0	0	1	1
1	1	0	1
1	0	1	1
0	1	1	1
1	1	1	1

(a)   (b)

Figure 3.13 (a) Symbol for an **OR** gate with three inputs. (b) Truth table for the **OR** gate.

A small circle placed at an input or output of a symbol for a digital block indicates a *logical negation*. If the small circle is placed at an input terminal, it has the following effect: If the signal entering the circle is **1**, the signal leaving the circle and entering the block is **0**, and vice versa. Similarly, if the circle is placed at the output of the block, it has the following effect: When the signal leaving the block (and entering the circle) is **0**, the signal leaving the circle is **1**, and vice versa.

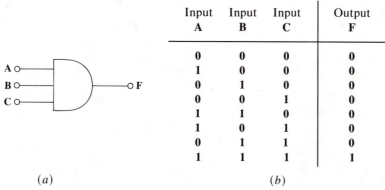

Input A	Input B	Input C	Output F
0	0	0	0
1	0	0	0
0	1	0	0
0	0	1	0
1	1	0	0
1	0	1	0
0	1	1	0
1	1	1	1

(a)   (b)

Figure 3.14 (a) Symbol for an **AND** gate with three inputs. (b) Truth table for the **AND** gate.

EXAMPLE 3.10
   Construct the truth table for the following block:

SOLUTION

This block consists of an **AND** gate with a logical negation at input **A** and another at the output. The truth table can be obtained from that of Figure 3.13(*b*), with the following changes: All **0**'s in column **A** are changed to **1**'s, and vice versa, and the same is done for the output column. The result is as follows:

Input A	Input B	Input C	Output F
1	0	0	1
0	0	0	1
1	1	0	1
1	0	1	1
0	1	0	1
0	0	1	1
1	1	1	1
0	1	1	0

∎

Two special cases in which logical negation is used occur often enough to deserve special mention. An **AND** gate with a negation at the output is known as a **NAND** gate. The logic symbol and truth table for the **NAND** gate are shown in Figure 3.15. The effect of the **NAND** gate is as follows: If and only if all inputs are **1**, the output is **0**. The second important special case is the **OR** gate with a negation at the output; this block, which is known as a **NOR** gate, is shown in Figure 3.16. Its effect is this: If one or more of the inputs is **1**, the output is **0**.

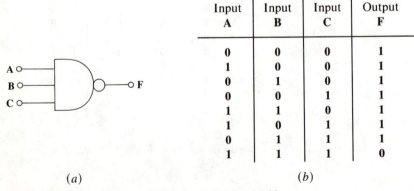

Input A	Input B	Input C	Output F
0	0	0	1
1	0	0	1
0	1	0	1
0	0	1	1
1	1	0	1
1	0	1	1
0	1	1	1
1	1	1	0

(*a*)   (*b*)

Figure 3.15   The **NAND** gate. (*a*) logic symbol; (*b*) truth table.

Two other logic operations occur frequently enough to be mentioned here: **EXCLUSIVE OR** and **COMPLEMENT**. The **EXCLUSIVE OR** gate, which as a rule has only two inputs, does the following: The output is **1** if one or the other of the inputs is **1**, but not if both inputs are **1**. The logic symbol and truth

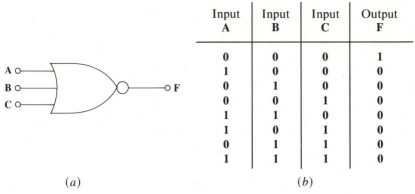

Input A	Input B	Input C	Output F
0	0	0	1
1	0	0	0
0	1	0	0
0	0	1	0
1	1	0	0
1	0	1	0
0	1	1	0
1	1	1	0

(a)   (b)

Figure 3.16 The **NOR** gate. (*a*) logic symbol; (*b*) truth table.

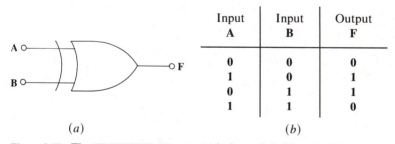

Input A	Input B	Output F
0	0	0
1	0	1
0	1	1
1	1	0

(a)   (b)

Figure 3.17 The **EXCLUSIVE OR** gate. (*a*) logic symbol; (*b*) truth table.

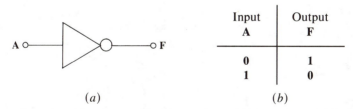

Input A	Output F
0	1
1	0

(a)   (b)

Figure 3.18 The **COMPLEMENT** operation. (*a*) logic symbol; (*b*) truth table.

table for the **EXCLUSIVE OR** are shown in Figure 3.17. The **COMPLEMENT** operation is a very simple one, performed on only a single input: When the input is **1**, the output is **0**, and vice versa. The logic symbol and truth table for this operation are shown in Figure 3.18. A circuit designed to perform the **COMPLEMENT** operation is often called an *inverter*.

In a typical logic system, the output of one logic block may be connected to the input of another. It is possible, then, to generate the truth table for a system of logic blocks.

EXAMPLE 3.11

Construct a truth table showing the values of the output **F** as a function of the four inputs **A, B, C,** and **D.**

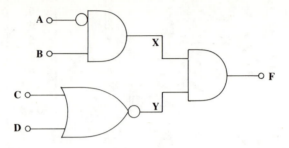

SOLUTION

It is convenient to proceed by first calculating the intermediate functions **X** and **Y.** The function **X** is found by modifying the truth table for an **AND** gate, interchanging **0** and **1** in column **A.** An **AND** gate gives an output of **1** if and only if all inputs are **1.** Hence the table for a two-input **AND** gate is

Input A	Input B	AND Output
0	0	0
0	1	0
1	0	0
1	1	1

Negating column **A,** we obtain the truth table for **X:**

A	B	X
1	0	0
1	1	0
0	0	0
0	1	1

The intermediate function **Y** is found by a **NOR** operation: If **C,** or **D,** or both are **1, Y** is **0.** Hence the truth table for **Y:**

C	D	Y
0	0	1
0	1	0
1	0	0
1	1	0

The final output is obtained by an **AND** operation on **X** and **Y.** There are now 16 input possibilities. The final truth table is

A	B	C	D	X	Y	F
1	0	0	0	0	1	0
1	1	0	0	0	1	0
0	0	0	0	0	1	0
0	1	0	0	1	1	1
1	0	0	1	0	0	0
1	1	0	1	0	0	0
0	0	0	1	0	0	0
0	1	0	1	1	0	0
1	0	1	0	0	0	0
1	1	1	0	0	0	0
0	0	1	0	0	0	0
0	1	1	0	1	0	0
1	0	1	1	0	0	0
1	1	1	1	0	0	0
0	0	1	1	0	0	0
0	1	1	1	1	0	0

This system selects out one of the 16 possible input states. It gives an output of **1** only in the case of **A = 0, B = 1, C = 0, D = 0.** ∎

A summary of the logical operations is given in Table 3.1.

Table 3.1

Logical Operation	Definition	Symbol in Logic Diagram
AND	F = 1 when A = B = C = 1 F = 0 otherwise	
OR	F = 0 when A = B = C = 0 F = 1 otherwise	
NAND	F = 0 when A = B = C = 1 F = 1 otherwise	
NOR	F = 1 when A = B = C = 0 F = 0 otherwise	
EXCLUSIVE OR	F = 1 if A or B = 1 but not if both = 1	
COMPLEMENT	F = 0 when A = 1 F = 1 when A = 0	

It should be noted that the foregoing discussion of logical operations is really a mathematical discussion. The logic blocks convert inputs of **0** and **1** into outputs of **1** or **0,** but nothing has been said about what voltages are actually present. A physical circuit, on the other hand, converts inputs of high and low into an output of high or low. Now, when we speak of high and low we are talking about voltages, but the mathematical significance is not clear unless one specifies whether high stands for **0** or **1.** It is helpful, in clarifying this rather subtle point, to point out that the same physical block can perform different logic functions, depending on whether positive or negative logic is used. Let us consider a physical block with two inputs, as shown in Figure 3.19(*a*). Let the *electrical* action of this block be as shown in Figure 3.19(*b*). Now suppose we adopt the positive logic conventions: then low voltage stands for **0** and high voltage for **1.** The truth table is then as shown in Figure 3.19(*c*); we recognize this as the truth table of the **OR** operation. Therefore when the positive logic convention is adopted, the physical block functions as an **OR** gate, and would be represented on a logic diagram by the symbol shown in Figure 3.12(*a*). Now, however, let us suppose that the negative logic convention is adopted: High stands for **0** and low stands for **1.** The truth table for the same circuit in this case is as shown in Figure 3.19(*d*). According to this truth table the output is **1** only if both inputs are **1;** hence the physical block is now performing the **AND** function! If negative logic were being used, this physical block would be represented by the **AND** symbol in the logic diagram.

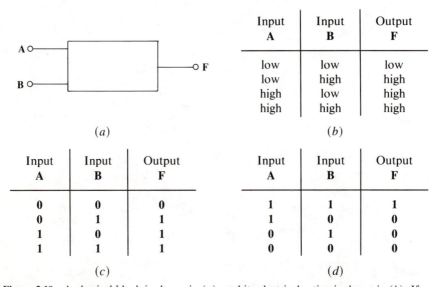

Input A	Input B	Output F
low	low	low
low	high	high
high	low	high
high	high	high

(*a*)

(*b*)

Input A	Input B	Output F
0	0	0
0	1	1
1	0	1
1	1	1

Input A	Input B	Output F
1	1	1
1	0	0
0	1	0
0	0	0

(*c*)

(*d*)

Figure 3.19 A physical block is shown in (*a*), and its electrical action is shown in (*b*). If we adopt the positive logic convention, the truth table for the block is as shown in (*c*); this truth table represents the **OR** operation. However, if we adopt the negative logic convention, the truth table is as shown in (*d*); this table represents the **AND** operation.

The functions of the logical blocks are given by their truth tables, just as the circuit model describes the operation of a linear block. The truth table, however, does not convey all possible information to the system designer. Since we have already considered the effects of loading on an amplifier, we may expect that the output voltage of a logic gate depends to some degree upon what load is connected there. The most common load connected to the output of a logic circuit is the input to another logic circuit, or several inputs in parallel. The number of inputs connected to the output of a block is known as the *fan-out*. In general there are limits on the permissible fan-out. If these limits are exceeded, the block may not be able to maintain its output voltage within the allowed ranges. Thus the fan-out of a block is a typical parameter which can be specified. Specifications for digital blocks are taken up in Chapter 8, where we consider their internal construction.

There is also much to be said about the analysis and design of systems of digital blocks. Special mathematical techniques for these purposes have been developed. This subject is treated in Chapter 9.

Arithmetic with Logic Blocks   To conclude this section, it is interesting to see how digital logic blocks can be used to perform arithmetic operations in a computer. Let us imagine that we wish to add two binary numbers. (For example, we might add 1011 to 1010. Their sum is 10101.) Just as in ordinary addition, we begin by adding the right-hand digits. Let us call the right-hand digit of the first number **A** and that of the second number **B**. Let the right-hand digit of the answer be called **C**. Then in order to add these digits, we wish to generate **C** according to the truth table shown in Figure 3.20(*a*). The asterisk in the last column means that a "carry" signal must be generated when **A** and **B** are both **1**; that is, since $1 + 1 = 10$, a 1 must in this case be added to the next column to the left.

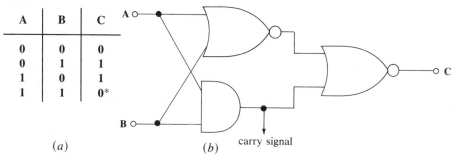

A	B	C
0	0	0
0	1	1
1	0	1
1	1	0*

(*a*)   (*b*)   carry signal

Figure 3.20   (*a*) Truth table for addition of two binary digits. The asterisk signifies that a carry signal must be generated when both **A** and **B** are **1**. (*b*) A system of logic operations which executes the functions shown in (*a*).

A system of logic operations which executes the desired functions is shown in Figure 3.20(*b*). [Verification that the truth table of Figure 3.20(*a*) does represent the system of (*b*) is left to a problem. The procedure is like that of Example 3.11.] We see that a "carry" signal is generated at the output of the **AND** gate: a **1** appears at the carry terminal if and only if both **A** and **B** are **1**. The carry signal would be connected as an input to another set of logic blocks, whose function is to add the second digits of the original numbers. The second-digit adder would have to be slightly different from Figure 3.19(*b*), because it would have to add three digits: one each from the original addends, plus the carry signal. The small system of Figure 3.20 is known as a *binary half-adder*. Half-adders are available commercially in integrated-circuit form.

Summary

- Active circuits are those circuits in which the output signal power can exceed the input signal power.

- An apparatus consisting of many electronic circuits working together is called a system. Individual electronic circuits can be regarded as building blocks for the system. Building blocks are described by simple rules which describe their function. The subject of system design deals with the interconnection of blocks to make up systems. Circuit design deals with the internal construction of the blocks. There are both analog and digital blocks.

- Dependent voltage and current sources are sources whose values cannot be specified independently. Instead, their values are determined by the value of voltage or current at a place somewhere else in the circuit.

- The most common analog building block is the amplifier. Amplifier blocks are conveniently described by models. A model is a collection of ideal linear circuit elements with the same number of terminals as the block it represents. When the model is substituted for the block in a system diagram, it imitates the action of the block. Substitution of the model facilitates analysis of the system. A typical amplifier model contains three parameters. These are the open-circuit voltage amplification A, the input resistance R_i, and the output resistance R_0.

- The operational amplifier is a special type of amplifier that is in very common use. "Op-amps" are inexpensive and extremely versatile, and are available in convenient integrated-circuit form.

- Digital building blocks perform functions quite different from those of analog blocks. Digital blocks usually are designed to process binary signals. Binary signals must fall into one or the other of two ranges, known as the high and low ranges. In positive logic, a signal in the high range represents the binary digit 1, and a signal in the low range represents the digit 0. In negative logic the reverse is true.

• Among the most common digital blocks are those designed to perform logical operations. The most important logical operations are **OR, AND, NOR, NAND, EXCLUSIVE OR,** and **COMPLEMENT.** A block designed to perform one of the first five operations is called a gate, for example, "**AND** gate." A block designed to perform the **COMPLEMENT** operation is called an inverter.

References

Treatments on approximately the same level as this book:

Durling, A. E. *An Introduction to Electrical Engineering.* New York: The Macmillan Company, 1969.

Brophy, J. J. *Basic Electronics for Engineers and Scientists.* New York: McGraw-Hill, 1966.

Problems

3.1   In Figure 3.21, let the input signal consist of a rectangular pulse whose amplitude is 2 V. The input current during the pulse is 3 mA. At the output terminals a rectangular pulse with the same duration appears across the load resistance R_L. During the pulse the output voltage v_L is 0.5 V and the output current i_L is 0.6 A. What is the power amplification of the block? Is it an active or a passive circuit?

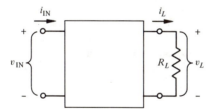

Figure 3.21

3.2   A man gives verbal instructions to an elephant, who responds by doing work with his trunk. Is the elephant an active device or a passive device? What is its "input signal"? Its "output signal"? Does this device turn raw power into useful output power? Where does the raw power come from?

3.3   In the circuit of Figure 3.22, the dependent current source depends on the current i_1, as shown. Let the value of β be 100. (What are the units of the constant β?) Find the current i_2.

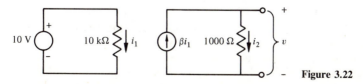

Figure 3.22

3.4   In Figure 3.22, find the voltage v. ANSWER: $v = 100$ V.

3.5   In the circuit of Figure 3.23, find the voltage v_2 and the current i_2. Does the value of R_1 have any influence on the answer?

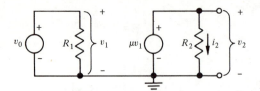

Figure 3.23

3.6   In Figure 3.24, calculate i_1. What is its numerical value if $v_1 = 2 \sin \omega t$ V; $r_\pi = 2500$ Ω; $R_L = 5000$ Ω; and $\beta = 100$?

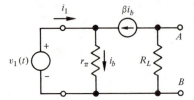

Figure 3.24

3.7   Let the circuit of Figure 3.24 be regarded as a subcircuit, with terminals A,B for external connection. Obtain the Norton equivalent of the subcircuit. It is not required that numerical values be substituted for the symbols.

3.8   Find the Thévenin resistance of the subcircuit shown in Figure 3.24, as seen from terminals A, B.

3.9   Find the Thévenin resistance of the subcircuit shown in Figure 3.24, using the alternative method in which independent voltage sources are set to zero. You may find it helpful to imagine an external voltage V_{TEST} connected to terminals A,B, in order to find the resistance that appears between these terminals.

3.10   Find the Thévenin equivalent of the subcircuit shown in Figure 3.25. The factor r_m is a constant.

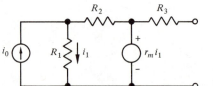

Figure 3.25

3.11   Check your answer to Problem 3.10 by using the alternative method (in which independent sources are set to zero) to find R_T.

3.12   Suppose there is a log cabin with two doors. A man stands at one of the doors. Snowballs come flying in the other door, and land on the floor. Every time one snowball flies in, the man picks up three snowballs from a large supply on the floor and flings them out his door. If snowballs are analogous to charge and the floor of the cabin is taken to be ground, devise a model for the system.

3.13 Suppose the device shown in Figure 3.26(*a*) is represented by the model shown in Figure 3.26(*b*). It is connected in a circuit as shown in (*c*). Calculate v_L. The quantity μ is a constant.

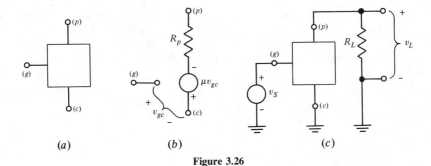

(a)   *(b)*   *(c)*

Figure 3.26

3.14 The active device shown in Figure 3.26(*a*) and (*b*) is connected as shown in Figure 3.27. Note that the orientation of the device is upside down compared with Figure 3.26(*a*) and (*b*).

(1) Substitute the device model into the circuit of Figure 3.26, thus obtaining a circuit model.

(2) Use the circuit model to calculate v_c. *Suggestion:* This problem is slightly tricky. Be careful in locating the reference voltage v_{gc}.

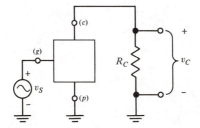

Figure 3.27

3.15 A block with one pair of input terminals and one pair of output terminals can often be represented by a model of the form shown in Figure 3.28(*a*). This model is known as a *hybrid model,* because it is not symmetrical, having instead a current source on one side and a voltage source on the other. The parameters are

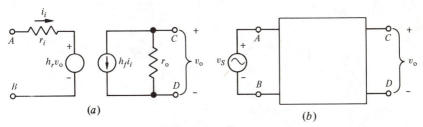

(a)   *(b)*

Figure 3.28

known as *hybrid parameters* or *h-parameters,* and the subscripts stand for "input," "output," "reverse," and "forward."

A block represented by the hybrid model is connected as shown in Figure 3.28(*b*). Find the output voltage v_0.

3.16   Show that any amplifier which is represented by the model of Figure 3.9(*b*) can also be represented by the more general hybrid model of Figure 3.28. Evaluate the four hybrid parameters in terms of R_i, R_0, and A in the amplifier model. *Suggestion:* Convert the output part of the hybrid model to its Thévenin equivalent.

3.17   The operational amplifier of Figure 3.9 is used in the circuit of Figure 3.29. Find v_{OUT} using, for simplicity, the approximations $R_i = \infty$, $R_0 = 0$. What limit does your answer approach as $A \to \infty$?

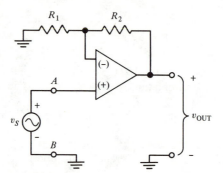

Figure 3.29

3.18   The op-amp of Figure 3.9 is used in the circuit of Figure 3.29. Calculate v_{OUT} for the circuit, without any simplifying assumptions concerning R_i, R_0, and A. Check your result by seeing if it reduces to the answer of the previous problem as $R_i \to \infty$, $R_0 \to 0$.

3.19   An amplifier represented by the model of Figure 3.8(*b*) is used as part of a larger circuit, as shown in Figure 3.30.
(1) Find the voltage amplification of this circuit.
(2) Find the input resistance of the circuit.
(3) Find the output resistance when an ideal voltage source v_S is connected across the input terminals.

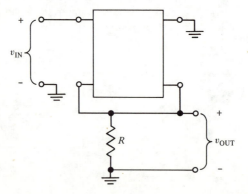

Figure 3.30

3.20   Calculate the input resistance of the op-amp circuit of Figure 3.29, as seen looking in to the right from terminals A,B. Use the op-amp model of Figure 3.9.
Assume that no external load is connected across the output terminals.

3.21   Find the output resistance of the op-amp circuit of Figure 3.29, using the op-amp
model of Figure 3.9. Assume that an ideal voltage source is connected to the
input terminals, as shown in Figure 3.29.

3.22   Two amplifiers which can be represented by the model of Figure 3.8(b) are connected *in cascade* (that is, head-to-tail) as shown in Figure 3.31. Their parameters
are A_1, R_{i1}, R_{01} and A_2, R_{i2}, R_{02}, respectively. Find v_{OUT}. Discuss how your
answer is influenced by the ratio R_{i2}/R_{01}, and explain what happens in the limits
as this ratio approaches zero and as it approaches infinity.

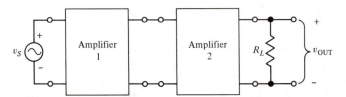

Figure 3.31

3.23   Let positive logic be used, and let the high range be 4 to 5 V and the low range
be 0 to 1 V.
(1) Sketch a possible waveform representing the binary number 10011010.
(2) Sketch a waveform representing this number if negative logic is used.

3.24   Explain the meaning of: (1) the binary digit 1; (2) the logical state **1**; (3) the high
voltage range.

3.25   Construct a truth table for the block shown in Figure 3.32.

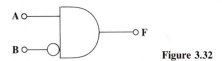

Figure 3.32

3.26   Construct a truth table for the combination of blocks shown in Figure 3.33.

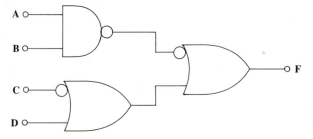

Figure 3.33

3.27 Show that in Figure 3.34, the output **F** is independent of the input **A**.

Figure 3.34

3.28 Prove that the block shown in Figure 3.35 is equivalent to an **AND** gate.

Figure 3.35

3.29 Suppose that a resistance is connected between the output terminal of an **AND** gate and ground. Would this be expected to influence operation of the gate? What if the resistance is made very small? How is this question related to the subject of output resistance in an analog block?

3.30 Let the electrical output of a block be related to the electrical inputs as shown in Figure 3.36. What logical block or combination of blocks describe this electrical operation, (1) if the negative logic convention is adopted? (2) if the positive logic convention is adopted?

Input A	Input B	Output F
low	low	high
low	high	·low
high	low	low
high	high	high

Figure 3.36

3.31 Verify the truth table of Figure 3.20.

3.32 Devise a connection of other gates which performs a function equivalent to that of an **EXCLUSIVE OR** gate.

Chapter 4

Sinusoidal Analysis and Phasors

In this chapter we shall consider the response of circuits to sinusoidal signals. We have previously encountered this subject in Section 2.3, where some basic ideas were introduced, and solutions to some simple problems of analysis were obtained. In this chapter we shall go further, developing powerful techniques for the analysis of circuit response to sinusoidal signals.

Before proceeding with the subject of circuit analysis, it may be useful to review the properties of sinusoidal functions and the parameters by which they are defined. This is done in Section 4.1. In Section 4.2 the quantities known as *phasors* are introduced. Phasors are complex numbers used to represent sinusoids in circuit calculations. As in the case of sinusoidal functions, it may be useful to review the properties of complex numbers; this is done at the start of Section 4.2, as an introduction to the material dealing specifically with phasors. In Section 4.3 we continue the discussion of sinusoidal analysis by introducing the quantity known as *impedance*. By means of the impedance concept, some of the techniques used in dc analysis of resistive circuits can be taken over, and made to serve for sinusoidal analysis as well.

The reader may well be wondering, at this point, why

so much attention is directed to the subject of response to sinusoidal signals. One reason is that there is an important class of signals that are nearly sinusoidal in form. Such signals are most often found in radio communications apparatus, and hence are called *radio-frequency* (or "rf") signals. The subject of rf circuits is introduced in Section 4.4. However, the usefulness of sinusoidal analysis is not confined to rf circuits. On the contrary, it is possible to use sinusoidal analysis to gain an idea of how circuits will perform with signals that are *not* sinusoidal. The basic idea here is that a signal of nonsinusoidal form can be expressed as a sum of numerous sinusoids. This statement, known as the *Fourier theorem*, enables us to estimate the response of a circuit to an arbitrary signal by testing the circuit's response to its sinusoidal components. Section 4.4 concludes with an introduction to the Fourier theorem and its use in circuit analysis.

4.1 Properties of Sinusoids

A *sinusoid* is a mathematical function of time having the form

$$\mathfrak{f}(t) = A \cos(\omega t + \phi) \tag{4.1}$$

This function is characterized by three parameters. The quantity A is the *amplitude;* ω is the *angular frequency;* and ϕ is the *phase* or *phase angle* of the sinusoid. Any sinusoid is completely described when the values of these three parameters have been given. The physical meanings of A, ω, and ϕ are best explained by referring to a graph of a sinusoid, such as Figure 4.1.

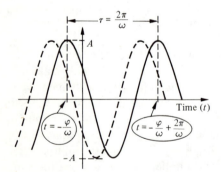

Figure 4.1 The solid curve represents the sinusoidal function $\mathfrak{f}(t) = A \cos(\omega t + \phi)$. The period is $2\pi/\omega$. Maxima occur at the time $t = -\phi/\omega$ and also at $t = -\phi/\omega \pm 2n\pi/\omega$ (where n is an integer). The dotted curve represents a sinusoid with identical A and ω but slightly larger ϕ. The effect of increasing ϕ is to displace the curve toward earlier times.

Amplitude The amplitude A of the sinusoid is simply equal to the value of the sinusoid at its maximum, as indicated in Figure 4.1. The minimum value of the sinusoid is $-A$. In general the units of the amplitude are the units of whatever physical quantity is represented by the sinusoid. For example, if $v(t)$ is a time-varying voltage and $v(t) = V_0 \cos(\omega t + \phi)$, then the amplitude V_0 must have the units of volts. To remind one of this, the letter symbol V would be used for the amplitude instead of A.

Frequency   The angular frequency ω specifies how often the maxima occur. To see this, we note that when the argument of the cosine function [Equation (4.1)] is zero, that is, at an instant of time $t = t_0$ such that

$$(\omega t_0 + \phi) = 0 \tag{4.2}$$

the sinusoid will be at its maximum. At times slightly later than t_0, the sinusoid has decreased; it goes to its minimum value, and then comes back to its maximum when the argument of the cosine function reaches 2π. Let us call the time when this happens t_1; then t_1 is defined by

$$(\omega t_1 + \phi) = 2\pi \tag{4.3}$$

Subtracting Equation (4.2) from Equation (4.3), we have

$$\omega(t_1 - t_0) = 2\pi \tag{4.4}$$

Let us define $\tau \equiv (t_1 - t_0)$. Then τ is the time which elapses between maxima of the sinusoid; τ is known as the *period*. From Equation (4.4) we have

$$\boxed{\tau = \frac{2\pi}{\omega}} \quad \text{(period of a sinusoid)} \tag{4.5}$$

The *ordinary frequency f* is equal to the number of repetitions of the sinusoid per second, and is given by

$$\boxed{f = 1/\tau} \tag{4.6}$$

It is not the same as the angular frequency, but is related to the latter by the formula

$$\boxed{\omega = 2\pi f} \tag{4.7}$$

Regrettably, in casual usage both ordinary frequency and angular frequency are often referred to simply as "frequency"; when the word "frequency" appears, one must make certain which of the two is meant. Angular frequency is given in units of radians per second. The units of ordinary frequency are hertz (Hz); units such as kilohertz (kHz) and megahertz (MHz) are also used.[1]

EXAMPLE 4.1
    The transmitter of a certain radio station produces a sinusoidal voltage at its antenna. The frequency of this sinusoid is 98.7 MHz. What is the time that elapses between voltage maxima?

---

[1] Until the 1960's ordinary frequencies were stated in cycles per second. This fine, self-descriptive old unit has now been dropped from standard usage.

SOLUTION

Since the frequency is stated in units of megahertz, we know that ordinary frequency f (rather than angular frequency ω) is being stated. The time that elapses between maxima is the period of the sinusoid τ. From Equation (4.5), $\tau = 2\pi/\omega$, and from Equation (4.7), $\omega = 2\pi f$. Thus

$$\tau = 1/f = (9.87 \times 10^7)^{-1} \text{ sec}$$
$$= 1.013 \times 10^{-8} \text{ sec} = 10.13 \text{ nsec} \qquad \blacksquare$$

Phase   The phase angle ϕ specifies at what instants of time the sinusoid reaches its maxima. For instance, we know there must be a maximum of the sinusoid when the argument of the cosine, $(\omega t + \phi)$, equals zero. Thus there must be a maximum when $t = -\phi/\omega$, and also, because of the periodicity of the sinusoid, at $t = -(\phi/\omega) \pm (2n\pi/\omega)$, where n is any integer. The phase angle ϕ may be thought of as specifying the right and left position of the sinusoid, in Figure 4.1, with respect to the time axis; increasing ϕ moves the sinusoid to the left, that is, toward earlier times. Since it is an angle, ϕ can be specified either in degrees or in radians. However, note that ω is usually stated in radians per second. If a sum such as $(\omega t + \phi)$ is formed, one must take care that ωt and ϕ are stated in the same units.

The sinusoid of Equation (4.1) happens to be expressed in terms of a cosine function. However, the use of the cosine function instead of the sine is an arbitrary choice; $\mathfrak{f}(t)$ could just as well be specified in terms of a sine function. There is a trigonometric identity which states that $\cos \alpha = \sin(\alpha + \pi/2$ radians), where α is any angle. Applying this identity to Equation (4.1), we can find an alternative expression for the same sinusoid in terms of the sine function:

$$\mathfrak{f}(t) = A \cos(\omega t + \phi) = A \sin\left(\omega t + \phi + \frac{\pi}{2}\right) \qquad (4.8)$$

EXAMPLE 4.2

It is desired to convert the function $\mathfrak{g}(t) = B \sin(\omega t - 37°)$ to the form $\mathfrak{g}(t) = B \cos(\omega t + \phi)$ where ϕ is in radians. Evaluate ϕ.

SOLUTION

We can use the trigonometric identity $\sin x = \cos[x - (\pi/2)$ radians$]$. Thus $\mathfrak{g}(t) = B \cos[\omega t - 37° - (\pi/2)$ radians$]$. The angle $37°$ must now be expressed in radians. One radian equals $57.3°$; therefore $37° = 0.65$ radian. Thus $\phi = -0.65 - (\pi/2) = -0.65 - 1.57 = -2.22$ radians. $\qquad \blacksquare$

There is a commonly used shorthand method for the description of sinusoids. In this method one simply specifies the amplitude A and the phase ϕ, using the notation $A \angle \phi$. For example, instead of stating $v(t) = 27 \; V \cdot \cos(\omega t + 60°)$, in this notation we would say $v(t) = 27 \; V \angle 60°$. The frequency of the sinusoid is not described by this notation, and must be specified separately.

This is not much of a shortcoming, as in sinusoidal steady-state situations all signals in the circuit usually have the same frequency.

EXAMPLE 4.3

A voltage $v_1(t)$ is described by the sinusoid 160 V $\angle 37°$. The ordinary frequency is 60 Hz. What is $v_1(t)$ at $t = 34$ msec?

SOLUTION

The desired voltage is

$$v_1(t = 34 \text{ msec}) = 160 \text{ V} \cdot \cos[2\pi \ (60) \ (0.034) \text{ radians} + 37°]$$

where the factor 2π has been inserted to convert ordinary frequency to angular frequency in radians per second. In computing the argument of the cosine it must be remembered that the first term is now in radians and the second in degrees. Converting the latter to radians, we find

$$\begin{aligned} v_1(t = 34 \text{ msec}) &= 160 \text{ V} \cdot \cos[12.82 \text{ radians} + 0.65 \text{ radian}] \\ &= 160 \text{ V} \cdot \cos[13.47 \text{ radians}] \\ &= 160 \text{ V} \cdot \cos[(13.47 - 4\pi) \text{ radians}] \\ &= 160 \text{ V} \cdot \cos[0.90 \text{ radian}] = 99.7 \text{ V} \quad \blacksquare \end{aligned}$$

4.2 Phasors

A *phasor* is a complex number which describes the amplitude and phase of a sinusoid. As in the case of the $A \ \angle\phi$ notation, the frequency of the sinusoid is not described by the phasor and must be specified separately. The phasor, however, is more than just a shorthand notation; it is the basis for a powerful technique of circuit analysis. Since phasor analysis involves the use of complex numbers, the terminology and properties of the complex number system will now be briefly reviewed.

Complex Numbers   Ordinary numbers, such as $2.3, -3/4$, or π, are known as *real numbers*. The square of any real number is always a positive real number. A second family of numbers are those whose squares are negative real numbers. Such numbers are called *imaginary numbers*. The particular imaginary number which is the square root of -1 is given the symbol j. Other imaginary numbers are expressed as multiples of j. For instance, $\sqrt{-4} = \sqrt{(4)\ (-1)} = (\sqrt{4})(\sqrt{-1}) = 2j$. Since $j^2 = -1$, we note that $1/j = -j$.

A number which is the sum of a real number and an imaginary number is called a *complex number*. For instance, if x and y are any two real numbers, $\mathbf{z} = x + jy$ is a complex number. In this book the symbols for complex numbers are printed in boldface type, for example, $\mathbf{z}$.

If $\mathbf{z} = x + jy$, we call x the *real part* of $\mathbf{z}$, abbreviated Re($\mathbf{z}$). We shall refer to y as the *imaginary part of* $\mathbf{z}$, abbreviated Im($\mathbf{z}$). Note that according to this definition the imaginary part of $\mathbf{z}$ is a real number.[2] To every complex

---

[2] The reader is cautioned that some writers use a different definition, and call jy the imaginary part of $\mathbf{z}$.

number there corresponds a second complex number known as its *complex conjugate.* If $z = x + jy$, the complex conjugate of z, whose symbol is z^*, is by definition equal to $x - jy$. As a practical matter, the complex conjugate of any complex number that is not expressed in simple form can be found by reversing the algebraic sign before every term in which j appears. For example, the complex conjugate of $(1 + 2j)/(3 - 4j)$ is equal to $(1 - 2j)/(3 + 4j)$. When a number is multiplied by its own complex conjugate, the product is a real number. This may be proven as follows: $zz^* = (x + jy)(x - jy) = x^2 + jxy - jxy - j^2 y^2 = x^2 + y^2$. Since x and y are both real, zz^* is also real. The number zz^* is given the symbol $|z|^2$ and is called the *absolute square* of z.

EXAMPLE 4.4

Find the real and imaginary parts of the complex number

$$z = \frac{2 + 3j}{4 + 5j}$$

What is its complex conjugate? Its absolute square?

SOLUTION

To find the real and imaginary parts it is necessary to express the number in the form $x + jy$. This can be done by multiplying the numerator and denominator of z by the complex conjugate of the denominator. Doing this

$$z = \frac{(2 + 3j)(4 - 5j)}{(4 + 5j)(4 - 5j)} = \frac{8 + 12j - 10j + 15}{16 + 25} = \frac{23 + 2j}{41} = \frac{23}{41} + \frac{2}{41}j$$

The real part of z is therefore $23/41$, and its imaginary part is $2/41$. The complex conjugate is $z^* = \frac{23}{41} - \frac{2}{41}j$. The absolute square is

$$|z|^2 = zz^* = \left(\frac{23}{41} + \frac{2}{41}j\right)\left(\frac{23}{41} - \frac{2}{41}j\right) = \left(\frac{23}{41}\right)^2 + \left(\frac{2}{41}\right)^2 = \frac{533}{1681}$$

An alternative way to obtain the complex conjugate (although in a less simple form) is to reverse the sign of all terms containing j in the original expression for z. Thus

$$z^* = \frac{2 - 3j}{4 - 5j} \qquad \blacksquare$$

The value of a complex number can be depicted graphically as a point in the *x-y* plane. One simply plots the value of x as the *x*-coordinate of the point and the value of y as the *y*-coordinate. One speaks of the *x-y* plane as the *complex plane.* Several examples of numbers in the complex plane are shown in Figure 4.2.

When the number z is expressed in the form $x + jy$, it is said to be expressed in *rectangular form,* because it is specified in terms of its rectangular coordinates in the complex plane. It is also possible to specify the same number by means of its polar coordinates. The relationship between the two

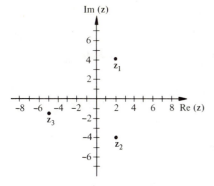

Figure 4.2 Locations of three complex numbers in the complex plane. $z_1 = (2 + 4j)$; $z_2 = (2 - 4j)$; $z_3 = -5 - 1.5 j$. Note that $z_2 = z_1{}^*$.

descriptions is shown in Figure 4.3. From trigonometry it is immediately clear that the length of the radius vector M is given by

$$M = \sqrt{x^2 + y^2} \qquad (4.9)$$

and that the polar angle θ is given by

$$\theta = \tan^{-1}\left[\frac{y}{x}\right] \qquad (4.10)$$

Conversely, it is also evident from the figure that

$$x = M \cos \theta \qquad (4.11)$$

and

$$y = M \sin \theta \qquad (4.12)$$

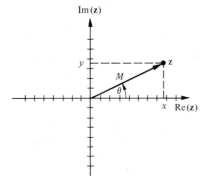

Figure 4.3 Relationship between rectangular and polar representations for the complex number z. By trigonometry we see that $z = x + jy = M(\cos \theta + j \sin \theta)$. The radius vector M is called the absolute value of the complex number, and is given the symbol $|z|$.

Thus the complex number $z = x + jy$ may be expressed in the form $z = M \cos \theta + jM \sin \theta$, or

$$z = M(\cos \theta + j \sin \theta) \tag{4.13}$$

When expressed as in Equation (4.13), z is said to be expressed in its *polar form*.

The radius vector M is known as the *absolute value* of the complex number z, and is given the symbol $|z|$. We have already seen that $zz^* = x^2 + y^2$. Thus, using Equation (4.9)

$$\boxed{|z| \equiv M = \sqrt{zz^*}} \tag{4.14}$$

Equation (4.14) is a very useful formula. The polar angle θ is called the *argument* of the complex number z. Its value is given by Equation (4.10).

EXAMPLE 4.5
What is the absolute value of the number $z = (6.4 - 5.6\,j)$?

SOLUTION
The absolute value $|z|$ is obtained from Equation (4.14). Here $z^* = 6.4 + 5.6\,j$. The product zz^* is given by

$$zz^* = (6.4 - 5.6\,j)(6.4 + 5.6\,j) = (6.4)^2 - (5.6\,j)(6.4) + (5.6\,j)(6.4) + (5.6)^2$$
$$= (6.4)^2 + (5.6)^2 = 72$$

Therefore

$$|z| = \sqrt{zz^*} = \sqrt{72} = 8.5 \qquad \blacksquare$$

Still a third way of expressing a complex number is by a mathematical identity known as *Euler's formula*. This identity states that if θ is any angle

$$\boxed{e^{j\theta} \equiv \cos \theta + j \sin \theta} \tag{4.15}$$

[Euler's formula may be proved by showing that the power series expansions for the left and right sides of Equation (4.15) are identical.] As important special cases of Equation (4.15), note that $e^{j(0)} = 1$, $e^{j(\pi/2)} = j$, $e^{j(\pi)} = -1$, and $e^{j(3\pi/2)} = e^{-j(\pi/2)} = -j$. Comparing Equation (4.15) with Equation (4.13), we now find that

$$z = Me^{j\theta} \tag{4.16}$$

where the values of M and θ are still as given by (4.9) and (4.10). When expressed as in Equation (4.16), z is said to be expressed in its *exponential form*.

We can show that if $z = Me^{j\theta}$, then $z^* = Me^{-j\theta}$. Referring to Equation (4.15), $z = M \cos \theta + jM \sin \theta$. Therefore $z^* = M \cos \theta - jM \sin \theta = M \cos$

$(-\theta) + jM \sin(-\theta)$. Now comparing again with (4.15), we see that $\mathbf{z}^* = Me^{j(-\theta)} = Me^{-j\theta}$.

The various definitions and relationships derived above are summarized in Table 4.1.

Table 4.1   Properties of Complex Numbers

Alternative Expressions for the Complex Number $\mathbf{z}$:

$$\text{Rectangular: } \mathbf{z} = x + jy$$
$$\text{Polar: } \mathbf{z} = M(\cos\theta + j\sin\theta)$$
$$\text{Exponential: } \mathbf{z} = Me^{j\theta}$$

Relationships Between Expressions:

To convert from rectangular to polar or exponential forms:

$$M = \sqrt{x^2 + y^2}$$
$$\theta = \tan^{-1}\left(\frac{y}{x}\right)$$

To convert from polar or exponential to rectangular form:

$$x = M\cos\theta$$
$$y = M\sin\theta$$

Complex Conjugate of $\mathbf{z}$:

$\mathbf{z} = x + jy$	$\mathbf{z}^* = x - jy$
$\mathbf{z} = \dfrac{A + jB}{C + jD}$	$\mathbf{z}^* = \dfrac{A - jB}{C - jD}$
$\mathbf{z} = M(\cos\theta + j\sin\theta)$	$\mathbf{z}^* = M(\cos\theta - j\sin\theta)$
$\mathbf{z} = Me^{j\theta}$	$\mathbf{z}^* = Me^{-j\theta}$

Absolute Value of $\mathbf{z}$:

$$|\mathbf{z}| \equiv M$$
$$|\mathbf{z}| = \sqrt{\mathbf{z}\mathbf{z}^*}$$

The manipulation of complex numbers follows the rules of ordinary algebra, with the added rule that $j^2 = -1$. Note that when two numbers expressed in exponential form are multiplied together, the absolute value of the product is the product of the absolute values, but the argument of the product is the *sum* of the arguments. For example, if $\mathbf{z}_1 = M_1 e^{j\theta_1}$, $\mathbf{z}_2 = M_2 e^{j\theta_2}$, then following the usual rules of algebra $\mathbf{z}_1\mathbf{z}_2 = (M_1 M_2)\, e^{j\theta_1 + j\theta_2} = M_1 M_2 e^{j(\theta_1 + \theta_2)}$.

EXAMPLE 4.6

Divide $3.1e^{j(1.8)}$ by $(-3.6 + 2.9j)$ and express the quotient in exponential form. What is its absolute value? Its argument?

SOLUTION

Let us first convert $(-3.6 + 2.9j)$ to exponential form. We use Equation (4.16), where M and θ are given by Equations (4.9) and (4.10). Thus

$$M = \sqrt{(3.6)^2 + (2.9)^2} = 4.6$$

$$\theta = \tan^{-1}\left(\frac{2.9}{-3.6}\right) = -39° = -0.68 \text{ radian}$$

Therefore

$$-3.6 + 2.9j = 4.6 \, e^{-0.68j}$$

The required quotient is

$$q = \frac{3.1 \, e^{1.8j}}{4.6 \, e^{-0.68j}} = \frac{3.1}{4.6} \cdot \frac{e^{1.8j}}{e^{-0.68j}}$$

$$= 0.67 \cdot e^{j(1.8 + 0.68)} = 0.67 \, e^{j(2.48)} \quad \blacksquare$$

EXAMPLE 4.7

Let $z_1 = 3.9 \, e^{j(4.2)}$ and $z_2 = 0.63 \, e^{-j(1.8)}$. Calculate $\text{Re}(z_1 z_2^*)$.

SOLUTION

The complex conjugate of z_2 is $z_2^* = 0.63 \, e^{+j(1.8)}$. The product of z_1 and z_2^* is

$$z_1 z_2^* = 3.9 \, e^{j(4.2)} \cdot 0.63 \, e^{j(1.8)} = (3.9)(0.63) \, e^{j(4.2+1.8)}$$

$$= 2.46 \, e^{j(6.0)}$$

To find the real part of this number, we can use Equation (4.11):

$$\text{Re}(z_1 z_2^*) = 2.46 \cos(6.0) = 2.46 \cos(-16°)$$

$$= (2.46)(0.96) = 2.36 \quad \blacksquare$$

Phasors We shall now introduce the complex numbers known as *phasors*, which are used to represent sinusoidal functions. Let us define the phasor **v**, representing the sinusoid $v(t) = V_0 \cos(\omega t + \phi)$, according to the formula

$$\boxed{v = V_0 e^{j\phi}} \tag{4.17}$$

where V_0 and ϕ are the amplitude and phase of the original sinusoid. The absolute value M of the phasor is set equal to the amplitude V_0 of the sinusoid, and the argument ϕ of the phasor is set equal to the phase angle ϕ of the sinusoid. Thus the phasor contains both amplitude and phase information.[3] The phasor representation is more, however, than just an arbitrary shorthand way of representing amplitude and phase, because there is a rather direct mathe-

---

[3] Note that a phasor is a *constant,* not a time-varying quantity. However, it *describes* a time-varying quantity (the sinusoid). Hence the use of a lower-case symbol.

matical relationship between the phasor and the sinusoid it represents. Let us multiply the phasor **v** by the complex number $e^{j\omega t}$ and take the real part of the product. We find that $\mathrm{Re}[ve^{j\omega t}] = \mathrm{Re}[(V_0 e^{j\phi})\,(e^{j\omega t})] = \mathrm{Re}[V_0 e^{j(\omega t + \phi)}] = \mathrm{Re}[V_0\cos(\omega t + \phi) + V_0\, j\,\sin(\omega t + \phi)] = V_0\cos(\omega t + \phi)$. Thus *to find a sinusoidal function when its phasor is known, one must multiply the phasor by $e^{j\omega t}$ and take the real part of the product.* The converse rule, which tells how the phasor is found when the sinusoidal function is known, follows from the definition of the phasor: *Let a sinusoid be expressed in the form $\mathcal{F}(t) = A\cos(\omega t + \phi)$; the phasor representing it is then given by $\mathcal{F} = A\,e^{j\phi}$.*

EXAMPLE 4.8
A certain current is known to be given by the formula $i(t) = I_0\sin(\omega t + \theta)$, where θ is a constant angle. What is the corresponding phasor **i**?

SOLUTION

The function $i(t)$ must first be put into the form of a cosine function. As in Example 4.2, this may be done using the trigonometric identity $\sin\alpha = \cos(\alpha - \pi/2)$. Thus

$$i(t) = I_0\cos\left[\omega t + \theta - \frac{\pi}{2}\right]$$

$$= I_0\cos\left[\omega t + \left(\theta - \frac{\pi}{2}\right)\right]$$

which now has the desired form. The required phasor is therefore

$$\mathbf{i} = I_0 e^{j(\theta - \pi/2)} = I_0 e^{j\theta - j\pi/2} \qquad\blacksquare$$

As has already been stated, the phasor corresponding to the sinusoid $\mathcal{F}(t) = A\cos(\omega t + \phi)$ is $\mathcal{F} = A e^{j\phi}$. Thus if a phasor is expressed in the form $A e^{j\phi}$, the coefficient of the exponential (A, in this case) is equal to the amplitude of the corresponding sinusoid. But A is equal to the absolute value of **f**. (Proof: $|\mathcal{F}| = \sqrt{\mathbf{ff}^*} = \sqrt{(A e^{j\phi})\,(A e^{-j\phi})} = A$.) Thus the amplitude of a sinusoid is equal to the absolute value of its phasor.

EXAMPLE 4.9
A sinusoid is represented by the phasor $\mathbf{f} = 2 - 3j$. Find A, the amplitude of the sinusoid.

SOLUTION

The amplitude is equal to the absolute value of **f**. Thus

$$A = |\mathbf{f}| = \sqrt{\mathbf{ff}^*}$$

$$= \sqrt{(2 - 3j)\,(2 + 3j)} = \sqrt{4 + 9} = \sqrt{13} \qquad\blacksquare$$

The phase angle of the sinusoid ϕ is simply equal to the argument of the phasor.

EXAMPLE 4.10

A sinusoid is represented by the phasor $\mathbf{f} = 2 - 3j$. Find ϕ, the phase angle of the sinusoid.

SOLUTION

The phase angle ϕ is equal to the argument of the phasor $\mathbf{f}$. According to Equation (4.10), this is given by $\phi = \tan^{-1}(y/x)$, where $y = \mathrm{Im}(\mathbf{f})$ and $x = \mathrm{Re}(\mathbf{f})$. In this case $y = -3, x = 2$. Thus $\phi = \tan^{-1}(-3/2) = -56.3°$. ■

The sum of any two sinusoids of the same frequency is another sinusoid of that frequency. We can show that *the phasor representing the sum of two sinusoids of the same frequency is the sum of the two phasors representing the parts.* Let $v_3(t) = v_1(t) + v_2(t)$, and let $v_1(t) = V_1 \cos(\omega t + \phi_1)$, $v_2(t) = V_2 \cos(\omega t + \phi_2)$. Then $\mathbf{v}_1 + \mathbf{v}_2 = V_1 e^{j\phi_1} + V_2 e^{j\phi_2}$. Let us give the sinusoid corresponding to the phasor $(\mathbf{v}_1 + \mathbf{v}_2)$ the name $g(t)$. Then we may evaluate $g(t)$:

$$\begin{aligned} g(t) &= \mathrm{Re}[(\mathbf{v}_1 + \mathbf{v}_2)\,e^{j\omega t}] \\ &= \mathrm{Re}[\mathbf{v}_1 e^{j\omega t}] + \mathrm{Re}[\mathbf{v}_2 e^{j\omega t}] \\ &= v_1(t) + v_2(t) \end{aligned} \qquad (4.18)$$

Thus $g(t) = v_3(t)$, which was to be proved.

EXAMPLE 4.11

Let $v_1(t) = 2.3 \cos(\omega t + 0.6)$, $v_2(t) = 1.9 \cos(\omega t - 1.0)$. Find the phasor representing $v_3(t) = v_1(t) + v_2(t)$.

SOLUTION

$$\mathbf{v}_3 = \mathbf{v}_1 + \mathbf{v}_2 = \{2.3e^{j(0.6)}\} + \{1.9e^{j(-1.0)}\}$$

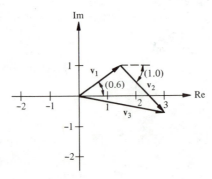

One way to simplify this expression is to convert the complex numbers to rectangular form by means of Equations (4.13), (4.11), and (4.12):

$$\begin{aligned} \mathbf{v}_3 &= 2.3[\cos(0.6) + j\sin(0.6)] + 1.9[\cos(-1.0) + j\sin(-1.0)] \\ &= [2.3\cos(0.6) + 1.9\cos(-1.0)] + j[2.3\sin(0.6) + 1.9\sin(-1.0)] \\ &= 2.9 - j(0.3) \end{aligned}$$

Graphically, the addition just performed appears as in the sketch (the numbers in parentheses are angles in radians). The complex number $\mathbf{v}_3 = 2.9 - j(0.3)$ is the answer, expressed simply in rectangular form. It may be desired to express it in polar form, so that the amplitude or the phase of the sinusoid $v_3(t)$ can be found. This is done by means of Equation (4.9):

$$M = \sqrt{(2.9)^2 + (-0.3)^2} = 2.92$$

$$\phi = \tan^{-1}\left(\frac{-0.3}{2.9}\right) = -0.10 \text{ radian}$$

Therefore $\mathbf{v}_3 = 2.92e^{j(-0.10)}$. The amplitude of the sinusoid which is the sum of v_1 and v_2 is 2.92, and its phase is -0.10 radian. ∎

If the sum of several sinusoids is zero, then the sum of the phasors representing the sinusoids must be zero. For example, let $f(t)$, $g(t)$, and $h(t)$ be three sinusoids of the same frequency whose phasors are $\mathbf{f}$, $\mathbf{g}$, and $\mathbf{h}$, and let us assume that $f(t) + g(t) + h(t) = 0$. Then we can show that $\mathbf{f} + \mathbf{g} + \mathbf{h} = 0$. To prove this, let us assume the contrary, that $(\mathbf{f} + \mathbf{g} + \mathbf{h}) = (x + jy)$. Then we would have $f(t) + g(t) + h(t) = \text{Re}[(\mathbf{f} + \mathbf{g} + \mathbf{h})\, e^{j\omega t}] = \text{Re}[(x + jy)\, e^{j\omega t}] = x \cos \omega t - y \sin \omega t$. But $f(t) + g(t) + h(t) = 0$, so $0 = x \cos \omega t - y \sin \omega t$. This equation can be true at all times only if $x = 0$, $y = 0$, which proves the theorem.

Phasor Representing the Derivative The derivative with respect to time of a sinusoid is another sinusoid of the same frequency. We can show that *if $\mathbf{v}$ is the phasor representing a sinusoid $v(t)$ with angular frequency ω radians per second, then the phasor representing the sinusoid $[dv(t)/dt]$ is $j\omega\mathbf{v}$.* This may be proved as follows. Let $v(t) = A \cos(\omega t + \phi)$. Then

$$\frac{dv}{dt} = -A\omega \sin[\omega t + \phi] \tag{4.19}$$

Using the trigonometric identity $\sin \alpha = -\cos(\alpha + \pi/2)$ (where α is any angle), we have

$$\frac{dv}{dt} = A\omega \cos[\omega t + (\phi + \pi/2)] \tag{4.20}$$

The phasor corresponding to this sinusoid is $A\omega e^{j(\phi + \pi/2)} = A\omega(e^{j\pi/2})e^{j\phi} = A\omega j e^{j\phi} = j\omega\mathbf{v}$, which proves this very important theorem.

The properties of phasors are summarized in Table 4.2.

Table 4.2 Properties of Phasors

Finding Phasor When Sinusoid is Known:

Express sinusoid in the form $\mathcal{F}(t) = M \cos(\omega t + \theta)$. The phasor representing $\mathcal{F}(t)$ is then equal to $Me^{j\theta}$.

Finding Sinusoid When Phasor is Known:

(1) The sinusoid corresponding to the phasor $\mathbf{f}$ is $\mathcal{F}(t) = \text{Re}(\mathbf{f}\, e^{j\omega t})$.
(2) If $\mathbf{f} = Me^{j\theta}$, $\mathcal{F}(t) = M \cos(\omega t + \theta)$.
(3) The amplitude of the sinusoid $\mathcal{F}(t)$ is equal to $|\mathbf{f}|$. Its phase angle is equal to the argument of $\mathbf{f}$.

Table 4.2   Properties of Phasors **(Continued)**

Sum of Sinusoids of the Same Frequency:

(1) If $\mathfrak{f}(t)$ and $\mathfrak{g}(t)$ are sinusoids of the same frequency whose phasors are **f** and **g**, respectively, the phasor representing the sinusoid $\mathfrak{f}(t) + \mathfrak{g}(t)$ is **f** + **g**.

(2) If the sum of several sinusoids of the same frequency is zero, the sum of their phasors is zero.

Phasor for Derivative

If $\mathfrak{f}(t)$ is a sinusoid, $d\mathfrak{f}/dt$ is a sinusoid with the same frequency. If the phasor representing $\mathfrak{f}(t)$ is **f**, the phasor for the sinusoid $d\mathfrak{f}/dt$ is $j\omega\mathbf{f}$, where ω is the angular frequency expressed in radians per second.

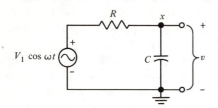

Figure 4.4   A circuit for demonstrating the use of phasors.

Use of Phasors in Circuit Analysis   We are now ready to make use of phasors to represent sinusoidal voltages and currents. The result obtained in the preceding paragraph is very useful because it provides a way to convert differentiation (such as occurs in the capacitor equation $i = C\ dv/dt$) into simple multiplication by $j\omega$, thus reducing differential equations to ordinary algebraic equations. Let us demonstrate the use of this phasor technique in a circuit problem, using the circuit of Figure 4.4. It is desired to find the amplitude and phase of the voltage v, indicated in the figure, as a function of the angular frequency ω. To do this we write a node equation for node x, where the voltage is v:

$$\frac{v}{R} - \frac{V_1 \cos \omega t}{R} + C \frac{dv}{dt} = 0 \tag{4.21}$$

Since the sum of the three sinusoidal functions on the left side of Equation (4.21) is zero, the sum of the phasors representing them must also be zero. The phasor representing the sinusoid $V_1 \cos \omega t$ is $V_1 e^{j(0)} = V_1$. We denote the (as yet unknown) phasor representing $v(t)$ by **v**. Then the phasor form of Equation (4.21) is

$$\frac{1}{R}(\mathbf{v} - V_1) + Cj\omega\mathbf{v} = 0 \tag{4.22}$$

Solving (4.22) algebraically for **v**, we have

$$\mathbf{v} = \frac{V_1}{1 + j\omega RC} = V_1 \cdot \frac{1 - j\omega RC}{1 + \omega^2 (RC)^2} \tag{4.23}$$

The amplitude of the sinusoid $v(t)$ is equal to the absolute value of the phasor **v**:

$$|\mathbf{v}| = (\mathbf{vv}^*)^{1/2}$$

$$= \frac{V_1}{\sqrt{1 + \omega^2 (RC)^2}} \tag{4.24}$$

The phase angle of $v(t)$ is equal to the argument of **v**, which is obtained from Equation (4.23) by means of (4.10):

$$\phi = \tan^{-1}(-\omega RC) \tag{4.25}$$

Using the results obtained in Equations (4.24) and (4.25), graphs can be constructed of the amplitude and phase of $v(t)$, as a function of frequency. This is done in Figure 4.5. As is customary, logarithmic scales are used for

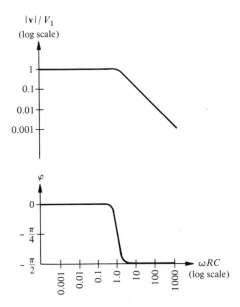

Figure 4.5 Graph of amplitude and phase of the output voltage v in Figure 4.4.

amplitude and frequency (but not the phase). We observe that the circuit just analyzed, that of Figure 4.4, is identical to that of Figure 2.14, and the results of the present analysis, Equations (4.24) and (4.25), are the same as the previous results, Equations (2.27) and (2.28). However, with the phasor method the results have been obtained without the need for solving any differential equations. Thus a considerable saving of computational effort results from use of the phasor method. When the circuit to be analyzed is more complicated, perhaps containing several capacitors and inductors, the simplification afforded by the phasor method is practically indispensable.

EXAMPLE 4.12

Find the current $i(t)$ which flows in the circuit shown below.

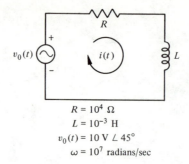

$$R = 10^4 \ \Omega$$
$$L = 10^{-3} \text{ H}$$
$$v_0(t) = 10 \text{ V} \angle 45°$$
$$\omega = 10^7 \text{ radians/sec}$$

SOLUTION

Writing a loop equation, we have

$$v_0(t) - i(t)R - L \frac{d}{dt} i(t) = 0$$

The phasor representing $v_0(t)$ is

$$\mathbf{v}_0 = 10 \ e^{j(\pi/4)} = \frac{10}{\sqrt{2}} (1 + j)$$

The phasor form of the loop equation is therefore

$$\frac{10}{\sqrt{2}} (1 + j) - \mathbf{i} R - L j\omega \, \mathbf{i} = 0$$

Solving for **i**

$$\mathbf{i} = \frac{10}{\sqrt{2}} \frac{1 + j}{(R + j\omega L)} = \frac{10}{\sqrt{2}} \cdot \frac{(R + \omega L) + j(R - \omega L)}{(R^2 + \omega^2 L^2)}$$

The amplitude and phase of the current are, from Equations (4.9) and (4.10),

$$M = \frac{10}{\sqrt{2}} \frac{[(R + \omega L)^2 + (R - \omega L)^2]^{1/2}}{R^2 + \omega^2 L^2} = \frac{10}{\sqrt{R^2 + \omega^2 L^2}} = \frac{10^{-3}}{\sqrt{2}} \text{ amperes}$$

$$\phi = \tan^{-1} \frac{(R - \omega L)}{(R + \omega L)} = \tan^{-1} (0) = 0$$

The current $i(t) = 7.07 \times 10^{-4} \text{ A} \angle 0°$. ■

Time Average of Product of Sinusoids   The power flowing in a circuit is given by the formula $P(t) = v(t) \, i(t)$. If $v(t)$ and $i(t)$ are sinusoids of the same frequency, the power $P(t)$ varies in time. Its value at any given instant is known as the *instantaneous power*. However, one is usually more interested in the time average of the power. For example, if a current flows through a resistor, the instantaneous power is $P(t) = v(t) \, i(t) = i^2(t)R$. Unless the frequency of $i(t)$ is extremely low, the temperature of the resistor will not be able to change fast enough to increase and decrease along with $i^2(t)$. Nonetheless, on the average, power is being dissipated in the resistor; the increase in its temperature will depend on the average value of $P(t)$. This is equal to the time average of $v(t)\,i(t)$, or, in this case, of $i^2(t)R$. We shall denote the time average by Avg $[P(t)] = $ Avg $[v(t)\,i(t)]$.

A theorem exists which allows us to calculate average power very simply when the phasors for $v(t)$ and $i(t)$ are known: *If $f(t)$ and $g(t)$ are two sinusoidal functions of the same frequency, and the phasors representing them are* **f** *and* **g** *respectively, then the time average of the product of $f(t)$ and $g(t)$ is given by* $1/2\,\mathrm{Re}(\mathbf{f}\,\mathbf{g}^*)$.

The theorem may be proved as follows. The time average of $f(t)\,g(t)$ is defined as

$$\text{Avg } (fg) = \lim_{T\to\infty} \frac{1}{T} \int_0^T f(t)\, g(t)\, dt \tag{4.26}$$

Let $f(t) = F\cos(\omega t + \phi_F)$, $g(t) = G\cos(\omega t + \phi_G)$. Then (4.26) becomes

$$\text{Avg } (fg) = \lim_{T\to\infty} \frac{1}{T} \int_0^T FG\cos(\omega t + \phi_F)\cos(\omega t + \phi_G)\, dt \tag{4.27}$$

Using the trigonometric identity $(\cos x)(\cos y) = 1/2\,[\cos(x+y) + \cos(x-y)]$, this becomes

$$\begin{aligned}
\text{Avg } (fg) &= \frac{FG}{2} \lim_{T\to\infty} \frac{1}{T} \int_0^T [\cos(2\omega t + \phi_F + \phi_G) + \cos(\phi_F - \phi_G)]\, dt \\
&= \frac{FG}{2} \left[\lim_{T\to\infty} \frac{1}{T} \int_0^T \cos(2\omega t + \phi_F + \phi_G)\, dt + \lim_{T\to\infty} \frac{1}{T} \int_0^T \cos(\phi_F - \phi_G)\, dt \right]
\end{aligned} \tag{4.28}$$

The first limit in the bracket is zero (because the integrand is a sinusoid, whose average value is zero). However the integrand of the second integral, $\cos(\phi_F - \phi_G)$, is a constant, and thus can be taken out of the integral. We then have

$$\begin{aligned}
\text{Avg } (fg) &= \frac{FG}{2} \cos(\phi_F - \phi_G) \lim_{T\to\infty} \frac{1}{T} \int_0^T dt \\
&= \frac{FG}{2} \cos(\phi_F - \phi_G) \lim_{T\to\infty} (1) \\
&= \frac{FG}{2} \cos(\phi_F - \phi_G)
\end{aligned} \tag{4.29}$$

Now let us find the value of $1/2\,\mathrm{Re}\,(\mathbf{f}\mathbf{g}^*)$. We know that $\mathbf{f} = F\,e^{j\phi_G}$, $\mathbf{g} = G\,e^{j\phi_G}$. Thus

$$\begin{aligned}
\tfrac{1}{2}\,\mathrm{Re}(\mathbf{f}\mathbf{g}^*) &= \tfrac{1}{2}\,\mathrm{Re}[FG e^{j(\phi_F - \phi_G)}] \\
&= \frac{FG}{2}\,\mathrm{Re}[\cos(\phi_F - \phi_G) + j\sin(\phi_F - \phi_G)] \\
&= \frac{FG}{2} \cos(\phi_F - \phi_G)
\end{aligned} \tag{4.30}$$

Since result (4.30) is identical to (4.29), the theorem is proved.

EXAMPLE 4.13

In the circuit of Example 4.12, the instantaneous power produced by the voltage source is the product of its instantaneous voltage times the instantaneous current directed outward from the (+) terminal. Find the time average of this power.

SOLUTION

$$\text{Avg}\,[i(t)\,v_0(t)] = \tfrac{1}{2}\,\text{Re}[\mathbf{i}\,\mathbf{v}_0^*]$$

From the result of Example 4.12

$$\mathbf{i} = 7.07 \times 10^{-4}\,e^{j(0)} = 7.07 \times 10^{-4}\ \text{amperes}$$

We have that

$$\mathbf{v}_0 = 10\,e^{j(\pi/4)}\ \text{volts}$$

therefore

$$\mathbf{v}_0^* = 10\,e^{-j(\pi/4)}\ \text{volts} = \frac{10}{\sqrt{2}}\,(1-j)\ \text{volts}$$

Now the average power can be obtained:

$$\text{Avg}\,[i(t)\,v_0(t)] = \frac{1}{2}\,\text{Re}\left[(7.07 \times 10^{-4})\left(\frac{10}{\sqrt{2}} - \frac{10}{\sqrt{2}}j\right)\right]$$

$$= \frac{1}{2}\,(7.07 \times 10^{-4})\left(\frac{10}{\sqrt{2}}\right) = 2.5 \times 10^{-3}\ \text{watts} \qquad \blacksquare$$

Before going on to further examples of the use of phasors, it is useful to introduce the concept of *impedance*. The use of the impedance concept provides further simplification of problem-solving procedures, and makes them almost identical to the methods used with purely resistive circuits.

4.3 Impedance

In the case of a resistance, if the voltage across the terminals is given, the current is determined by Ohm's law. The ratio of voltage to current is equal to the constant R: $v(t)/i(t) = R$. The existence of this fixed ratio of v to i gives the *I-V* relationship of the resistor a very simple form. Through the use of phasors the *I-V* equations of inductors and capacitors can be reduced to an equally simple form.

Suppose that the phasor **v** represents the sinusoidal voltage across the terminals of a circuit element, and that the phasor representing the sinusoidal current through the element is **i**. Then *we define the impedance* **Z** *of the circuit element as the ratio of the phasor representing the sinusoidal voltage across it to the phasor representing the sinusoidal current flowing through it:*

$$\frac{\mathbf{v}}{\mathbf{i}} = \mathbf{Z} \tag{4.31}$$

Not every circuit element possesses an impedance; an ideal voltage source does not, since the voltage across a voltage source is not functionally related to the current through it, and the ratio $\mathbf{v}/\mathbf{i}$ has no fixed value. The elements for which the definition of impedance is most useful are resistance, capacitance, and inductance. The impedances of these elements will now be found.[4]

For the resistance, the constitutive equation is $v = iR$. Therefore $\mathbf{v} = \mathbf{i}R$, and the impedance is $\mathbf{v}/\mathbf{i} = R$:

$$\boxed{\mathbf{Z}_R = R \qquad \text{(impedance of a pure resistance)}} \qquad (4.32)$$

Similarly, for the capacitance, $i = C\ dv/dt$. Therefore $\mathbf{i} = Cj\omega\mathbf{v}$ and

$$\boxed{\mathbf{Z}_C = \frac{1}{j\omega C} \qquad \text{(impedance of a pure capacitance)}} \qquad (4.33)$$

For the inductance, $v = Ldi/dt$. Therefore $\mathbf{v} = Lj\omega\mathbf{i}$, and

$$\boxed{\mathbf{Z}_L = j\omega L \qquad \text{(impedance of a pure inductance)}} \qquad (4.34)$$

It should be noted that the symbol for an impedance ($\mathbf{Z}$) is boldface because in general an impedance is a complex number. However, the use of boldface type does *not* mean that impedance is a phasor. Impedances describe the properties of circuit elements; phasors describe sinusoidal voltages or currents.

When the phasor voltage across a resistance, capacitance, or inductance is known, the current through it can immediately be found by means of Equations (4.32), (4.33), or (4.34).

EXAMPLE 4.14

The sinusoidal voltage across a 10^{-6} F capacitor is represented by the phasor $\mathbf{v} = 6\ e^{j\phi}$ V, where $\phi = 0.6$ radian. The frequency ω is 10^4 radians/sec. What is the amplitude of the sinusoidal current $i(t)$ through the capacitor?

SOLUTION

Using Equation (4.33), the impedance of the capacitor is given by

$$\mathbf{Z}_C = \frac{1}{j\omega C}$$

From Equation (4.31) we have

$$\mathbf{i} = \frac{\mathbf{v}}{\mathbf{Z}_C} = \mathbf{v}j\omega C$$
$$= 6\omega C e^{j(\phi+\pi/2)} = 6(10^4)(10^{-6})\ e^{j(\phi+\pi/2)}$$
$$= 6 \times 10^{-2}\ e^{j(\phi+\pi/2)}\ \text{A}$$

The absolute value of $\mathbf{i}$ is 6×10^{-2} A; therefore this is the amplitude of the sinusoidal current through C. ∎

---

[4] The definition of impedance given in this paragraph is not the most general definition possible. More general definitions of impedance are used in advanced work.

When circuit elements are connected in series or in parallel, the impedance of the combination is obtained from rules identical in form to those for series or parallel resistances:

$$\mathbf{Z}_S = \mathbf{Z}_1 + \mathbf{Z}_2 \qquad \text{(impedances in series)} \tag{4.35}$$

$$\mathbf{Z}_P = \mathbf{Z}_1\mathbf{Z}_2/(\mathbf{Z}_1 + \mathbf{Z}_2) \qquad \text{(impedances in parallel)} \tag{4.36}$$

(The derivations of these formulas are similar to those for resistances in series and parallel. The proofs are deferred to a problem.) Formulas (4.35) and (4.36) are very useful in ac circuit analysis.

EXAMPLE 4.15

Find the impedance of the combination of elements shown in the sketch.

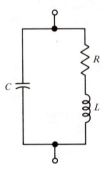

SOLUTION

The impedance of the combination is, using Equations (4.35) and (4.36)

$$\begin{aligned}
\mathbf{Z} &= \mathbf{Z}_C \| (\mathbf{Z}_R + \mathbf{Z}_L) \\
&= \frac{\mathbf{Z}_C(\mathbf{Z}_R + \mathbf{Z}_L)}{\mathbf{Z}_C + \mathbf{Z}_R + \mathbf{Z}_L} \\
&= \frac{\left(\dfrac{1}{j\omega C}\right)(R + j\omega L)}{\dfrac{1}{j\omega C} + R + j\omega L} \\
&= \frac{R + j\omega L}{1 + j\omega RC - \omega^2 LC}
\end{aligned}$$
∎

EXAMPLE 4.16

Suppose a current source whose phasor is $\mathbf{i}_0$ is connected to the terminals of the network in Example 4.15, as shown in sketch (a).

(1)   What is the phasor $\mathbf{v}$ representing the voltage at the terminals indicated in the diagram?

(2)   What is the amplitude of the sinusoidal voltage $v(t)$ at these terminals?

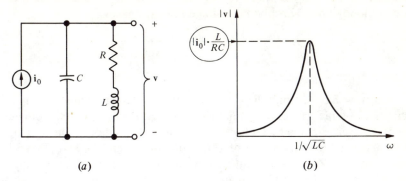

(a)   (b)

SOLUTION

(1)   By the definition of impedance, $\mathbf{v} = \mathbf{i}_0\mathbf{Z}$, where $\mathbf{Z}$ is the impedance of the network found in Example 4.15:

$$\mathbf{v} = \mathbf{i}_0 \bullet \frac{R + j\omega L}{(1 - \omega^2 LC) + (j\omega RC)}$$

(2)   The amplitude of the sinusoidal voltage $v(t)$ is equal to the absolute value of the phasor $\mathbf{v}$. Since the absolute value of the product of two complex numbers is the product of their absolute values, $|\mathbf{v}| = |\mathbf{i}_0| \, |\mathbf{Z}|$.

The value of $|\mathbf{Z}|$ can be found most easily by means of the identity $|\mathbf{Z}| = \sqrt{\mathbf{Z}\mathbf{Z}^*}$. Thus

$$|\mathbf{Z}| = \left\{ \frac{R^2 + (\omega L)^2}{[1 - \omega^2(LC)]^2 + [\omega(RC)]^2} \right\}^{1/2}$$

The amplitude of $v(t)$ is then equal to this quantity multiplied by $|\mathbf{i}_0|$. For instance, if $i_0(t)$ were 20 mA $\angle 60°$, $|\mathbf{i}_0|$ would be 20 mA, and $|\mathbf{v}| = (20 \text{ mA}) \bullet |\mathbf{Z}|$ volts.

The result obtained in this example illustrates the phenomenon of *resonance* in an RLC circuit. Due to the form of the first term in the denominator of $|\mathbf{Z}|$, the absolute value of the impedance of the network, regarded as a function of frequency, has a maximum at a certain frequency. This frequency is called the *resonant frequency*. At the resonant frequency, the voltage across its terminals is a maximum. An example of $|\mathbf{v}|$ versus ω, for the case of $(RC)^2 \ll (LC)$, is given in sketch (b). We see that the value of the resonant frequency is

$$\omega_{RES} \cong \frac{1}{\sqrt{LC}}$$ ∎

Phasor Form of Kirchhoff's Laws   In Section 4.2 it was pointed out that if the sum of several sinusoids is zero, the sum of their phasors likewise must be zero. We can make use of this fact to express Kirchhoff's two laws in phasor form. Let $i_1(t)$, $i_2(t)$,..., $i_n(t)$ be the sinusoidal currents entering a node. Then Kirchhoff's current law, in phasor form, is

$$\sum_{k=1}^{n} \mathbf{i}_k = 0 \qquad \binom{\text{Kirchhoff's}}{\text{current law}}$$ (4.37)

Similarly, let $v_1(t)$, $v_2(t)$,..., $v_n(t)$ be the sinusoidal voltage drops around a complete loop. Then Kirchhoff's voltage law, in phasor form, is

$$\sum_{k=1}^{n} \mathbf{v}_k = 0 \qquad \begin{pmatrix} \text{Kirchhoff's} \\ \text{voltage law} \end{pmatrix} \tag{4.38}$$

We now have at our disposal everything needed to analyze sinusoidal ac circuits in a manner exactly analogous to that for dc circuits. Voltages and currents are represented by their phasors. The node method or the loop method can be used, taking advantage of Equations (4.37) and (4.38). Ohm's law is replaced by $\mathbf{v} = \mathbf{iZ}$. Thus, except that complex quantities are used, sinusoidal analysis can now be performed in just the same way as the dc analysis of Chapter 1.

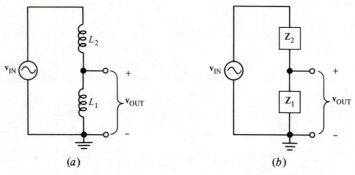

(a)   (b)

Figure 4.6 Generalized voltage dividers. (a) With two inductors; (b) with arbitrary impedances.

As an example of the use of the node method in phasor analysis, let us consider the circuit of Figure 4.6(a). This circuit is similar to the voltage divider, except that inductors are present instead of resistors. To obtain $\mathbf{v}_{OUT}$, we write a phasor node equation for the node between L_1 and L_2. The phasor current through each of the inductors is equal to the phasor voltage across the inductor, divided by its impedance. Thus

$$\frac{\mathbf{v}_{OUT} - \mathbf{v}_{IN}}{\mathbf{Z}_{L_2}} + \frac{\mathbf{v}_{OUT}}{\mathbf{Z}_{L_1}} = 0 \tag{4.39}$$

Since $\mathbf{Z}_{L_1} = j\omega L_1$, $\mathbf{Z}_{L_2} = j\omega L_2$, we have

$$(\mathbf{v}_{OUT} - \mathbf{v}_{IN})\, L_1 + \mathbf{v}_{OUT}\, L_2 = 0$$

$$\mathbf{v}_{OUT} = \mathbf{v}_{IN} \cdot \frac{L_1}{L_1 + L_2} \tag{4.40}$$

The reader will note that no differential equations were needed for this solution, and the algebra was much like that used in solving the resistive voltage divider, except that complex voltages and currents are used.

We can distinguish a general family of circuits, characterized by the form of Figure 4.6(*b*). This general circuit resembles the resistive voltage divider (Figure 1.24), except that now the boxes designated $\mathbf{Z}_1$ and $\mathbf{Z}_2$ may represent any resistance, capacitance, or inductance, or two-terminal combinations of these elements. The impedance between the terminals of box $\mathbf{Z}_1$ is determined by whatever elements or combination of elements happens to be in the box, and the same is true for $\mathbf{Z}_2$. (For example, if the box marked $\mathbf{Z}_2$ contains a resistor R and inductance L in series, then $\mathbf{Z}_2 = R + j\omega L$.) Proceeding as in Equation (4.39), we find

$$\frac{v_{OUT} - v_{IN}}{\mathbf{Z}_2} + \frac{v_{OUT}}{\mathbf{Z}_1} = 0$$

$$\boxed{v_{OUT} = v_{IN} \cdot \frac{\mathbf{Z}_1}{\mathbf{Z}_1 + \mathbf{Z}_2}} \qquad (4.41)$$

This result closely resembles that for the resistive voltage divider, Equation (1.35), except that resistances are now replaced by impedances. We may call the circuit of Figure 4.6(*b*) the *generalized voltage divider*. The case of Figure 4.6(*a*) is now seen to be just a specific example of the general case, Figure 4.6(*b*), and Equation (4.40) can be obtained from Equation (4.41).

EXAMPLE 4.17
Find the phasor v_{OUT} in the following circuit:

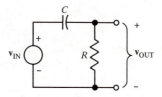

SOLUTION
This circuit has the form of a generalized voltage divider, as may be seen by comparison with Figure 4.6(*b*). Accordingly we may use Equation (4.41). In this case $\mathbf{Z}_1 = R$, $\mathbf{Z}_2 = (1/j\omega C)$. Thus

$$v_{OUT} = v_{IN} \cdot \frac{R}{R + \dfrac{1}{j\omega C}}$$

$$= v_{IN} \cdot \frac{j\omega RC}{1 + j\omega RC}$$

This circuit is the same as that of Figure 2.16; the results previously found in Equation (2.32) can also be found from v_{OUT}. For instance, the amplitude of the output sinusoid is given by

$$|v_{\text{OUT}}| = |v_{\text{IN}}| \cdot \left| \frac{j\omega RC}{1 + j\omega RC} \right|$$

$$= |v_{\text{IN}}| \cdot \sqrt{\frac{j\omega RC}{1 + j\omega RC} \left(\frac{j\omega RC}{1 + j\omega RC} \right)^*}$$

$$= |v_{\text{IN}}| \cdot \frac{\omega RC}{\sqrt{1 + (\omega RC)^2}}$$

in agreement with (2.32).   ∎

EXAMPLE 4.18

In the sketch, find the phasor **v** representing the voltage at the output terminals.

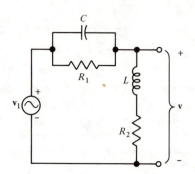

SOLUTION

This is a special case of the generalized voltage divider of Figure 4.6(*b*), with

$$\mathbf{Z}_2 = \mathbf{Z}_C \| \mathbf{Z}_{R_1} = \frac{R_1 / j\omega C}{R_1 + \dfrac{1}{j\omega C}} = \frac{R_1}{1 + j\omega R_1 C}$$

$$\mathbf{Z}_1 = \mathbf{Z}_{R_2} + \mathbf{Z}_L = R_2 + j\omega L$$

Using Equation (4.41)

$$\mathbf{v} = \mathbf{v}_1 \cdot \frac{R_2 + j\omega L}{R_2 + j\omega L + \left(\dfrac{R_1}{1 + j\omega R_1 C} \right)}$$ ∎

The following example demonstrates the use of the loop method in phasor analysis.

EXAMPLE 4.19

In the sketch, find the phasor representing the current flowing through R (as shown) in terms of the input current phasor i_1.

SOLUTION

We introduce two phasor loop currents as shown. The left loop current is controlled by the current source and must have the value i_1. The second phasor loop current equals the unknown **i**. Writing a loop equation for the right-hand loop

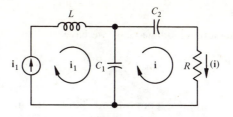

$$(\mathbf{i} - \mathbf{i}_1)\mathbf{Z}_{C_1} + \mathbf{i}\mathbf{Z}_{C_2} + \mathbf{i}\mathbf{Z}_R = 0$$

Inserting the values of the three impedances

$$\frac{(\mathbf{i} - \mathbf{i}_1)}{j\omega C_1} + \frac{\mathbf{i}}{j\omega C_2} + \mathbf{i}R = 0$$

$$\mathbf{i} = \mathbf{i}_1 \frac{1}{1 + \dfrac{C_1}{C_2} + j\omega R C_1}$$ ■

When dependent sources are present in a circuit, circuit analysis can proceed as usual, but with the value of the dependent source and its reference both regarded as phasors. For instance, suppose the value of a dependent current source is $\beta i_1(t)$, where $i_1(t)$ is a sinusoidal reference current. Then if $i_1(t)$ is represented by the phasor $\mathbf{i}_1$, the sinusoidal current through the dependent control must be represented by the phasor $\beta\mathbf{i}_1$.

EXAMPLE 4.20

Consider the following circuit. This circuit contains a dependent current source whose value is a constant β times the current through $\mathbf{Z}_1$, as shown. Find the phasor representing $v(t)$ in terms of the phasor representing the sinusoidal input voltage $v_1(t)$.

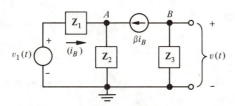

SOLUTION

The phasor representing the current through the dependent source equals β times the phasor for the current through $\mathbf{Z}_1$. The problem may be solved by writing node equations for $\mathbf{v}_A$ and $\mathbf{v}$. The node equation at node A is

$$\frac{\mathbf{v}_A - \mathbf{v}_1}{\mathbf{Z}_1} \cdot (1 + \beta) + \frac{\mathbf{v}_A}{\mathbf{Z}_2} = 0$$

Similarly, the node equation at node B is

$$\frac{\beta(\mathbf{v}_1 - \mathbf{v}_A)}{\mathbf{Z}_1} + \frac{\mathbf{v}}{\mathbf{Z}_3} = 0$$

Solving these two equations simultaneously, we have

$$\mathbf{v} = -\mathbf{v}_1 \cdot \frac{\beta \, \mathbf{Z}_3}{\mathbf{Z}_2(1 + \beta) + \mathbf{Z}_1} \qquad \blacksquare$$

With phasor techniques, generalized Thévenin and Norton equivalent circuits can be found for subcircuits containing voltage and current sources, resistance, capacitance, and inductance. The procedure is the same as that described in Section 2.1 for purely resistive circuits, with the following changes: (1) phasors representing voltages and currents replace the voltages and currents themselves, and (2) the Thévenin and Norton resistances are replaced by Thévenin and Norton impedances. The Thévenin and Norton parameters thus are found by using the following rules:

$$
\begin{array}{ll}
\mathbf{v}_T = \mathbf{v}_{OC} \\
\mathbf{Z}_T = -\mathbf{v}_{OC}/\mathbf{i}_{SC} & \text{(general Thévenin equivalent)}
\end{array}
\qquad (4.42)
$$

$$
\begin{array}{ll}
\mathbf{i}_N = -\mathbf{i}_{SC} \\
\mathbf{Z}_N = -\mathbf{v}_{OC}/i_{SC} & \text{(general Norton equivalent)}
\end{array}
\qquad (4.43)
$$

The sign conventions of $\mathbf{v}_{OC}$, $\mathbf{i}_{SC}$, $\mathbf{v}_T$, and $\mathbf{i}_N$ are as shown in Figure 4.7.

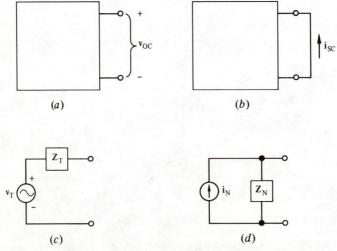

Figure 4.7 Sign conventions for finding generalized Thévenin and Norton equivalent circuits. (a) shows the sign convention for open-circuit voltage; (b) shows the sign convention for short-circuit current. Figure (c) shows the generalized Thévenin equivalent, and (d) the generalized Norton equivalent.

EXAMPLE 4.21

Find the Thévenin equivalent of the circuit shown in sketch (a).

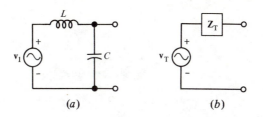

(a)   (b)

SOLUTION

This will be recognized as a generalized voltage divider of the type of Figure 4.6(b). Using Equation (4.41) we have

$$\mathbf{v}_{OC} = \mathbf{v}_1 \frac{\mathbf{Z}_C}{(\mathbf{Z}_C + \mathbf{Z}_L)} = \mathbf{v}_1 \frac{\dfrac{1}{j\omega C}}{\dfrac{1}{j\omega C} + j\omega L} = \frac{\mathbf{v}_1}{1 - \omega^2 LC}$$

The short-circuit phasor current is given by

$$\mathbf{i}_{SC} = -\mathbf{v}_1/j\omega L$$

(The minus sign results from the sign convention for $\mathbf{i}_{SC}$.) Thus from Equation (4.42)

$$\mathbf{v}_T = \frac{\mathbf{v}_1}{1 - \omega^2 LC}; \qquad \mathbf{Z}_T = \frac{j\omega L}{1 - \omega^2 LC}$$

The Thévenin equivalent circuit is shown in sketch (b).

It is interesting to note that the Thévenin parameters take on infinite values when $\omega^2 = 1/LC$. As in Example 4.16, this is again the phenomenon of resonance. When $\omega^2 = 1/LC$, the impedance of the series combination of L and C is

$$\mathbf{Z}_L + \mathbf{Z}_C = j\omega L + \frac{1}{j\omega C}$$

$$= \frac{jL}{\sqrt{LC}} - \frac{j\sqrt{LC}}{C} = j \left[\sqrt{\frac{L}{C}} - \sqrt{\frac{L}{C}} \right] = 0$$

Thus at this frequency the impedance across v_1 would seem to become zero and consequently the current which flows from v_1 through the series combination would seem to become infinite. If such a circuit were constructed, the current would indeed have a maximum when $\omega = 1/\sqrt{LC}$, but of course could not be infinite. The actual physical circuit would inevitably contain some resistance (especially in the coil windings), and this resistance, if included in the equations, would prevent the impedance from becoming zero for any real value of ω.

Note that the parallel combination of L and C in Example 4.16 had a maximum of impedance at the resonant frequency, while in this example, with a series combination of L and C, there is a minimum of impedance at the resonant frequency. ■

EXAMPLE 4.22

In the subcircuit shown in sketch (a), the dependent current source is controlled by the voltage at node A. Find the Norton equivalent for this subcircuit.

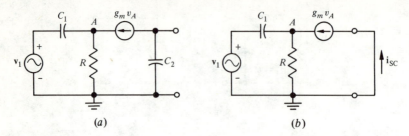

(a)   (b)

SOLUTION

We first find the short-circuit current. When the terminals are short-circuited, no current flows through C_2. The situation is shown in sketch (b).

Let us refer to the value of v_A when the terminals are shorted as $v_{A,\text{SC}}$. Writing a node equation for node A, we have

$$\frac{v_{A,\text{SC}} - v_1}{\dfrac{1}{j\omega C_1}} + \frac{v_{A,\text{SC}}}{R} - g_m\, v_{A,\text{SC}} = 0$$

Solving, we have

$$v_{A,\text{SC}} = v_1 \cdot \frac{j\omega C_1}{\left(j\omega C_1 + \dfrac{1}{R} - g_m\right)}$$

Clearly $i_{\text{SC}} = g_m\, v_{A,\text{SC}}$, so

$$i_{\text{SC}} = g_m\, v_1 \frac{j\omega C_1}{j\omega C_1 + \dfrac{1}{R} - g_m}$$

Next we shall find the open-circuit voltage. We shall call the value of v_A with the terminals open-circuited $v_{A,\text{OC}}$. Again writing a node equation for node A

$$\frac{v_{A,\text{OC}} - v_1}{\dfrac{1}{j\omega C_1}} + \frac{v_{A,\text{OC}}}{R} - g_m\, v_{A,\text{ OC}} = 0$$

This equation, as it happens, is identical with the equation for $v_{A,\text{SC}}$, and hence has the solution

$$v_{A,\text{OC}} = v_1 \cdot \frac{j\omega C_1}{\left(j\omega C_1 + \dfrac{1}{R} - g_m\right)}$$

To find v_{OC}, we observe that

$$v_{\text{OC}} = -g_m\, v_A Z_{C_2}$$

where Z_{C_2}, the impedance of C_2, is $1/j\omega C_2$. Thus

$$\mathbf{v}_{OC} = -\frac{g_m}{j\omega C_2} \cdot \frac{j\omega C_1}{\left(j\omega C_1 + \dfrac{1}{R} - g_m\right)} \cdot \mathbf{v}_1$$

$$= -\frac{g_m C_1 \,\mathbf{v}_1}{C_2\left(j\omega C_1 + \dfrac{1}{R} - g_m\right)}$$

Now using Equation (4.43), we have

$$\mathbf{i}_N = -\mathbf{i}_{SC} = -g_m \,\mathbf{v}_1 \frac{j\omega C_1}{\left(j\omega C_1 + \dfrac{1}{R} - g_m\right)}$$

$$\mathbf{Z}_N = -\mathbf{v}_{OC}/\mathbf{i}_{SC} = \frac{1}{j\omega C_2}$$

The Norton equivalent circuit is as shown in sketch (c).

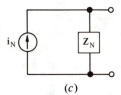

(c)

4.4 Applications of Sinusoidal Analysis

In this section we shall consider some applications of phasor techniques to problems of circuit analysis. In some cases circuits are designed to handle signals which are sinusoidal, or nearly so. The application of sinusoidal analysis to this type of circuit is relatively straightforward. However, even when the circuit is intended for use with non-sinusoidal signals, sinusoidal analysis is a useful technique. We shall consider the two cases separately, beginning with the former, simpler case.

RF Circuits In electronics, sinusoidal signals are most often encountered in communications equipment, where specific radio frequencies are used. For example, the carrier wave of a radio station ("1100 kilohertz on your dial") passes through the station's transmitter as a very nearly sinusoidal signal, and it enters your receiver in the form of a small but nearly sinusoidal antenna voltage. This kind of signal is called an *rf signal*. (The abbreviation is for "radio frequency.") Circuits designed to handle such signals are known as *rf circuits*.

One structure very characteristic of rf circuits is the *resonant circuit*. Resonant circuits are circuits which exhibit special properties at frequencies near a special frequency characteristic of the circuit, known as its *resonant frequency*.

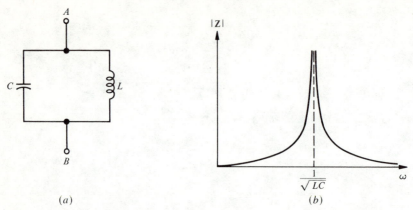

Figure 4.8   (a) A parallel LC resonant circuit. (b) Absolute value of impedance as a function of frequency.

Although other kinds of resonant circuits can be made, the most common type contains both capacitance and inductance. Let us consider the *parallel resonant circuit* shown in Figure 4.8(a). The impedance that appears between terminals A and B of this parallel combination is

$$\mathbf{Z} = \frac{(j\omega L)\left(\dfrac{1}{j\omega C}\right)}{j\omega L + \dfrac{1}{j\omega C}} = \frac{j\omega L}{1 - \omega^2 LC} \tag{4.44}$$

Often one is interested in the absolute value $|\mathbf{Z}|$. (For example, if a current source $\mathbf{i}_0$ were connected across terminals A,B, the amplitude of the voltage between the terminals would be $|\mathbf{v}| = |\mathbf{i}_0| \cdot |\mathbf{Z}|$.) The absolute value of the parallel impedance given in Equation (4.44) is

$$\boxed{|\mathbf{Z}| = \frac{\omega L}{1 - \omega^2 LC} \qquad \begin{pmatrix} \text{parallel} \\ \text{resonance} \end{pmatrix}} \tag{4.45}$$

This function of ω is graphed in Figure 4.8(b). We see that the function $|\mathbf{Z}|$ has a singularity at the frequency $\omega_R = 1/\sqrt{LC}$; its value appears to become infinite when $\omega = \omega_R$. This frequency ω_R given by

$$\boxed{\omega_R \equiv \frac{1}{\sqrt{LC}} \qquad \begin{pmatrix} \text{resonant} \\ \text{frequency} \end{pmatrix}} \tag{4.46}$$

is called the *resonant frequency* of the circuit.

In a practical circuit the impedance would not actually become infinite at any frequency. One reason for this is that actual physical circuit elements would contain some resistance in their wires. For example, the windings of the

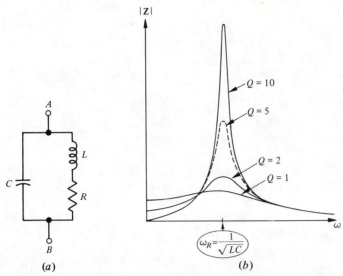

Figure 4.9 (*a*) A parallel resonant circuit with resistance in the inductive branch. (*b*) Plots of the absolute value of the impedance versus ω for several values of Q.

inductor may contain substantial resistance. If this resistance is taken into consideration, the circuit has the form shown in Figure 4.9(*a*). This circuit has already been considered in Examples 4.15 and 4.16. There it was found that the impedance between terminals A and B is

$$\mathbf{Z} = \frac{R + j\omega L}{1 - \omega^2 LC + j\omega RC} \tag{4.47}$$

so that the absolute value of the impedance is

$$|\mathbf{Z}| = \left[\frac{R^2 + (\omega L)^2}{(1 - \omega^2 LC)^2 + (\omega RC)^2} \right]^{1/2} \tag{4.48}$$

This quantity is plotted in Figure 4.9(*b*) for several cases, corresponding to the same resonant frequency but to different values of R. We shall define a dimensionless parameter Q, known as the *quality factor* of the circuit (or more frequently, simply as its "Q"), according to the formula

$$Q = \frac{\omega_R L}{R} \tag{4.49}$$

The curves of Figure 4.9(*b*) are designated by their values of Q. We see that when Q is large (that is, when R is small) the curves have a sharply peaked form which approaches that of Figure 4.8(*b*) as Q approaches infinity. On the other hand, when $Q = 1$, the curve becomes broad, and in fact the maximum of $|\mathbf{Z}|$ occurs at a frequency slightly less than ω_R.

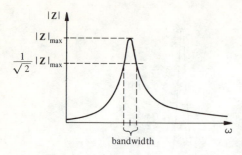

Figure 4.10 The bandwidth of a high Q circuit is given by $BW = \omega_R/Q$.

As the value of Q increases, the width of the peak becomes "narrower." Let us define the *bandwidth* of the resonant circuit as the range of frequencies over which $|\mathbf{Z}|$ is larger than $1/\sqrt{2}$ times its maximum value. This definition is illustrated by Figure 4.10. Then it can be shown from Equation (4.48) that if $Q \gg 1$, the bandwidth is given by

$$BW \cong \frac{\omega_R}{Q} \tag{4.50}$$

The above definition of bandwidth makes quantitative our previous statement that larger values of Q imply a "narrower" resonance curve.

EXAMPLE 4.23

A parallel-resonant circuit of the type shown in Figure 4.9(a) is to be used as a "filter" to remove a signal of an unwanted frequency. The circuit contains two current sources of equal magnitude but with different frequencies, 1 MHz and 1.01 MHz, as shown below. Assume that ω_R for the resonant circuit equals 1 MHz, that $Q \gg 1$, and that $C = 3 \times 10^{-10}$ F. Find the value of R which will cause the amplitude of 1.01 MHz voltage at the voltmeter to be 3 db less than that of the 1 MHz voltage.

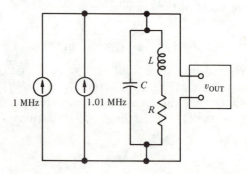

SOLUTION

As can be seen from Figure 4.10, we need a circuit for which the bandwidth extends approximately from 0.99 MHz to 1.01 MHz, so that the 1.01 MHz signal will lie at one end of the bandwidth and be attenuated by $1/\sqrt{2}$, or 3 db. Thus we wish $BW = 0.02$ MHz. From Equation (4.50) we have

$$Q \cong \frac{\omega_R}{BW} = \frac{2\pi \cdot (1)}{2\pi \cdot (0.02)} = 50$$

The value of Q is related to that of R by Equation (4.49). However, before (4.49) can be used we must find L. This can be done using the formula for resonant frequency, Equation (4.46):

$$L = \frac{1}{\omega_R^2 C} = \frac{1}{(2\pi)^2 (10^6)^2 \cdot 3 \cdot 10^{-10}} = 8.5 \times 10^{-5} \text{ H}$$

Now from (4.49)

$$R = \frac{\omega_R L}{Q} = \frac{(2\pi) \cdot 10^6 \cdot (8.5 \cdot 10^{-5})}{50} = 10.6 \ \Omega$$

Thus we find that an inductor with resistance of 10.6 Ω or less is required to make the filter. In practice a lower limit exists for the bandwidth of LC filters, because of the inevitable resistance of the wire in the inductor. ∎

Another important resonant circuit is the *series resonant circuit,* shown in Figure 4.11(a). Assuming for the moment that $R = 0$, the impedance of this connection is

$$\mathbf{Z} = j\omega L + \frac{1}{j\omega C} = \frac{1 - \omega^2/\omega_R^2}{j\omega C} \tag{4.51}$$

where ω_R, as before, is defined by $\omega_R = 1/\sqrt{LC}$. We see that for the series resonant circuit, the impedance decreases to zero when $\omega = \omega_R$. The absolute value of $\mathbf{Z}$ [as given by Equation (4.51)] is graphed as the solid curve in Figure 4.11(b), as a function of frequency. Of course, resistance must also be present

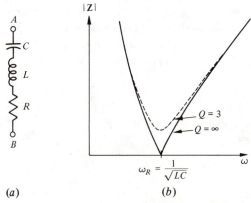

(a)   (b)

Figure 4.11 (a) A series resonant circuit. (b) For the case of $Q = \infty$ $(R = 0)$, the impedance vanishes at the resonant frequency, as shown by the solid curve. For comparison the dashed curve shows the case of $Q = 3$.

in the series resonant circuit, and in practice this prevents the impedance from actually becoming zero. However, provided that Q, as defined in Equation (4.49), is much greater than unity, the absolute value of the series impedance will still have a well-defined minimum at the resonant frequency. For comparison, a curve with $Q = 3$ is also shown in Figure 4.11.

Typical rf systems contain resonant circuits and amplifiers. Another block that occurs in rf circuits is the *oscillator*. An oscillator is basically an unstable amplifier which has infinite amplification at a certain frequency. If $\mathbf{v}_{OUT} = A\mathbf{v}_{IN}$, and $A(\omega) = \infty$ for some value of ω, then one may expect an output signal to be produced at the frequency ω, even if $\mathbf{v}_{IN} = 0$. Oscillators are used to generate sinusoids in electronics. For example, the nearly sinusoidal signals transmitted by every radio and TV station originate in oscillator circuits.

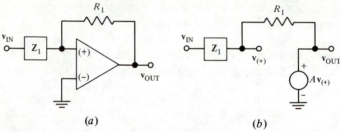

(*a*)   (*b*)

Figure 4.12   Conceptual illustration of an oscillator. (*a*) Oscillator circuit. (*b*) Circuit obtained after replacement of op-amp by its model. For simplicity it is assumed that the op-amp input resistance is infinite and its output resistance is zero.

We shall not present an extensive discussion of oscillator circuits here, but it may be interesting to see in a general way how such a circuit could come about. Let us consider the circuit of Figure 4.12(*a*), which contains an op-amp of the kind discussed in Chapter 3. Let us use the op-amp model of Figure 3.9(*b*); for simplicity we shall assume $R_i = \infty$ and $R_0 = 0$. Upon substitution of the model, we have the circuit of Figure 4.12(*b*). Writing a node equation for $\mathbf{v}_{(+)}$, the voltage at the (+) input terminal, we have

$$\frac{\mathbf{v}_{(+)} - \mathbf{v}_{IN}}{\mathbf{Z}_1} + \frac{\mathbf{v}_{(+)} - A\mathbf{v}_{(+)}}{R_1} = 0 \tag{4.52}$$

Solving for $\mathbf{v}_{(+)}$, using the relationship $\mathbf{v}_{OUT} = A\mathbf{v}_{(+)}$, and approximating by means of $A \gg 1$, we have

$$\mathbf{v}_{OUT} = \mathbf{v}_{IN} \cdot \frac{A}{1 - A \cdot \dfrac{\mathbf{Z}_1}{R_1}} \tag{4.53}$$

We see that the overall amplification $\mathbf{v}_{OUT}/\mathbf{v}_{IN}$ can become infinite if $\mathbf{Z}_1 = R_1/A$. One possibility would be to choose for $\mathbf{Z}_1$ the parallel resonant circuit of Figure 4.9. If the Q is chosen just right, one can have $|\mathbf{Z}_1| < R_1/A$ at every frequency except the resonant frequency. At the resonant frequency, where $\mathbf{Z}_1$ is real, the denominator of (4.53) would be made to vanish. Thus one would expect oscillation at the resonant frequency $\omega_R = 1/\sqrt{LC}$. The output frequency of the oscillator could then be varied by changing the value of L or C. It should be realized that the foregoing is a very crude description of an oscillator circuit, intended only to convey the idea. Considered in more detail, the behavior of oscillators is a complicated (but interesting) subject.

As a final example of the application of phasors to rf circuits, let us discuss a useful result known as the *power transfer theorem*. Suppose that we have an oscillator circuit, which may be represented by a Thévenin equivalent. The Thévenin parameters $\mathbf{v}_T$ and $\mathbf{Z}_T$ of the oscillator are not adjustable. We inquire, what load impedance $\mathbf{Z}_L$ should be connected across the oscillator terminals, as shown in Figure 4.13(a), so that the time-averaged power transmitted out of the oscillator into the load is maximum? The answer to this question is as follows.

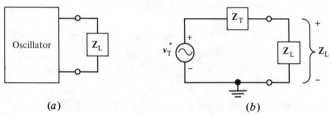

(a)   (b)

Figure 4.13 Illustrating the power transfer theorem. (a) Nonadjustable oscillator connected to adjustable load. (b) Circuit after replacement of the oscillator by its Thévenin equivalent. For maximum power transfer we choose $\mathbf{Z}_L = \mathbf{Z}_T{}^*$.

When the oscillator is replaced by its Thévenin equivalent, the situation is as shown in Figure 4.13(b). Using the generalized voltage-divider formula, we see that the voltage across the load is given by

$$\mathbf{v}_L = \frac{\mathbf{Z}_L}{\mathbf{Z}_L + \mathbf{Z}_T} \bullet \mathbf{v}_T \tag{4.54}$$

The current flowing through the load is given by

$$\mathbf{i}_L = \frac{\mathbf{v}_T}{\mathbf{Z}_L + \mathbf{Z}_T} \tag{4.55}$$

The time-averaged power [as proved in Equations (4.26) to (4.30)] is given by

$$\tfrac{1}{2} \operatorname{Re}(\mathbf{v}_L \, \mathbf{i}_L^*) = \frac{|\mathbf{v}_T|^2}{2} \operatorname{Re} \frac{\mathbf{Z}_L}{|(\mathbf{Z}_L + \mathbf{Z}_T)|^2} \tag{4.56}$$

Since $\mathbf{v}_T$ and $\mathbf{Z}_T$ are not adjustable, our problem is to adjust $\mathbf{Z}_L$ so that the quantity on the right of Equation (4.56) is maximized. Let us set $\mathbf{Z}_L = R_L + jX_L$ and $\mathbf{Z}_T = R_T + jX_T$. Then we must maximize

$$\text{Re} \frac{Z_L}{|(\mathbf{Z}_L + \mathbf{Z}_T)|^2} = \frac{R_L}{(R_L + R_T)^2 + (X_L + X_T)^2} \tag{4.57}$$

It is clear by inspection that we should choose $X_L = -X_T$. This reduces the problem to maximization of $R_L/(R_L + R_T)^2$. To maximize, we differentiate with respect to R_L and set the derivative equal to zero:

$$\frac{d}{dR_L}\left[\frac{R_L}{(R_L + R_T)^2}\right] = \frac{(R_L + R_T)^2 - 2R_L(R_L + R_T)}{(R_L + R_T)^4} = 0 \tag{4.58}$$

Solving, we find $R_T = R_L$. Thus for maximum power transfer, we choose $R_L = R_T$, $X_L = -X_T$; that is, we choose

$$\boxed{\mathbf{Z}_L = \mathbf{Z}_T^* \quad \begin{bmatrix} \text{power transfer} \\ \text{theorem} \end{bmatrix}} \tag{4.59}$$

EXAMPLE 4.24

For the circuit of Figure 4.13, $\mathbf{Z}_T$ consists of a 10-Ω resistance and a 20-μH inductance in series. The angular frequency $\omega = 10^6$. Design a load $\mathbf{Z}_L$ which will receive the maximum power obtainable from the oscillator.

SOLUTION

In this case the value of $\mathbf{Z}_T$ is $10 + j(2 \times 10^{-5})(10^6) = (10 + j20)$ Ω. Therefore we must have $\mathbf{Z}_L = (10 - j20)$ Ω. There are two simplest solutions to the problem of designing a load with this impedance: We may use a series combination of elements for $\mathbf{Z}_L$ or a parallel combination. Choosing the former, we see that a series combination of a resistance and an inductance will not work, because the series impedance $R + j\omega L$ of such a combination will have a positive imaginary part, and a negative imaginary part is needed. However, a series combination of a resistance and a capacitance can be used. Such a combination has series impedance

$$\mathbf{Z}_L = R + \frac{1}{j\omega C} = R - \frac{j}{\omega C}$$

We set $R = 10$ Ω and choose C such that

$$\frac{1}{\omega C} = 20$$

$$C = \frac{1}{20\omega} = 5 \times 10^{-8}\ \text{F}$$

The required load is shown in the sketch.

Response of Circuits to Non-sinusoidal Signals With the exception of rf circuits, most electronic circuits are made to handle signals which are by no means sinusoidal. However, phasor techniques often can be used to infer the response of a circuit to non-sinusoidal signals. Some ways in which this can be done will now be considered.

A principle of basic importance is the following: *Any signal of finite duration may be regarded as a sum of sinusoidal signals.* In general, sinusoids of many frequencies will be needed; in fact, in general one must add up sinusoids of *all* frequencies to make up an arbitrary signal. Each sinusoidal component must also have a certain phase and amplitude, so that they may add up to the original signal.

Like any other sinusoids, the sinusoidal components of a non-sinusoidal signal can be described in terms of their phasors. Since many frequencies are present, we must introduce a notation which specifies the phasor for each component frequency. If the non-sinusoidal signal is $f(t)$, let us give the phasor representing the sinusoidal component which has frequency ω the name $\mathbf{f}(\omega)$. A single sinusoidal component at the frequency ω is then given, as with all phasors, by $\text{Re}[\mathbf{f}(\omega)e^{j\omega t}]$. Now, however, to obtain the total signal $f(t)$, we must add up all the sinusoidal components. Thus

$$f(t) = \int_0^\infty \text{Re}[\mathbf{f}(\omega)\ e^{j\omega t}]\ d\omega \tag{4.60}$$

Here an integration over ω has been performed, in order to include contributions from sinusoids of all frequencies.

Interestingly, there is a convenient formula by which the phasors $\mathbf{f}(\omega)$ representing the sinusoidal components of the signal may be found. This formula is

$$\mathbf{f}(\omega) = \frac{1}{\pi}\int_{-\infty}^\infty f(t)\ e^{-j\omega t}\ dt \tag{4.61}$$

Equations (4.60) and (4.61) together are known as the *Fourier integral theorem*. The phasors $\mathbf{f}(\omega)$ are known as the *Fourier components* of the sinusoidal function $f(t)$.

For example, let $v(t)$ be a rectangular voltage pulse of amplitude V_0 and duration $2t_0$, as shown in Figure 4.14(a). We shall find its Fourier components. Using Equation (4.61)

$$\mathbf{v}(\omega) = \frac{1}{\pi}\int_{-\infty}^\infty v(t)\ e^{-j\omega t}\ dt$$

$$= \frac{V_0}{\pi}\int_{-t_0}^{t_0} e^{-j\omega t}\ dt = -\frac{V_0}{\pi j\omega}[e^{-j\omega t_0} - e^{j\omega t_0}]$$

$$= \frac{2V_0}{\pi}\frac{\sin(\omega t_0)}{\omega}$$

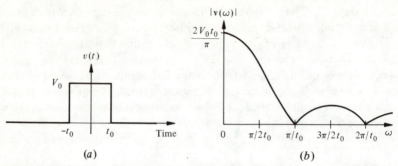

Figure 4.14 Fourier components of a non-sinusoidal signal. (*a*) A rectangular pulse. (*b*) Graph of the amplitudes of its Fourier components, as a function of frequency.

In this simple example, the Fourier components $\mathbf{v}(\omega)$ all happen to be real; in general they are complex. Whether real or complex, the absolute value of the Fourier component indicates the amplitude of the sinusoidal component of the corresponding frequency. That is, $|\mathbf{v}(\omega)|$ indicates how much of the frequency ω is present in the pulse. The quantity $|\mathbf{v}(\omega)|$ is graphed as a function of ω in Figure 4.14(b).

It can now be seen how sinusoidal analysis can allow one to estimate the performance of a circuit with nonsinusoidal signals. The nonsinusoidal signal is decomposed into its Fourier components. By phasor analysis, it is determined what frequencies are correctly processed by the circuit. If the frequencies present in the signal are among those processed satisfactorily, then the circuit will perform adequately with this signal.

EXAMPLE 4.25
    Find the range of frequencies transmitted from input to output, without significant attenuation, through the circuit in sketch (*a*).

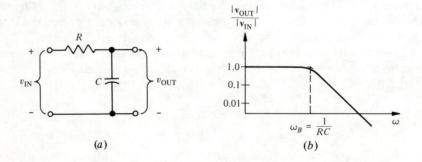

SOLUTION
    Using the generalized voltage-divider formula, we find that

$$\mathbf{v}_{OUT} = \mathbf{v}_{IN} \, \frac{1}{1 + j\omega(RC)}$$

$$|\mathbf{v}_{OUT}| = |\mathbf{v}_{IN}| \cdot \frac{1}{\sqrt{1 + \omega^2(RC)^2}}$$

A log-log plot of $|\mathbf{v}_{OUT}|/|\mathbf{v}_{IN}|$ (that is, of the ratio of the output amplitude to input amplitude) as a function of frequency is as shown in sketch (b).

We recognize the break frequency ω_B which is equal to the reciprocal of the time constant RC. It is seen that so long as the frequency is less than $(RC)^{-1}$, the output sinusoid has an amplitude of at least $1/\sqrt{2}$ times the amplitude of the input sinusoid. ∎

EXAMPLE 4.26

Pulses such as those of Figure 4.14(a) are to be passed through the filter circuit of Example 4.25. How short can the pulses be without being significantly degraded by the filter?

SOLUTION

The Fourier amplitudes of the pulse are shown in Figure 4.14(b). We see that most of the components lie in the range of frequencies $\omega < \pi/t_0$. [The power in any given component is proportional to the square of its Fourier amplitude. This makes the components at frequencies greater than π/t_0 less important than they appear at first glance in Figure 4.14(b).] If the filter transmits frequencies less than π/t_0 well, the pulse will go through the filter without too much degradation. Referring to the plot of $|\mathbf{v}_{OUT}|$ versus $|\mathbf{v}_{IN}|$ in Example 4.25, we see that frequencies satisfying $\omega < (RC)^{-1}$ are transmitted well. Thus one should make

$$\frac{\pi}{t_0} < \frac{1}{RC}$$

or

$$t_0 > \pi \, (RC)$$

The duration of the pulse in Figure 4.6(a) is actually $2t_0$; therefore

$$\text{pulse length} > 2\pi(RC)$$

A similar result was obtained in Chapter 2, Figure 2.13, by quite a different approach. The difference of a numerical factor results from the imprecision of the methods; these are only estimates.

It should be noted that the prescription $2t_0 > 2\pi(RC)$ is not sufficient to prevent some distortion of the pulse. Not only will the higher frequency components of the pulse (small, but not completely absent) be lost, there is also the possibility that the frequency components at the output will emerge with different *phases* than when they entered the input. Alteration of phases will distort the outcoming pulse just as will alteration of amplitudes. Both kinds of distortion can be minimized by making the pulse somewhat longer than the minimum length estimated in this example, thus leaving a margin of safety. The effect of using pulses which are too short was illustrated in Figure 2.13. ∎

In a practical case, a signal would not consist of a single pulse, but rather of an ongoing, time-varying voltage or current. In a digital system the signal consists of a procession of pulses, perhaps with varying height or spacing. In an analog system the signal would consist of a continuously varying voltage or current. To apply the ideas of this section, one may think in terms of the average Fourier amplitudes of the signal. Here the idea is that one estimates what the frequency composition of the time-varying signal is on the average, and designs the circuit to handle that.

In the case of a train of well-spaced pulses, ascertaining the average power spectral density is fairly easy. If all pulses are of the same duration, they all contain the same mixture of frequencies, and the average spectrum has approximately the same form as the spectrum of an individual pulse. In the case of analog signals, one must somehow have knowledge about the average range of frequencies present in the signal. For example, if the apparatus being considered is a phonograph amplifier, only those frequencies within the range of human hearing (about 20 Hz to 20,000 Hz) need be transmitted by the amplifier. Phasor analysis of the amplifier circuit can then determine the performance of the amplifier over this range of frequencies.

EXAMPLE 4.27

A certain amplifier is represented by the circuit model in sketch (a). With regard to frequency response, discuss the suitability of this amplifier for use in a telephone system.

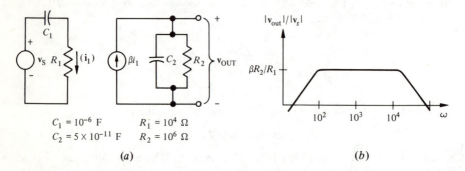

$$C_1 = 10^{-6} \text{ F} \qquad R_1 = 10^4 \ \Omega$$
$$C_2 = 5 \times 10^{-11} \text{ F} \qquad R_2 = 10^6 \ \Omega$$

(a)   (b)

SOLUTION

If the input is a sinusoid whose phasor is v_s, we find the phasor for i_1 to be

$$i_1 = \frac{v_s}{R_1 + 1/j\omega C_1} = \frac{j\omega C_1 v_s}{1 + j\omega R_1 C_1}$$

The phasor representing v_{out} is

$$v_{out} = \beta i_1 \cdot (Z_{R_2} \parallel Z_{C_2}) = \beta i_1 \cdot \frac{R_2}{1 + j\omega R_2 C_2}$$

Substituting for i_1,

$$v_{out}/v_s = \frac{\omega C_1 R_2 \beta}{(1 + j\omega R_2 C_2)(1 + j\omega R_1 C_1)}$$

The ratio of the amplitude of the output sinusoid to the amplitude of the input sinusoid is given by

$$|\mathbf{v}_{out}|/|\mathbf{v}_s| = \frac{\omega\beta R_2 C_1}{[1 + \omega^2 (R_2 C_2)^2]^{1/2} [1 + \omega^2 (R_1 C_1)^2]^{1/2}}$$

The functional appearance of this quantity, plotted against frequency on log-log scales, is shown in sketch (b). We observe that in the range $100 < \omega < 20,000$ (that is, 16 Hz $< f <$ 3200 Hz), the amplification is nearly constant at the value $\beta R_2/R_1$. This quantity is called the *midband gain* of the amplifier. Frequencies below 16 Hz or above 3200 Hz will be amplified by a smaller factor, or perhaps attenuated; hence these Fourier components will be lost. The range of frequencies over which the amplification is within 3 db of its maximum value is called the *passband*.

Human speech varies, but in general contains frequencies in the range 50 to 15,000 Hz. The proposed circuit will remove the components at frequencies higher than 3200 Hz. Actually, it is found that speech is still quite intelligible if only components in the range 100 to 3000 Hz are retained, so the circuit is adequate for telephone use. For high-fidelity amplifier use, however, it would be inadequate.   ∎

Summary

- A sinusoidal signal is characterized by three parameters: amplitude, frequency, and phase.

- A phasor is a complex number which represents the amplitude and phase of a sinusoid.

- When a sinusoid is known it is possible to find the corresponding phasor, and vice versa.

- The phasor representing the sum of several sinusoids of the same frequency is the sum of their individual phasors.

- If the phasor corresponding to a sinusoid $f(t)$, with frequency ω, is $\mathbf{f}$, the phasor corresponding to the sinusoid df/dt is $j\omega\mathbf{f}$.

- If $v(t)$ and $i(t)$ are a sinusoidal voltage and current with the same frequency, the instantaneous power is $i(t)v(t)$. The time-averaged power is given by Avg $[i(t)\,v(t)] = \frac{1}{2}$ Re $(\mathbf{iv}^*)$, where $\mathbf{i}$ and $\mathbf{v}$ are the phasors representing $i(t)$ and $v(t)$.

- The impedance of a circuit element is the ratio of the phasor for the voltage across it to the phasor for the current flowing through it. Impedances are complex numbers which play a role similar to that of resistance in dc calculations.

- The node and loop methods can be used for sinusoidal analysis of ac circuits. Voltages and currents are represented by their phasors in the calculation. The use of phasors eliminates the need for solving differential equations.

- Thévenin and Norton equivalents can be constructed for sinusoidal ac circuits. The Thévenin voltage is represented by its phasor, and the Thévenin resistance is replaced by a Thévenin impedance.

- Rf circuits are circuits designed to handle signals which are nearly sinusoidal.

- Resonant circuits are circuits that have singular behavior at or near a special frequency which is characteristic of the circuit. This frequency is called the resonant frequency. A parallel combination of inductance and capacitance makes a parallel resonant circuit; such a connection has a maximum of impedance at the resonant frequency. A series combination of inductance and capacitance makes a series-resonant circuit; such a connection has a minimum of impedance at the resonant frequency. If there is little resistance present in the circuit, the resonant frequency in either case is approximately $1/\sqrt{LC}$.

- Non-sinusoidal signals can be regarded as being sums of sinusoidal signals. Response of circuits to a non-sinusoidal signal can be studied by decomposing the signal into its component sinusoids. The process of decomposing a signal into its sinusoidal components is known as Fourier analysis.

References

Extensive treatments of phasor analysis are found in many books, including:

Thompson, H. A. *Alternating-Current and Transient Circuit Analysis.* New York: McGraw-Hill, 1955.

Close, C. M. *The Analysis of Linear Circuits.* New York: Harcourt, Brace & World, 1966.

Skilling, H. H. *Electrical Engineering Circuits.* New York: John Wiley & Sons, 1957.

A large number of worked examples are to be found in:

Edminster, J. A. *Electric Circuits.* In Schaum's *Outline Series.* New York: McGraw-Hill, 1965.

Rf circuits are treated in:

Brophy, J. J. *Basic Electronics for Scientists.* New York: McGraw-Hill, 1966.

Most books treat Fourier analysis at a more advanced level. Comparatively simple treatments will be found in:

Cooper, G. R., and C. D. McGillen. *Methods of Signal and System Analysis*. New York: Holt, Rinehart and Winston, 1967.

Goldman, S. *Frequency Analysis, Modulation, and Noise*. New York: McGraw-Hill, 1948.

Problems

4.1   The ordinary frequency of a certain power-line voltage is 60 Hz.

(1) Find the angular frequency and the period.

(2) If a second sinusoidal voltage has the same frequency as the first but a phase angle 18° larger, how much earlier will the maxima of the second occur than the maxima of the first?

4.2   A sinusoid described by $f(t) = A \cos(\omega t + \phi)$ has two maxima at $t = 0.014$ sec and at $t = 0.018$ sec, with no other maxima in between these two. Evaluate ω and ϕ. Is there more than one possible answer for ϕ?

4.3   A sinusoid with ordinary frequency 10 kHz has a maximum at $t = 0$. What is the least number of radians its phase angle ϕ can be increased to have a zero at $t = 0$?

4.4   It is desired to express the function $f(t) = 27 \sin(18t + 47°)$ in the form $A \cos(\omega t + \phi)$. Find A, ω, and ϕ.

4.5   A sinusoid with ordinary frequency 1500 Hz is described by the notation 27 V $\angle 60°$. Express this sinusoid in the form $A \cos(\omega t + \phi)$, where ϕ is in radians and ω is in radians per second.

4.6   A quantity known as the *root-mean-square*, or *RMS*, value of a sinusoid is often used. If $v(t) = V_0 \cos(\omega t + \phi)$, the RMS value of $v(t)$ is defined by

$$v_{\mathrm{RMS}} = \sqrt{\frac{1}{\tau} \int_{t_1}^{t_2} [v(t)]^2 \, dt}$$

where τ is the period of the sinusoid and the time integral is over any time interval of duration τ. Calculate v_{RMS}. ANSWER: $v_{\mathrm{RMS}} = V_0 / \sqrt{2} = 0.707 \, V_0$.

4.7   Consider the complex number $z = 12 + 17j$.

(1) What is z^*?

(2) Express z in exponential form.

(3) Express z^* in exponential form.

(4) What is $|z|$?

4.8   Let $z_1 = 0.6e^{-0.8j}$.

(1) Find z_1^*.

(2) Express z_1 in rectangular form.

(3) Express z_1^* in rectangular form.

(4) What is $|z_1|$?

4.9   Prove that $\exp\{j(2.23 \text{ radians})\} = j \exp\{j(0.66 \text{ radian})\}$.

4.10   Let $z_1 = 3 - 2j$ and $z_2 = -1 + 6j$. Find $z_1 + z_2$. Illustrate the summation by means of vectors in the complex plane.

4.11   Let $z_1 = 2 + 3j$, $z_2 = 4 - 2j$.

    (1) Calculate their product z_1z_2 in rectangular form.

    (2) Convert the result to exponential form.

    (3) Convert z_1 and z_2 to their exponential forms.

    (4) Multiply together the exponential forms obtained in (3). The result should agree with (2).

4.12   Let $z_3 = 0.2 - 0.3j$, $z_4 = 6 + 5j$. Consider their quotient $z_4/z_3 = (6 + 5j)/(0.2 - 0.3j)$.

    (1) Convert this fraction to a form in which the denominator is real. *Suggestion:* Multiply denominator and numerator by the complex conjugate of the denominator.

    (2) Convert the result obtained in (1) to exponential form.

    (3) Convert z_3 and z_4 to exponential form.

    (4) Compute z_4/z_3 using the exponential forms obtained in (3). The result should agree with (2).

4.13   If $z_1 = a + bj$, $z_2 = c + dj$, find $|z_1/z_2|$.

    ANSWER: $|z_1/z_2| = [(a^2 + b^2)/(c^2 + d^2)]^{1/2}$

4.14   Use the power series expansions of the sine, cosine, and exponential functions to verify that $e^{j\theta} = \cos \theta + j \sin \theta$.

4.15   Prove that if z_1 and z_2 are any two complex numbers, $\mathrm{Re}(z_1z_2^*) = \mathrm{Re}(z_2z_1^*)$.

4.16   Find the phasors that represent the following sinusoidal functions and express them in exponential form:

    (1) $v_1(t) = 17 \cos(\omega t + 10°)$ V

    (2) $v_2(t) = 27.6 \sin(\omega t - 54°)$ V

    (3) $i_1(t) = 3.4 \cos(\omega t - 0.42 \text{ radian})$ mA

4.17   Obtain the sinusoids corresponding to the following phasors and express them in the form $A \cos(\omega t + \phi)$:

    (1) $v_1 = 12 \, e^{j(34°)}$ V

    (2) $v_2 = (7 + 9j)$ V

    (3) $i_1 = (6 - 2j)$ mA

4.18   The sinusoid $v_1(t)$ is represented by the phasor $v_1 = 10 + 12j$, and another sinusoid of the same frequency $v_2(t)$ is represented by $v_2 = -7 - 9j$.

    (1) Find the phasor corresponding to the sinusoid $v_1(t) + v_2(t)$.

    (2) Use the phasor found in (1) to express $v_1(t) + v_2(t)$ in the form $A \cos(\omega t + \phi)$ (that is, find A and ϕ).

4.19   Find the sum of $3 \sin(\omega t + 28°)$ and $4 \cos(\omega t - 71°)$.

4.20   The voltage across a 20 μF capacitor is $v(t) = 160 \cos(377t \text{ radians})$ volts, where t is in seconds.

    (1) What is the phasor representing $v(t)$?

    (2) What is the phasor representing $i(t)$, the current through the capacitor?

    (3) Express $i(t)$ in the form $A \cos(\omega t + \phi)$.

    (4) Do the maxima of $i(t)$ come earlier in time or later than the maxima of $v(t)$? (If the former is true, one says the current "leads" the voltage; if the latter, one says the current "lags" the voltage.)

4.21 The current through a circuit element is $i(t) = I_1 \cos \omega t$, and the voltage across it is $v(t) = V_1 \cos(\omega t + \phi)$. Calculate the average power entering the element as a function of ϕ. What does negative power mean, physically?

4.22 A sinusoidal voltage represented by the phasor v_1 is applied to the terminals of a capacitor with value C. How much time-averaged power enters the capacitor?

4.23 A sinusoidal voltage whose phasor is v is applied across the terminals of an impedance Z. (1) What is i, the phasor representing the current through Z? (2) What does $|i|$ equal?

4.24 Let Z_L be the impedance of a 10 μH inductance and Z_C be the impedance of a 10 μF capacitance. (1) Over what range of ordinary frequencies is $|Z_L| > 1000 \ \Omega$? (2) Over what range of ordinary frequencies is $|Z_C| > 1000 \ \Omega$?

4.25 Prove that if two elements with impedances Z_1 and Z_2, respectively, are connected in series, the impedance of the combination is $Z_1 + Z_2$.

4.26 Prove that if two elements with impedances Z_1 and Z_2, respectively, are connected in parallel, the impedance of the combination is $Z_1 Z_2/(Z_1 + Z_2)$.

4.27 Find the impedances between terminals A and B for each of the combinations shown in Figure 4.15, at an arbitrary frequency ω. Express your answers in their simplest rectangular form.

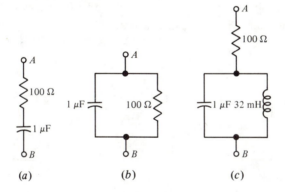

(a)   (b)   (c)

Figure 4.15

4.28 A sinusoidal voltage of amplitude 100 V and ordinary frequency 1000 Hz is applied across each of the impedances shown in Figure 4.15. Calculate the amplitude of the current that flows through terminals $A \ B$ in each case.

4.29 In the circuit of Figure 4.16, write a phasor loop equation and solve for v.

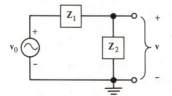

Figure 4.16

4.30   Write a phasor node equation for the circuit of Figure 4.16 and solve for **v**.

4.31   In Figure 4.16 find the numerical value of **v**. Let $\mathbf{v}_0 = 10\,e^{j(20°)}$, $f = 1000$ Hz, and let $\mathbf{Z}_1$ and $\mathbf{Z}_2$ be the compound elements shown in Figures 4.15(a) and 4.15(b), respectively.

4.32   Write a set of loop equations sufficient to analyze the circuit of Figure 4.17. It is not asked that you solve the equations.

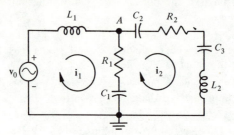

Figure 4.17

4.33   Write a node equation for the circuit of Figure 4.17 and solve for the voltage at node A.

4.34   It is desired to find a Thévenin equivalent for the subcircuit of Figure 4.18(a) in the form shown in Figure 4.18(b). The angular frequency is 230 radians/sec.

(1) Find the numerical values of $\mathbf{v}_T$ and $\mathbf{Z}_T$.
(2) Devise a connection of ideal circuit elements which has impedance equal to $\mathbf{Z}_T$.

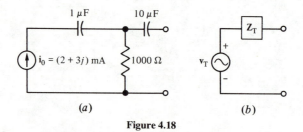

(a)   (b)

Figure 4.18

4.35   Calculate the phasor voltage **v** in the circuit of Figure 4.19.

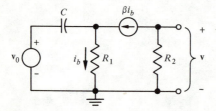

Figure 4.19

4.36   A 10 mH inductance, a 3-Ω resistance, and a variable capacitance are connected in series as shown in Figure 4.20. (The arrow through the capacitor symbol indicates that its value is adjustable.) A voltage source with ordinary frequency $f = 800$ Hz is connected across the combination, as shown. At what value of the capacitance is the circuit resonant?

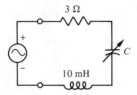

Figure 4.20

4.37 An inductance L, a capacitance C, and a resistance R are connected in parallel. Calculate $|\mathbf{Z}|$, the absolute value of the impedance of this parallel combination.

(1) Make a graph, on linear scales, of $|\mathbf{Z}|$ as a function of ω, over the range $0 < \omega < 25{,}000$ radians per second. Take $L = 1$ mH, $C = 10$ μF, $R = 2$ Ω.

(2) Repeat Part (1), with $C = 5$ μF.

(3) Repeat Part (1), with $L = 1$ mH, $C = 10$ μF, $R = 1$ Ω.

4.38 A parallel resonant circuit consists of a fixed L and C and a variable R, all connected in parallel. Let $\mathbf{Z}$ be the parallel impedance. Calculate the value of $|\mathbf{Z}|$ when the frequency is such that $|\mathbf{Z}|$ is maximum. Express this maximum value of $|\mathbf{Z}|$ as a function only of L, C, and the quality factor Q of the circuit.

4.39 Determine the 3-db passband, in terms of ordinary frequency, for the circuit of Figure 4.21. (That is, the range of frequencies for which $|v_{OUT}/v_{IN}|$ is within three db of its maximum value.) Express your answer in terms of ordinary frequency (Hz).

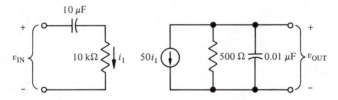

Figure 4.21

4.40 The passband of a certain amplifier extends from zero to ω_M. It is desired that rectangular pulses of duration T be amplified with only slight distortion. Estimate the smallest suitable value for ω_M if (1) $T = 1$ msec;   (2) $T = 1$ μsec;   (3) $T = 1$ nsec.

4.41 Use equation (4.48) to verify Equation (4.50). Assume $Q \gg 1$.

4.42 Devise a load consisting of two elements in parallel which will accept maximum power from the Thévenin source of Example 4.24.

4.43 Show that if the amplitude of the pulse of Figure 4.14 is changed, all the Fourier amplitudes are changed in the same proportion, so that the relative frequency distribution is unchanged. If a circuit which contains only ideal R, L, and C transmits the pulse of Figure 4.14 with little distortion, would you expect it to do the same for a pulse of ten times greater amplitude?

4.44   Prove that a pulse identical to that of Figure 4.14, but occurring at a different time, has the same Fourier amplitudes. (Note that only the amplitudes of the Fourier components are the same. The phases are different. If two pulses have the same Fourier amplitudes, however, they contain the same amount of signal at each frequency.) Would you expect a circuit that transmits the pulse of Figure 4.14 with little distortion to also transmit it if the pulse occurs at a later time?

Chapter 5

Properties and Applications of Diodes

The foregoing chapters have dealt with techniques for calculating the currents and voltages of a given circuit. Kirchhoff's current and voltage laws, combined with the properties (constitutive relationships) of the circuit elements, lead to equations sufficient to calculate any desired voltage or current. We now apply the techniques we have developed to electronic circuits. In addition to the elements already introduced in Chapter 1, practical electronic circuits contain a number of semiconductor devices, especially diodes and transistors; hence it is necessary to have available in some form the constitutive relationships for these devices. Reflecting the trend in modern electronics, the emphasis is on semiconductor devices rather than, say, vacuum tube devices. The desirable electronic properties — small size, low cost, low power dissipation, and high reliability of semiconductor devices — have led to almost complete disappearance of other electronic devices in modern practice. Semiconductor integrated circuits, in which the entire circuit is contained within a tiny semiconductor "chip," represent the culmination of this trend. Because of their low cost and small size, integrated circuits are steadily taking on greater importance in the electronic market.

The first semiconductor device to be considered is

the *pn* junction diode. This device, in addition to being an important circuit element in its own right, is a basic building block for more complex devices, such as the transistor (discussed in Chapter 6) and the integrated circuit (Chapter 7). Hence it is important not only to discuss the terminal properties of the *pn* junction, but the internal processes as well, in order that the more complex devices may be understood.

It is necessary to begin with a discussion of the fundamental processes of current flow in semiconductors, the subject of Section 5.1. We then proceed to the specific case of the *pn* junction diode in Section 5.2. Reasoning from the principles set out in Section 5.1, the basic features of *pn* junction operation are outlined, and the *I-V* characteristic is presented. For those who wish to explore the internal processes in greater detail, a mathematical theory of *pn* junctions is given in Appendix II.

In the subsequent sections we examine the elementary circuit properties of diodes. First, in Section 5.3, the various aspects of the basic diode non-linearity are considered, and a number of idealizations and simplifying circuit models are developed. The applications of the diode which involve its rectification properties, that is, as a one-way conductor of current, are then examined in Section 5.4, making use of the idealizations. Finally, in Section 5.5, we examine the extremely important application of the diode as a logic element.

5.1   Electrons and Holes

To determine the relationship between the currents and voltages at the various terminals of an electronic device, namely, a *pn* junction, we look inside the device, that is, take a microscopic point of view. On a microscopic scale, a current consists of a flow of charged particles, called *carriers*. The flow of electrons in a metal in response to an applied electric field is a familiar example. Metals have an abundance of *free electrons,* that is, electrons which can readily move about within the material. Consequently they conduct a current when a field is applied. In other solids, such as mica and glass, the electrons are bound into more or less fixed positions; they cannot, for example, respond to an applied electric field. Hence these materials are insulators. The name *semiconductor* refers to a class of materials intermediate between metals and insulators. In semiconductors, some of the electrons are bound and others are free. Furthermore it is possible to control the relative number of free and bound electrons. This control, in fact, is the basis of semiconductor device operation. Let us look more closely then at the nature of free and bound electrons in semiconductors to see how this control comes about.

Bonding Electrons and Free Electrons   The two most important semiconductors, silicon and germanium, have the same atomic arrangement (see Figure 5.1). Such an orderly repetitive arrangement is called a *crystal* structure. The spheres in Figure 5.1 represent the atoms in the crystal. Of course

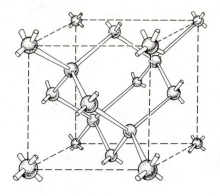

Figure 5.1 A model of a silicon crystal. The spheres represent the atom including its core electrons. The lines, or sticks, represent the bonding electrons that hold the atoms together. The broken lines are added only to indicate the cubic nature of the crystal.

any real crystal has many more atoms than shown in the figure. However the small portion shown represents the basic unit from which larger crystals may be constructed by stacking this basic cube. Each atom has four nearest neighbors; when the figure is visualized correctly each atom is seen to be at the center of a tetrahedron, with its four nearest neighbors at the vertices. The crystal is held together by the bonding of the outer electrons between nearest neighbor atoms. The "sticks" connecting the atoms in Figure 5.1 represent then the bonding electrons. In silicon, for example, there are 14 electrons per atom. Of these, only four are involved in bonding; the other ten are within the atom itself. Consequently each stick represents two electrons, one contributed by each atom at either end.

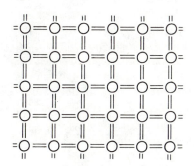

Figure 5.2 A two-dimensional representation of a silicon crystal. Each atom is connected to its four nearest neighbors by two electrons, as in the real three-dimensional crystal.

To simplify the figure for further discussion, this bonding arrangement may be drawn schematically as a two-dimensional array, as shown in Figure 5.2. The circles represent the atoms, and the lines represent the bonding electrons which hold the atoms together. Each atom contributes four bonding electrons; consequently every atom is bound to each of its nearest neighbors with two electrons. The bonding electrons represented by the short lines in Figure 5.2 are tightly held in place, and hence are unavailable for carrying a

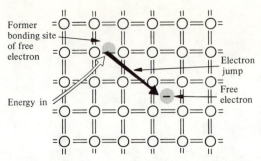

Figure 5.3   The creation of a free electron and a hole. Energy is supplied to break the bond, and the bonding electron is freed. The empty site, formerly occupied by the bonding electron, is called a hole.

current. To make current flow possible, it is necessary to add extra free electrons not involved in bonding, or else to free some of the electrons from their bonds by some means.[1] To break a bond, energy, in the amount of the bond energy, must be supplied. If sufficient energy is supplied, for example as heat or light, a bonding electron may be set loose, as shown in Figure 5.3. The resulting *free* electron, because it is not tied up in local bonds, moves about freely in the crystal. If for example an electric field is applied, the electron responds, giving rise to conduction. Hence free electrons are also called *conduction* electrons.

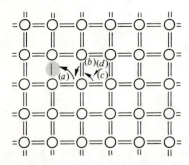

Figure 5.4 The transport of a hole. Electron (*a*) jumps into the hole, moving the hole into the position formerly occupied by (*a*). Electron (*b*) may then proceed to move into the hole, and the process continues. In this way the hole may be transported, for example, from its initial position, into the position initially occupied by electron (*d*).

Holes   It is evident in Figure 5.3 that when a bonding electron is converted to a free electron, by breaking it free from its bonding position, a vacant site remains. These vacant bonding sites are called *holes*. It is possible for a neighboring bonding electron to jump from its bonding position into the vacant position. For example, electron (*a*) in Figure 5.4 could jump

---

[1] The application of an electric field is generally not sufficient to break a bonding electron free for conduction. Only very large fields, on the order of 10^7 V/cm can pull electrons out of their bonding positions.

into the hole shown. After such a jump there is still a hole, that is, a vacant site; however, it has moved. In the example depicted, electron (*a*) and the hole essentially change places. Further motion can occur, as illustrated in Figure 5.4, by electron (*b*) jumping into the hole since the latter occupies the former site of electron (*a*). Electrons (*c*) and (*d*) may follow in turn, resulting in the transport of the hole to the original site of electron (*d*). In this process there is a slight motion of numerous bonding electrons in one general direction; consequently there is a net motion of charge. Such motion of bonding electrons via holes constitutes a second mechanism for current flow.

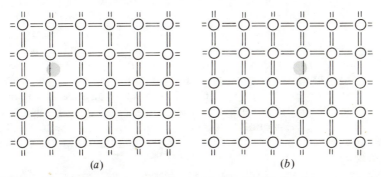

Figure 5.5   The electronic arrangement (*a*) before and (*b*) after the sequence of bonding electron jumps shown in Figure 5.4. The net change, from (*a*) to (*b*), may be described as either the motion of one bonding electron, two atomic units to the left, or the motion of one hole, two atomic units to the right.

Let us look more closely at the process depicted in Figure 5.4 to see how the current might be computed. The electronic arrangement before and after the motion of Figure 5.4 is shown in Figure 5.5. If we are not concerned with all the details of the electron jumps shown in Figure 5.4, it appears from Figure 5.5 that the entire process may be described in terms of *hole motion*. The hole has simply moved two positions to the right. It is convenient to think of the holes themselves as a kind of particle that can move through the lattice. As may be seen from Figure 5.5, the motion of a hole constitutes a flow of charge. As the hole moves to the right, negative charge, in the form of bonding electrons, moves to the left. This motion may equally well be described as the motion of positive charge to the right. There is in fact a positive charge associated with a hole, as one electron is missing from a normally neutral medium. Figure 5.5(*b*) differs from Figure 5.5(*a*) in that there has been net motion of one electron two positions to the left, or in an alternative description, a net motion of one positive charge two positions to the right. Whereas either description gives the same current, it is much easier to follow the motion of the hole than the combined motion of all the bonding electrons. Whenever there are missing bonding electrons in a semiconductor, we describe any current flow in terms of hole motion, treating the hole as a kind of free particle, with one unit of positive charge.[2] For example, the electron motion depicted in Figure 5.4 would be produced by an electric field directed to the right (producing electron flow to the left). Thinking only of the hole rather than the bonding electrons yields the same result; a positively charged hole moves to the right. Thinking of either the motion of the hole, or the more complicated motion of the bonding electrons yields the same answer, a current flow toward the right.

Returning then to Figure 5.3, we see that when a bond is broken two kinds of carriers are produced. Both free electrons and holes are created, and both are able to move through the crystal and thus carry a current. The process of creation of electrons and holes (whether by thermal, optical, or some other process) is called *generation*. It should be realized that there is a corresponding reverse process, called *recombination,* in which a free electron hops into a hole, annihilating both particles. At room temperature, bonds are constantly being broken by thermal excitations, and free electrons and holes are constantly recombining. The processes just balance in equilibrium. There is a characteristic parameter, called the *lifetime,* which describes the average time a newly generated carrier spends in a semiconductor before recombining. As will be seen in Chapter 6, the lifetime is important in transistor design.

Whereas holes are always simply referred to as holes, various writers refer to free electrons as "conduction electrons," "conduction band electrons," "free electrons," or just "electrons." The latter terminology is preferred because it is the simplest. Thus one often says, for example, that the electron concentration in a certain crystal is $10^{19}/cm^3$; this simply means that the free electron concentration equals $10^{19}/cm^3$. Obviously the total electron concentration is always much higher (since the atomic concentration in any solid is on the order of $10^{23}/cm^3$).

Electron and Hole Concentrations   The current carried by electrons and holes in a semiconductor depends upon how many of these particles are present. It is convenient to deal with the particle concentration (that is, density) rather than the absolute number present in some given volume. The concentration of electrons, that is, free electrons, is given the standard symbol n and the concentration of holes is given the symbol p.

A comparison of the concentration of charge carriers brings out quite dramatically the differences among metals, insulators, and semiconductors. Copper, the most common metal used when high conductivity is desired, has about 8×10^{22} free electrons per cm^3. In contrast pure silicon has about 2×10^{10} electrons and holes per cm^3 at room temperature. A good insulator like quartz has fewer than even one electron or hole per cubic centimeter.

---

[2] The simple discussion given here is intended only to introduce the concept of the hole as a particle. With a rigorous quantum-mechanical treatment it may be shown that the current is correctly computed using the hole-particle concept in essentially all situations of practical interest.

The carrier concentrations in semiconductors are quite variable, and the numbers above are given only to contrast semiconductors with metals and insulators. There are many ways to alter the concentration of electrons and holes in semiconductors. Just shining a light on a crystal, for example, increases the electron and hole concentration (if the photon energy exceeds the bond energy). Clearly it is impossible to predict the electron or hole concentration without specifying the forms and amount of energy flowing into and out of the crystal. The easiest possible condition to specify is the condition of *thermal equilibrium* in which the same amount of energy flows in as out, no matter what the form of the energy. To be in thermal equilibrium, a semiconductor should be at the same temperature as its surroundings and be receiving no energy from external sources such as batteries. In practice, thermal equilibrium usually means no voltage applied to the device in question.

Under thermal equilibrium conditions, some electrons and holes will be present just from thermal excitations, provided the semiconductor is at some nonzero temperature. Thermal energy will break a few bonds, producing equal numbers of electrons (free electrons) and holes. Carriers produced in this way are quite scarce in germanium and silicon at room temperature, owing to a relatively large bond energy. A semiconductor in which the electrons and holes are produced purely by thermal excitations is said to be *intrinsic*. The concentration of electrons equals that of holes, and is called the intrinsic electron concentration. It is given a special symbol, n_i. At room temperature in germanium, $n_i = 2.5 \times 10^{13}/\text{cm}^3$, whereas in silicon, with its larger bond energy, $n_i = 1.5 \times 10^{10}/\text{cm}^3$.

Impurities   We have not yet considered the effect of impurities in semiconductors. It will be shown in the following discussion that impurities can supply electrons and holes. A semiconductor in which the carrier concentrations are determined by the impurities present is said to be *extrinsic*. In most cases of practical importance, intentionally added impurities, called *dopants*, are used to control the concentration of electrons and holes. In particular, elements from groups III and V of the periodic table are normally included in germanium and silicon to achieve the desired carrier concentrations.

Suppose, for example, an element such as phosphorus (group V) is added to silicon, and that the phosphorus atoms occupy normal atomic posi-

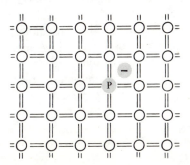

Figure 5.6   A model of a silicon lattice with a phosphorus atom replacing one silicon atom. The phosphorus atom has one more electron than silicon. This extra electron "escapes" to become a free, or conduction, electron.

tions in the lattice (see Figure 5.6). Phosphorus has one more electron than silicon, thus one electron beyond those needed for bonding. This extra electron, if left undisturbed, would orbit the phosphorus atom, much in the same way a single electron orbits in a hydrogen atom. However, at room temperature, thermal excitations are sufficient to separate this electron from the phosphorus atom, and it becomes a free electron, available for carrying a current. Because each phosphorus atom incorporated into the crystal donates one free electron, phosphorus is called a *donor*. Other elements from group V of the periodic table, notably arsenic and antimony, behave similarly to phosphorus and act as donors in silicon and germanium. The similarity stems of course from the fact that all the group V elements have five outermost electrons which are the electrons involved in chemical combinations with other elements. Thus arsenic and antimony, like phosphorus, have four electrons for bonding and one to donate as a free electron. If the donors are spread uniformly in the lattice, then at room temperature a uniform concentration of electrons (free electrons) results. Figure 5.7 illustrates a small region of

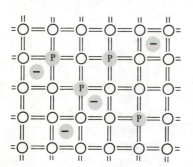

Figure 5.7   A region of a silicon crystal containing four donors. Each of the donors has given up one electron that is free to wander through the crystal. (The relative number of dopant atoms is exaggerated here for illustration. A more typical ratio might be one dopant atom for every 10^6 silicon atoms.)

a crystal containing four phosphorus donors and four electrons. A material which has donors added is said to be *n-type*, and the electron concentration n just equals[3] the donor concentration N_D. In an *n*-type semiconductor the hole concentration is much less than the electron concentration. A hole is not created with each free electron donated by a donor. For this reason alone the electrons would be expected to outnumber holes. (Actually, the hole concentration in *n*-type material is even less than in undoped material because of recombination. In essence, the greater number of electrons in *n*-type material results in a larger number of recombination events with holes, so that p is actually less than n_i. A quantitative discussion is given in Appendix II.) Even though the densities of electrons and holes are unequal, an *n*-type semiconductor may be electrically neutral. The negative charges of the free electrons are compensated by the positive charges of the phosphorus ions

---

[3] This statement is true provided the donor concentration greatly exceeds n_i, the very small concentration of electrons which is present in the absence of any donors because of thermal excitations.

that have each lost one electron. These positive charges, unlike the free electrons, are immobile and cannot contribute to any flow of current.

Now let us consider group III impurities, which have one fewer electron available for bonding than does silicon. For example, in Figure 5.8 one

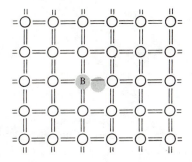

Figure 5.8 A silicon crystal in which a silicon atom has been replaced by a boron atom. The boron atom is deficient in one electron compared to its neighbors; hence a hole, that is, an empty bonding site, results.

silicon atom is replaced by a boron atom. For each such impurity atom added, one valence electron disappears from the crystal, that is, one hole is created. This hole is only loosely associated with the boron atom, and at room temperature moves off and becomes an ordinary hole, available to carry current. The boron atom accepts a neighboring bonding electron to complete its bonds; consequently boron, and other group III elements, are said to be *acceptors*. One hole is created per acceptor atom added into the crystal; hence in a material doped with acceptors the hole concentration p just equals[4] the acceptor concentration N_A. A material that has a greater number of holes than electrons, produced by incorporating acceptors in the lattice, is said to be *p-type*.

From the preceding discussion we see that it is possible to produce *n*- or *p*-type semiconductors simply by incorporating the appropriate concentration of the desired impurity into the crystal. In an *n*-type crystal, $n > p$ and electrons are said to be the *majority carriers* and holes the *minority carriers*. "Carrier" refers to the fact that the electrons and holes carry electrical charges. Hence, their motions can give rise to current flow.

5.2 *PN Junctions*

With this brief introduction to the nature of semiconductors, we proceed to a discussion of the most basic semiconductor device, the *pn* junction. Only a very simple description of *pn* junction operation is presented here. While it is necessarily oversimplified, the description introduces the basic physical mechanisms responsible for *pn* junction behavior, and makes the observed *I-V* characteristics plausible. Also it further prepares the groundwork for

---

[4] Again, this statement assumes that the dopant concentration is much greater than the intrinsic electron or hole concentration.

understanding the functioning of the transistor, the subject of Chapter 6. A more detailed and quantitative treatment of *pn* junction operation is given in Appendix II.

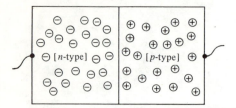

Figure 5.9 Schematic drawing of a *pn* junction.

Fundamentally a *pn* junction is a structure which contains a *p*-type semiconductor in contact with an *n*-type semiconductor (Figure 5.9). (In practice junctions usually arise from an abrupt change of doping, from donors to acceptors, inside an otherwise uniform crystal; however, one may imagine a junction as being formed when a *p*-type semiconductor and an *n*-type semiconductor are brought into intimate contact.) As we have seen in Section 5.1 the *n*-type material contains many mobile electrons, while the *p*-type material contains mobile holes. Wires are connected to each of the regions to connect the device into a circuit. It is convenient to consider separately the two possible polarities of applied voltage.

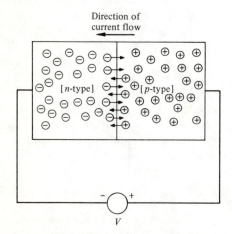

Figure 5.10 Schematic illustration of a *pn* junction showing the effect of forward-bias voltage. Substantial current flows from right to left.

Forward Bias   Let us suppose that an external battery is connected to the wires, with polarity positive on the *p* side and negative on the *n* side (Figure 5.10). Then the holes, which carry a positive charge, are repelled by the positive voltage on their side of the junction and tend to flow toward the *n* side. Similarly, the electrons, with negative charge, are repelled by the negative voltage applied to their side of the junction and tend to flow toward the *p* side. As fast as the electrons flow into the *p* side, or as the holes flow into the *n* side, they recombine with their opposite types; thus there is no buildup of minority carriers on either side. This condition is known as *forward bias*. The positive charges moving to the left and negative charges moving to the right both contribute to the total current, which flows to the left. In some cases it is of interest to know which current is greater, the current of electrons or the current of holes. The primary consideration in answering this question is the relative doping on the *p* and *n* sides. If, for example, the *n* side is heavily doped compared to the *p* side, then there are more electrons available to flow to the right than holes to the left. Consequently, in forward bias the electron current would exceed the hole current. Such a device is known as an *n⁺p* junction. In contrast, in a *p⁺n* junction, in which the *p* side is more heavily doped, the current in forward bias consists primarily of holes flowing into the *n* side. In practice, most junctions are *n⁺p* or *p⁺n*, not *pn*. The reasons, which are related to ease of fabrication and to desired device behavior, will become clear in Chapters 6 and 7.

Although not obvious from this discussion, it is also important to realize that most of the applied voltage is dropped across a narrow region quite close to the interface of the *p*- and *n*-type sides. Very little voltage drop (typically only a few millivolts) is required to move the electrons through the *n* region and holes through the *p* region. On the other hand, some fraction of a volt, say 0.75 V, is normally required to send the electrons and holes across the interface.

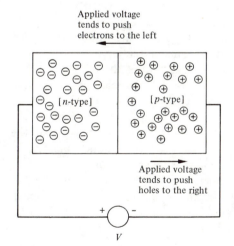

Figure 5.11 Schematic illustration of a *pn* junction showing the effect of reverse-bias voltage. The applied voltage tends to push electrons to the left and holes to the right. Little current flows.

Reverse Bias   We now consider the situation in which the external voltage is applied with the opposite polarity, as shown in Figure 5.11. In this case the electrons tend to flow to the left and the holes to the right. However there is almost no source of electrons on the right nor holes on the left. Thus the main result of applying a voltage with this polarity is a slight separation of negative and positive charge, which occurs at the interface. There are so few electrons on the *p* side available to flow over to the left that only a very small electron current results. Similarly the lack of holes on the left limits the hole current. Consequently there is practically no movement of charge, and practically no current flow at all. This condition is known as *reverse bias*.

The foregoing description of the *pn* junction operation is very crude, and many details have been sacrificed for the sake of simplicity. Nonetheless, two aspects of *pn* junction behavior are revealed that are most important for the understanding of semiconductor devices. The first is that the junction conducts current readily when a voltage of one sign is applied (positive on the *p* side) but conducts practically no current when voltage of the opposite sign is applied. Note that none of the ideal circuit elements introduced in Chapters 1 to 3 has this property; a resistor, for example, conducts equally well in either direction. A second aspect of junction behavior may be seen from Figure 5.10. When the junction is forward biased, carriers leave their "home territory" and are injected into the opposite kind of material, where they are in the minority. This effect of the forward voltage is known as *minority carrier injection*, and is of great significance in the operation of transistors.

A third aspect of *pn* junction behavior should also be mentioned, as it is important in the operation of devices, such as transistors, which contain *pn* junctions. Suppose that an electron is introduced into the *p*-type material at a place near the junction. (Another junction in the neighborhood might be injecting minority carriers, for example.) We wish to show here that the action of the *pn* junction is to sweep the electron from the *p* side, over to the *n* side of the interface. In other words the *pn* junction acts as a *collector* of minority carriers, transporting them from the region where they are minority carriers, to a region where they are majority carriers. This collecting action of the *pn* junction is caused by the presence of an electric field at the interface of the *p* and *n* regions. Let us briefly examine how this electric field comes to exist. First of all it is easily shown that such a field must be present in a reverse-biased junction. Suppose, for example, that the voltage source in Figure 5.11 is a 10-V battery. Then 10 V must be dropped in going from the *n* side to the *p* side. The bulk *n* and *p* regions contain many free carriers; hence a large current would flow if the voltage were dropped across these regions. Therefore the voltage must be dropped across a very narrow region just at the interface of the *p*- and *n*-type sides. Thus there must be an electric field at the interface as proposed, and in such a direction as to sweep electrons to the left and holes to the right (Figure 5.11).

It is interesting, and important for transistor operation, that there is an electric field present even in the absence of an external bias. To see how such a "built-in field" comes about let us imagine that two uniform and electrically neutral pieces of semiconductor are brought together to form a *pn* junction. Both the electrons (in the *n* side) and the holes (in the *p* side), being mobile particles and possessing thermal energy, move randomly about. (This random motion, like the motion of molecules in a gas, is a form of diffusion and does not require the presence of any electric field.) Some electrons will diffuse to the *p* side (where they recombine with the numerous holes present). Similarly some holes will diffuse to the *n* side. In fact there is a tendency for the electron and hole concentrations to equalize throughout the material, analogous to the way the concentration of gas molecules equalizes throughout a closed container. However, unlike gas molecules, electrons and holes are charged, so when an electron, for example, flows over to the *p* side, the *p* side becomes more negatively charged. The diffusion of electrons over to the *p* side and holes over to the *n* side produces a separation of charge as shown in Figure 5.12. In other words an electric field is produced. This field continues to grow until it is of sufficient magnitude to repel any further flow of electrons and holes. When a steady state is reached, the tendency of an electron on the *n* side to diffuse to the *p* side is opposed exactly by the force of the electric field which tends to push the electron back to the *n* side.

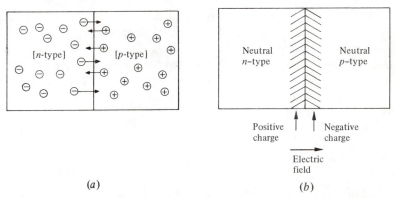

(a)   (b)

Figure 5.12   The origin of the electric field in a *pn* junction. (*a*) When the junction is formed, there is a diffusion of electrons toward the *p* side and holes toward the *n* side. (*b*) The movement of electrons to the right and holes to the left produces a separation of charge as illustrated. This charge gives rise to an electric field that opposes further motion.

This electric field, which is present in the absence of any external bias, is called the built-in field. It can act to collect excess minority carriers, as noted in the previous discussion.

As was already shown above, the collecting field is increased when a reverse bias is applied to the junction. Forward biasing the junction tends to decrease the electric field; however, as shown in Appendix II, the built-in field is large enough that forward biasing the junction only reduces the magnitude of the field, but does not eliminate it.

The major significance of the electric field under discussion is that it is a collector for minority carriers. No matter whether a *pn* junction is forward biased or reverse biased, minority carriers approaching the interface are swept across to the region where they are majority carriers. It will be shown in the next chapter that this collection action plays an essential role in the operation of a transistor.

The Ideal Diode Equation   In the preceding discussion we have indicated qualitatively why *pn* junctions conduct more easily for one polarity of bias than for the opposite. A quantitative treatment of the relationship between the diode current and the applied voltage is given in Appendix II. It is shown there that for one polarity (forward bias) the current increases exponentially with applied voltage, whereas for the other polarity (reverse bias), the current is limited to a small constant value. The complete *I-V* characteristic, valid for either bias polarity, may be written as

$$i = I_S \left(\exp\left\{ \frac{qv}{kT} \right\} - 1 \right)$$

(5.1)

The sign of the current is defined as positive into the *p* side and the applied voltage *v* is defined as positive on the *p* side, as shown in Figure 5.13. Here *q* is the electronic charge, *k* is Boltzmann's constant, and *T* is the temperature. It is useful to note that $kT/q = 0.026$ V at room temperature. The quantity I_S depends upon the constructional parameters of the diode (such as doping and area), and is also a function of temperature. For silicon diodes the value of I_S is typically in the range of 10^{-8} to 10^{-14} A.

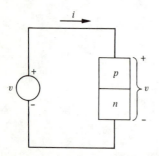

Figure 5.13   The sign conventions for *pn* junctions. The voltage is defined as positive when the potential of the *p* side is higher than that of the *n* side. The current is defined as positive when directed into the *p* side.

The current-voltage characteristic described by Equation (5.1) is displayed graphically in Figure 5.14(*a*). In this figure a value of I_S of 10^{-13} A, typical of a small silicon diode at room temperature, is assumed. The current-voltage characteristic of an actual silicon diode is shown for comparison in Figure 5.14(*b*). Figure 5.14(*a*) and Equation (5.1), from which it is drawn, manifest the polarity-dependent conduction described in the previous discussion. When *v* is positive, that is, in forward bias, the current increases rapidly with increasing voltage. For any positive voltage greater than ~100 mV, the (−1) term is negligible in comparison with the exponential term. Hence the current is simply given by $I_S \exp\{qv/kT\}$. For the characteristic illustrated in Figure 5.14(a), I_S equals 10^{-13} A and the forward current equals 10^{-3} A at 0.6 V. Furthermore, the current doubles for every 18-mV increase in the voltage.

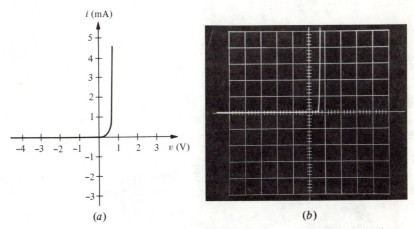

Figure 5.14 The current-voltage characteristic of a *pn* junction. (*a*) According to Equation (5.1) with $I_S = 10^{-13}$ A. (*b*) The actual characteristics of a silicon diode (curve tracer photograph). Vertical scale: 1 mA per large division; horizontal scale: 1 V per large division.

When v is made negative, that is, when the junction is reverse biased, the exponential term becomes quite small in comparison with the (-1) term. Consequently the current approaches a constant value of $-I_S$ in reverse bias. The constant I_S is called the *saturation current,* and as noted above is in the nanoampere range for silicon diodes. In most situations such a small current is entirely negligible, and the reverse current may be regarded as zero.

EXAMPLE 5.1

A certain *pn* junction has the characteristics given by Equation (5.1), where $I_S = 10^{-14}$ A at room temperature and $I_S = 10^{-9}$ A at 125°C. The diode is forward biased with a constant-current source of 1 mA at room temperature. Assuming that the current does not change, find the diode voltage at room temperature and at 125°C. ($kT/q = 0.026$ V at room temperature and $= 0.035$ V at 125°C.)

SOLUTION

We may solve the ideal diode equation, Equation (5.1), for v as a function of i. Rearranging the equation

$$\frac{i}{I_S} + 1 = \exp\left\{\frac{qv}{kT}\right\}$$

Taking the natural logarithm of both sides and solving for v, we have

$$v = \frac{kT}{q} \ln\left(\frac{i}{I_S} + 1\right)$$

Now at room temperature

$$v = 0.026 \ln\left(\frac{10^{-3}}{10^{-14}} + 1\right)$$
$$= 0.026 \times 25.3$$
$$= 0.66 \text{ V}$$

At 125°C

$$v = 0.036 \ln\left(\frac{10^{-3}}{10^{-9}} + 1\right)$$
$$= 0.50 \text{ V}$$

The temperature dependence of the diode characteristics that is displayed in this example is typical of silicon diodes. Often this temperature dependence leads to complications in circuit design; however it can also be useful, leading to such applications as diode temperature sensors. ∎

Just as the characteristics of the ideal circuit elements described in Chapter 1 are never quite realized in practice, the characteristics of a real *pn* junction diode are never exactly given by Equation (5.1). Consequently Equation (5.1) is known as the *ideal diode equation*.

Let us examine carefully the characteristics of the real device of Figure 5.14(*b*) and see to what degree it obeys the ideal diode equation. For this purpose we replot the actual characteristics along with the ideal characteristics in Figure 5.15.

To examine a wide range of currents and voltages, several plots are given, with different current and voltage scales. In each case the actual characteristics are plotted with circles and the ideal characteristics as a solid line. A typical range of currents and voltages in an electronic circuit includes currents from 0 to 10 mA and voltages from 0 to 20 V. The diode characteristic in this range is given in Figure 5.15(*a*). This figure shows that under normal operating conditions, the ideal diode equation describes the real characteristics quite well. The only deviation occurs at higher current levels, where the current increases with voltage more slowly than theoretically predicted.

When the characteristics are examined on a greatly magnified scale, as in Figure 5.15(*b*), considerable deviations from the ideal diode equation are seen. The magnitude of the current, both in forward and reverse bias, is larger than predicted. Furthermore the current does not "saturate" in the reverse direction, but increases slowly with reverse voltage. However, these deviations occur in the range of very low currents and are of no consequence in most applications.

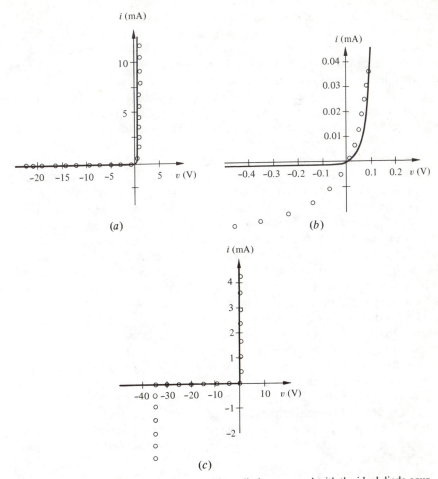

Figure 5.15 The characteristics of a real silicon diode compared with the ideal diode equation (with $I_S = 10^{-12}$ A). The actual characteristics are plotted as circles and the theoretical characteristics are drawn with a solid line. (*a*) The characteristics plotted over a typical operating range of current and voltage. (*b*) The same characteristics, greatly magnified to show the low current behavior. (*c*) The same characteristics over an extended voltage range to show reverse breakdown.

A more significant deviation of the actual characteristics from the theoretical may be seen when the voltage range is greatly expanded, as in Figure 5.15(*c*). As the reverse bias voltage is increased there comes a point, called the *breakdown voltage*, beyond which the current rapidly increases. Often the breakdown is so sharp that almost any increase in the voltage beyond the breakdown voltage results in very large currents. The diode of Figure 5.15(*c*) has a breakdown voltage of 35 V. The precise value of a diode's breakdown voltage depends mainly on the doping, and may be designed to be anywhere in the range from a few to thousands of volts. In designing diode circuits, one must keep in mind the maximum reverse voltage that will appear across each diode, and select diodes with appropriate breakdown voltage.

It should be emphasized that breakdown is not destructive (unless of course the diode is overheated) and may be used to obtain accurate voltages. For example, if a current source of −1 mA were connected to the diode of Figure 5.15, the voltage across the diode terminals would be 35 V. Diodes especially made to be operated in this fashion are called Zener diodes. Zener diodes are used as the voltage references in all kinds of modern electronic circuits.

Diode Time Constants   We have seen that when a *pn* junction is forward biased minority carriers are injected into the bulk regions in the vicinity of the junction. We wish to consider what would happen if a reverse voltage were suddenly applied. The current would not immediately drop to zero because the *pn* junction, in its role as a collector for minority carriers, would gather back some of the carriers it injected in the forward-bias condition. The minority carriers injected in forward bias are essentially stored until they are removed — either by recombination or by collection at the junction. When the diode voltage is reversed, these stored carriers are collected, and the resulting current does not cease until they are all collected or recombined.

In "switching" diodes, used in logic circuits, this effect is undesired, as it causes a delay before the desired new condition of the circuit is achieved. Therefore every effort is made to reduce the so-called storage time, both by reducing the lifetime of the minority carriers and by reducing the thickness of the region in which the carriers are stored. The better, that is, faster, switching diodes may have storage times in the nanosecond region.

If it is desired to include these storage effects in circuit analysis, then the diode is modeled as an ideal diode in parallel with a capacitor. Similar effects take place in transistors, so we shall defer further consideration of this matter to Chapters 6 and 10.

5.3   The Diode as a Circuit Element

Diodes are important elements in many kinds of electronic circuits. Their importance stems primarily from their nonlinear characteristics and, in particular, from their property of conducting current more easily in one direction than in the other. A standard circuit symbol reflecting this basic property has been adopted for the *pn* junction diode. The circuit symbol and a schematic illustration of the *pn* junction that it represents are given in Figure 5.16. The current and voltage conventions are also shown in the figure. The circuit symbol derives from the direction of "easy current flow" in a *pn* junction. As we have seen, the current flows easily from *p* side to *n* side in forward bias; hence the arrow points in this direction. In forward bias both *i* and *v*, as defined, are positive. When *v* is made negative, the diode is in reverse bias, and essentially

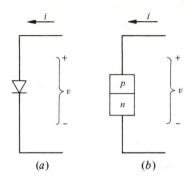

Figure 5.16   The circuit symbol for a *pn* junction diode. The standard symbol is shown in (*a*) and the structure it represents is given in (*b*). Forward bias corresponds to positive voltage, and produces a current flow in the direction of the arrow.

(*a*)   (*b*)

no current flows. What little current there is flows in the direction against the arrow.

Unlike the circuit elements discussed in Chapter 1, the diode is a non-linear device; that is, the current through the device is not linearly proportional to the voltage across it. This nonlinearity imposes some restrictions on the tools used in circuit analysis; for instance, superposition and the Norton and Thévenin theorems cannot be used on circuits containing diodes. Of course the basic laws of Kirchhoff retain their validity, and just as in the circuit problems already encountered, either the loop or nodal analysis method yields sufficient equations to determine all the circuit unknowns. However, the presence of diodes in the circuit causes the equations to be nonlinear, and thus complicates finding the solution. Consequently, graphical or iterative numerical techniques very often are used to find the solutions for the desired unknowns.

EXAMPLE 5.2

Find the current I_D flowing through the diode in the circuit shown. (Note: $kT/q = 0.026$ V at 300° K.)

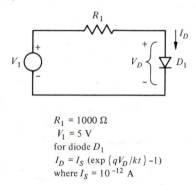

$R_1 = 1000\ \Omega$
$V_1 = 5$ V
for diode D_1
$I_D = I_S\ (\exp\ (qV_D/kt\ ) - 1)$
where $I_S = 10^{-12}$ A

SOLUTION

Writing Kirchhoff's voltage law around the single loop, and adding the voltage drops, we have

$$-V_1 + I_D R_1 + V_D = 0$$

where V_D is the voltage across the diode. We may obtain an expression for V_D using the diode equation [(5.1)]:

$$V_D = \frac{kT}{q} \ln\left(\frac{I_D}{I_S} + 1\right)$$

Substituting for V_D in the loop equation

$$-V_1 + I_D R_1 + \frac{kT}{q} \ln\left(\frac{I_D}{I_S} + 1\right) = 0$$

This equation has only one unknown, I_D; nonetheless we cannot solve directly for I_D because of the presence of the term $\ln(I_D/I_S + 1)$. Probably the most straightforward way of solving for I_D is to guess values for I_D, and to keep changing the guess until the equation is satisfied. If a very high degree of accuracy were desired, we would use a computer, applying the Newton–Raphson algorithm described in Chapter 2. However a simple iterative guessing scheme will be employed here.

Let us define the function $F(I_D)$ directly from the last equation by

$$F(I_D) = -V_1 + I_D R_1 + \frac{kT}{q} \ln\left(\frac{I_D}{I_S} + 1\right)$$

We wish then to find the value of I_D for which $F(I_D)$ equals 0. An orderly procedure can be devised based on the fact that $F(I_D)$ is positive when I_D is too large, and negative when I_D is too small. We make a guess for I_D, and if $F(I_D)$ is positive, we know we must lower our guess, and vice versa. The maximum value of I_D may be first estimated by finding the current that would flow in the circuit if the diode voltage were zero. Similarly, the minimum current may be estimated by calculating the current in the circuit for a diode drop of 1 V. Using these limits on the diode drop, we obtain a maximum current of $V_1/R = 5 \times 10^{-3}$ A and a minimum current of $(V_1 - 1)/R = 4 \times 10^{-3}$ A.

Let us then use 4.5×10^{-3} A as a first guess for I_D. The following table shows the value of $F(I_D)$ for the two limits and for the first guess.

I_D (A)	$F(I_D)$
5×10^{-3}	+0.58
4×10^{-3}	−0.42
4.5×10^{-3}	+0.08

Our first guess, 4.5×10^{-3} A, is too high because $F(I_D)$ is positive, so the answer lies between 4.0×10^{-3} and 4.5×10^{-3} A. Furthermore, from the size of $F(I_D)$, the answer appears to be closer to 4.5×10^{-3} than to 4×10^{-3} A. Thus a good next guess might be 4.4×10^{-3} A. We proceed in this fashion until the answer is known with whatever accuracy is desired. Four further guesses are given below:

I_D (A)	$F(I_D)$
4.4×10^{-3}	−0.023
4.45×10^{-3}	+0.027
4.42×10^{-3}	−0.003
4.43×10^{-3}	+0.007

The value of I_D must lie between 4.42 and 4.43×10^{-3} A.   ■

The iterative method of solving nonlinear equations is quite simple, but tedious. Such repetitive calculations are best carried out on a digital computer. Often it is sufficient to find a quick estimate of the answer using a graphical method of solution. Then, if necessary, more accurate calculations can be made on the computer.

EXAMPLE 5.3

Solve the problem of Example 5.2 using the graphical load-line technique introduced in Chapter 2.

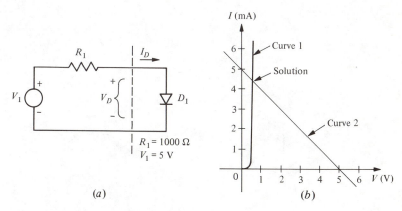

(a)   (b)

SOLUTION

The load-line technique is based on the concept of identifying two parts of the circuit whose characteristics may be separately graphed. An imaginary break is indicated by the dotted line in the circuit in sketch (a). We sketch, on one set of axes, I_D versus V_D for the subcircuits to the left and right of the dotted line; the solution is found at the intersection of the two curves. For the diode

$$I_D = I_S\left(\exp\left\{\frac{qV_D}{kT}\right\} - 1\right)$$

This characteristic is graphed as curve 1 in sketch (b). For the resistor and voltage source (as seen from the diode terminals)

$$V_D = V_1 - I_D R_1$$

This characteristic is graphed as curve 2 in the same figure. (Note that the positive sense of I_D must be the same for both curves – clockwise in the circuit diagram, in this example.) It is seen that the diode current is about 4.5×10^{-3} A. ■

The difficulties involved in the analysis of circuits containing nonlinear elements prompt the use of idealizations that simplify the analysis. There are two common idealizations for the pn junction diode that are routinely used in the analysis of diode circuits. One is a new ideal element, called the perfect rectifier; and the other is a device model, which we call the large-signal diode model.

The Perfect Rectifier   Various degrees of idealization of the diode may be used to simplify diode circuits. The more the actual behavior of the diode is simplified, the easier it is to visualize the operation of the diode in the circuit. However, the greater the idealization, the less accurate and the less widely applicable it becomes. For example, if one were to proceed to the extreme of saying that a diode, since it conducts current, acts like a resistor, then the most essential property—that of conducting in only one direction—would be lost.

The term *rectifier* refers to a device which passes current in only one direction; hence the *pn* junction is a kind of rectifier. A *perfect rectifier* would be a device which freely conducts current in one direction, with no voltage drop, but allows no current whatsoever to pass in the other direction. In other words, for positive currents the perfect rectifier is just a short circuit, whereas for negative currents it is an open circuit. The *I-V* characteristic shown in Figure 5.17(*a*) is obtained by combining portions of the *I-V* characteristics of a zero resistance and an infinite resistance, using the former in the forward-biased region and the latter in the reverse-biased region.

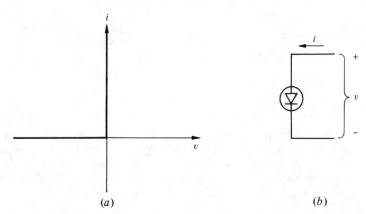

(*a*)   (*b*)

Figure 5.17   The perfect rectifier. (*a*) The *I-V* characteristic. The vertical portion, corresponding to forward bias, is the same characteristic as that of a short circuit. The horizontal portion, corresponding to reverse bias, is the same characteristic as that of an open circuit. (*b*) The circuit symbol for the perfect rectifier.

A comparison of the *I-V* characteristic of the perfect rectifier with the characteristic of a typical *pn* junction diode, for example, Figure 5.14, shows that the perfect rectifier is an idealization of the diode. In many cases in the analysis of diode circuits, it is useful to replace the diodes by perfect rectifiers. The operation of the circuit may be quickly visualized, and an approximate analysis for the desired circuit parameters may be made. The circuit symbol for the perfect rectifier is given in Figure 5.17(*b*). It should be kept in mind that in using an idealization, some features of the real device have been lost. One must be sure that in the problem under consideration, the omitted features do not make a vital difference. Since the essential difference between the perfect rectifier and an actual diode is the small but nonzero forward voltage drop of the actual diode, the effect of this voltage drop should be considered.

EXAMPLE 5.4

The input voltage v_1 in the circuit in sketch (a) is a pure sinusoid. Find the output voltage v_2 (that is, graph v_2 versus time). Treat the problem approximately by replacing the diode by a perfect rectifier.

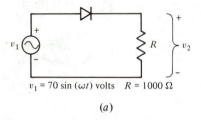

$v_1 = 70 \sin (\omega t)$ volts $R = 1000 \ \Omega$

(a)

SOLUTION

Whenever the voltage v_1 is positive, the junction is forward biased; whenever v_1 is negative, it is reverse biased. Replacing the diode by a perfect rectifier is equivalent to replacing the diode by a short circuit for positive v_1 and an open circuit for negative v_1 [see sketch (b)].

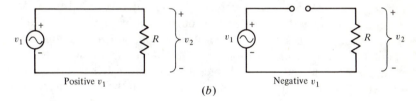

Positive v_1 Negative v_1

(b)

Accordingly, whenever v_1 is positive, $v_2 = v_1$. When v_1 is negative the model requires that the diode current, and consequently the current through the resistor, be zero. In this case v_2 must be zero. The graphs of the input voltage and output voltage are given in sketch (c).

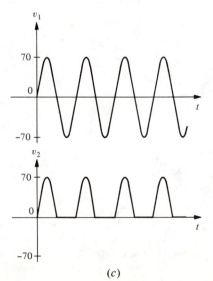

(c)

Since the voltages present in the circuit during most of the cycle are large compared to the actual diode forward drop of about 0.7 V, the answer as obtained with the perfect rectifier model is quite reasonable. For example, at the instant the input voltage is +70 V, the output voltage in this approximation is +70 V. The actual output voltage at this instant must be somewhat less, say 69 or 69.5 V, because of the voltage drop which occurs across a forward-biased diode. Whereas this error is very small, the error is relatively larger for smaller input voltages. For example, at the instant the input voltage equals +2 V, we indicate an output voltage of +2 V. The actual output voltage would be closer to 1.3 V, again because of the small but nonzero voltage drop of a real diode.   ■

The Large-Signal Diode Model   The essential approximation that is made in treating a *pn* junction diode as a perfect rectifier amounts to neglecting the forward voltage drop. It is often the case that the error thus introduced is unacceptably large; yet we hesitate to use the actual diode characteristic because of the analytical difficulties introduced by its nonlinearity. In such instances a better approximation to the actual diode characteristic can be made. In a real diode the current increases so rapidly with voltage that the forward voltage drop in silicon diodes is almost always in the range of 0.5 to 0.8 V for practical currents. For example, we see that for the diode of Figure 5.14 the forward voltage drop is between 0.6 and 0.7 V for currents in the range of 0.5 to 5 mA. This observation prompts an improved idealization, shown in Figure 5.18(*a*). This characteristic more closely approximates the actual diode characteristic, and at the same time retains much of the simplicity of the perfect rectifier characteristic. However, in the improved idealization the diode has a constant forward voltage drop of 0.7 V.

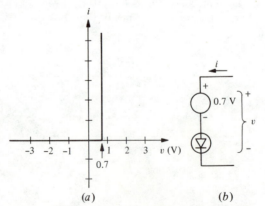

(*a*)   (*b*)

Figure 5.18   The large-signal diode model. (*a*) The *I-V* characteristic. For reverse voltages, and forward voltages less than 0.7 V, the current is zero. The current rises without limit when the forward voltage reaches 0.7 V. (*b*) The circuit model. The combination of a 0.7-V voltage source and a perfect rectifier produces the *I-V* characteristics shown in (*a*).

The characteristic displayed in Figure 5.18(*a*) is produced by a combination of two ideal elements, as shown in Figure 5.18(*b*). This circuit, consisting of a perfect rectifier in series with a 0.7-V voltage source, will be called the *large-signal diode model.*[5]

Solving diode circuit problems with the large-signal diode model is almost as straightforward as using the perfect rectifier. The constant voltage of 0.7 V merely adds to the signal appearing at the terminals of the perfect rectifier.

EXAMPLE 5.5

Find the current I_D in the circuit in sketch (*a*) using the large-signal diode model. This is the same problem solved in Example 5.2 using the numerical method. It is interesting to compare the accurate answer obtained there with the approximate result to be obtained here.

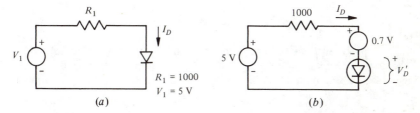

(*a*)   (*b*)

SOLUTION

The circuit is first redrawn, replacing the diode by the large-signal diode model [sketch (*b*)]. The voltage drops around the single loop circuit must sum to zero, hence

$$-5 + 1000 \, I_D + 0.7 + V_D' = 0$$

It may be seen, either from this equation or directly from the circuit, that 4.3 V must be dropped across the combination of resistor and the perfect rectifier. Furthermore, the polarity is such that the rectifier can only be forward biased. Since in forward bias there is no voltage drop across a perfect rectifier, the 4.3 V must be dropped across the resistor. From Ohm's law

$$I_D = 4.3/1000 = 4.3 \text{ mA}$$

This approximate answer is to be compared with the more exact result (4.42 mA) obtained in Example 5.2.   ∎

The principal use of the large-signal diode model stems from the rapid circuit analysis made possible by its simplicity. However, its limitations should be kept in mind. The idealized characteristics given in Figure 5.18 are not exactly like any real diode characteristics. For example, the model incorrectly implies that the current is zero whenever the forward bias is less than 0.7 V. Actually the change in conduction is more gradual. Furthermore, according to the model the voltage drop in forward bias never exceeds 0.7 V, whereas

---

[5] The 0.7-V forward drop is appropriate for silicon diodes. In germanium diodes the forward drop is more typically 0.3 V and in gallium arsenide diodes, 1 V.

in real diodes the voltage drop increases gradually with current and may exceed this value. The model is used then as a tool in determining quickly approximate values of desired circuit unknowns. In the analysis of some diode circuits, one may substitute the perfect rectifier alone, ignoring completely the voltage drop across the forward-biased diode. Such a substitution may be made if the voltages in the circuit are so large that a correction of about 0.7 V would make no difference. This is usually the case in rectifier circuits, to be discussed in the next section. The 0.7-V drop across the forward-biased diode does become important in diode logic circuits, as will be discussed in Section 5.5 and Chapter 8. In logic circuits 0.7 V is significant; moreover the effects of this voltage tend to accumulate from stage to stage. Finally, there are infrequently some cases in which neither approximation is warranted, and the exact diode characteristics, along with numerical or graphical techniques, must be used.

EXAMPLE 5.6

The input v_1 in the circuit in sketch (*a*) is a time-varying signal, as shown. Find and sketch the output v_2 in the time interval for which the input is given. Make a first estimate of v_2, assuming that the diodes are perfect rectifiers; then make a more accurate estimate using the large-signal diode model.

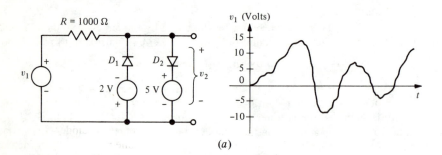

(*a*)

SOLUTION

Let us first imagine, without bothering to redraw the circuit, that the diodes are perfect rectifiers. Suppose that $v_1 = 0$. Then both D_1 and D_2 are reverse biased, D_1 by 2 V and D_2 by 5 V. In reverse bias no current flows, that is, the diodes behave as open circuits. The circuit may be visualized with the diodes removed, in which case it is seen that $v_2 = v_1$. This situation persists, with v_2 equal to v_1 until v_1 becomes large enough, positive or negative, to forward bias one of the diodes. This happens when $v_1 > 5$ V or when $v_1 < -2$ V. When v_1 increases to 5 V, the 5 V source in series with D_2 is then canceled (as can be seen by considering the loop starting at the top of v_1, passing through R, D_2, the 5 V source, and coming back through v_1), and D_2 becomes forward biased. Since a perfect rectifier is a short circuit in forward bias, the 5 V battery is connected to the output terminals, and $v_2 = 5$ V. The difference between v_1 and v_2 is dropped in the resistance R, through which a current flows. (The value of R is not important, but should be large enough to prevent excessive current from passing through D_2.) Meanwhile, diode D_1 is now reverse biased by 7 V, and still has no effect on the circuit. Similar behavior occurs when the sign of v_1 is negative; when $v_1 < -2$ V, diode D_1 becomes forward biased and $v_2 = -2$ V. The output voltage v_2 obeys the following rules:

$$\text{If } -2 \text{ V} \le v_1 \le +5 \text{ V} \qquad \text{then } v_2 = v_1$$
$$\text{If } v_1 < -2 \text{ V} \qquad \text{then } v_2 = -2 \text{ V}$$
$$\text{If } v_1 > +5 \text{ V} \qquad \text{then } v_2 = +5 \text{ V}$$

A graph of the output for the given input is given in sketch (b).

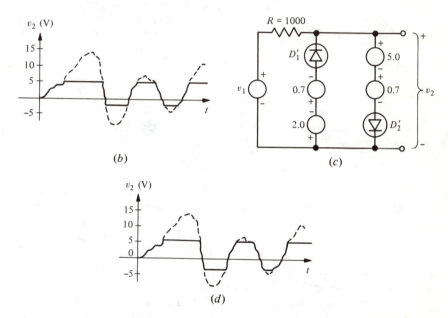

(b)

(c)

(d)

Let us now reexamine circuit operation using the large-signal diode model. The circuit is shown in sketch (c). This circuit is identical to the one previously considered, except that each voltage source is increased by 0.7 V. The behavior is similar, the only difference being that an input of 5.7 V or greater (rather that 5 V) is required to forward bias D_2', and an input of -2.7 V or less, to forward bias D_1'. In this case the output is shown in sketch (d).

If an exact analysis (or a measurement) were carried out, it would show a slight rounding of the sharp corners of v_2 versus t, but would closely correspond to sketch (d) in other respects. The circuit of this example is known, for obvious reasons, as a limiter. Any desired limiting voltages may be selected, by choosing appropriate voltage sources. Such circuits are useful, for example, in limiting the input voltages in digital circuits. ∎

Small-Signal Behavior   When we deal with very large signals, the real diode behaves nearly as a perfect rectifier: an open circuit for one polarity and a short circuit for the other. On the other hand, for very small signals the behavior is much like that of a resistor, as shown in Figure 5.19. Suppose the diode is biased at current I_b and voltage V_b; then if the voltage is increased by a small amount Δv, the current increases by the amount Δi, in proportion to

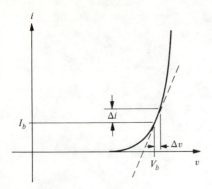

Figure 5.19 The response of a diode to small signals. For very small signals, the current increases approximately in proportion to the increase in voltage Δv. The constant of proportionality is the slope di/dv at the bias point.

Δv. As shown in Figure 5.19, the constant of proportionality is the slope of the *I-V* curve at the point of interest. Taking an analytical, rather than graphical point of view, the current as a function of the voltage may be expanded in a Taylor series about the bias point V_B:

$$i = I_b + a \, \Delta v + b \, (\Delta v)^2 + c \, (\Delta v)^3 + \ldots \qquad (5.2)$$

where $a = di/dv$ evaluated at the bias point; $b = (1/2) \, d^2i/dv^2$ evaluated at the bias point; $c = (1/6) \, d^3i/dv^3$ evaluated at the bias point. For the very smallest signals only the term $a \, \Delta v$ is important, and changes in i are proportional to changes in v. However as the signal size increases, it is necessary to take into account more and more terms to calculate the current accurately. For large signals, greater than about a few tenths of a volt, so many terms must be taken into account that the Taylor series approach becomes impractical. Fortunately, the large-signal diode model already developed is then appropriate.

EXAMPLE 5.7

One important application of diodes is that of a mixer, which is a circuit that combines signals of two different frequencies to form new signals at the sum and difference frequencies. Show that the output in the circuit shown has components at the frequencies $\omega_1 + \omega_2$ and $\omega_1 - \omega_2$. Assume that the diode characteristic is given by the first three terms in the Taylor series expansion, Equation (5.2), and that the resistance R is so small that it has no effect on the current.

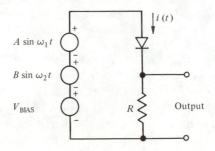

SOLUTION

If R is sufficiently small, the signal Δv applied to the diode is given directly by $A \sin \omega_1 t + B \sin \omega_2 t$. Assuming that the diode characteristic may be represented by

$$i = I_b + a \, \Delta v + b \, \Delta v^2$$

we have

$$i = I_b + a(A \sin \omega_1 t + B \sin \omega_2 t)$$
$$+ b(A \sin \omega_1 t + B \sin \omega_2 t)^2$$

the third term may be multiplied out and then simplified through the identity

$$\sin \theta_1 \sin \theta_2 = \frac{\cos(\theta_1 - \theta_2) - \cos(\theta_1 + \theta_2)}{2}$$

Thus

$$i = I_b + aA \sin \omega_1 t + aB \sin \omega_2 t - \frac{b}{2} A^2 \cos 2\omega_1 t$$

$$- \frac{b}{2} B^2 \cos 2\omega_2 t + bAB \cos(\omega_1 - \omega_2)t$$

$$- bAB \cos(\omega_1 + \omega_2)t + \frac{b}{2}(A^2 + B^2)$$

The output signal is iR. It is clear that the nonlinearity of the diode leads to output components at a number of different frequencies, including the sum and difference frequencies. If the nonlinearity is not present, that is, $b = 0$, then the last five terms disappear and no mixing occurs. ∎

5.4 Rectifier Circuits

One of the oldest and most familiar applications of the diode is as a rectifier. This name is applied to a device for converting ac power to dc. Inasmuch as almost all commercial electric power is ac, while almost all electronic circuits require dc, there is a rectifier circuit in nearly every piece of electronic equipment.

A simple rectifier circuit is shown in Figure 5.20(a). This circuit is familiar from the examples already given. The ac voltage source (which in this case might represent the ac power line) supplies the voltage v_1 shown in Figure 5.20(b). The effect of v_1 is to apply forward and reverse bias to the diode. When the diode is forward biased a current flows through the resistance, which is the load. Since the voltage drop across a forward-biased diode is usually small compared with v_1, the current which flows on this part of the cycle is essentially the input voltage divided by the value of the resistance. On the other half of the cycle, when the diode is reverse biased, negligible current flows. Thus the current flowing through the resistance has the form of the top half of the sine wave, as shown in Figure 5.20(c). This resulting current is called pulsating dc. It is considered dc rather than ac because the current always

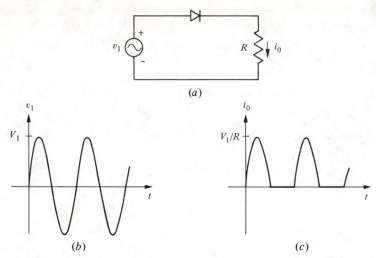

(a)

(b)   (c)

Figure 5.20   The operation of a half-wave rectifier circuit. The circuit is given in (a). In (b) the input voltage, such as might come from an ac power line, is given. The amplitude of the sinusoid is v_1. The rectified current (pulsating dc) flowing in resistance R is shown in (c). Its amplitude is v_1/R, approximately.

flows in the same direction and thus never changes sign. A rectifier circuit of this type is known as a half-wave rectifier because current flows to the load, at most, on only one-half of the operating cycle.

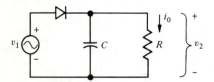

Figure 5.21   A rectifier circuit with filter capacitor. The function of the capacitor is to supply current to the load during the period when the diode is reverse biased.

In practical circuits it is usually necessary to obtain smooth dc, that is, nearly constant current, through R; thus, the pulsations of the rectified current must be smoothed out. This may be done by adding a capacitor, known as a filter capacitor,[6] as shown in Figure 5.21. To understand the effect of C on i_0, let us imagine that v_1 is turned on at time $t = 0$, as shown in Figure 5.20(b), and imagine that the diode is a perfect rectifier. Until v_1 reaches its first maximum, the diode is forward biased and acts as a short circuit, so that v_2, the voltage across the capacitor, equals v_1 during this period (Figure 5.22). If a sufficiently large capacitor is provided, then after the first maximum of v_1 the capacitor remains charged nearly to the value V_1, while v_1 decreases. This makes the diode reverse biased, and it conducts no current. The load resistance across

---

[6] The conventional name filter capacitor refers to the filtering of ac (that is, time-varying) components out of the dc currents. The circuit here, although different in function, is related to the filter circuits discussed in Chapter 2.

the capacitor causes the voltage v_2 to decay exponentially during this period with an exponential time constant equal to RC, as described in Chapter 2. When v_1 returns to its next positive maximum, the capacitor is recharged to the value V_1, and the process repeats. If the time constant RC is chosen to be long compared with the period of the sinusoid, the current i_0 is nearly constant.

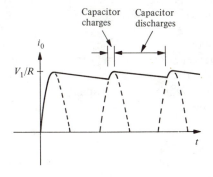

Figure 5.22 Output current of a rectifier with filter capacitor (Figure 5.21). The solid line indicates the current through the load resistance R. For comparison the dotted line shows the unsmoothed pulsating dc of Figure 5.20(c).

EXAMPLE 5.8

Sketch the load current for the rectifier circuit, Figure 5.21, for the case $C = 50 \times 10^{-6}$ F, $R = 1000 \ \Omega$, and $v_1 = 165 (\sin 377 \ t)$. (The latter is approximately the standard U.S. line voltage.) Assume for simplicity that the diode is a perfect rectifier.

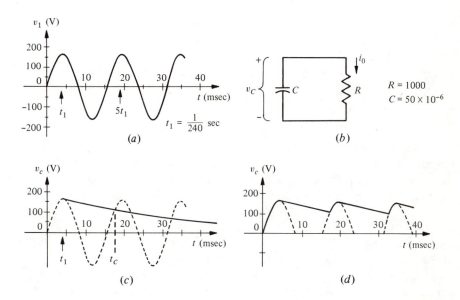

SOLUTION

We first sketch the input voltage [see sketch (a)]. According to the discussion already given, the capacitor is charged fully to 165 V at time t_1. In the time interval starting at t_1 and ending just before $5t_1$, the input and the diode are effectively removed from the circuit. During this time the capacitor discharges into the load, supplying the load current, and the circuit reduces to a resistance in parallel with a capacitance [see sketch (b)]. We know from the discussion of RC circuits in Chapter 2 that the voltage v_C decays exponentially with time constant RC. The voltage is thus of the form $v_C = 165 \exp\{-t'/RC\}$ where $t' = 0$ is the instant when the input was disconnected, that is, when $t = t_1$. In other words, $v_C = 165 \exp\{-(t-t_1)/RC\}$. This equation is plotted in sketch (c).

This equation only describes the output up to the time t_e shown in the figure because the charging of the capacitor starts again at this instant and the cycle is repeated. Thus the actual waveform is as given in sketch (d). The load current has the same form as the output voltage, because $i_0 = v_C/R$.   ■

In a practical rectifier further refinements are generally made. A resistance is often placed between the diode and the capacitor to prevent large currents (which might damage the diode) from flowing into the capacitor when the circuit is first turned on. Additional capacitors may be added to improve the smoothing (multistage filtering).

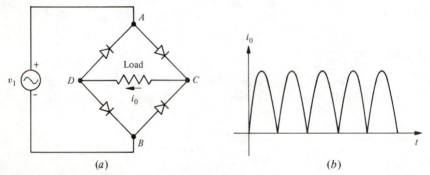

(a)   (b)

Figure 5.23   A full-wave rectifier. (a) A four-diode circuit known as a "bridge rectifier." The current through the load can only flow from C to D because of the "steering" action of the diodes. (b) The current waveform in the load.

A basic modification in the rectifier circuit can simplify the filtering problem by improving the shape of the unfiltered output. A suitable circuit, and its output (for a sinusoidal input), are shown in Figure 5.23. The circuit functions by always steering the current in one direction through the load, thus always producing a positive load current. For example, during the half cycle when $v_A > v_B$ there is a current path from A to B via A–C–D–B. Treating the diodes as perfect rectifiers, we see that $v_C = v_A$ and $v_D = v_B$ during this half cycle. The diodes between A–D and C–B are reverse biased, so current can only flow along the route already stated. Now during the opposite half cycle, $v_B > v_A$. During this time there is a current path from B to A via B–C–D–A. Again the current in the load flows from C to D. The diodes between A–C and B–D are reverse biased, blocking any current flow except via B–C–D–A. The current waveform shown in Figure 5.23(b) results. Because both half cycles of the input sinusoid are rectified and appear at the output, this circuit is called a full-wave rectifier.

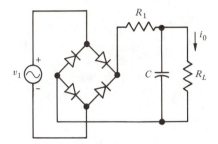

Figure 5.24 A full-wave rectifier circuit with filtering.

A complete full-wave rectifier with output filter is shown in Figure 5.24. It functions much like the simpler circuit of Figure 5.21, except that the time lapse between charging cycles of the filter capacitor is halved with full-wave rectification, and therefore the filtering is improved.

The kinds of diodes which are used in rectifier applications are often subjected to large reverse voltages. Furthermore, in some circuits they must conduct quite sizable forward currents. Power supplies yielding tens of amperes are very common; some high-power installations handle thousands of amperes.

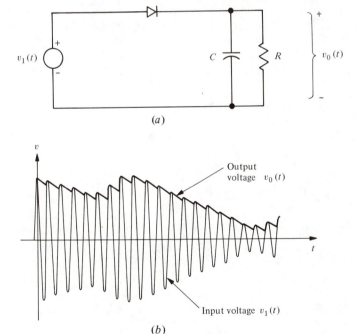

(a)

(b)

Figure 5.25 A diode demodulator. The capacitor charges up to the peak value of the sine wave. The RC time constant is chosen so that the output can vary in response to slow changes in the amplitude of the sine wave. However, RC is made large compared with the period of the sinusoid, so that the sine wave itself does not appear in the output.

The voltage drop in a forward-biased diode is in the neighborhood of 0.5 to 1 V, so the power dissipated may be considerable when large currents flow. For example, at 10 A, 5 to 10 W of heat is generated within the diode. Diodes used in rectifier applications must therefore be chosen with appropriate power and reverse voltage capability.

Diode Peak Detectors and Clamps   Aside from applications in power supplies, diode rectifier circuits are very useful in many kinds of signal processing applications. The same basic circuit shown in Figure 5.21 may be used as a peak detector. When the load resistance is large, the output voltage charges up to the peak positive signal voltage. If the magnitude of the peak voltage further varies with time the circuit may be designed to follow this variation. The principle is illustrated in Figure 5.25. The RC time constant of the filter is chosen so that the output can follow the variations of the amplitude of the signal being rectified. The input signal shown in Figure 5.25(b) is an amplitude modulated sine wave and the detector output is proportional to the amplitude. In this application the circuit is therefore called a demodulator.

EXAMPLE 5.9

The input to the circuit in sketch (a) is of the form $v = (A + B \sin 2\pi f_s t) \sin 2\pi f_c t$ (v is called an amplitude-modulated sine wave). In this case f_s is the signal frequency (1000 Hz) and f_c is the carrier frequency (10^6 Hz). Choose a capacitor value such that the output closely follows the 1000 Hz signal, but removes most of the 10^6 Hz ripple.

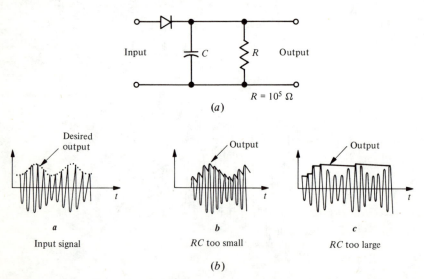

(a)

(b)

SOLUTION

The input signal has the form shown below in sketch (b), part **a**. If a very small capacitor is chosen, then the output signal will have the form shown in part **b**. The "droop" that occurs between charging cycles is excessive, with the result that the output contains not merely the desired $\sin 2\pi f_s$ signal, but also ripple at the carrier frequency. On the other hand, as shown in part **c**, if the capacitor is very large, the output signal is nearly constant and does not follow the desired signal. Thus to achieve the desired operation, C must be chosen as follows: (1) $RC \gg 1/2\pi f_c$; this condition assures that the capacitor voltage remains nearly constant during the period between charging peaks $(1/f_c)$. (2) $RC \ll 1/2\pi f_s$; this condition assures that the capacitor discharges in a time which is short compared to the period of the signal $(1/f_s)$. Thus C is chosen so that

$$\frac{1}{2\pi f_c} \ll RC \ll \frac{1}{2\pi f_s}$$

With $R = 10^5 \ \Omega$, $f_c = 10^6$ Hz, and $f_s = 1000$ Hz, we have

$$1.6 \times 10^{-12} < C < 1.6 \times 10^{-9} \ \text{F}$$

A value of $C = 50 \times 10^{-12}$ F, for example, would satisfy the requirement. ■

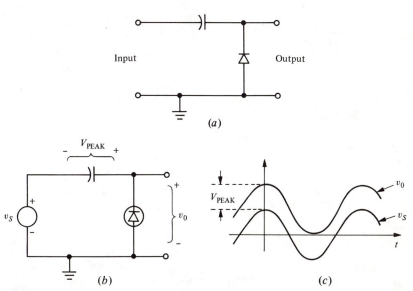

Figure 5.26 The diode clamp. The capacitor charges up to the peak negative input voltage V_{PEAK}. The output voltage follows the input, but is shifted by the amount V_{PEAK}.

The rectifying property of a *pn* junction diode can also be used to advantage in the application known as *clamping* or *level restoring*. The so-called diode clamp circuit is shown in Figure 5.26, along with the input and output waveforms. The basic circuit is shown in Figure 5.26(*a*) and is redrawn in Figure 5.26(*b*) for analysis. The circuit is recognized as the half-wave rectifier circuit, with the output taken across the diode instead of across the capacitor. Assuming for simplicity that the diode is a perfect rectifier, the capacitor is charged to the peak negative value of the input, and since there is no resistor in parallel to drain off charge, it remains charged to this value. For example, if the input is a 10 V peak-to-peak sine wave, the capacitor charges up to 5 V, with the polarity indicated in Figure 5.26(*b*). The output signal is obtained by adding the capacitor voltage to the input signal. Since the capacitor voltage is constant, the output has the same form as the input, but is shifted upward by the amount of the capacitor voltage. The capacitor voltage equals the peak negative signal voltage; hence the output is shifted upward just enough that it never goes negative. The output is said to be clamped to ground.

In a real circuit with a diode with a nonzero forward voltage drop, the capacitor voltage is reduced by the amount of the drop. Hence the voltage shift is reduced, and the output voltage may go slightly negative. In the analysis of a practical circuit, the effect of the load resistance (which slowly discharges the capacitor) should also be taken into account.

5.5 Diode Logic Circuits

We now move from one of the oldest applications of diodes (as rectifiers) to one of the newest, as logic elements which are used, for example, in digital computers. In Chapter 3 several logic blocks, or gates, were described. We now show how diode circuits that perform the **AND** and **OR** logic functions may be constructed.

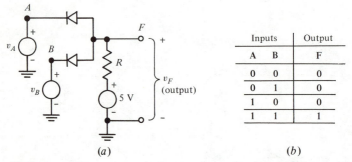

Figure 5.27   A diode **AND** gate. (*a*) The circuit. The inputs are terminals *A* and *B*, and the output is taken at terminal *F*. (*b*) The truth table (see text).

The first logic circuit to be considered is the **AND** gate, shown in Figure 5.27. We wish to determine the value of the output for each of the various possible input combinations, that is, construct the truth table for the circuit. For the purposes of the discussion here, the voltage range of 0 to 1 V is taken as logic **0** and the voltage range of 4 to 5 V as logic **1.** Let us first determine the approximate circuit behavior, assuming that the diodes are perfect rectifiers, and then come back and compute the output more accurately using the large-signal diode model. In a digital circuit, when the output is expected to be in one of two logical states, there is a simple method of determining the state. The method is essentially trial and error; a state is assumed, and it is determined whether or not this state is consistent with the basic circuit laws. The other logical state is then assumed, and again a self-consistency check is made. If the logic circuit functions properly, only one of the two states will be found to be consistent with the circuit laws.

The circuit is redrawn in Figure 5.28, once for each possible combination of logic inputs. In Figure 5.28(a) both inputs are grounded. Let us first suppose that the output at F is logical **1,** that is, the voltage v_F is in the range of 4 to 5 V. In this case both diodes would be forward biased by a voltage of 4 to 5 V. Such a large forward bias would lead to very large currents; in fact, for perfect rectifiers the current would become infinite. Since there is no source of infinite current, the assumption that $\mathbf{F} =$ logical **1** must be false. On the other hand, suppose $\mathbf{F} = \mathbf{0}$, that is, v_F is in the range 0 to 1 V. In this case a current of $4/R$ to $5/R$ flows up through R and must flow to ground through one or both of the diodes. According to Figure 5.17 any positive current may flow through a perfect rectifier (in the direction of the arrow); hence the state $\mathbf{F} = \mathbf{0}$ is consistent with the inputs. If the diodes are perfect rectifiers then $v_F = 0$. The first line in the truth table for this circuit is thus $\mathbf{A} = \mathbf{0}, \mathbf{B} = \mathbf{0}, \mathbf{F} = \mathbf{0}.$

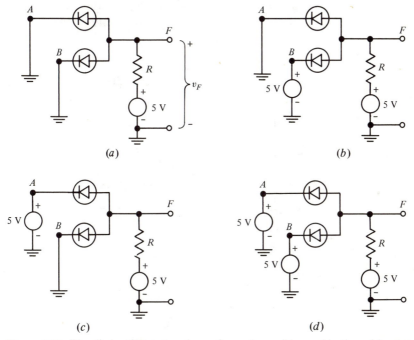

Figure 5.28 The diode **AND** gate redrawn for each possible combination of inputs. The diodes are assumed to be perfect rectifiers.

This circuit is redrawn for a second set of input conditions in Figure 5.28(b). The logical values of the inputs are $\mathbf{A} = \mathbf{0}, \mathbf{B} = \mathbf{1}$. If an output of logical **1** is assumed, the previous inconsistency, a positive voltage of 4 to 5 V across a diode, reappears. On the other hand, if the output is assumed to be zero volts, that is, logical **0**, then a current of $5/R$ flows up through R and to ground through diode A. Diode B is then reverse biased by 5 V, and is effectively an open circuit. The second line in the truth table thus reads $\mathbf{A = 0}$, $\mathbf{B = 1}$, $\mathbf{F = 0}$.

The third input combination, illustrated in Figure 5.28(c), is analogous to the case just examined, with one input high and one low. The same reasoning leads us to conclude that the output must be at zero volts. Consequently the third line in the truth table reads $\mathbf{A = 1}$, $\mathbf{B = 0}$, $\mathbf{F = 0}$. The last possible combination of inputs is illustrated in Figure 5.28(d). Let us again first suppose that $\mathbf{F = 1}$, that is, that v_F is in the range of 4 to 5 V. If, in particular, $v_F = 5$ V, then both diodes as well as the resistor have zero volts across their terminals. No current would flow through any part of the circuit. There is no inconsistency in this state of affairs; hence, the output is $\mathbf{1}$ if both inputs are $\mathbf{1}$. As a check, let us instead assume that the output is $\mathbf{0}$, that is, in the range 0 to 1 V. If this were the case, a current of magnitude $4/R$ to $5/R$ flows up through the resistor R, yet both diodes are reverse biased by 4 to 5 V, and as a result can conduct no current. Clearly the assumption that the output is low when both inputs are high leads to an inconsistency; a current would flow up through R, yet no current can flow through either diode. The last line in the truth table must therefore read $\mathbf{A = 1}$, $\mathbf{B = 1}$, $\mathbf{F = 1}$. The completed table is given in Figure

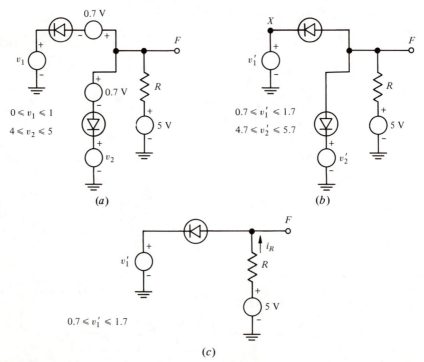

(a)

(b)

(c)

Figure 5.29 The diode **AND** gate redrawn using the large-signal diode model. (a) The circuit with input **A** assumed to lie in the range of logical $\mathbf{0}$, and input **B** in the range of logical $\mathbf{1}$. (b) Circuit of (a), simplified. (c) Circuit of (b), simplified.

5.27(*b*). This truth table is characteristic of an **AND** gate; the output is high only when both inputs **A** *and* **B** are high.

Having determined that the circuit of Figure 5.27 is an **AND** circuit, we should now go back and compute more carefully the output voltage for at least one of the various possible input combinations. Two refinements must be made in the analysis: First, the full range of possible input values, rather than just the extremes, of 0 and 5 V should be considered; second, a more realistic diode model should be used. The large-signal diode model is particularly appropriate for digital circuit calculations, because generally diodes in such circuits are either fully on or fully off, avoiding the region of low forward bias where the model is weakest.

Let us for example reconsider the case represented by line 2 in the truth table. The circuit is redrawn, using the large-signal diode model in Figure 5.29(*a*). A simplification is immediately possible; the voltage sources may be combined as in Figure 5.29(*b*). Now since we know the circuit is intended to be an **AND** gate, we strongly suspect that the output is low. We saw in the approximate analysis given previously that the diode whose input is low prevents the output from rising above the input potential; otherwise a voltage would appear across a forward-biased perfect rectifier. Similarly, it is clear from Figure 5.29(*b*) that the potential of point F cannot rise above the potential of point X. The exact potential of the second input plays no role; the diode in this leg is always reverse biased if the input is high. Consequently the circuit may be further simplified, and takes on the form shown in Figure 5.29(*c*). The current i_R, which equals $(5-v_F)/R$, flows up through the resistor R and to ground through the forward-biased diode. Since zero volts appears across the perfect rectifier, the output voltage falls somewhere in the range of 0.7 to 1.7, according to the exact value of the input voltage.

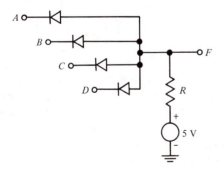

Figure 5.30 A four-input diode **AND** gate.

It is noteworthy from this analysis that the output does not always lie in the allowed logical **0** range of 0 to 1 V; if the low input is at the top end of the low range (1 V), then the output is close to 1.7 V. It is a characteristic property of diode logic circuits that the signal levels are degraded in this manner. For this reason an active element, namely, a transistor, is always included in logic circuits to restore the proper logic levels. Diode-transistor logic circuits are taken up in detail in Chapter 8 and an exact computation is made there for the output under all possible input conditions. The discussion of the effects of any loading on the output is also deferred until Chapter 8, because the transistors normally included in logic circuits modify loading effects.

The circuit shown in Figure 5.27 is a two-input **AND** gate. With only the addition of extra diodes, the number of inputs may be increased to three, four, or even more. A four-input **AND** gate is shown, for example, in Figure 5.30. This circuit functions much like the two-input circuit already considered. If any one of the four inputs is low, then the output must be low, or else an impossible forward voltage appears across the diode connected to the low input. As a consequence *all* of the inputs must be high if the output is high and the circuit functions as an **AND** gate.

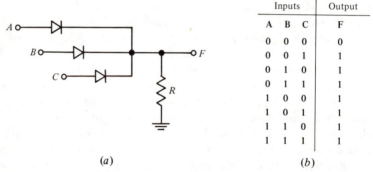

Inputs			Output
A	B	C	F
0	0	0	0
0	0	1	1
0	1	0	1
0	1	1	1
1	0	0	1
1	0	1	1
1	1	0	1
1	1	1	1

(a)   (b)

Figure 5.31   A diode **OR** gate. (a) The circuit. (b) The truth table. The output is 1 if any of the inputs is 1.

It is also possible to construct an **OR** gate using diodes as logic elements; an example of such a circuit is shown in Figure 5.31. Let us briefly examine the operation of this **OR** gate. Suppose one or all of the inputs is in the range of logical **1,** that is, 4 to 5 V. If the output is assumed to be low, a contradiction results; whichever diode (or diodes) is connected to the high input would have an impossibly large (3 to 5 V) forward bias. On the other hand, if the output is assumed to be high, no inconsistency results; whichever diode (or diodes) is forward biased supplies the current flowing down through R which maintains the 4 to 5 V output voltage. The output is low only when all of the inputs are low; in this case no currents flow in any part of the circuit.

EXAMPLE 5.10
    Find the output voltage of the **OR** circuit of Figure 5.31 under the condition that one input equals 4.5 V and the other two inputs are at ground potential.

SOLUTION

The circuit is redrawn below with the diode replaced by the large-signal diode model, and with the stated input voltages. We know from the approximate analysis that the output is high under these conditions. Thus the two lower diodes are reverse biased;

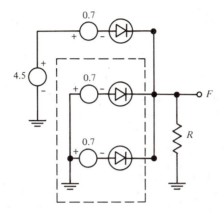

hence the part of the circuit within the dotted line is effectively an open circuit and may be ignored. It is seen immediately then that the upper diode is forward biased. The voltage drop across the perfect rectifier is zero, and the output voltage equals 3.8 V. This voltage is not strictly within the range of logical **1**. We see that the diode **OR** gate, like the **AND** gate, somewhat degrades the signal voltages.   ■

Most logic circuits operate in the range of a few to ten volts maximum signal, and currents are restricted to the range of a fraction of a milliampere to a few milliamperes. Thus, in contrast with rectifier diodes, power dissipation and voltage breakdown play a minor role in the design or selection of diodes for logic applications. The most important parameter for logic diodes is the speed at which they may be turned on and off. Naturally the faster that switching occurs, the faster the logic operations may be performed.

Summary

- Semiconductor devices are widely used in modern electronic circuits. Silicon is the most important semiconductor.

- Current flow in semiconductors arises from the motion of either electrons or holes. The inclusion of donor impurities, from group V of the periodic table, results in an *n*-type semiconductor, in which the electron concentration exceeds the hole concentration. Conversely, the inclusion of acceptor impurities, from group III of the periodic table, gives rise to *p*-type material, in which the hole concentration exceeds the electron concentration.

- A *pn* junction is a semiconductor structure in which a *p*-type region is immediately adjacent to an *n*-type region. The *pn* junction is important both as a device, the *pn* junction diode, and as a subunit of more complex devices, such as transistors.

- The *pn* junction diode is a nonlinear device which conducts current more easily for one bias polarity than for the opposite polarity. Applying a positive voltage on the *p* side with respect to the *n* side produces forward bias, the polarity of easy current flow. In forward bias electrons and holes flow across the junction interface. In an n^+p junction the flow consists primarily of electrons, injected as minority carriers, into the *p*-type region. In reverse bias, in which a negative voltage is applied to the *p* side with respect to the *n* side, practically no current flows.

- The *I-V* characteristic of a *pn* junction is closely approximated by the ideal diode equation. Several idealizations are useful in circuit analysis. When very large signals are considered, the *pn* junction diode may be treated as a perfect rectifier, in which case the voltage drop in forward bias is ignored. A more accurate analysis is possible using the large-signal diode model which assumes a constant 0.7 V forward drop. For very small signals, the diode characteristic may be expanded as a Taylor series about the operating point.

- Rectifier circuits convert ac currents to dc. A major application of rectifiers is in power supplies which produce constant dc supply voltages for electronic equipment. Diode clamps and peak detectors are related circuits that perform special operations on ac signals.

- Simple logic circuits, performing operations such as **AND** and **OR,** can be constructed using diodes. However, diode logic circuits have a deficiency in that degradation of the signal levels may occur.

References

Treatments at approximately the same level as this book:

Brophy, J. J. *Basic Electronics for Scientists.* New York: McGraw-Hill, 1966.

Durling, A. E. *An Introduction to Electrical Engineering.* New York: The Macmillan Co., 1967.

At a more advanced level:

Gray, P. E. *Introduction to Electronics.* New York: John Wiley & Sons, 1967.

Gibbons, J. F. *Semiconductor Electronics.* New York: McGraw-Hill, 1966.

Problems

5.1 In the growth of a certain crystal of silicon two atoms of phosphorus are added for every 10^6 atoms of silicon. Find the conductivity type and the concentration of majority carriers. (A silicon crystal contains about 5×10^{22} atoms/cm³.)

5.2 It is shown in Appendix II that the equilibrium minority carrier concentration in a semiconductor equals n_i^2 divided by the majority carrier concentration. (1) Suppose a silicon crystal is doped with 5×10^{15} boron atoms/cm³. Find n and p. (2) Suppose a germanium crystal is doped with 5×10^{15} boron atoms/cm³. Find n and p.

5.3 Consider the apparatus shown in Figure 5.32. A cylindrical tank has a powerful fan in the center which keeps the gas on the right side of the tank from crossing to the left side. The left side, furthermore, is being exhausted by a vacuum pump. (1) Comment on the suitability of this apparatus as a model for the motion of electrons in a pn junction. Let the gas molecules represent electrons and ignore the holes. To what is the fan analogous? The vacuum pump? (2) A squirt of gas is injected into the system at point B, near the fan. What happens to it? What is this analogous to? What might be different if the squirt occurs at A instead of B?

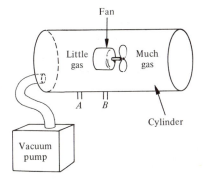

Figure 5.32

5.4 A certain diode has a saturation current equal to 10^{-13} A at room temperature. Calculate the current for applied forward voltages of -10, -1, -0.1, 0, $+0.1$, $+0.5$, and $+1.0$ V. Plot on linear graph paper. It is sometimes said that a diode conducts with positive forward bias and does not conduct significantly with negative forward bias. How well is this statement justified by your graph? Can you suggest a better statement?

5.5 It is shown in Appendix II that I_S varies strongly with temperature. In fact, for silicon diodes near room temperature I_S increases by about a factor of 6 for every 10°C increase in temperature. The current flowing in a diode is also dependent on temperature through the factor qV/kT in the exponential. (1) A diode with $I_S = 10^{-13}$ A at 300°K is forward biased by a constant voltage of 0.6 V. Does the current increase or decrease with temperature? By what factor does the current change for a 10°C change in temperature? (2) The same diode is forward biased by a constant-current source of value 1 mA. Does the forward voltage drop increase or decrease with temperature? What is the change in the forward voltage drop (in mV) for a 10°C change in temperature?

5.6   The diode in the circuit of Figure 5.33 has the *I-V* characteristic shown in Figure 5.15. (1) Calculate v_2 for the following values of v_1 (all positive): 0, 5, 10, 20, 30, 35, 36, 40, 50 V. (2) Plot v_2 versus v_1 based on the above calculation. (3) For what purpose might this circuit be useful?

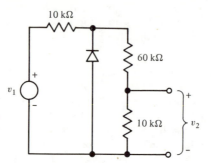

Figure 5.33

5.7   Find the solution, to an accuracy of at least three significant figures, to the problem of Example 5.2 using the Newton–Raphson algorithm. Use $i_D = 4.3 \times 10^{-3}$ A for an initial guess.

5.8   Find the voltage drop across the diode in the circuit of Figure 5.33 if $v_1 = -10$ V. Assume the diode has the characteristic given in Figure 5.14(*a*) ($I_S = 10^{-13}$ A) and use a graphical technique. *Suggestion:* Make a Thévenin equivalent of the voltage source and the three resistors.

5.9   Find the output of the circuit of Example 5.4 if the input is a "triangular wave" that has a peak positive value of 100 V, a peak negative value of -100 V, and a period of 10 msec. Comment on the suitability of the perfect rectifier model.

5.10   In a certain diode logic circuit, a diode with saturation current of 10^{-14} A is used. In normal operation, the diode is either reverse biased by 5 V or forward biased in the range of 0.5 to 3 mA. Compute the exact diode reverse current and forward voltage drop, and comment on the suitability of the large-signal diode model for use in analyzing such circuits.

5.11   Find v_2 in the circuit of Figure 5.33 for the case of $v_1 = -20$ V. (1) Use a graphical technique, assuming $I_S = 10^{-13}$ A. (2) Use the large-signal diode model.

5.12   Find the output of the circuit of Example 5.6, assuming the input is a sinusoidal signal of amplitude 100 V and frequency 100 Hz. Use the large-signal diode model.

5.13   Suppose a diode that obeys the ideal diode equation and has a saturation current of 10^{-14} A is used in the circuit of Example 5.7. Let $V_{BIAS} = 0.6$ V, $R = 10$ Ω, $A = 0.05$ V, $\omega_1/2\pi = 2 \times 10^6$ Hz, $B = 0.02$ V, and $\omega_2/2\pi = 3 \times 10^6$ Hz. Find the amplitudes of the components of the output voltage: (1) at 10^6 Hz, (2) at 2×10^6 Hz, (3) at 3×10^6 Hz, (4) at 4×10^6 Hz, (5) at 5×10^6 Hz, and (6) 6×10^6 Hz.

5.14   Suppose the voltage V_{BIAS} in Problem 5.13 were reduced to zero; how are the answers affected?

5.15   Find and sketch the output voltage v_2 of the circuit of Figure 5.21, as a function of time, for the case $R = 4k\Omega$, $C = 10^{-6}$ F, $v_1 = 50 \sin(2\pi ft)$. (1) Let $f = 50$ Hz (the European standard line frequency). (2) Let $f = 400$ Hz (a typical frequency used in portable electronic equipment). Comment on the relative difficulty of the filtering problem for inputs of different frequency.

5.16   Find the output waveform for the conditions of Problem 5.15 if the full-wave rectifier circuit of Figure 5.23 is used. (The same 10^{-6} F capacitor is placed in parallel with the load.)

5.17   The circuit of Example 5.9 is to be used to demodulate a signal. The signal frequency f_s lies in the range of 100 to 5000 Hz. The carrier frequency f_c is 110 kHz. Choose a suitable value for the capacitor C.

5.18   The input to the clamp circuit of Figure 5.26 is a "triangular wave" of amplitude 5 V (10 V peak-to-peak) and a period of 1000 Hz. Find the output. (1) Use the perfect rectifier. (2) Use the large-signal diode model.

5.19   A simple diode clipper circuit is constructed by paralleling the diode in the circuit of Example 5.5 with a second diode, connected oppositely. (The circuit is much like the circuit of Example 5.6, except that the voltage sources are removed.) Find and plot the output voltage of a clipper (the voltage across the diodes) as a function of the input voltage, as the latter varies over the range -10 volts to $+10$ volts. (1) Use the perfect rectifier concept. (2) Use the large-signal diode model. (3) Use a graphical method (at several values of the input voltage), assuming the diode characteristic of Figure 5.14(a) and a 1000-Ω resistance value.

5.20   Find the value of the output voltage of the **AND** gate of Figure 5.27 for the following input combinations: (1) $v_A = 0$ V, $v_B = 0$ V; (2) $v_A = 1$ V, $v_B = 0$ V; (3) $v_A = 5$ V, $v_B = 1$ V; (4) $v_A = 5$ V, $v_B = 4$ V; (5) $v_A = 5$ V, $v_B = 5$ V. Use the large-signal diode model, and assume $R = 1000$ Ω.

5.21   Find the output voltage for case (3) and case (4) of Problem 5.20, using a graphical technique. Assume $I_S = 10^{-13}$ A. Compare your results with the approximate results obtained using the large-signal diode model.

5.22   Find the truth table for the diode **AND** gate of Figure 5.30.

5.23   Find the output voltage of the diode **OR** gate of Figure 5.31 for the following input combinations: (1) $v_A = 0$ V, $v_B = 0$ V, $v_C = 1$ V; (2) $v_A = 1$ V, $v_B = 1$ V, $v_C = 1$ V; (3) $v_A = 4$ V, $v_B = 1$ V, $v_C = 5$ V; (4) $v_A = 5$ V, $v_B = 5$ V, $v_C = 5$. Use the large-signal diode model.

5.24   Find and plot the current i in the circuit of Figure 5.34 over the input voltage range of -20 to $+20$ V. Assume the diodes are identical. (1) Use the perfect rectifier concept. (2) Use the large-signal diode model. (3) Use the ideal diode equation. (This simple circuit is very important in integrated circuits.)

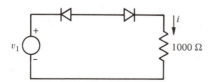

Figure 5.34

5.25   An "improved" large-signal diode model would include a series resistance to account for the slight increase of forward voltage drop with current. Suppose we make an improved model consisting of a perfect rectifier, a voltage source (not necessarily 0.7 V), and a resistor, all in series. Find a suitable value for the resistance and for the voltage source to model accurately the current-voltage characteristic of a diode with $I_S = 10^{-14}$ A over the range of 0 to 1 mA forward bias. *Suggestion:* A graphical approach is suggested in which the characteristic of the diode and the characteristic of the model are plotted.

5.26   A more complicated diode logic circuit is shown in Figure 5.35(a). A possible set of inputs is given in Figure 5.35(b). (1) Find the output voltage v_F over the time interval shown, assuming that the switch S_1 is open and assuming that the diodes are perfect rectifiers. (2) Find the output voltage v_F with the switch closed, again assuming that the diodes are perfect rectifiers. (3) Find v_F with the switch closed, using the large-signal diode model. (4) Construct the truth table for this circuit, assuming 0 to 1 V is logical **0**, and 4 to 5 V is logical **1**. (Ignore the degradation of the signal levels.)

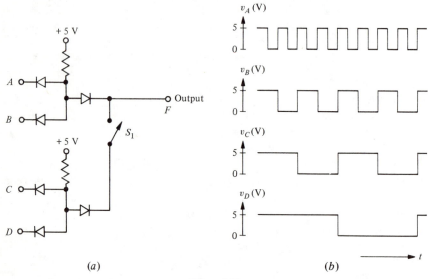

(a)   (b)

Figure 5.35

5.27   It is sometimes useful to define a "small-signal resistance" of a *pn* junction by the relationship $r \equiv dv/di$. (1) Find an exact expression for r in terms of the total diode current i (not the voltage) and other constants. (2) Simplify the expression in the case $i \gg I_S$. (3) Compute r in a diode at a forward bias of 1 mA. (If the expression derived in (2) is correct, r does not depend on I_S.) Assume $T = 300°$K. (4) What is the approximate value of r at a *reverse* bias of several volts? (ANSWER: (3) 26 Ω)

5.28   Figure 5.36 shows a famous electronic "trick" circuit in which rectifiers are concealed in the base of bulbs B_1 and B_2 as well as in switches S_1 and S_2. (1) Describe the action of the circuit for all combinations of switch positions. What happens when one bulb is unscrewed? When the bulbs are reversed in their sockets? (2) What happens if one bulb is screwed in an ordinary light socket?

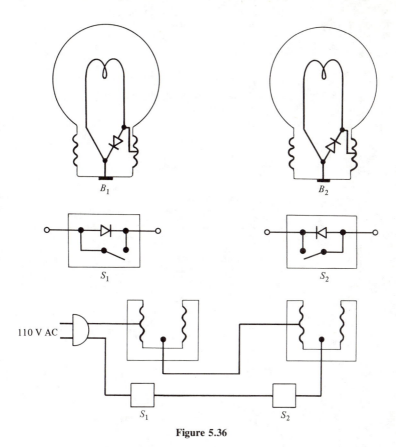

Figure 5.36

5.29   The circuit of Figure 5.37 is known as a "voltage doubler." Sketch the output voltage as a function of time if the capacitors are initally uncharged, and $v_1 = 100 \sin \omega t$. Explain the action of the circuit. *Suggestion:* Treat the diodes as perfect rectifiers.

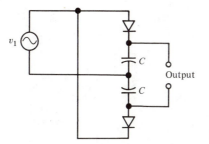

Figure 5.37

5.30   The diode whose characteristics are given in Figure 5.38(a) is known as a tunnel diode. Using a graphical technique, find the diode voltage when operating in the circuit of Figure 5.38(b). (1) Let $v_1 = 0.7$ V, $R_1 = 200$ Ω. (2) Let $v_1 = 5$ V, $R_1 = 10\text{k}\Omega$. If more than one answer is obtained in (2), the answers should be examined to see if they are stable. (3) How might the tunnel diode be used as a "memory" device?

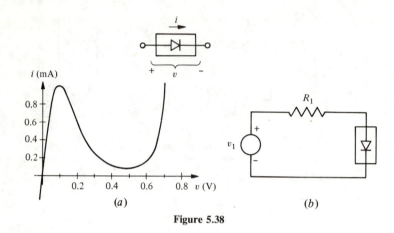

(a)   (b)

Figure 5.38

Chapter 6

Junction Transistors

The principal active circuit element in modern electronics is the bipolar junction transistor, or "BJT," more familiarly known simply as the transistor. The transistor is small, reliable, and cheap; most important, it can be integrated into circuits that are built entirely within a small piece of silicon (see Chapter 7). Although other types of transistor devices are of growing importance (see, for example, Chapter 13), the conventional transistor is, and will remain for some time, the single most important electronic device. In applications ranging from phonographs to computers, transistor circuits play a dominant role.

This chapter develops those relationships necessary for the analysis of circuits containing transistors. The required relationships, which we have called the constitutive relationships (or *I-V* relationships), are the equations relating the various currents and voltages at the transistor terminals. The transistor is remarkable in that the *I-V* relationships may be easily developed from a very elementary understanding of its operation. Thus we begin in Section 6.1 with a physical description of the transistor and a discussion of the principles of transistor action. In Section 6.2 the *I-V* relationships for the transistor are derived, and a graphical presentation of the *I-V* characteristics is made. The range of applied bias is restricted in this section to that corresponding to the application of transistors in typical analog circuits such as amplifiers. In such cases the transistor is said to be operating in the active mode.

In digital circuits the transistor generally operates over a wider range of current and voltage, and it is necessary to derive *I-V* relationships that are valid over the full operating range. Thus the subject of Section 6.3 is the so-called large-signal transistor operation. The terminal *I-V* relationships are developed, and the more important *I-V* characteristics are graphed. Finally, in Section 6.4, the characteristics of real transistors are presented, and compared with the theoretical results of Sections 6.2 and 6.3. The most significant limitations of the equations in describing the terminal properties of transistors are discussed.

6.1   Transistor Principles

The bipolar transistor consists of a semiconductor crystal containing two junctions in close proximity, as depicted in Figure 6.1(*a*). (This drawing is only schematic; the various regions are not shown in correct proportion.) To see how operation takes place, let us recall from Chapter 5 that *pn* junctions can both inject and collect minority carriers.

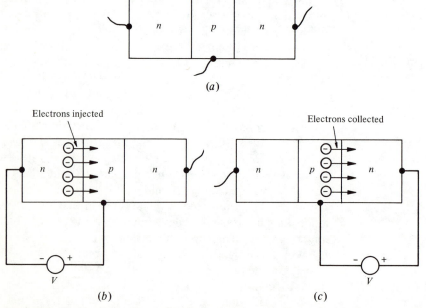

Figure 6.1 An *npn* transistor and the basic processes involved in its operation. (*a*) The transistor consists of a crystal containing two junctions in close proximity. (*b*) When one of the junctions is forward biased, it can inject electrons into the *p* region. (*c*) When one of the junctions is reverse biased, it can collect the electrons.

Let us assume (for reasons which will become clear) that the *n* region to the left in Figure 6.1(*a*) is more heavily doped than the *p* region. If such an n^+p junction is forward biased, the more heavily doped *n* side *injects* (or "emits") carriers into the lightly doped *p*-type material, where they become minority carriers. This behavior is illustrated in Figure 6.1(*b*).

A *pn* junction can also act as a collector for minority carriers that approach the boundary of the *p* and *n* regions. Minority carriers, wandering up to the interface of a *pn* junction, are swept across the interface by the electric field present there. Figure 6.1(*c*) illustrates a situation in which minority electrons somehow have been introduced into the *p* region, and are swept into the *n* region. Not all of the minority carriers that are introduced into the *p* region are necessarily collected, as they may instead recombine with any of the numerous holes present there. If the junction is reverse biased, then essentially the only current that flows consists of minority carriers, which, having somehow gotten into the *p* region, wander up to the interface and are swept across. Even if the junction is forward biased, there is still a sufficient field in the junction to collect any minority carriers that approach it. However, in forward bias, the usual forward current flows in addition to (and in the opposite direction from) any current of collected minority carriers.

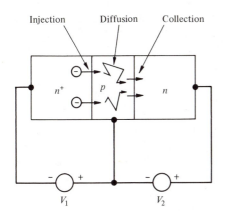

Figure 6.2 The operation of a transistor. An n^+p junction is forward biased and therefore injects electrons into the *p* region. The electrons wander about in the *p* region by diffusion. Some of them reach the opposite junction (collecting junction) before recombining with holes, and therefore are swept on into the *n* region.

A transistor is made by placing two junctions close enough together that carriers injected by one may be collected by the other. Referring to Figure 6.2, electrons injected into the *p* region by the emitting junction diffuse to the collecting junction and are collected (swept into the *n* region). The source of minority carriers (the n^+ region on the left in this example) is called the *emitter*. The other *n* region is called the *collector*. The region through which the minority carriers must diffuse to pass from emitter to collector (the *p* region in this example) is called the *base*. The base is essentially a "hostile territory" through which the electrons must journey before reaching the collector. Subsequent analysis will show that it is desirable to collect as large a fraction as possible of the emitted carriers. With a very narrow base, practically all of the injected minority carriers are collected; therefore the base is made as narrow as possible. Its width is typically in the range of 10^{-5} to 10^{-3} cm, which is small enough so that the minority carriers can diffuse across it before they die by recombination with holes.[1]

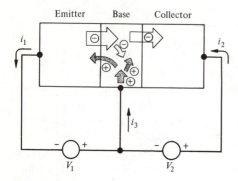

Figure 6.3 Current flow in the transistor of Figure 6.2. The electrons flowing from emitter to collector produce a current i_2 into the collector and i_1 out of the emitter. Because most of the emitter current consists of electrons that transit the base and flow out the collector, i_2 is only slightly smaller than i_1. Consequently, the current i_3 flowing into the base is quite small. It consists of a small flow of holes, some of which are injected into the emitter and some of which recombine with electrons in the base.

Most of the current through the emitter-base junction consists of electrons injected into the base, and most of these electrons reach the collector-base junction and flow out through the collector. Consequently, when the currents are measured in the external wires connected to the three regions (Figure 6.3), it appears that most of the current in the transistor flows directly from collector to emitter (in the direction opposite to that of the electron motion). Large currents can flow in the circuitry connected to the collector and emitter, while only a relatively small current flows in the base circuit. A base current that is only 1% of the collector current is typical. The very useful control, or amplification, action of the transistor arises out of the fact that a small base current may be used to control much larger currents in the emitter or collector circuits. Moreover, the voltage required to produce the emitter current (the emitter-base voltage) may be much smaller than the voltage produced by the collector current as it flows through an external load resistance. Thus the bipolar transistor can produce both current gain and voltage gain.

The structure which has just been described is called an *npn transistor,* meaning that the emitter and collector regions are *n*-type and the base *p*-type. Normally, the mode of construction is not a simple arrangement of parallel planes, as suggested by Figure 6.1, although this is a useful approximation. A more typical configuration of an *npn* transistor is shown in Figure 6.4. The methods of planar processing discussed in Chapter 7 may be used to fabricate such a structure: A *p*-type base region and an *n*-type emitter region are in-

---

[1] The distance that minority carriers can diffuse before dying by recombination, called the diffusion length, is normally in the range of 10^{-4} to 10^{-2} cm for silicon, hence the requirement for a very narrow base width. However, in most transistors a small electric field of appropriate sign is provided in the base to assist, that is, speed up, the movement of the electrons through the base.

corporated into an *n*-type crystal, and suitable contacts are applied. An oxide layer covers and protects the top surface, except in the region of the contacts. Although the structure of Figure 6.4 is more realistic, we shall continue to use the simpler diagram of Figure 6.1 where we are concerned only with the principles of operation.

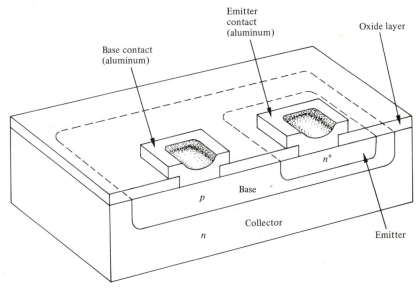

Figure 6.4   A cross section of a planar *npn* silicon transistor. The base region is incorporated into the collector region, and the emitter is incorporated into the base. A protective oxide covers the surfaces, except where contacts are made to the emitter and base. The collector contact is applied to the bottom.

The complementary structure to the *npn* device is called a *pnp transistor,* which operates analogously. In both cases a forward-biased emitter injects minority carriers into the base. The minority carriers diffuse across the base and are collected by the collector. In a *pnp* transistor, the heavily doped *p*-type emitter emits holes into the *n*-type base. The holes diffuse across the base and are collected by the *p*-type collector. As in the *npn* transistor, most of the emitter current flows through the collector, with only a very small base current. However it is noteworthy that the direction of current flow in *npn* transistors is opposite to that in *pnp* transistors. In the former the current is by a flow of negatively charged electrons from emitter to collector, while in the latter the current is by a flow of positively charged holes, again from emitter to collector.

Having discussed briefly the physical operation of the transistor, we now consider its operation in electronic circuits. It is desirable to adopt some conventions for the signs of the various circuit parameters, that is, the terminal currents and voltages, and to adopt a circuit symbol.

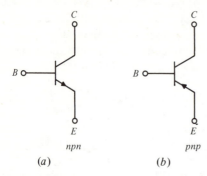

Figure 6.5 The standard circuit symbols for bipolar transistors. The transistor type, (a) *npn* or (b) *pnp*, is identified by an arrow on the emitter that indicates the direction of current flow in normal operation.

Transistor Conventions and Symbols   The standard circuit symbols for bipolar transistors are given in Figure 6.5. The emitter has an arrow indicating the actual current direction under normal operation, in which the emitter injects minority carriers into the base. For example when the emitter-base junction of an *npn* transistor, Figure 6.5(*a*), is forward biased, the electrons flowing from the emitter toward the collector constitute a current flowing out of the emitter. Hence the symbol for an *npn* transistor has an arrow pointed outward from the emitter. Similarly, in normal operation of a *pnp* transistor, there is a flow of holes from emitter to collector. Since holes carry a positive charge, this flow constitutes a current flowing into the emitter, and the symbol of Figure 6.5(*b*) is adopted.

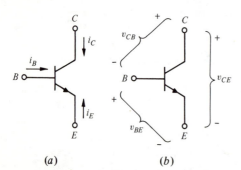

Figure 6.6 The terminal variables for transistors. (*a*) There are three terminal currents that are always defined as positive when directed inward for both *npn* and *pnp* transistors. (*b*) There are three interterminal voltages whose signs are defined by the order of the subscripts.

To specify completely the circuit currents and voltages of a three-terminal device, three terminal currents and three interterminal voltages must be specified. For the transistor, then, the emitter, base, and collector currents, as well as the emitter-to-base, emitter-to-collector, and base-to-collector voltages, must be specified. Independent of the actual direction of current flow, all the terminal currents are conventionally defined as *positive when directed inward,* as illustrated in Figure 6.6(*a*). Voltages between the terminals are named according to the formula $v_{ij} = v_i - v_j$. (For example, v_{BE} stands for the potential of the base minus that of the emitter.) The interterminal voltages are labeled according to this notation in Figure 6.6(*b*). As in the case of the current definitions, the *defined* polarities are identical for *pnp* and *npn* transistors. Of course the *actual* direction of current flow and the polarity of the interterminal voltages do differ for the two transistor types. Consequently, the signs of the variables differ; for example, i_E is normally positive for *pnp* transistors and negative for *npn* transistors. We see that one must specify six variables (three currents and three voltages) to describe the precise operating point of a transistor. However, two relationships between the variables can be obtained directly from the basic circuit laws discussed in Chapter 1. From Kirchhoff's current law

$$i_E + i_B + i_C = 0 \qquad (6.1)$$

and from Kirchhoff's voltage law

$$v_{EB} + v_{BC} + v_{CE} = 0 \qquad (6.2)$$

These equations are valid for both *npn* and *pnp* transistors. The problem of finding the operating point of a transistor has been reduced to finding the values of four of the six terminal variables. Equations (6.1) and (6.2) then specify the values of the remaining two variables.

The analysis to determine the transistor operating point is easiest when certain restrictions are placed on the range of the terminal variables. These restrictions are not arbitrarily invented, but correspond to particular modes of operation that occur in real circuits. For example, in amplifier circuits (as well as many other kinds of circuits) transistors are generally operated in the so-called active mode.

6.2 Transistor Operation in the Active Mode

The active mode of transistor operation is the mode in which only the emitter injects carriers into the base. In other words, in the active mode only the emitter-base junction is forward biased. Let us consider specifically the *npn* transistor.[2] For the active mode of operation v_{BE} must be positive so that electrons are emitted into the *p*-type base region. (Remember, the *p* side of a junction is positive with respect to the *n* side in forward bias.) The biasing polarity, and the corresponding particle flow, are indicated in Figure 6.7(*a*). The actual current directions are indicated by the signs of i_E, i_B, and i_C in Figure 6.7(*b*). The voltage source V_1 must be sufficiently positive to cause a current of usable magnitude to flow through the emitter-base junction. The emitter current consists almost entirely of electrons injected into the base,

---

[2] The *npn* transistor is chosen as the principal vehicle for explanation of the transistor because in practice, primarily for reasons related to ease of fabrication, it is used more often than the *pnp* transistor.

which subsequently diffuse across to the collector-base junction [Figure 6.7(a)]. The collector-base junction is either zero biased or reverse biased ($v_{CB} \geq 0$), and therefore in the absence of an emitter current the collector current would be essentially zero.[3] However, in the presence of an emitter current, the electrons injected into the base by the emitter diffuse across the base and are collected by the collector. Hence, as illustrated in Figure 6.7(a), negative charges enter the collector, whence they must flow out the collector terminal of the transistor. Therefore a collector current flows. The collector current is positive in the active mode of operation.

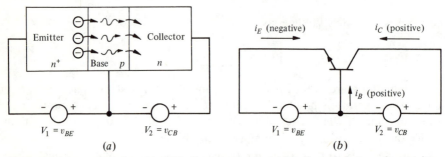

(a)   (b)

Figure 6.7   An *npn* transistor biased in the active mode. (*a*) The flow of electrons from emitter to collector. (*b*) The circuit shown with appropriate polarity for active mode operation. The currents are all defined as positive inward; however the actual emitter current direction is outward, hence i_E is negative.

Transistor Relationships in the Active Mode   We shall now obtain mathematical expressions relating the various terminal currents and voltages of a transistor operating in the active mode. From these expressions it is possible to plot the transistor characteristics, in other words, to display the relationship between the various currents and voltages graphically. In a graphical display the restricted range of currents and voltages corresponding to the active mode is referred to as the active region.

We consider first the collector current, which as we have seen originates at the emitter-base junction. The major fraction of the emitter current consists of minority carriers that cross the base and flow out the collector. This fraction is conventionally given the symbol α. Remembering that current flow is opposite to electron flow, and that all currents are defined as inward [Figure 6.7(b)], we may write

$$\boxed{i_C = -\alpha \, i_E}$$
(6.3)

---

[3] Recalling *pn* junction operation in reverse bias, we realize that for zero emitter current a small "leakage" current (typically 10^{-8} to 10^{-11} A) would flow in the reverse-biased, collector-base junction of a silicon transistor. We are neglecting this leakage.

This equation reflects the fact that in the active mode the collector current is positive, while the emitter current is negative. The quantity α depends primarily upon geometrical and doping parameters. It also depends on the magnitude of i_E, as well as on the temperature of the transistor. Typical values of alpha in modern transistors are in the range of 0.98 to 0.999.

Equation (6.3) holds for both *npn* and *pnp* transistors, so long as operation is in the active region. Another form of Equation (6.3) can be obtained. Using Equation (6.1) to substitute for i_E in Equation (6.3), we have

$$\boxed{i_C = \beta \, i_B} \tag{6.4}$$

where we have defined a new quantity β by

$$\boxed{\beta = \frac{\alpha}{1 - \alpha}} \tag{6.5}$$

Again, Equations (6.4) and (6.5) are valid for both *npn* and *pnp* transistors in the active region. Beta, like alpha, is a characteristic parameter of each transistor. As Equation (6.4) suggests, it may be easily measured by finding the ratio of collector to base current when the transistor is operating in the active mode.

It was stated previously that alpha, for silicon transistors, is typically in the range of 0.98 to 0.999. Hence, from Equation (6.5), the value of beta is typically in the range of 50 to 1000.

Let us now turn our attention to the emitter. Since the emitter-base junction is forward biased in the active mode, the base-to-emitter voltage is just the voltage drop across a forward-biased *pn* junction. From the *pn* junction diode equation, Equation (5.1), we know that in forward bias the emitter current must be related to the emitter-base voltage by a relationship of the form

$$i_E = -I_{ES}\left(\exp\left\{\frac{q \, v_{BE}}{kT}\right\} - 1\right) \qquad [npn]^4 \tag{6.6}$$

where I_{ES} is some constant. The negative sign arises from the peculiar definition of current direction, namely, inward into the emitter, Figure 6.6. The constant I_{ES} is then positive. It has a typical value on the order of 10^{-14} A; hence v_{BE} must be about 0.7 V in order to produce 1 mA of emitter current (at room temperature). It is a simple matter to determine experimentally I_{ES}: the emitter current is measured at a given voltage and Equation (6.6) is then solved for I_{ES}.

Often it is desired to have an expression for the base current as a func-

---

[4] Certain equations are valid for transistors of only one type, for example, Equation (6.6), which is valid for *npn* transistors. The analogous equation to Equation (6.6) for *pnp* transistors is given in Table 6.2, and involves only sign changes.

tion of v_{BE}. It is simplest to relate the base current to the emitter current, and use Equation (6.6). From Equations (6.1) and (6.4)

$$i_B = -\frac{i_E}{\beta + 1} \tag{6.7}$$

In other words, the magnitude of the base current is just the emitter current, divided by the large factor $(\beta + 1)$. Again, the minus sign arises from the current conventions.

EXAMPLE 6.1

Establish that the transistor in the circuit in sketch (a) is operating in the active mode, and find the base current. Assume that β equals 99 for this transistor.

SOLUTION

In the active mode the emitter-base junction is forward biased and the collector-base junction reverse biased. The current source I_1 causes a negative emitter current to flow. This current, which flows in the direction of the emitter arrow, has the proper sign for forward biasing the emitter-base junction. [The resulting forward voltage drop may be obtained from Equation (6.6), if desired.] The collector-base voltage is simply V_2 (see figure). Hence the collector junction is reverse biased by 15 volts. With the emitter-base junction forward biased, and the collector-base junction reverse biased, the transistor is indeed operating in the active region. We may solve for i_B using Equation (6.7):

$$i_B = -\frac{i_E}{\beta + 1}$$

$$i_B = -\frac{-10^{-3}}{100} = 10^{-5}\ \text{A}$$

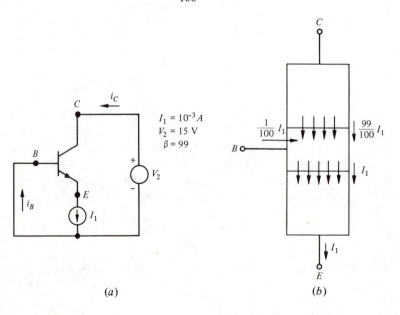

$I_1 = 10^{-3}\,A$
$V_2 = 15\ \text{V}$
$\beta = 99$

(a)   (b)

We may also find i_C from Equation (6.4):

$$i_C = \beta\, i_B$$
$$i_C = 99 \times 10^{-5} = 0.99 \text{ mA}$$

Sketch (*b*) shows the *currents* flowing in the transistor. The manner in which i_E divides into i_B and i_C should be particularly noted. (Since the carriers are negatively charged electrons, the motion of the carriers is opposite to the direction of the arrows.) ∎

In Example 6.1 it is demonstrated that in the active mode of operation, the emitter current gives rise to a base current many times smaller (in the example, 100 times smaller). This result is not surprising, since the base current is related to the emitter current by Equation (6.4). Therefore the reverse statement is also true: In normal active operation, a small base current implies an emitter current that is many times larger. Similarly, the collector current is much larger than the base current. Thus we see that in the active mode of operation, the transistor can be used as a current amplifier by means of which a small base current can be used to control a larger current in the emitter or collector circuits. When used this way the transistor is said to have current gain.

Another interesting point about operation in the active mode is that a voltage gain is possible. It requires only about 0.7 V to forward bias the emitter-base junction of a silicon transistor. Furthermore, to produce large changes in the emitter current requires only a fraction of this forward bias. From the fact that the emitter-base junction has an exponential *I-V* characteristic, that is, $i_E \sim \exp\{qv_{BE}/kT\}$, we see that emitter current doubles for every 18-mV increase in v_{BE} at room temperature. Thus a small change in the emitter-base voltage results in a larger change in the emitter current, and therefore also the collector current. If the collector current flows through a large load resistance, a large change in load voltage results from the change in collector current. Therefore an appreciable voltage gain (the change in load voltage divided by the change in the input voltage) will result.

EXAMPLE 6.2

The transistor in the following circuit is operating in the active mode. Find the base current, collector current, and collector-to-emitter voltage, assuming that I_{ES}

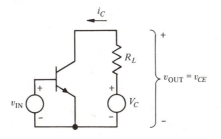

$= 10^{-14}$ A. Find the change in collector-to-emitter voltage assuming that the input voltage increases by 0.018 V. $v_{IN} = 0.7$ V; $R_L = 1000\ \Omega$; $V_C = 15$ V; $\beta = 200$; $kT/q = 0.026$ V. (This value corresponds to $T = 300°$ K, that is, room temperature.)

SOLUTION

The emitter-base junction is forward biased by 0.7 V. According to Equation (6.6) this would give

$$i_E = -10^{-14} \exp \left\{ \frac{0.7}{0.026} \right\} = -5 \times 10^{-3}\ \text{A}$$

Since $\beta = 200$, alpha is close to one, in fact from Equation (6.5), $\alpha = \beta/(1 + \beta) = 0.995$. Thus i_C is nearly equal in magnitude to i_E, that is, $i_C \simeq 5 \times 10^{-3}$ A. We can solve for v_{CE} by a single loop equation:

$$v_{CE} - V_C + i_C R_L = 0$$

Thus

$$v_{CE} = 15 - 5 \times 10^{-3} \times 1000 = 10\ \text{V}$$

This voltage is sufficient to assure that the transistor is indeed biased in the active region because $v_{CB} = v_{CE} - v_{BE} = 9.3$ V, and the collector-base junction is therefore reverse biased. The base current is given by

$$\frac{i_C}{\beta} = 2.5 \times 10^{-5}\ \text{A}$$

We now recompute the output voltage with an 18-mV increase in the input voltage. Such an increase approximately doubles the emitter current.

$$i_E = -10^{-14} \exp \left\{ \frac{0.718}{0.026} \right\} = -10^{-2}\ \text{A}$$

Therefore the base current and collector current double. From the loop equation given above $v_{CE} = 15 - 10^{-2} \times 10^3 = 5$ V. (Note that the transistor remains in the active mode.) It is interesting that for a change in v_{BE} of 0.018 V, there is a change in v_{CE} of -5 V. That is, input voltages are multiplied by a factor of $-5/0.018 = -280$. ∎

For future reference, we summarize in Table 6.1 the polarity of the various transistor terminal parameters. The voltage polarities are obtained from the requirement that the emitter-base junction be forward biased and the collector-base junction be reverse biased. The emitter current direction is obtained from the emitter-base forward bias requirement, and the base and collector current directions are obtained from Equations (6.3) and (6.4).

Relationships between emitter current, or base current, and emitter-base junction voltage are called *input relationships*. The equations relating the collector current to the emitter or base current, or relating the emitter current to the base current, are called *output relationships*. The input and output relationships, as well as Equations (6.1) and (6.2), are summarized in Table 6.2 for transistors operating in the active mode.

Table 6.1 Bias Polarities for a
Transistor Operating in the Active
Mode

Parameter	Polarity	
	npn	*pnp*
v_{CE}	Positive	Negative
v_{BE}	Positive	Negative
v_{CB}	Positive	Negative
i_E	Negative	Positive
i_B	Positive	Negative
i_C	Positive	Negative

It should be realized that all of these equations are not independent. It will be shown in the next section that there are only four independent equations for *npn* transistors (and similarly only four for *pnp*). The remaining equations are variations, stated for convenience.

Table 6.2 Input and Output Relationships for Transistors Operating in the Active Mode

	npn	*pnp*
Input Relationships	$i_E = -I_{ES}(\exp\{qv_{BE}/kT\} - 1)$ $i_B = -i_E/(\beta + 1)$	$i_E = I_{ES}(\exp\{qv_{EB}/kT\} - 1)$ $i_B = -i_E/(\beta + 1)$
Output Relationships	$i_C = \beta\, i_B = -\alpha i_E$ $i_E = -(\beta + 1)\, i_B$	$i_C = \beta\, i_B = -\alpha i_E$ $i_E = -(\beta + 1)\, i_B$
Current law Voltage law	$i_E + i_B + i_C = 0$ $v_{EB} + v_{BC} + v_{CE} = 0$	$i_E + i_B + i_C = 0$ $v_{EB} + v_{BC} + v_{CE} = 0$

The Graphical *I-V* Characteristics   The results expressed by the equations in Table 6.2 may also be presented in a graphical form, much in the same way as the diode *I-V* characteristics were given graphically as well as analytically in Chapter 5. However, the fact that the transistor has three terminals makes the presentation considerably more complicated than for two-terminal devices. It is simplest to specify the conditions at one of the terminals and then proceed with the other two terminals as if they were the terminals of a two-terminal device. Suppose, for example that the base current were set to some fixed value, say 1 μA, as shown in Figure 6.8(*a*). We might then inquire about the collector current as a function of the collector-to-emitter voltage, in other words the i_C versus v_{CE} characteristic. Certainly this characteristic could be measured; it may also be calculated from Table 6.2. Alternatively, we might set the emitter-to-base voltage to some fixed value as in Figure 6.8(*b*), and similarly calculate or measure the i_C versus v_{CE} characteristics. There is nothing special about the two circuit arrangements shown in Figure 6.8; obviously many other combinations are possible. Which curves then are desired?

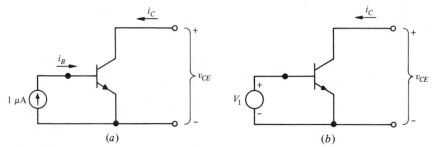

Figure 6.8   (*a*) An *npn* transistor with the base current fixed at 1 μA. (*b*) An *npn* transistor with base-to-emitter voltage fixed at the value V_1.

The answer depends upon the transistor application. For example, in many applications the input signal is applied between base and emitter, and the output signal is taken between collector and emitter. In such applications the characteristics of primary interest are the base current versus base-to-emitter voltage, and the collector current versus collector-to-emitter voltage.

The *input characteristic*, i_B versus v_{BE}, is plotted for constant v_{CE}, whereas the *output characteristic*, i_C versus v_{CE}, is plotted for constant i_B. Let us see how such plots are constructed, considering first the output characteristic. A typical value of base current is selected, and i_C is calculated (and plotted) as a function of v_{CE}. A different value of i_B is selected, and another curve i_C versus v_{CE} is plotted. In this way a whole family of such curves, each curve corresponding to a different value of i_B, is plotted on a single graph. We now draw such a graph for an *npn* transistor with typical values of $\beta = 100$, $I_{ES} = 10^{-14}$. Suppose i_B is fixed at 5 μA. Then from Equation (6.4) the collector current equals βi_B, or 0.5 mA. It happens that the collector current is independent of v_{CE}; hence the curve i_C versus v_{CE} amounts to a straight line, parallel to the voltage axis. This "curve" is plotted in Figure 6.9, and is labeled $i_B = 5$ μA. Also shown in the figure are plots of i_C versus v_{CE} for other assumed values of i_B. For example, when $i_B = 15$ μA, i_C is constant at 1.5 mA, again directly from Equation (6.4).

We are dealing only with transistor operation in the active region and the curves must not extend out of this region. When operating in the active mode, the transistor emitter-base junction is forward biased; hence v_{BE} is typically some fraction of a volt, say 0.6 to 0.8 V. The collector-base junction must be reverse biased, that is, v_{CB} must be positive. Now since v_{BE} may be as much as 0.8 V, v_{CE}, which equals $v_{CB} + v_{BE}$, must be at least 0.8 V. Thus the active region of operation consists of most of the first quadrant in Figure 6.9; only the region $v_{CE} < 0.8$ V is excluded.

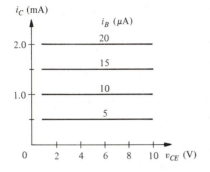

Figure 6.9 The output characteristics of an *npn* transistor biased in the active mode. The collector current is plotted versus the collector-to-emitter voltage for four different values of base current: 5, 10, 15, and 20 μA.

Figure 6.9 may be used to determine the collector current for any collector-to-emitter voltage in the range of 0.8 to 10 V, for any one of four base currents. In fact, by interpolating between the curves, the collector current may be computed for any base current in the range of 0 to 20 μA. Such a graph as Figure 6.9 is called a *parametric graph,* because an extra parameter (i_B), which is not one of the coordinates, is introduced.

We now consider the transistor input characteristics. In particular, let us plot i_B versus v_{BE} for various (fixed) values of v_{CE}. From the first two lines of Table 6.2 we have that for *npn* transistors,

$$i_B = \frac{I_{ES}}{1+\beta}\left(\exp\left\{\frac{q\,v_{BE}}{kT}\right\} - 1\right) \quad (npn) \tag{6.8}$$

This result is somewhat surprising in that i_B is independent of v_{CE}. Thus, instead of a family of curves, the input characteristics (in the active mode) consist of a single curve obtained directly from Equation (6.8). This single characteristic is plotted in Figure 6.10, again assuming the typical values of $\beta = 100$ and $I_{ES} = 10^{-14}$.

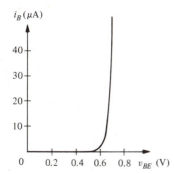

Figure 6.10 The input characteristic of an *npn* transistor biased in the active mode.

The base-to-emitter *I-V* characteristic in the active mode has the form of a typical *pn* junction characteristic, with negligible current below about 0.5 V and a steeply rising current in the neighborhood of 0.7 V. Therefore it is often appropriate to assume in circuit calculations that the base-emitter voltage drop is 0.7 V. Even when this approximation is unsatisfactory, it is certainly safe to assume that v_{BE} lies in the range of 0.5 to 0.8 V.

EXAMPLE 6.3

The circuit in sketch (*a*) is used to measure the input and output characteristics of an *npn* transistor. Suppose that the transistor has $\beta = 250$ and $I_{ES} = 10^{-15}$ A. Plot the following in the active region: (1) i_B versus v_{BE} (over the range $0 < i_B < 10\ \mu$A). (2) i_C versus v_{CE}, for $i_B = 10^{-6}$A, 2×10^{-6}A, and 3×10^{-6}A.

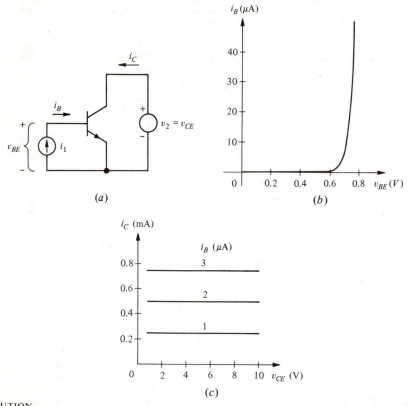

(*a*)

(*b*)

(*c*)

SOLUTION

The i_B versus v_{BE} characteristic is found from Table 6.2 or Equation (6.8). It is plotted in sketch (*b*) for the case $I_{ES} = 10^{-15}$ A. The curve is plotted only in the active region, that is, only for positive v_{BE}. The output characteristics, i_C versus v_{CE} for various values of i_B, may be plotted from Equation (6.4):

$$i_C = \beta i_B$$

The three curves, for the three specified values of i_B, are plotted in sketch (*c*). Again, the curves are restricted to the active region. To assure that v_{CB} is always positive, v_{CE} must be greater than or equal to v_{BE}, because

$$v_{CE} = v_{BE} + v_{CB}$$

From the input characteristic above, v_{BE} may be as large as 0.75 V; hence only the range $v_{CE} > 0.75$ V is covered in the plot of the output characteristics.   ■

EXAMPLE 6.4

The transistor of Example 6.1 was shown to be operating in the active mode. The circuit is repeated in sketch (a). We now consider varying V_2 and plotting i_C versus V_2, that is, i_C as a function of v_{CB}. Make such a plot over the active region for three values of I_1: 1, 2, and 3 mA.

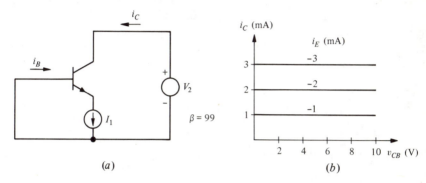

(a)   (b)

SOLUTION

V_2 is applied directly between collector and base. Since I_1 forward biases the emitter-base junction, it is only necessary that V_2 be positive to assure that the transistor is operating in the active mode. Hence the active region is the entire first quadrant of the i_C versus v_{CB} plot. Table 6.2 gives the collector current as a function of the emitter current.

$$i_C = -\alpha\, i_E$$

Alpha is obtained from Equation (6.5), and equals 0.99. Suppose $I_1 = 1$ mA, then $i_E = -1$ mA; hence i_C must equal 0.99 mA independent of v_{CB}. Similarly, when I_1 equals 2 mA, $i_C = 1.98$; and when $I_1 = 3$ mA, $i_C = 2.97$ mA. The graph is given in sketch (b).   ■

This section has described transistor operation in the active mode. A large class of circuits, such as linear amplifiers, use transistors operating in this mode. However, in digital circuits a wider range of operating conditions is used. Hence, in the next section a more general set of equations will be derived. These equations, known as the Ebers-Moll equations, describe the transistor under general operating conditions, and include the active mode as a special case.

6.3 Large-Signal Operation

In the previous section we restricted the transistor operating mode to one in which the emitter-base junction was forward biased and the collector-base junction was reverse biased. In this section we extend the analysis to either bias condition (forward or reverse) on either junction. The equations we derive are conventionally called large-signal equations because they apply to signals of any size or polarity. (There are always some restrictions, of course. For example, the equations do not apply beyond the breakdown voltage of either junction.) Again in this section the *npn* transistor will be considered.

The Large-Signal Equations for *npn* Transistors   The large-signal equations are a direct consequence of the dual roles a *pn* junction can play. A junction can both inject minority carriers and collect minority carriers. Consider, for example, the collector-base junction in the *npn* transistor. In the first place, it is a collector of minority carriers (electrons) from the base, and it plays this role whether it is reverse biased, zero biased, or even forward biased. However, if forward bias were applied, it could also act to inject minority carriers into the base, just as any forward-biased junction would do.

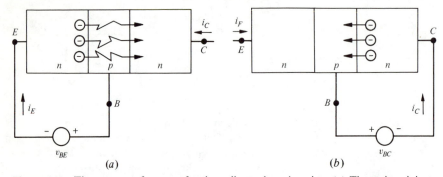

Figure 6.11   The sources of current for the collector-base junction. (*a*) The emitter injects electrons into the base, a fraction of which is collected by the collector. A collector current of magnitude αI_{ES} (ecp $\{qv_{BE}/kT\} - 1$) results. (*b*) The collector-base junction acts like an ordinary diode, conducting a current equal to $-I_{CS}$ (exp $\{qv_{BC}/kT\} - 1$).

The first source of collector current, as illustrated in Figure 6.11(*a*) is the emitter-base junction. It was shown in Section 6.2 that the emitter-base junction, when biased by the voltage v_{BE}, is a source of current of magnitude I_{ES} (exp $\{q v_{BE}/kT\} - 1$). Most of this current consists of electrons injected into the base, which then, for the most part, pass through the base and are collected by the collector. The collector current resulting from injection at the emitter is thus proportional to I_{ES} (exp $\{q v_{BE}/kT\} - 1$), with a constant of proportionality α. The constant α, as defined in Section 6.2, is the fraction of the current from the emitter-base junction that is collected. The sign of the collector current is positive when the emitter current is negative. Electrons are moving into the emitter terminal and out the collector terminal; thus current enters the collector terminal and flows out from the emitter terminal.

The second possible source of collector current is the collector-base junction itself, as illustrated in Figure 6.11(*b*). Although a reverse-biased, collector-base junction produces a negligible current on its own, a forward-biased, collector-base junction can be a source of considerable current. In fact, just as for a simple *pn* junction, the *I-V* characteristics of the collector base junction are of the form $i_C = -I_{CS} (\exp \{q\, v_{BC}/kT\} - 1)$. (The minus sign, as in the case of the emitter, arises from the fact that the current i_C is defined as positive when directed into the n region.) Summing the current from both sources, we have for the total collector current

$$i_C = \alpha\, I_{ES} \left(\exp \left\{ \frac{q\, v_{BE}}{kT} \right\} - 1 \right) - I_{CS} \left(\exp \left\{ \frac{q\, v_{BC}}{kT} \right\} - 1 \right) \quad [npn]^5 \quad (6.9)$$

Three parameters other than circuit parameters appear. Alpha and I_{ES} have already been introduced in the previous section and retain their meanings. The new parameter I_{CS} is the collector-base saturation current. As may be readily seen from Equation (6.9), I_{CS} may be obtained by studying the collector-base diode characteristic. It would be simplest to short the emitter-base junction, in which case the first term in Equation (6.9) is zero. Then a measurement of i_C for any value of v_{BC} determines I_{CS}.

EXAMPLE 6.5

It may be shown from Equation (6.9) that when $v_{BE} = 0$ and v_{CE} is large and positive, $i_C \cong -I_{CS}$. In principle, then, one could measure I_{CS} directly by shorting the emitter-base junction and reverse biasing the collector-base junction. In practice this procedure is not a good method for determining I_{CS}. It is such a small quantity (typically less than 10^{-12} A) that small currents arising from other mechanisms mask it. A better method is suggested here. A graph of the collector current versus collector-base voltage for a certain npn transistor is given. The emitter-base voltage was set to zero in obtaining this graph. Determine I_{CS}.

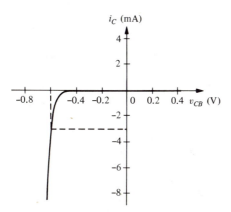

SOLUTION

According to Equation (6.9), when $v_{BE} = 0$, then $i_C = -I_{CS} (\exp \{q\, v_{BC}/kT\} - 1)$. From the graph, $i_C = -3$ mA at $v_{BC} = 0.6$ V, and from this point we will calculate I_{CS}. At room temperature, $kT/q = 0.026$ V; thus $\exp \{q \times 0.6/kT\} = \exp \{23\} = 1 \times 10^{10}$.

---

[5] As in Section 6.2, equations that are valid only for transistors of a given type will be indicated. The complete set of equations for both npn and pnp will be summarized in Table 6.3.

Solving for I_{CS}

$$I_{CS} = \frac{-i_C}{\exp\left\{\dfrac{q\,v_{BC}}{kT}\right\} - 1} = \frac{3 \times 10^{-3}}{10^{10}} = 3 \times 10^{-13}\ \text{A}$$

■

Before proceeding further with the discussion of Equation (6.9), let us derive a second, analogous equation, this time for the emitter current. Equation (6.9) expresses the dual roles of the collector-base junction, first as a collector and then as a junction that in itself can conduct a current in forward bias. The emitter likewise can play both roles, and a similar equation can be derived for it. We are already familiar with the role of the emitter-base junction as a source of current; when biased by the voltage v_{BE}, it conducts according to $i_E = -I_{ES} (\exp\{qv_{BE}/kT\} - 1)$. Furthermore, the emitter, like the collector, can collect minority carriers from the base. If the collector acts as a source of magnitude $-I_{CS} (\exp\{q\,v_{BC}/kT\} - 1)$, then some fraction of this collector current consists of electrons injected into the base, and some fraction of these electrons is collected by the emitter. The overall success of the collector-base junction in producing an emitter current is expressed by a constant called "alpha reverse," and is given the symbol α_R. The emitter current that results from carriers injected by the collector thus equals $\alpha_R I_{CS} (\exp\{v_{BC}/kT\} - 1)$. Summing the two contributions to the emitter current, we have

$$\boxed{\ i_E = \alpha_R I_{CS} \left(\exp\left[\frac{q\,v_{BC}}{kT}\right] - 1\right) - I_{ES} \left(\exp\left[\frac{q\,v_{BE}}{kT}\right] - 1\right)\ } \quad [npn] \qquad (6.10)$$

One new parameter has been introduced, α_R. Alpha reverse is the transistor alpha when the normal roles of the emitter and collector are interchanged, that is, when the collector-base junction injects minority carriers into the base and the emitter-base junction collects them. Alpha reverse can be measured by forward biasing the collector-base junction, reverse biasing the emitter-base junction, and measuring the ratio of emitter current to collector current. Equation (6.10) is exactly analogous to Equation (6.9), reflecting the basic symmetry of the transistor (a p-type base sandwiched between two n-type regions). However, the symmetry is not exact in a practical transistor, such as the one shown in Figure 6.4. The emitter and collector dopings and areas are usually quite different. As a consequence, the coefficients α and α_R are not equal. Typically α_R is in the range of 0.01 to 0.5, a good deal less than α. Furthermore I_{ES} is generally smaller than I_{CS} by 1 or 2 orders of magnitude because of doping differences, and because the emitter is smaller than the collector. It is important to realize that α, α_R, I_{ES}, and I_{CS} are "known" quantities (either they are specified by the manufacturer or they are measured), but they are a property of the transistor, not of the circuit in which it is used. On the other hand, quantities i_E, i_C, v_{BE}, and v_{BC} are "unknowns," and depend on the circuit in which the transistor finds itself.

Equations (6.9) and (6.10) are one form of the famous *Ebers-Moll equations* for *npn* transistors. Together with the current and voltage laws given in Equations (6.1) and (6.2) they completely describe transistor operation for large or small signals of any polarity. The meaning of this powerful statement may be clarified by examining the unknowns and the equations governing their values in an arbitrary transistor connection. For example, suppose the emitter is grounded, a signal in the form of a voltage is applied to the base, a battery is connected to the collector, and the current flowing out of the collector is the output signal. Then of the six electrical variables (v_{EB}, v_{BC}, v_{CE}, i_E, i_B, i_C) two — v_{BE}, the input signal, and v_{CE}, the collector battery voltage — are already specified. To find the values of the other four it is only necessary to solve the four equations (6.1), (6.2), (6.9), and (6.10) for the four unknowns. In general, then, when two parameters are specified, that is, determined by the circuit, the four equations allow the solution for the four remaining unknowns.

EXAMPLE 6.6

The transistor below has voltage sources applied between emitter and base and between base and collector; thus v_{BE} and v_{BC} are specified. Determine v_{CE}, i_E, i_B, and i_C.

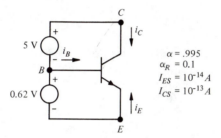

$$\alpha = .995$$
$$\alpha_R = 0.1$$
$$I_{ES} = 10^{-14}\,A$$
$$I_{CS} = 10^{-13}\,A$$

SOLUTION

First, by a direct addition of voltages, or using Equation (6.2), $v_{CE} = 5.62$ V. To solve for i_C and i_E, we simply evaluate Equations (6.9) and (6.10) for the values $v_{BC} = -5$ V, $v_{BE} = 0.62$ V. At room temperature $kT/q = 0.026$ V. From Equation (6.9)

$$i_C = \alpha I_{ES}\left(\exp\left\{\frac{q\,v_{BE}}{kT}\right\} - 1\right) - I_{CS}\left(\exp\left\{\frac{q\,v_{BC}}{kT}\right\} - 1\right)$$

Note that the latter term is negligible for large negative v_{BC}.

$$i_C = 0.995 \times 10^{-14} \times \left(\exp\left\{\frac{0.62}{0.026}\right\} - 1\right) - 10^{-13}\left(\exp\left\{\frac{-5.0}{0.026}\right\} - 1\right)$$
$$= 0.995 \times 10^{-14} \times 2.3 \times 10^{10} + 10^{-13}$$
$$i_C = 2.3 \times 10^{-4}\,A$$

From Equation (6.10)

$$i_E = -I_{ES}\left(\exp\left\{\frac{q\,v_{BE}}{kT}\right\} - 1\right) + \alpha_R I_{CS}\left(\exp\left\{\frac{q\,v_{BC}}{kT}\right\} - 1\right)$$
$$= -10^{-14}\,(2.3 \times 10^{10}) - 10^{-14}$$
$$i_E = -2.3 \times 10^{-4}\,A$$

We obtain i_B by subtraction, that is, from Equation (6.1).

$$i_B = -i_C - i_E$$
$$\cong 0$$

Now in reality i_B is not zero; the error stems from the approximation that $0.995 \times 2.3 \times 10^{-14} \times 10^{10}$ equals 2.3×10^{-4} in calculating i_C. Thus

$$i_B = -0.995 \times 2.3 \times 10^{-4} + 2.3 \times 10^{-4}$$
$$i_B = 1.15 \times 10^{-6} \qquad \blacksquare$$

Equations (6.9) and (6.10) give the emitter and collector currents as a function of the junction voltages. Many other forms of these equations are possible and are used. However, the other forms can be derived from Equations (6.1), (6.2), (6.9), and (6.10) by algebraic manipulation. If, for example, it were desired to express the emitter current as a function of v_{BE} and i_C, rather than as a function of v_{BE} and v_{BC}, we could solve Equation (6.9) for v_{BC}, and substitute into Equation (6.10) to obtain i_E as a function of v_{BE} and i_C. Other examples appear in the subsequent discussion and in the problems.

EXAMPLE 6.7
   Find i_B in terms of v_{BC} and i_C.

SOLUTION
   It is desired to find an expression for i_B that contains only v_{BC} and i_C as terminal variables. By combining Equations (6.1), (6.9), and (6.10), it is possible to find such a relationship. We first eliminate the term I_{ES} $(\exp \{q\, v_{BE}/kT\} - 1)$ from Equations (6.9) and (6.10) by solving Equation (6.9) for this factor, and substituting into Equation (6.10):

$$i_E = \alpha_R I_{CS} \left(\exp \left\{ \frac{q\, v_{BC}}{kT} \right\} - 1 \right) - \frac{i_C + I_{CS} \left(\exp \left\{ \frac{q\, v_{BC}}{kT} \right\} - 1 \right)}{\alpha}$$

We substitute this value for i_E into Equation (6.1) and gather terms to obtain

$$i_B = -i_E - i_C = \left(\frac{1 - \alpha}{\alpha} \right) i_C + \frac{1 - \alpha\alpha_R}{\alpha} I_{CS} \left(\exp \left\{ \frac{q\, v_{BC}}{kT} \right\} - 1 \right) \qquad \blacksquare$$

pnp Transistors   While we have focused our attention on *npn* transistors, equations analogous to Equations (6.9) and (6.10) may be derived for *pnp* transistors. However, the direction of current flow and the polarities of applied bias are opposite to those in the *npn* transistor. Consequently all of the signs of the terminal parameters in the Ebers-Moll equations are opposite for *npn* and *pnp* transistors. The complete set of four equations describing transistor operation is summarized in Table 6.3.

From the discussion thus far it appears that the values of four independent parameters (α, α_R, I_{ES}, I_{CS}) must be given to specify completely the characteristics of a given transistor. However these parameters are not really completely independent; there is a relationship between them that makes the

Table 6.3 The Characteristic Equations for *npn* and *pnp* Transistors

<table>
<tr><td align="center">npn</td></tr>
</table>

$$i_C + i_B + i_E = 0$$
$$v_{EB} + v_{BC} + v_{CE} = 0$$

$$i_C = \alpha\, I_{ES} \left(\exp\left\{ \frac{qv_{BE}}{kT} \right\} - 1 \right) - I_{CS} \left(\exp\left\{ \frac{qv_{BC}}{kT} \right\} - 1 \right)$$

$$i_E = \alpha_R I_{CS} \left(\exp\left\{ \frac{qv_{BC}}{kT} \right\} - 1 \right) - I_{ES} \left(\exp\left\{ \frac{qv_{BE}}{kT} \right\} - 1 \right)$$

<table>
<tr><td align="center">pnp</td></tr>
</table>

$$i_C + i_B + i_E = 0$$
$$v_{EB} + v_{BC} + v_{CE} = 0$$

$$i_C = -\alpha\, I_{ES} \left(\exp\left\{ \frac{qv_{EB}}{kT} \right\} - 1 \right) + I_{CS} \left(\exp\left\{ \frac{qv_{CB}}{kT} \right\} - 1 \right)$$

$$i_E = -\alpha_R I_{CS} \left(\exp\left\{ \frac{qv_{CB}}{kT} \right\} - 1 \right) + I_{ES} \left(\exp\left\{ \frac{qv_{EB}}{kT} \right\} - 1 \right)$$

specification of only three necessary. The fourth is then completely determined by the following relationship:

$$\alpha\, I_{ES} = \alpha_R\, I_{CS} \tag{6.11}$$

Let us see the physical basis for Equation (6.11). We consider a typical structure, such as the one shown in Figure 6.4 and repeated in Figure 6.12. In this structure, which is typical of a transistor made by diffusion, the emitter-base junction area is only a small fraction, say one-tenth, of the collector-base junction area. Just from this area difference we expect a factor of 10 difference in

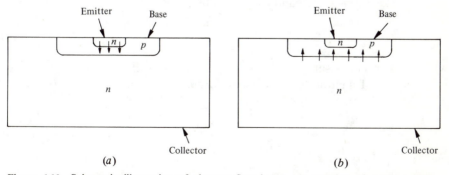

Figure 6.12 Schematic illustration of electron flow in an *npn* transistor. (*a*) Most of the electrons injected by the emitter traverse the base and are collected by the collector. (*b*) Many of the electrons injected by the collector are injected a large distance from the emitter junction.

I_{ES} and I_{CS}, that is, $I_{CS} \approx 10 \, I_{ES}$. Furthermore the collector junction, because of its larger area, would have greater success in collecting minority carriers injected into the base from the emitter than would the emitter in collecting minority carriers injected from the collector into the base. Referring to Figure 6.12(a) we see that most of the electrons injected by the emitter-base junction travel only a short distance before being collected by the collector-base junction. On the other hand, if, as in Figure 6.12(b), electrons are injected by the collector-base junction, only a small fraction of them is injected immediately opposite the emitter. Thus about 90% of these injected electrons would "miss" the emitter-base junction. Consequently the fraction α, describing transit from emitter to collector, is about a factor of 10 higher than α_R, which describes transit from collector to emitter. Thus we see that decreasing the emitter area relative to the collector area decreases α_R/α, and increases I_{CS}/I_{ES} in such a way that Equation (6.11) is always satisfied. Asymmetries in doping in the emitter and collector similarly affect the parameters; when one parameter is increased, another is decreased in precisely the proportion described by Equation (6.11).

The Ebers-Moll equations, being general large-signal equations, include operation in the active mode as a special case. Thus Equations (6.4) and (6.6) may be derived from Equations (6.9) and (6.10) under the condition that the emitter-base junction is forward biased and the collector-base junction is reverse biased. When v_{BC} is negative and v_{BE} is positive, Equation (6.9) may be written as

$$i_C = \alpha \, I_{ES} \left(\exp \left\{ \frac{q \, v_{BE}}{kT} \right\} - 1 \right) \qquad [npn] \qquad (6.12)$$

Similarly, Equation (6.10) simplifies to

$$i_E = -I_{ES} \left(\exp \left\{ \frac{q \, v_{BE}}{kT} \right\} - 1 \right) \qquad [npn] \qquad (6.13)$$

The latter is the same as Equation (6.6). To derive Equation (6.4) we first combine Equations (6.12) and (6.13) to obtain

$$\boxed{i_C = -\alpha \, i_E} \qquad (6.14)$$

which, again, is the same as Equation (6.3). By substituting for i_E in terms of i_B and i_C from Equation (6.1), we obtain Equation (6.4):

$$\boxed{i_C = \beta \, i_B} \qquad (6.4)$$

where, as before, β is defined by

$$\boxed{\beta = \frac{\alpha}{1 - \alpha}} \qquad (6.5)$$

It is noteworthy that in almost all cases of practical interest, the term -1, which appears after the exponentials in Equations (6.9) and (6.10), is insignificant. This circumstance arises because of the small size of the coefficients I_{CS} and I_{ES} in silicon transistors; it takes the product of such a coefficient with a large exponential to produce a term worthy of our attention. Only in very large transistors (designed for handling large amounts of power) are the coefficients I_{ES} and I_{CS} in themselves significant.

Large-Signal I-V Characteristics The Ebers-Moll equations, taken together with Equations (6.1) and (6.2), completely describe transistor operation. Let us use these equations to construct a plot of the transistor characteristics. The set of four equations (6.1), (6.2), (6.9), and (6.10) contains in all six variables: i_E, i_C, i_B, v_{EB}, v_{BC}, and v_{CE}. Since there are four equations and six variables, two of the variables may be specified arbitrarily. As was demonstrated in Example 6.6, when values are assigned to any two of the six variables, the other four values may be found by solving the remaining four equations in four unknowns. If, instead of arbitrarily fixing two variables, we allow one to vary over some desired range, then a plot may be obtained of any of the four unknowns as a function of the parameter that is varied. For example, suppose v_{BE} is fixed at some assumed value, and i_C is computed for several different values of v_{CB}. Then a graph of i_C versus v_{CB} could be constructed. Furthermore, the curves could be recomputed for several values of v_{BE}, and a parametric graph made of the function i_C versus v_{CB}, with v_{BE} as a parameter.

EXAMPLE 6.8
Plot the dependence of i_C on v_{CB} over the range $-10 < v_{CB} < 10$ V, $-1.0 < i_C < 1.0$ mA. Obtain three curves corresponding to $v_{BE} = 0.60$, 0.62, and 0.64 V.

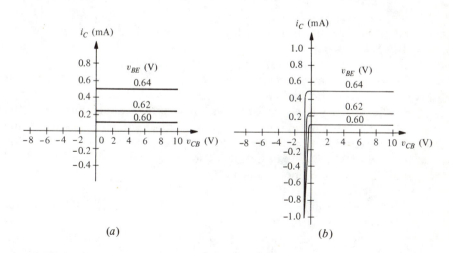

(a)   (b)

SOLUTION
Equation (6.9) may be used to compute i_C for any specified values of v_{BE} and v_{CB}. Thus, constructing the plot is only a matter of computing i_C at a sufficient number

of points. Let us consider first the region $v_{CB} > 0$. As shown in Example 6.6, in this range Equation (6.9) reduces to $i_C \cong \alpha I_{ES} \, (\exp \{q \, v_{BE}/kT\} - 1)$. Furthermore $\alpha \cong 1$, and the -1 is negligible if $v_{BE} \gg kT/q$, as it is here. Thus for $v_{CB} > 0$

$$i_C \approx I_{ES} \exp \left\{ \frac{q \, v_{BE}}{kT} \right\}$$

Clearly, in this range of v_{CB} the collector current i_C is independent of v_{BC}. If $v_{BE} = 0.62$ V, $i_C = 2.3 \times 10^{-4}$ A, as already calculated in Example 6.6. For $v_{BE} = 0.6$ V, we obtain $i_C = 1 \times 10^{-4}$ A and for $v_{BE} = 0.64$ V, $i_C = 4.9 \times 10^{-4}$ A. The graph may now be partially constructed as in sketch (a).

To complete the graph we must consider the range $v_{CB} < 0$. In this case the second term in Equation (6.9) is no longer negligible. However, it is no problem to include it. For example, if $v_{CB} = -0.5$ V, then

$$-I_{CS} \left(\exp \left\{ \frac{q \, v_{CB}}{kT} \right\} - 1 \right) = -10^{-13} \times (2.3 \times 10^8)$$
$$= -2.3 \times 10^{-5} \text{ A}$$

Thus the total value of i_C (for $v_{CB} = -0.5$ V, $v_{BE} = 0.6$ V) is the sum

$$1 \times 10^{-4} \text{ A} - 2.3 \times 10^{-5} \text{ A} = 7.7 \times 10^{-5} \text{ A}$$

This forms one point on the curve of i_C versus v_{CB} for $v_{BE} = 0.6$ V. If $v_{CB} = -0.6$ V, then the second term contributes a current of -10^{-3} A, and a second point, with $i_C = 1 \times 10^{-4} - 10^{-3} = -9 \times 10^{-4}$, is obtained. Carrying this out for many values of v_{CB}, we obtain the complete set of curves [see sketch (b)]. ∎

Specifying two terminal variables determines all the others; hence we may say that any variable is a function of any two others. In the previous example, we solved for i_C as a function of v_{CB} and v_{BE}. Such a relationship may be written as $i_C(v_{CB}, v_{BE})$. This relationship is particularly easy to find, because it is given directly by Equation (6.9). In general, any of the six parameters may be regarded as a function of any other two, and a parametric graph may be drawn to display the relationship. There are numerous possible functional relationships that might be computed from the Ebers-Moll equations; however there are two relationships of special interest because they display the terminal properties of a transistor as seen from the input and output of the most common transistor circuit. The relationships of greatest interest are the input *I-V* characteristics and output *I-V* characteristics for a transistor in the so-called *common-emitter* connection. The name common-emitter derives from the circuit configuration, shown in Figure 6.13, in which the input appears at the base and the output is taken at the collector. The emitter is connected to one of the input terminals and one of the output terminals. Thus the emitter terminal is common to both input and output.

The *common-emitter output characteristic* is the function $i_C(v_{CE}, i_B)$. This characteristic was displayed, in part, in Figure 6.9. (However, we are now able to compute i_C versus v_{CE} over the full range, not restricting v_{CE} to be greater than 0.7 V, as in Figure 6.9.) The other relationship of great interest is $i_B(v_{BE}, v_{CE})$, which is known as the *common-emitter input characteristic*.

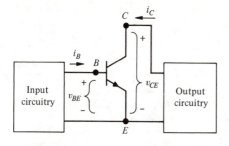

Figure 6.13 The common-emitter circuit configuration. The input is applied between base and emitter and the output is taken between collector and emitter. The emitter terminal is common to both circuits.

Again, this graph was partially presented in Figure 6.10, under the restriction $v_{CB} > 0$. Using the Ebers-Moll equations we may now compute and plot the input and output characteristics of a typical silicon *npn* transistor. While no two transistors are exactly alike, we can give parameter values that are representative of common low-power devices. Let us adopt the values given in Table 6.4 as characteristic of a typical silicon transistor at room temperature.

Table 6.4 The Parameters for a
Typical Silicon *npn* Transistor

$$\alpha = 0.99$$
$$\alpha_R = 0.1$$
$$I_{ES} = 10^{-14} \text{ A}$$
$$I_{CS} = 10^{-13} \text{ A}$$

As shown in Figure 6.13, in the common-emitter connection the input terminals are the base and the emitter. The desired input characteristic is the functional dependence of the base current on the emitter-base voltage for various fixed values of v_{CE}. This value of i_B is obtained by adding $-i_C$, from Equation (6.9), to $-i_E$, from Equation (6.10), according to Equation (6.1). Writing out the analytic form of i_B from these equations, and gathering common terms, we have

$$i_B = (1 - \alpha)I_{ES}\left(\exp\left\{\frac{q\,v_{BE}}{kT}\right\} - 1\right)$$
$$+ (1 - \alpha_R)I_{CS}\left(\exp\left\{\frac{q\,v_{BC}}{kT}\right\} - 1\right) \qquad [npn] \qquad (6.15)$$

We may now evaluate the function $i_B(v_{BE}, v_{CE})$ directly from Equation (6.15). The fact that the unknown voltage v_{BC} appears in the equation poses little problem; it may simply be obtained by subtracting v_{CE} from v_{BE}. Figure 6.14 is the resulting graph. Only values of $v_{CE} > 0$ have been considered, as negative values are seldom encountered. For all values of v_{CE} greater than about 0.3 V, all the curves lie one on top of the other. Furthermore this common curve is the same curve as that given previously for the active mode in Figure 6.10.

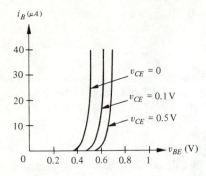

$i_B (\mu A)$

40

30    $v_{CE} = 0$

20    $v_{CE} = 0.1\,V$

10    $v_{CE} = 0.5\,V$

v_{BE} (V)

0   0.2   0.4   0.6   0.8   1

Figure 6.14 The input characteristics of a typical *npn* silicon transistor in the common emitter configuration. If $v_{CE} \geq 0.3$ V, the curves all merge into the curve shown for $v_{CE} = 0.5$ V.

The output characteristics are described by the function $i_C(v_{CE}, i_B)$. Again, making use of Table 6.3, we solve for i_C. For the transistor parameters given in Table 6.4, we obtain the curves given in Figure 6.15. A careful examination of the characteristics displayed in Figure 6.15 can give considerable insight into transistor operation. In particular, three modes of transistor operation can be identified, each of which is important in circuit applications. The three modes are defined as regions on the graph of the output characteristics in Figure 6.16. The first of these modes, the active mode, has already been discussed at some length in the previous section. The active mode of operation is characterized by reverse bias on the collector-base junction and forward bias on the emitter-base junction. The base-emitter voltage v_{BE} is positive, on the order of 0.6 to 0.8 V (a forward-biased diode drop). Since v_{BC} may be any negative voltage, even as small as zero, v_{CE}, which equals $v_{BE} - v_{BC}$, must be positive and greater than 0.8 V. Hence the active region is found in the first quadrant of Figure 6.16 for v_{CE} greater than 0.8 V. Of course the curves in this region correspond exactly to the curves of Figure 6.10. Because i_C is independent of v_{CE}, each of the curves is horizontal. The collector current in the active region of Figure 6.16 is directly proportional to the base current, with a constant of proportionality equal to β (100 in the case of this figure). Any variations in current applied to the base give rise to collector current varia-

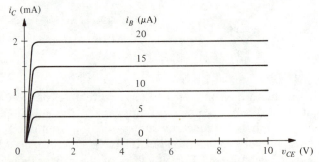

i_C (mA)

i_B (μA)

20

2

15

10

1

5

0

0   2   4   6   8   10 v_{CE} (V)

Figure 6.15 The output characteristics of a typical *npn* transistor in the common-emitter configuration.

tions β times larger; in other words there is a current gain from input to output. Furthermore, as was demonstrated in Example 6.2, a transistor operating in the active region can also produce a voltage gain. Clearly the transistor is active in the sense that a power gain from input to output occurs, hence the label active mode.

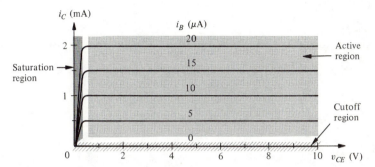

Figure 6.16 The output characteristics of a typical *npn* transistor. The regions of operation are indicated.

Saturation   Let us now consider the area in Figure 6.16 labeled the saturation region. In this region the collector-to-emitter voltage is quite small, and the collector current is less than β times the base current. Both junctions are forward biased; hence the voltage drop from collector to emitter is the small difference between the two forward-biased diode drops. For example, when $v_{CE} = 0.1$ V, the base-to-collector bias is only 0.1 V less than the base-to-emitter bias. The base current, for a given collector current, is larger than in the active mode. The relative increase in base current stems from the presence of forward bias on the collector-base junction. Typically, forward biasing a collector-base junction leads to almost equal base and collector currents. Consequently, when these currents are added to the base and collector currents arising from the emitter-base bias, the ratio of collector current to base current becomes less than beta. Summarizing the most important features of operation in the saturation region: (1) the collector-to-emitter voltage is less than the base-to-emitter voltage, and is typically less than a few tenths of a volt; (2) the collector current is less than beta times the base current.

Cutoff   The third mode of operation (really of nonoperation) of a transistor is called cutoff and is identified in Figure 6.16. In this mode the emitter-base junction is not sufficiently forward biased to cause any significant current to flow. As may be seen in Figure 6.14, if v_{BE} is small, say 0.1 V or less, then essentially no base current flows. If no base current flows, then we see from Figure 6.15 that no collector current flows. The transistor is said to be cut off; the collector-to-emitter circuit is essentially an open circuit.

The saturation and cutoff modes of transistor operation are very important in digital applications, where it is desired to have a device that operates like a switch. In the saturation mode, the voltage across the transistor (collector to emitter) may be quite small, even for large currents, just as in a closed switch. In the cutoff mode the transistor passes no current between collector and emitter, just as in an open switch. The operating mode of a transistor is determined by a combination of its constructional parameters, the circuit parameters, and the signal values. Example 6.9 illustrates the importance of the signal in determining the operating mode.

EXAMPLE 6.9

Find the collector current of the transistor in sketch (*a*) for the following three different values of i_1: 0, 5, and 20 μA. Assume the transistor is the "typical" transistor of Table 6.4, whose characteristics are illustrated in Figures 6.14 and 6.15.

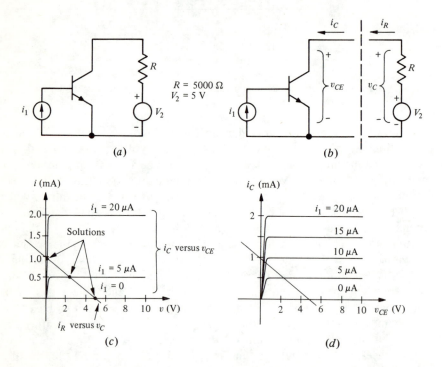

SOLUTION

To solve this problem we will use the graphical load-line technique. The method consists of matching the *I-V* characteristics of the transistor, as seen from the collector to emitter terminals, to the *I-V* characteristics of the collector circuit. An imaginary break is made in the circuit as shown in sketch (*b*). The break in the circuit is purely imaginary, and in reality i_C equals i_R and v_C equals v_{CE}. We plot the curve of i_C versus v_{CE} and the curve of i_R versus v_C on the same graph; the point of intersection of the curves is the desired solution. A set of curves of i_C versus v_{CE} for this transistor is already given in Figure 6.15; we need only choose the appropriate curve from the known base current ($i_B = i_1$). The characteristic i_R versus v_C is obtained from Kirchhoff's voltage law: $v_C = V_2 - i_R R$. We now plot i_C versus v_{CE} and i_R versus v_C on the same graph, as in sketch (*c*), for the specified values of R, V_2, and i_1. For the case $i_1 = 5$ μA, the point at which the conditions $i_C = i_R$ and $v_{CE} = v_C$ are satisfied is the point $i_C = 0.5$ mA, $v_{CE} = 2.5$ V. The transistor is operating in the active region. For the case $i_1 = 20\ \mu A$, the solution is $i_C = 0.95$ mA, $V_{CE} = 0.25$ V, and the transistor is in saturation. When $i_1 = 0$, then $i_C = 0$ and $v_{CE} = 5$ V and the transistor is cut off.

If the entire set of curves of Figure 6.15 is superimposed on the i_R versus v_C characteristics, as in sketch (d), then i_C may be quickly determined for any value of i_1. It is noteworthy that the transistor behavior is quite different in the different regions of operation. In the active region (that is, for $v_{CE} > 0.8$ V) the collector current and collector-emitter voltage depend strongly on i_1. In the saturation region the collector current becomes almost independent of i_1 and the collector-emitter voltage is nearly zero. In the cutoff region, when i_1 becomes very small, the collector current goes to zero and the collector-emitter voltage equals the battery voltage.   ■

The common-emitter input and output characteristics given in Figures 6.14 and 6.15 are by no means the only characteristics that may be plotted. However they are the most important characteristics, both for digital and analog circuit applications. When some other characteristic is desired, it may be plotted from the Ebers-Moll equations, as has already been demonstrated in Example 6.8.

6.4 Transistor Characteristics: Comparison with Theory

Let us examine the I-V characteristics of an actual transistor, and note the degree to which it obeys the theoretical characteristics just derived. A suitable transistor is the 2N3568, which is a general-purpose npn silicon device. Curve-tracer photographs of the room temperature common-emitter characteristics of a 2N3568 transistor are shown in Figure 6.17. We also reproduce in Table 6.5 part of the manufacturer's data sheet describing the device.

The experimental characteristics given in Figures 6.17(a) and (b) may be compared with the theoretical characteristics given in Figures 6.14 and 6.15. The basic similarity of the theoretical and experimental plots speaks well for the simple theory; however there are some noticeable differences in the output characteristics that deserve discussion.

The Early Effect It is evident from Figure 6.17(b) that the collector current is not really independent of the collector-to-emitter voltage in the active region as predicted theoretically. For example, with a base current of 15 μA, the collector current increases from 1.44 to 1.48 mA, as the collector voltage increases from 3 to 10 V. Though this increase in current is not large, we shall see in Chapter 10 that the effect is of sufficient importance that it must be taken into account in amplifier design. This dependence of collector current on collector voltage stems from the effect of collector-base voltage on base width, and is known as the Early effect. When the collector-base reverse voltage is increased, the magnitude of the electric field in this junction increases and the field reaches farther into the base region. The effect of narrowing the base can be analyzed exactly using the diode theory given in Appendix II. Reducing the

Table 6.5

2N3567 • 2N3568 NPN GENERAL PURPOSE TYPES

DIFFUSED SILICON PLANAR* EPITAXIAL TRANSISTORS

The 2N3567 and 2N3568 are NPN silicon **PLANAR*** epitaxial transistors designed primarily for amplifier and switching applications over a wide range of voltage and current. These devices feature a useful beta range to 500 mA and low saturation voltage. High collector-to-emitter voltage allows operation to 60 volts for the 2N3568 and 40 volts for the 2N3567.

ABSOLUTE MAXIMUM RATINGS

Maximum Temperatures
Storage Temperature   −55°C to +125°C
Operating Junction Temperature   +125°C Maximum
Lead Temperature (Soldering,
10 sec. time limit)   +260°C Maximum

Maximum Power Dissipation
Total Dissipation
at 25°C Case Temperature   0.8 Watt
at 25°C Ambient Temperature   0.3 Watt

Maximum Voltages   2N3567   2N3568
V_{CBO} Collector to Base Voltage   80 Volts   80 Volts
V_{CEO} Collector to Emitter Voltage   40 Volts   60 Volts
V_{EBO} Emitter to Base Voltage   5.0 Volts   5.0 Volts

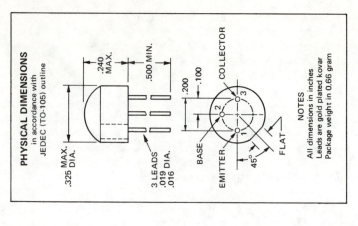

PHYSICAL DIMENSIONS
in accordance with
JEDEC (TO-105) outline

.240 MAX.
.500 MIN.
.325 MAX. DIA.
3 LEADS .019 DIA. .016 DIA.
.200
.100
COLLECTOR
BASE
EMITTER
45°
FLAT

NOTES
All dimensions in inches
Leads are gold plated kovar
Package weight in 0.66 gram

FAIRCHILD
SEMICONDUCTOR
A DIVISION OF FAIRCHILD CAMERA AND INSTRUMENT CORPORATION

Table 6.5 (continued)

ELECTRICAL CHARACTERISTICS (25°C Free Air Temperature unless otherwise noted)

Symbol	Characteristic	2N3567 Min.	Typ.	Max.	2N3568 Min.	Typ.	Max.	Units	Test Conditions
h_{FE}	DC Pulse Current Gain	40	80	120	40	80	120		$I_C = 150$ mA, $V_{CE} = 1.0$ V
h_{FE}	DC Pulse Current Gain	40			40				$I_C = 30$ mA, $V_{CE} = 1.0$ V
v_{CE}(sat)	Collector Saturation Voltage		0.15	0.25		0.15	0.25	Volts	$I_C = 150$ mA, $I_B = 15$ mA
v_{BE}(sat)	Base Saturation Voltage		0.9	1.1		0.9	1.1	Volts	$I_C = 150$ mA, $I_B = 15$ mA
h_{fe}	High Frequency Current Gain (f = 20 MHz)	3.0			3.0				$I_C = 50$ mA, $V_{CE} = 10$ V
C_{obo}	Open Circuit Output Capacitance		13	20		13	20	pF	$I_E = 0$, $V_{CB} = 10$ V
C_{ibo}	Open Circuit Input Capacitance		63	80		63	80	pF	$I_C = 0$, $V_{EB} = 0.5$ V
I_{CBO}	Collector Cutoff Current			50			50	nA	$I_E = 0$, $V_{CB} = 40$ V
I_{CBO}(75 C)	Collector Cutoff Current			5.0			5.0	μA	$I_E = 0$, $V_{CB} = 40$ V
I_{EBO}	Emitter Cutoff Current			25			25	nA	$I_C = 0$, $V_{EB} = 4.0$ V
BV_{CBO}	Collector to Base Breakdown Voltage	80			80			Volts	$I_C = 100$ μA, $I_E = 0$
v_{CEO}(sust)	Collector to Emitter Sustaining Voltage	40			60			Volts	$I_C = 30$ mA (pulsed), $I_B = 0$
BV_{EBO}	Emitter to Base Breakdown Voltage	5.0			5.0			Volts	$I_E = 10$ μA, $I_C = 0$

° Planar is a patented Fairchild process.

SMALL SIGNAL CHARACTERISTICS (f = 1.0 kHz)

Symbol	Characteristic	Typical	Units	Test Conditions	
h_{ie}	Input Resistance	1800	Ohms	$I_C = 1.0$ mA,	$V_{CE} = 5.0$ V
h_{oe}	Output Conductance	8.0	μmhos	$I_C = 1.0$ mA,	$V_{CE} = 5.0$ V
h_{re}	Voltage Feedback Ratio	2.1	$\times 10^{-4}$	$I_C = 1.0$ mA,	$V_{CE} = 5.0$ V
h_{fe}	Small Signal Current Gain	60		$I_C = 1.0$ mA,	$V_{CE} = 5.0$ V

distance that the injected carriers must diffuse has the effect of increasing the current at a given junction bias, in essence by shortening the journey across the base. Thus at a given emitter-base bias, when the base width is decreased, the flow of electrons from emitter to collector is increased. A simple method of taking this effect into account in the analysis of analog transistor circuits will be given in Chapter 10. The effect is not of much consequence in switching circuits.

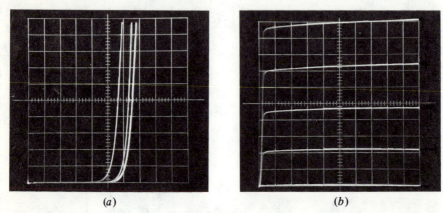

(a)   (b)

Figure 6.17   The input and output characteristics (photographs from curve tracer) of a 2N3568 *npn* silicon transistor. (*a*) The input characteristics i_B versus v_{BE}. Vertical scale: 10 μA per large division. Horizontal scale: 0.1 V per large division. The three curves, from left to right: $v_{CE} = 0$ V, $v_{CE} = 0.1$ V, $v_{CE} \geq 0.3$ V. (*b*) The output characteristics i_c versus v_{CE}. Vertical scale: 0.2 mA per large division. Horizontal scale: 1 V per large division. The five curves from top to bottom: $i_B = 20, 15, 10, 5,$ and 0 μA.

Variation of Parameters with Bias   It may be seen from a careful examination of Figure 6.17(*b*) that the ratio of collector current to base current in the active region is actually not constant as theoretically predicted. That is, beta is not a true constant but depends somewhat on collector current. This dependence is particularly pronounced at lower currents; in fact all but special transistors operate very poorly, that is, have a low value of β, at collector currents in the microampere range.

However, in the normal operating range of a given transistor type, the variation of β with current is far smaller than the variation from device to device.[6] The behavior of other transistor parameters is similar; generally the variation from one device to the next is of greater concern than the slight variation with current or voltage. For example, the manufacturer of the 2N3568 transistor guarantees only that β lies in the range of 40 to 120 at 150 mA collector current. This seemingly large variation in beta does not really amount

---

[6] To specify the ratio i_C/i_B, transistor manufacturers introduce a quantity h_{FE} (rather than β) in their data sheets. This is not to be confused with h_{fe}, another parameter which they specify. The latter is defined as $\Delta i_C/\Delta i_B$, the ratio of the change in i_C to the change in i_B for small changes in i_B.

to much variation in alpha (0.976 to 0.992), and is simply unavoidable in the manufacturing process. We shall see in later chapters that in a good circuit design, such variations of beta do not affect the performance. In consequence, the relatively small variations of beta with current can generally be ignored.

Junction Breakdown If the output characteristics are measured over a greater range of collector-to-emitter voltage, a region in which the collector current sharply increases with voltage is encountered. The characteristics of the same 2N3568 transistor of Figure 6.17 are displayed in Figure 6.18, but over a range of v_{CE} from 0 to 100 V rather than 0 to 10 V. It is seen that the behavior deviates quite strongly from the predictions of the simple Ebers-Moll theory for collector-to-emitter voltages beyond about 50 V. Just as in simple *pn* junctions, there is in transistor junctions a reverse voltage, called the *breakdown voltage,* beyond which the current increases drastically with voltage. Thus there is a limit to the voltage that may be applied between any two of the three transistor terminals.

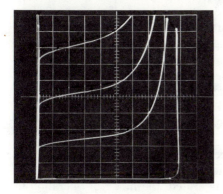

Figure 6.18 The output characteristics of a 2N3568 transistor. Vertical scale: 0.2 mA per large division. Horizontal scale: 10 V per large division. The four curves from top to bottom: $i_B = 15$, 10, 5, and 0 μA (curve-tracer photograph).

The emitter-base junction is forward biased in the active region. Thus in analog circuits such as amplifiers, in which the transistor is always biased in the active region, the emitter-base junction is never reverse biased. However, in many digital circuits the emitter-base junction may be reverse biased, and care must be taken in circuit design to assure that the reverse voltages do not exceed the emitter-base breakdown voltage. A typical value for this breakdown voltage is 6 to 7 V. The breakdown voltage of the collector-base junction is much higher, typically 20 to 50 V, and may be as high as several thousand volts in special transistors.

The most important voltage limitation in bipolar transistor circuits is the collector-to-emitter breakdown. In most circuits there is a voltage source, usually in series with a resistor, applied between collector and emitter (as in Example 6.2). The collector-to-emitter voltage, which tends to reverse bias the collector-base junction and forward bias the emitter-base junction, must not exceed the collector-to-emitter breakdown voltage. The exact value of this breakdown voltage may be shown to depend on the base current, as is evident from Figure 6.18. Typical values are 20 to 50 V in general-purpose transistors. However, for transistors in integrated circuits values are often lower, hence they must operate at restricted voltages.

Speed Limitations of Transistors   There are several important limitations on the speed of response of transistors to input signals. The most obvious of these effects is the transit time of minority carriers from emitter to collector. Suppose a large signal is suddenly applied to the emitter-base junction of an *npn* transistor, injecting a large number of electrons into the base. The output signal does not appear immediately at the collector; it is necessary to wait for the electrons to diffuse across to the collector. This transit-time delay imposes one upper limit on the maximum frequency response of transistors. In modern "thin-base" transistors, the transit time is readily made as low as 10^{-9} sec, and with special techniques even lower values are achieved.

Another effect causing a delay in the response arises from the charging time of a capacitance associated with the collector-base junction. When a junction is reverse biased there is a separation of charge in the vicinity, just as in a parallel-plate capacitor. A reverse-biased collector-base junction is in effect a capacitor, and when a signal is applied to the base, it takes some time to charge this capacitance. The charging time is determined by the resistance R through which the capacitor is charged, and is given by the usual RC time constant. The charging rate may, for example, be limited by the simple bulk resistance of the base, typically 100 Ω. Since the collector-base capacitance is usually very small, say 10^{-12} F (or 1.0 pF) the RC time constant is small, here 10^{-10} sec. External circuit resistances can further increase the RC time constant.

Finally another limitation on the speed of response of transistors operating as digital switches is worthy of mention. This effect, known as *collector storage,* occurs only when the device is saturated, the mode of operation in which both junctions become forward biased. In particular, when the collector-base junction becomes forward biased, we know that some minority carriers are injected from the collector into the base, but it is also true that minority carriers of the opposite type may be injected from the base into the collector. For example, in the *npn* transistor, holes may be injected from base into collector. As long as the forward bias is maintained on the collector-base junction, these holes wander about and eventually recombine with electrons. However, if the junction bias is reversed, they are collected by the now reverse-biased junction. This collection process takes some time, as many of the holes are quite deep in the collector region and take their time in diffusing back to the junction. In other words, current flow does not cease (the transistor does not turn off) until the wandering minority carriers in the collector have been eliminated. For this reason special impurities, called recombination centers, are included in the transistor collector region, hastening the removal of minority carriers by recombination. Gold is the usual impurity, and it may be put into silicon in such concentrations that minority carriers live on the average less than a few nanoseconds.

Summary

- A transistor is a semiconductor structure that contains two *pn* junctions in close proximity. In an *npn* transistor, an *n*-type emitter injects electrons into a *p*-type base region. The electrons diffuse across the base and are collected by an *n*-type collector.

- The transistor circuit symbol reflects the normal direction of current flow — out of the emitter for an *npn* transistor and into the emitter for a *pnp* transistor.

- A transistor operating in the active mode has the emitter-base junction forward biased and the collector-base junction reverse biased. In the active mode, the collector current is related to the base current by a constant factor, beta. Beta is typically in the range of 50 to 250 for silicon transistors.

- The Ebers-Moll equations describe the general transistor *I-V* relationships. The input characteristics (i_B versus v_{BE} with v_{CE} as a parameter) and the output characteristics (i_C versus v_{CE} with i_B as a parameter) may be plotted from the Ebers-Moll equations.

- Three important modes of transistor operation may be identified from the output characteristics. In the active region, the collector current is proportional to the base current. In the cutoff region, the base current and the collector current are zero. In the saturation region, the collector-to-emitter voltage is close to zero.

- The characteristics of real transistors may deviate somewhat from the predictions of the simple Ebers-Moll theory. Especially important is the effect of junction breakdown.

References

Treatments at approximately the same level as this book:

Gray, P. E., D. DeWitt, A. R. Boothroyd, and J. F. Gibbons. *Physical Electronics and Circuit Models of Transistors* (SEEC Notes, Vol. 2). New York: John Wiley & Sons, 1964.

Angelo, E. J. *Electronics: BJTs, FETs, and Microcircuits.* New York: McGraw-Hill, 1969.

At a more advanced level:

Gibbons, J. F. *Semiconductor Electronics.* New York: McGraw-Hill, 1966.

Valdes, L. B. *The Physical Theory of Transistors.* New York: McGraw-Hill, 1961.

Problems

6.1   Consider a certain *npn* transistor that is biased as shown in Figure 6.3. Suppose that of the current flowing across the emitter-base junction, 99.5% consists of electrons injected into the base and that the other 0.5% consists of holes injected from the base into the emitter. Assume further that an electron injected into the base has, on the average, a 99.7% chance of being collected (by the collector-base junction) and only a 0.3% chance of recombining with a hole in the base. Find α and β for this transistor.

6.2   In *npn* transistor operation, electrons are introduced into the *p* region by injection from an *n* region. It is also possible to introduce excess electrons by shining a light on the crystal. If the photon energy is sufficient to break an electron bond, the absorption of a photon produces an electron-hole pair. Suppose the emitter region in Figure 6.1 is removed. Find the current flowing in the collector circuit when 10^{12} photons/sec are absorbed in the *p* region. (Assume that there is no recombination in the base.) Is the current a function of the reverse bias *V*? Why or why not?

6.3   Explain how Equation (6.2) follows from Kirchhoff's voltage law.

6.4   Why must Equation (6.1) be valid? (For example, could not some electrons flow into the base and recombine, thus disappearing?)

6.5   Show a diagram of electron and hole flow (analogous to Figure 6.3) for the *pnp* transistor.

6.6   A certain *npn* transistor has a reverse-bias voltage of 5 V applied to the collector-base junction. It is found that the emitter current equals −0.8 mA and the base current equals 5 μA at an emitter-base bias of 0.65 V. Find β and I_{ES}.

6.7   Suppose the transistor of Example 6.2 were a *pnp* transistor with $\beta = 150$ and $I_{ES} = 3 \times 10^{-15}$ A. Let $V_C = -15$ V. (1) Find the value of v_{IN} such that $v_{OUT} = -10$ V. (2) Find v_{IN} such that $v_{OUT} = -11$ V. (3) What is the voltage gain, $\Delta v_{OUT}/\Delta v_{IN}$? (4) What is the current gain, $\Delta i_C/\Delta i_B$?

6.8   Find the voltage gain of the circuit of Example 6.2 if R_L is halved and all other parameters remain as stated in the example.

6.9   Noting that $i_C \cong -i_E$: (1) Derive an expression for v_{OUT} in the circuit of Example 6.2 as a function of v_{IN}. The fixed parameters V_C, R_L, I_{ES}, β, and kT/q may appear in the expression, but i_C, i_B, and i_E should not. (2) Find an expression for the voltage gain $\Delta v_{OUT}/\Delta v_{IN}$ (or dv_{OUT}/dv_{IN}).

6.10   Plot the input characteristic i_B versus v_{BE} in the active mode for a *pnp* transistor with $I_{ES} = 2 \times 10^{-14}$ A, $\beta = 75$ at room temperature.

6.11   Use a graphical method to solve Example 6.2. (1) Plot the input characteristic for the transistor of this example, and find i_B for the two values of v_{IN} graphically. (2) Plot the output characteristics for the two values of i_B determined in (1), and solve for v_{OUT} graphically.

6.12   Plot the input characteristic of an *npn* transistor operating in the active mode if $I_{ES} = 10^{-8}$ A and $\beta = 50$. (These figures are typical of germanium transistors.) Also plot the output characteristics for four values of base current from 25 to 100 μA.

6.13   Equation (6.8) expresses i_B as a function of v_{BE}. (1) Derive an expression for the change in i_B for a small change in v_{BE}. *Hint:* If $y = F(x)$, then $\Delta y = (df/dx)\Delta x$. (2) Eliminate v_{BE} from the expression derived in (1) by substituting from Equation (6.8). (3) Simplify the expression obtained in (2) by noting that in the active mode $i_B \gg I_{ES}$. (4) Assuming $i_B = 10\ \mu A$ and $\beta = 100$, find Δi_B for a 1-m V increase in v_{BE}. (ANSWER: 0.38 μA.)

6.14   The circ⸱⸱⸱ of Example 6.2 is a basic amplifier circuit. (1) Derive an expression for the change in v_{CE} (or v_{OUT}) for a small change in i_B. The expression should be in terms of the load resistance R_L and the transistor parameters. (2) Use the results of Problem 6.13(3) to find an expression relating Δv_{OUT} to Δv_{IN}. (3) Evaluate the expression of part (2) to verify the numerical calculation made in Example 6.2.

6.15   Plot the input characteristics of an *npn* transistor with $\beta = 200$, $I_{ES} = 10^{-15}$ A, and $I_{CS} = 10^{-13}$ A. (1) Assume $v_{CE} = 5$ V, (2) $v_{CE} = 0$ V, and (3) $v_{CE} = 0.1$ V.

6.16   A so-called lateral *pnp* transistor in integrated circuits often has a value of α in the neighborhood of 0.7. Plot the input characteristic of such a transistor at a collector-to-emitter bias of -5 V, assuming $I_{ES} = 10^{-13}$ A.

6.17   Find I_{ES} for the transistor of Figure 6.17.

6.18   Derive the Ebers-Moll equations for *pnp* transistors (Table 6.3). Follow the procedure used in the text for deriving the *npn* equations, paying close attention to the signs of the various terms.

6.19   In the so-called reverse active mode of transistor operation, the emitter-base junction is reverse biased, and the collector-base junction forward biased. For an *npn* transistor with $I_{ES} = 10^{-14}$ A, $I_{CS} = 10^{-13}$ A, $\alpha = 0.99$, and $\alpha_R \cong 0.1$; (1) find a simple expression for i_B as a function of v_{BC}. (2) Find and plot i_E as a function of v_{EB} with i_B as a parameter ($i_B = 100, 200, 300\ \mu A$).

6.20   In germanium transistors, I_{CS} and I_{ES} may not be negligible compared to currents flowing in a circuit. Derive, from the large-signal equations for an *npn* transistor, i_B as a function of i_C in the active mode, including I_{CS} and I_{ES}, if they appear.

6.21   A certain silicon power transistor has $\beta = 45$, $I_{ES} = 10^{-10}$, and $I_{CS} = 10^{-9}$. Plot (1) the input characteristics in the active region over the range $i_B = 0$ to 100 mA, and (2) the output characteristics in the active region for $i_B = 5, 10, 15$, and 20 mA.

6.22   (1) Derive, from the equations of Table 6.3, an expression for v_{CE} as a function of i_C and i_B for an *npn* transistor. *Suggested procedure:* First eliminate i_E, using Equation (6.1). Then solve for v_{BC} and v_{BE}, and combine to find v_{CE}. (2) Show that v_{CE} is small whenever $i_B > i_C/\beta$. (Neglect I_{CS} and I_{ES} in comparison with i_B and i_C.) This is the saturation region. (3) Plot v_{CE} versus i_C for the transistor of Table 6.4, assuming $i_B = 1$ mA. The graph should cover the range $v_{CE} = 0$ to 1 V, $i_C = 0$ to 1 mA.

6.23   Near room temperature (300°K), the values of I_{ES} and I_{CS} typically increase by a factor of 6 for every increase of 10°C in temperature. Beta, however, is only very slightly temperature dependent. Assume the transistor of Table 6.4 is biased at a constant emitter current of 1 mA. Find the change in v_{BE} if the temperature is increased by 10°C (in active mode operation). (ANSWER: -27 mV)

6.24   Find the exact collector current for the transistor of Table 6.4 biased at $v_{CE} = 10$ V, $i_B = 0$.

6.25   Sketch v_{CE} versus i_1 for the circuit of Example 6.9. (The graph should cover the range $v_{CE} = 0$ to 5 V, $i_1 = 0$ to 20 μA.) Identify the three regions of transistor operation.

6.26   In integrated circuits, transistors are often used as diodes. Some possible connections are shown in Figure 6.19. Assume the transistor parameters are $I_{ES} = 10^{-15}$ A, $I_{CS} = 10^{-14}$ A, $\alpha = 0.99$, $\alpha_R = 0.1$. (1) Derive the I-V relationships for the three connections shown, that is, i_{AB} as a function of v_{AB}. (2) Plot the I-V characteristics on the same graph over the range of 0 to 2 mA.

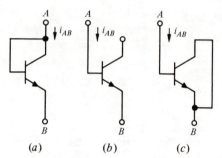

(a)   (b)   (c)   **Figure 6.19**

6.27   The transistor is often used together with an operational amplifier to produce an amplifier whose output is proportional to the log of its input. Show that when the collector-base junction is shorted, v_{BE} is proportional to the log of the collector current. How does the temperature effect the slope, $dv_{BE}/d(\ln i_C)$?

6.28   Find h_{fe} and h_{FE} for the transistor of Figure 6.17 when operating at a collector current of 1 mA and a collector-to-base voltage of 5 V.

6.29   The two circuits shown in Figure 6.20 represent two different methods for biasing an *npn* transistor. Find the range of i_C and v_{CE} in both circuits, assuming β can lie anywhere in the range of 50 to 150.

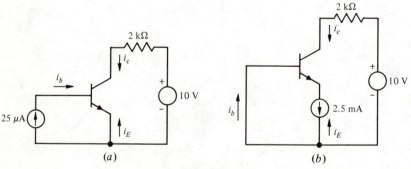

(a)   (b)

Figure 6.20

6.30 A simple biasing circuit is shown in Figure 6.21. The transistor has the properties given in Table 6.4 and displayed in Figures 6.14 and 6.15. (1) Assuming
that the transistor is operating in the active mode, find the base current using a
graphical technique. (2) Also find the base current approximately by noting that
v_{BE} is always close to 0.7 V in the active mode of operation. (3) Find the collector current. (4) What happens if β increases to 250?

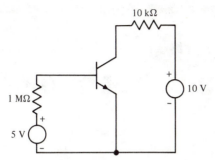

Figure 6.21

6.31 Two diodes are connected as shown in Figure 6.22. Assume that the diodes
obey the ideal diode equation with $I_S = 10^{-13}$ A. We might falsely imagine that
such a connection constitutes a transistor. (1) Compute the "input characteristics" (i'_B versus v'_{BE}) for this "transistor" for $v'_{CE} = 0$, 0.1, and 0.5 V. Compare with Figure 6.14. (2) Compute the output characteristics (i'_C versus v'_{CE})
for $i'_B = 5$, 10, 15, 20 μA. Compare with Figure 6.15.

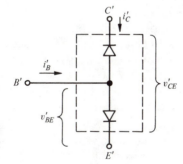

Figure 6.22

6.32 (1) Using the large-signal equation for *npn* transistors, derive an expression for
i_C as a function of i_E and v_{BC}. (2) Find an expression for i_E as a function of i_C
and v_{BE}. These two equations are a common alternative form for the Ebers-Moll
equations.

Chapter 7

Integrated Circuit Technology

In the last two chapters we investigated the electrical properties of two elementary semiconductor devices, the *pn* junction and the transistor. Both of these devices are constructed in a single crystal of silicon, by incorporating suitable impurities in different regions of the crystal. Resistors may also be constructed from silicon. Furthermore, a large number of diodes, transistors, and resistors may be fabricated within the same single crystal using the techniques to be described in this chapter. If these devices are appropriately interconnected, a circuit is formed, all of whose elements are contained within a tiny crystal of silicon. The term *integrated circuit,* or IC, has been adopted to describe such a structure.

It is instructive to compare integrated circuits to the simpler devices. For example, a transistor and an integrated circuit are shown in Figure 7.1. In outward appearance, Figure 7.1(*a*), they are similar, except that the IC has more leads. Furthermore, when the devices are opened up, as in Figure 7.1(*b*), the similarity is maintained. In each case, a tiny piece of semiconductor, called a *chip,* is found within the much larger container, which is called a *header.* The semiconductor chip contains the complete transistor or integrated circuit; the function of the header is merely

(a)

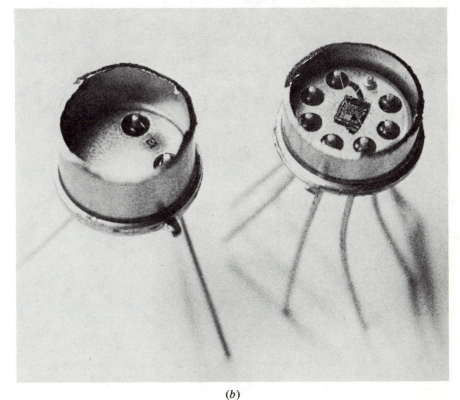

(b)

Figure 7.1 Microphotograph of a transistor and an integrated circuit. (a) The encapsulated devices. (b) The devices with the packages opened. The transistor and the integrated circuit are the tiny "chips" within the packages.

to protect the semiconductor and act as a support for the external wires.

Now let us consider the relative cost of producing these structures. The cost of producing a typical diode or transistor chip is a few pennies or less, and has steadily decreased. However, the cost of the header generally results in a much higher cost to the user. Thus, until the advent of integrated circuits, the user of semiconductor devices was largely unable to take advantage of either their inherent small size or small cost, since the basic device must be enclosed in a large and expensive header.

Integrated circuits have altered the entire picture. The integrated circuit is built within a single chip, and is contained within a single header. As a result, in many cases the entire circuit is as cheap as a single transistor.

Additional savings accrue from not having to assemble the circuit components and solder them together. Moreover, the elimination of numerous

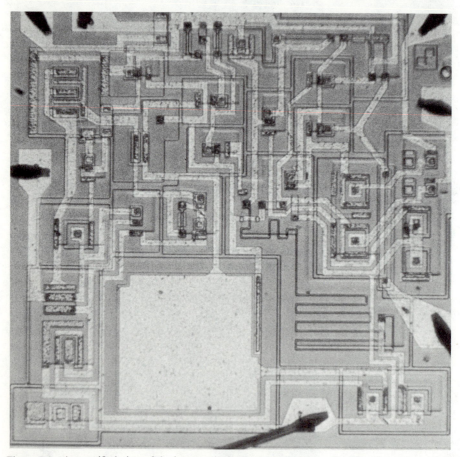

Figure 7.2   A magnified view of the integrated circuit shown in Figure 7.1. The dimensions of the silicon chip are about 1.5 by 1.5mm. The circuit diagram may be found in Chapter 12, Figure 12.4. (Courtesy of SGS – Societa Generale Semiconductori, Milan, Italy)

external soldered interconnections that are prone to failure results in higher circuit reliability in integrated form.

In comparison with ordinary circuits, integrated circuits have at least four important advantages: the individual components are cheaper; the components are already assembled as a circuit, saving the cost of assembly; the circuit is more reliable; and the circuit is smaller. While all this seems almost too good to be true, it is true, and has led to a revolution in electronics that is far from complete.

The greater number of present integrated circuits, such as that in Figure 7.1, typically contain a few dozen transistors and perhaps one hundred components in all. However, much larger circuits are available, and are growing in importance. The industry has coined the term large-scale integrated circuits to describe those circuits that perform the function of a hundred or so conventional IC's. Circuits containing thousands of transistors are now becoming available, making possible miniature and inexpensive electronic systems that were undreamed of a decade ago.

A magnified view of the IC shown in Figure 7.1 is given in Figure 7.2. This circuit contains about 30 components, and occupies an area of about 2×10^{-2} cm^2. The thickness of the thickest component of the structure is about 10^{-3} cm. Hence, we may estimate the effective volume of a typical integrated circuit component, with average present technology, as about 10^{-6} cm^3. (A reduction of about 2 orders of magnitude is presently attainable.) In principle, then, a packing density of about 10^6 components/cm^3 is possible; all the electronics in a modern computer might be packed into a briefcase. There are many problems, such as interconnections, heat dissipation, and servicing that are yet to be solved before we see such a degree of miniaturization, but the way of the future is clear; electronic systems are going to get smaller, cheaper, and more reliable.

Now we take a closer look at what constitutes an integrated circuit. A cross section of a small portion of an IC is shown schematically in Figure 7.3(a). It consists of a silicon crystal partially covered with silicon dioxide (an insulator) and aluminum. The same view is given in Figure 7.3(b), with the aluminum removed. The electrical circuit integrated in this structure is shown in Figure 7.3(c). Let us examine the correspondence between this circuit and the actual structure. First of all, the resistor R consists of the thin p region toward the right, which connects the aluminum conductors labeled C and D. The extent of this p region may best be seen in Figure 7.3(b), where it is outlined by a broken line. Toward the left-hand side of the crystal is an npn transistor, which may be identified by the three leads, A, B, and C connected to the emitter, base and collector, respectively. Again, the extent of the emitter and base regions is indicated in Figure 7.3(b). Comparison of Figures 7.3(a) and 7.3(c) shows that the aluminum layer serves as the wires that interconnect the circuit elements.

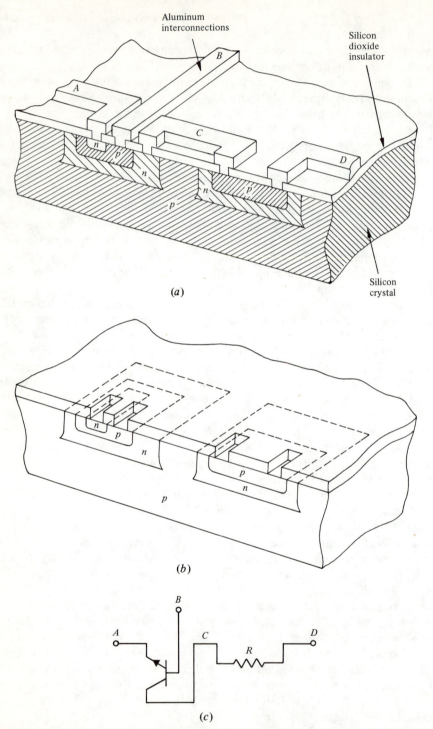

Figure 7.3 A small portion of an integrated circuit. (*a*) The physical structure. (*b*) The physical structure with the aluminum layer removed. (*c*) The electrical circuit.

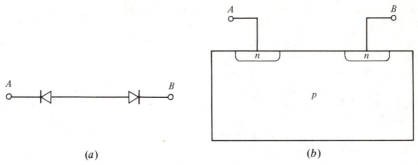

(a)   (b)

Figure 7.4 Two *pn* junction diodes connected back-to-back. (*a*) The electrical circuit. (*b*) The physical structure.

In reality the structure of Figure 7.3(*a*) is much more complicated than the circuit shown in Figure 7.3(*c*). In particular, the resistor and the transistor are connected, within the silicon crystal, by a series of *pn* junctions. How is it that these other junctions are neglected in Figure 7.3(*c*)? To answer this question we should examine the electrical characteristics of two *pn* junctions connected back-to-back, as in Figure 7.4. In such a connection, no current can flow either from *A* to *B* or from *B* to *A*, because for either polarity of bias, one of the junctions is reverse biased. Since no current can flow from *A* to *B*, independent of the voltage, the points may be regarded as unconnected, or isolated. The two large *n*-type regions in Figure 7.3(*a*) are isolated in this manner. This technique, called junction isolation, is the principal means by which all of the components of an integrated circuit are isolated. A more complete scheme of the partial IC of Figure 7.3(*a*) is shown in Figure 7.5. The parts of the circuit shown as dotted lines are the interconnections

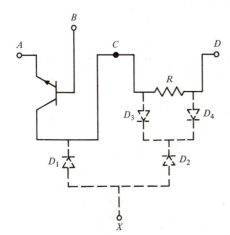

Figure 7.5 The complete electrical circuit of the structure shown in Figure 7.3(*a*). The portions of the circuit drawn with dotted lines may be deleted, because diodes D_1 to D_4 prevent any current flow in this subcircuit.

through the isolation junctions. Point X, which is the p-type substrate, is generally connected to the most negative potential in the circuit; thus diodes D_1 and D_2 can never become forward biased. Therefore diodes D_1 and D_2 can pass no current, and may be deleted from the circuit diagram.[1] Furthermore, diodes D_3 and D_4 are back-to-back; hence no current can flow from C to D via these diodes. They also may be deleted, and the simpler circuit diagram of Figure 7.3(c) is obtained.

Figure 7.3 represents only a small part of an integrated circuit, yet it displays the basic elements sufficient for constructing much larger, complete circuits: Transistors, resistors, and diodes are contained within a single crystal of silicon, but isolated from one another by reverse-biased pn junctions. The circuit elements are interconnected by an aluminum pattern on the surface. A thin silicon dioxide layer insulates the aluminum from the underlying silicon, except where connections are deliberately made. A complete circuit such as the one shown in Figure 7.2 is typically about 10 μm thick and a few square millimeters in area. All of the components essentially lie in one plane; such a structure is called *planar*.

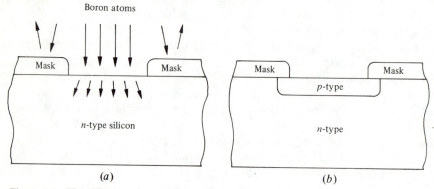

Figure 7.6 The diffusion of a p-type impurity, boron, into silicon. (a) A small region of a silicon wafer, greatly magnified, partly covered by a mask that selectively protects the wafer from the impurities. At a high temperature the impurities diffuse into the crystal, producing the structure shown in (b).

We consider next how an integrated circuit might be constructed. It would certainly be most economical if all of the components could be formed during a few basic processing steps. Owing to the fact that the structure is planar, it is possible to fabricate an entire circuit in only a few steps. For example, in Figure 7.3(b), the p-type region that functions as a resistor is formed simultaneously with the p-type region that functions as the transistor base. This step in the process is illustrated schematically in Figure 7.6. The desired p-type impurities are introduced into the crystal by a process known

---

[1] If two back-to-back diodes are very closely spaced, transistor action can occur, and the components may no longer be isolated. This effect may be used to advantage in fabricating pnp transistors, as discussed in Section 7.2.

as *solid state diffusion*. In this process the crystal is maintained at a high temperature, and the desired impurities are placed on the surface. They move into a crystal by diffusion. A *mask* is used to protect those areas where no impurities are desired, and openings are provided in the mask in the areas where a *p*-type region is desired. The diffusion process is described more fully in Section 7.1; it is mentioned here only as an example of a typical process step in which many components are fabricated in a single step. In the *p*-type diffusion step, all of the resistors and all of the transistors may be formed simultaneously. The silicon crystal is typically in the form of a wafer about 5 to 8 cm in diameter and 0.01 cm thick. A typical circuit takes up only a few square millimeters of area on this wafer; accordingly, many circuits can be fabricated simultaneously.

The cost of processing a wafer is almost independent of the number of circuits that it is to contain; hence the smaller the circuits, the more can be fabricated on one wafer and the cheaper they are to produce. This relationship between area and cost is referred to as "area economics." It is always desirable to make the circuits as small as possible; hence the individual components within the circuit should themselves be as small as possible. An estimate of circuit cost based on these economics is most revealing. It is generally agreed that the average cost of processing one silicon wafer using planar technology lies in the range $10 to $100. This includes purchasing the crystal and carrying out all the processes required to fabricate the finished circuits. To this cost must be added the cost of the header and mounting the individual circuits on the header, once they are separated. It should also be recognized that not all of the circuits will function. At any rate, the basic processing cost is about $1 to $10/cm². Since dozens of complete integrated circuits, or thousands of individual transistors, are fabricated in 1 cm², the cost per circuit or device is indeed very small.

In this chapter we examine both the processes and the components used to fabricate integrated circuits. By looking carefully into the construction of IC's, we can see, on the one hand, what considerations must go into the design of a circuit to be integrated, and, on the other hand, what properties we may expect of integrated circuits. We shall begin in Section 7.1 by discussing the technology—the processes by which the components of an IC are fabricated. In Section 7.2 we summarize the properties of typical components available to the IC designer and show how a simple integrated circuit might be laid out.

7.1   Modern Semiconductor Technology

In the early development of semiconductor electronics in the 1950s, germanium was the most important semiconductor. It is relatively easy to purify and to grow large crystals of germanium, and a number of processes for fabricating transistors and diodes were quickly developed. However, by 1960 it was clear that silicon was to replace germanium in almost all applications.

The most significant advantages of silicon are: (1) it is a common element (20 percent of the earth's crust) and is therefore cheap; (2) it has a larger bonding energy than germanium, which makes it superior for higher temperature operation;[2] and (3) it has a stable oxide which may be used to advantage in processing. When silicon is heated in the presence of oxygen, a layer of silicon dioxide (fused quartz) forms on its surface. This oxide may be used as a mask in the fabrication process and as an insulator in the final device or circuit. The latter is the most significant reason for the importance of silicon; the existence of silicon dioxide makes possible the entire integrated circuit technology of modern electronics.

The general term processing technology is used to describe the physical and chemical processes used to fabricate semiconductor devices. The most important steps in the technology are (1) crystal growth from the melt, (2) crystal growth from the vapor (epitaxy), (3) oxidation, (4) photolithography, (5) diffusion, (6) metallization, and (7) encapsulation. In addition, there is an important collateral process, mask-making, which is used to manufacture masks used in this processing. Each of these processes is now examined in turn.

Crystal Growth   To make reliable semiconductor devices and circuits, it is necessary to begin with a single crystal of the semiconductor. The quality of a good semiconductor crystal is comparable to that of the best gem crystals of diamond and ruby. Various chemical processes are used to prepare chemically pure silicon from raw ingredients, such as sand. Finally, pure silicon, in the form of powder or chunks, is placed in a crucible and heated to 1420°C, the melting point of silicon. An inert atmosphere is maintained over the molten silicon to prevent the formation of oxides. If, as is the usual case, some doping is desired, the appropriate amount of impurity is included with the silicon in the crucible. A small single crystal "seed" is inserted into the melt, which is maintained just at the melting point. The seed is slowly rotated and withdrawn, as shown schematically in Figure 7.7. The liquid silicon that clings to the seed freezes as it is withdrawn from the melt, and a crystal is grown. For economic reasons, it is desirable to grow as large a crystal as possible. In modern practice, 2- to 3-in. diameter crystals about 1 ft long are prepared.

The large crystals, known as ingots, are sawed into wafers typically 250 μm thick (250 μm $\approx$ 0.01 in.). Thus the silicon at this stage is in the form of round wafers 250 μm thick and 5 to 8 cm in diameter. One surface of these wafers is polished with successively finer abrasives until a mirror-smooth surface is achieved. Often an acid etching solution is applied as a final treatment to

---

[2] As noted in the discussion of Chapter 5, the intrinsic carrier concentration is larger in germanium than in silicon. At higher temperatures the intrinsic carrier concentration becomes so high in germanium that the doping no longer controls the carrier density. The device becomes essentially "swamped" with carriers and ceases to function properly.

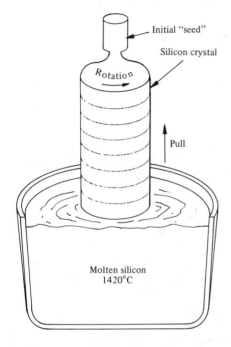

Figure 7.7 The growth of a silicon crystal. The raw silicon is melted, and maintained just at the melting point. A single crystal is inserted and slowly withdrawn, forming a large crystal.

remove any surface material whose crystal structure was damaged in the polishing steps. The wafer is ready at this point for subsequent processing steps.

Epitaxy As will be seen in Section 7.2, it is often convenient to have a thin layer of silicon of different conductivity type over the whole wafer, as illustrated in Figure 7.8.[3] Such a layer is called an *epitaxial* layer. (Epitaxy stems from the Greek language, and indicates that the layer is a continuation of the crystal structure of the substrate.)

Silicon may be grown epitaxially by several processes. The most prominent of these is $SiCl_4$ pyrolysis in which $SiCl_4$ gas is heated in the presence of hydrogen. At 1100° or 1200°C, silicon is deposited on the desired wafer surface. Sometimes another process, called the silane process, is used in which gaseous SiH_4 is heated and simply decomposes to form hydrogen and deposit silicon. In either case, a thin silicon layer of the desired thickness and doping is deposited on the substrate crystal.

---

[3] In this and subsequent figures, vertical distances are not drawn to the same scale as horizontal distances. Otherwise it would be impossible to show, for example, a 10-μm layer on a wafer that is greater than 10^4 μm in diameter. For the same reasons, distances in even one dimension are often distorted. For example, in Figure 7.8 the 10-μm layer is not in proportion to the 250-μm wafer thickness.

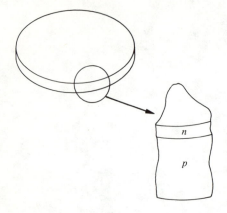

Figure 7.8   The cross section of a portion of a silicon wafer after the deposition of a thin expitaxial layer on the top surface. In this example an *n*-type layer is grown on a *p*-type substrate.

Oxidation   The principal oxide of silicon is silicon dioxide, a hard, clear glassy material, also called fused quartz. This oxide plays a major role in the fabrication and in the stability of silicon devices. It is a sort of super tough glass that is useful both as a mask in subsequent high-temperature processes, and as a permanent surface protection and insulator covering the devices within the silicon.

In semiconductor processing silicon dioxide is generally grown by the process of *thermal oxidation*. The silicon wafers are exposed to either oxygen or water vapor in a furnace at a high temperature (800° to 1300°C). This results in the formation of a layer of silicon dioxide on all exposed surfaces of the slices. A diagrammatic representation of the process is shown in Figure 7.9. As indicated in the figure, oxygen or water from the surrounding atmosphere diffuses through the existing silicon dioxide layer to combine with silicon at the silicon dioxide-silicon interface, thereby increasing the thickness of the oxide. By controlling the time and temperature of the oxidation, the device engineer can grow an oxide layer whose thickness is anywhere in the range of 0.01 to 10 μm.

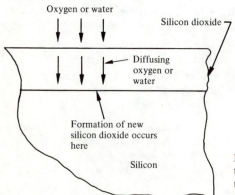

Figure 7.9   The oxidation of silicon. At a high temperature oxygen or water vapor diffuses through silicon dioxide to react with silicon, causing the oxide to thicken.

Oxidation Kinetics The simplicity of the oxidation process makes possible an accurate analysis and precise control of the oxide thickness. The oxidation process is almost completely controlled by the rate of diffusion of the oxidant through the oxide to the surface. Suppose silicon is being oxidized in steam. Then some form of water molecule or ion is present in high concentration C_1 at the oxide-vapor interface and in low concentration C_2 at the oxide-silicon interface (where its concentration is reduced because it reacts with the surface to form silicon dioxide). Since the concentration is lower at the oxide-vapor interface, the water molecules will flow toward the oxide-silicon interface by diffusion, as illustrated in Figure 7.9. (This flow is analogous to the diffusive flow of electrons and holes in semiconductors.) The magnitude of the flow is proportional to the slope in the concentration $(C_1 - C_2)/W$, where W is the thickness of the oxide. Furthermore, the rate of growth of the oxide is proportional to the flow, which may be expressed as

$$\frac{dW}{dt} = K \cdot \frac{C_1 - C_2}{W} \tag{7.1}$$

where K is the constant of proportionality. Equation (7.1) is a simple differential equation that may be solved by direct integration, as shown in Example 7.1. If the thickness W equals zero at $t = 0$, then at any time t

$$W = Rt^{1/2} \tag{7.2}$$

where R is a temperature-dependent constant.

EXAMPLE 7.1
Find the constant R in terms of the constant K by solving the differential equation.

SOLUTION
Equation (7.1) is first rearranged by multiplying both sides by W and dt:

$$W\,dW = K(C_1 - C_2)\,dt$$

Both sides are now integrable:

$$\frac{W^2}{2} = K(C_1 - C_2)\,t + \text{Constant}$$

If $W = 0$ at $t = 0$, then the constant must be zero. Clearly

$$W = Rt^{1/2}$$

where

$$R = \sqrt{2\,K(C_1 - C_2)} \qquad\blacksquare$$

Thus the oxidation kinetics are quite simple; at any given temperature, the oxide thickness is proportional to the square root of the oxidation time.[4]

---

[4] In the very first stages of oxidation the oxidation process may be complicated by such factors as the presence of initial oxide films. Thus very thin films require a more complex theory.

The device engineer can choose the appropriate combination of time and temperature to yield the desired oxide thickness. An experimental plot of oxide thickness versus time is given in Figure 7.10 for several temperatures. These curves are valid only for steam. If another oxidant, such as oxygen, is used, the oxidation rate is of course different.

EXAMPLE 7.2

In some processes it is desired to carry out the oxidation step at as low a temperature as possible. Suppose that a 1-μm oxide is desired, and that 3 hr is as long a time as may be allotted for the oxidation step. Find the minimum oxidation temperature.

SOLUTION

Because the oxidation rate increases with temperature, the minimum temperature for a given thickness is achieved by using the longest possible time, 3 hr. A 1.2-μm film is grown at 1100°C in 3 hr and a 0.8-μm film is grown at 1000°C in 3 hr. Interpolating between these values, a temperature of about 1050°C would yield a 1-μm oxide layer in 3 hr. ∎

It should be noted that the growth of silicon dioxide on silicon uses up silicon atoms, and thus reduces the thickness of the wafer. For every 1 μm of silicon dioxide grown, about 0.3 μm of silicon is used. Typical oxide thicknesses are in the range of 0.2 to 1.0 μm. Thus the effect on the overall wafer thickness is very slight.

Photolithography   As we have suggested in the introduction, hundreds of circuits are fabricated simultaneously in a single silicon wafer by a sequence of processing steps. For example, in a typical step, p-type impurities are introduced wherever transistor-base regions are desired.

It is necessary to have a means of defining those areas on the wafer to receive the particular treatment. In most cases, thermally grown silicon dioxide is used as a mask to select those areas into which impurities are introduced. Originally the oxide is grown over the whole surface; thus it must be selectively removed from appropriate areas. Photolithography offers an economical means of doing this. The entire surface of an oxide-covered wafer is

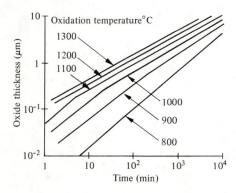

Figure 7.10   The oxidation of silicon in steam. The oxide thickness is a function of the temperature and time of oxidation.

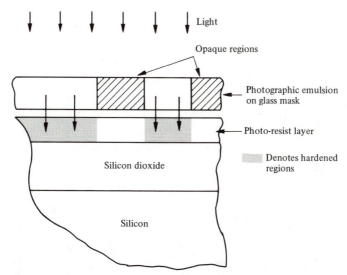

Figure 7.11 The exposure of photoresist through a photographic mask. Wherever the mask is transparent, the photoresist hardens, causing it to be insoluble in a developer. After exposure and development, the photoresist layer has the appearance shown in Figure 7.12.

coated with a photosensitive material called photoresist. This material is a lacquer that polymerizes or hardens when exposed to light, thus changing its solubility in a developing fluid.[5] A photographic mask containing the desired pattern is placed over the coated wafer. The wafer is illuminated by an intense light source through the mask, effectively making a contact print. The exposure process is illustrated in Figure 7.11. After exposure, the mask is removed and the photoresist is developed by dissolving the unexposed resist in a special solvent. The result is shown in Figure 7.12.

An important property of the photoresist is its insolubility in certain etching solutions. The wafer that is partially covered with photoresist (Figure 7.12) is placed in an acid solution capable of dissolving the silicon dioxide layer, and the oxide is etched away wherever there is no protective photoresist layer. The solution does not attack silicon; only the oxide is dissolved, and only at openings in the photoresist layer. This process of removing selectively silicon oxide is called making an *oxide cut.*

In another step, the remaining photoresist is removed in special solvent, leaving the wafer and oxide surfaces ready for further processing, as shown in Figure 7.13. The oxide layer now has open areas, called windows, corre-

---

[5] Both negative and positive photoresists exist, that is, resists that harden where struck by light and those that harden where not struck by light. In all of the examples here a negative photoresist is assumed.

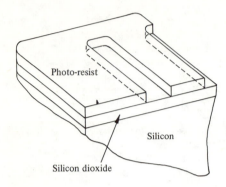

Photo-resist

Silicon

Silicon dioxide

Figure 7.12   The photoresist after development.

sponding to the dark areas of the photographic mask. In the next step impurities are introduced into the underlying silicon through these windows.

Diffusion   To fabricate an IC component, for example a transistor, it is necessary to introduce selectively impurities into the silicon crystal. Solid state diffusion offers a convenient and economical means of doing just this. At sufficiently high temperature, typically 1000° to 1200°C, impurity atoms can move through silicon by diffusion. The wafer is placed in a high-temperature furnace that also contains a source of the desired impurity atoms. A "boat" containing about 30 wafers in a diffusion furnace is shown in Figure 7.14. In those regions where no silicon dioxide protective layer exists, the dopant (typically boron or phosphorus) may diffuse into the wafer. This process is illustrated schematically in Figure 7.15. In this example boron is diffused into an n-type wafer, creating a pn junction wherever there is an opening in the oxide. Many variations exist in the details of diffusion, but all the variations have in common the principle of heating the wafer in the presence of the desired impurity. The depth of the diffused layer (the p-type layer in Figure 7.15) is controlled by the time and temperature of the diffusion.

Like the oxidation process, the diffusion process is readily amenable to precise analysis and control. Although the details of the mathematical analy-

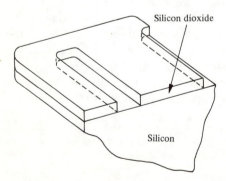

Silicon dioxide

Silicon

Figure 7.13   A silicon wafer with windows in the oxide. The originally uniform oxide is etched away in selected areas using photoresist as a protective film. Following the oxide etch, the photoresist is removed in a special solvent. The windows in the silicon dioxide layer correspond to the dark areas in the photographic mask, Figure 7.11.

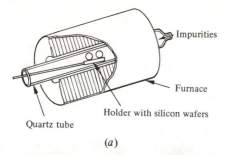

Impurities

Furnace

Holder with silicon wafers

Quartz tube

(a)

Figure 7.14 Doping of silicon by solid state diffusion. (a) A schematic view of a diffusion furnace, showing the position of the wafers. (b) A photograph showing about 30 wafers in a diffusion furnace. (Courtesy of Texas Instruments Inc.)

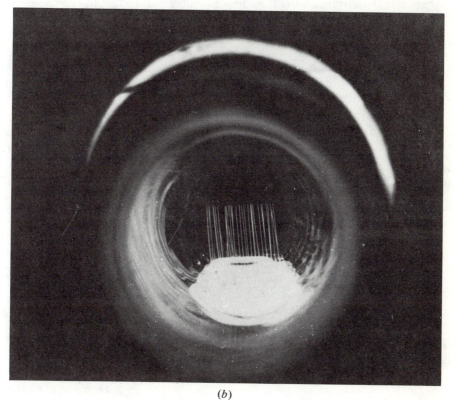

(b)

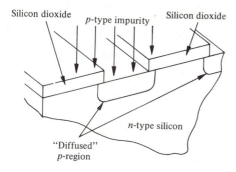

Silicon dioxide   p-type impurity   Silicon dioxide

n-type silicon

"Diffused"
p-region

T = 800 – 1200°C

Figure 7.15 The diffusion of a p-type impurity into an n-type silicon wafer. The impurity cannot diffuse through the silicon dioxide layer; therefore p-type regions are formed only beneath the windows in the oxide.

sis need not concern us here, we may state the principal result, namely, the concentration profile. A typical process results in an impurity distribution that varies as a Gaussian function of the depth into the crystal. That is, the impurity concentration $C(x)$ is given by

$$C(x) = C_0 \exp\left\{-\frac{x^2}{4Dt}\right\} \qquad (7.3)$$

where x is the distance from the surface (Figure 7.16), C_0 is the concentration at the surface ($x = 0$), t is the time duration of the diffusion, and D is a constant, called the diffusion constant. The constant C_0 depends on the duration of the diffusion, and both D and C_0 depend on the diffusion temperature. "Recipes," analogous to the oxidation recipe given in Figure 7.10, are used to select the time and temperature required for the required impurity concentration profile. It is important to realize that a p-type region may be produced in a crystal originally doped n-type, if sufficient p-type impurities are introduced. Suppose the original crystal contains 10^{15} phosphorus atoms/cm³, and in a certain region 10^{16} boron atoms/cm³ are introduced. The boron atoms accept the 10^{15} electrons from the phosphorus atoms, leaving 9×10^{15} boron atoms to accept bonding electrons, creating 9×10^{15} holes. Similarly, n-type impurities in sufficient concentration can "overdope" a p-type crystal, producing n-type regions wherever desired.

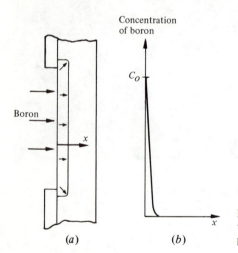

Concentration of boron

C_0

Boron

x

(a)   (b) x

Figure 7.16   The concentration of boron atoms versus depth following a solid state diffusion process.

EXAMPLE 7.3

A junction is formed by the diffusion of phosphorus into a uniformly doped p-type silicon wafer containing 10^{16} boron atoms/cm³.

The phosphorus has a concentration of $C_0 \exp -\{x/L\}^2$ where $L = 10^{-4}$ cm and $C_0 = 10^{20}$/cm³. (Here x is the distance into the wafer from the surface.) Find the junction depth and plot the concentration of majority carriers in the crystal.

SOLUTION

The junction occurs at the point where the phosphorus concentration just equals the background boron concentration; thus it occurs at the value of x satisfying

$$10^{20} \exp -\left\{\frac{x}{10^{-4}}\right\}^2 = 10^{16}$$

or

$$x = 3 \times 10^{-4} \text{ cm}$$

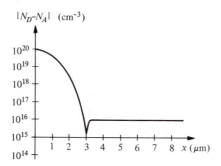

For $x < 3 \times 10^{-4}$ cm, the material is n-type and for $x > 3 \times 10^{-4}$ cm, the material is p-type.

Calling the phosphorus concentration N_D and the boron concentration N_A, a drawing of

$$\ln|N_D - N_A| = \ln\left|10^{20} \exp -\left\{\frac{x}{10^{-4}}\right\}^2 - 10^{16}\right|$$

is given in the figure shown.   ∎

The concentration profile resulting from a diffusion is not a step profile, that is, the concentration does not jump from one value to another. Nonetheless it may often be approximated as a step profile. A typical case is examined in Figure 7.17. An n-type impurity is diffused into a p-type wafer with a background acceptor doping of $10^{16}/\text{cm}^3$. The profile is sketched from Equation (7.3) in Figure 7.17, assuming $Dt = 10^{-8}$ cm^2. Both a linear concentration versus x plot and a log concentration versus x plot are shown. Also shown for comparison is the step approximation to the concentration profile. In an approximate analysis, the structure might be taken as an n^+p step junction with junction depth of 6 μm, and doping of about $10^{19}/\text{cm}^3$ and $10^{16}/\text{cm}^3$ on the n and p sides, respectively.

Following a diffusion step, the wafer is usually reoxidized in preparation for the next masking step. Whereas diodes require only a single diffusion step, transistors require two, and, as we shall see, most integrated circuits require three. The appearance of a wafer after a second oxidation is shown schematically in Figure 7.18. The oxide that is grown over the former windows is of

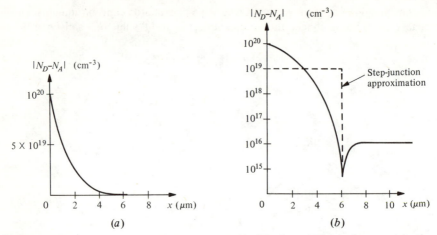

Figure 7.17   The concentration profile of a typical *pn* junction. (*a*) The doping versus depth. (*b*) The log of the doping versus *x*. Also shown is the step junction approximation to the profile.

course thinner than the remaining parts of the original oxide, the latter also thickening during the second oxidation.

Suppose we wish to make an *npn* transistor such as the one illustrated in Figure 6.4. Then it is necessary to form an *n*-type region (to serve as an emitter) within the confines of the *p*-type region. This is accomplished by a second sequence of photolithographic, oxide etching, and diffusion steps, as shown in Figure 7.19. A photoresist layer is applied and masked; the result

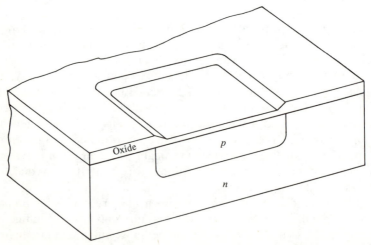

Figure 7.18   A small area of a silicon wafer that has undergone a *p*-type diffusion, and is reoxidized.

is illustrated in Figure 7.19(*a*). The size of the openings in the resist layer is different than in the first photolithographic step, and a second mask is required. Of course a very careful alignment must be made to assure that the second diffusion will be correctly placed with respect to the first. After oxide etching and photoresist removal, the second diffusion is carried out, in this case in a separate furnace with *n*-type impurities. The transistor structure appears as in Figure 7.19(*b*) at this stage.

If further diffusions are required, as in integrated circuit processes, subsequent oxidation, photolithographic, oxide etching, and diffusion steps are carried out in the fashion already described. When the diffusions are com-

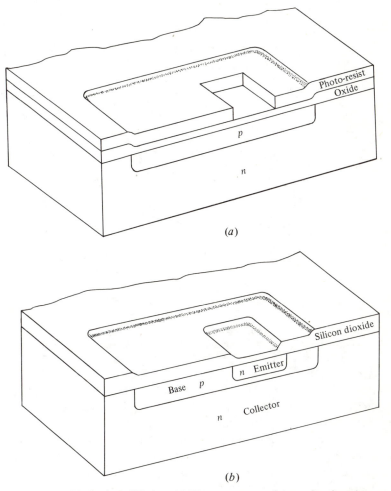

Figure 7.19   The emitter diffusion. (*a*) The appearance of the wafer after photoresist development. (*b*) The appearance after the oxide cut, photoresist removal, *n*-type diffusion, and second reoxidation steps.

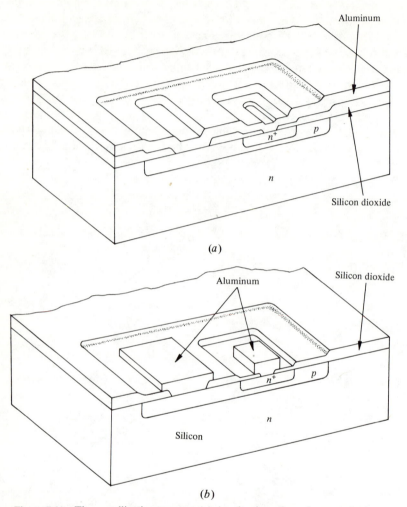

Figure 7.20   The metallization process. (*a*) An aluminum layer is evaporated over the entire structure. (*b*) The aluminum is selectively removed in a photolithographic and etching step.

plete, then the device, or devices, is finished and ready to be connected into a circuit. It is first necessary to make electrical contact to the various regions, such as the emitter and base of the transistor.

Metallization   Evaporated aluminum is generally used to make connections to desired areas on a semiconductor surface. The surface is first protected by growing an oxide and removing it (using the photoresist technology) only where contact to the bare silicon surface is desired. Next aluminum is evaporated over the entire surface of the wafer, and is then selectively removed, again using a photoresist process. A metallized wafer is shown in Figure 7.20(*a*) after aluminum evaporation, and in Figure 7.20(*b*) after unwanted areas of the aluminum are removed. The aluminum pattern may consist of simple contacts to which wires are to be attached, or it may be much more complex and interconnect a number of devices to form a complete circuit. The aluminum pattern on a large wafer may be quite complex as shown in Figure 7.21. The pattern for an individual circuit on the wafer is shown in a magnified view. The relatively large aluminum-covered regions around the borders of the circuit are the contact *bonding pads* — the places where external wires are to be connected to the circuit.

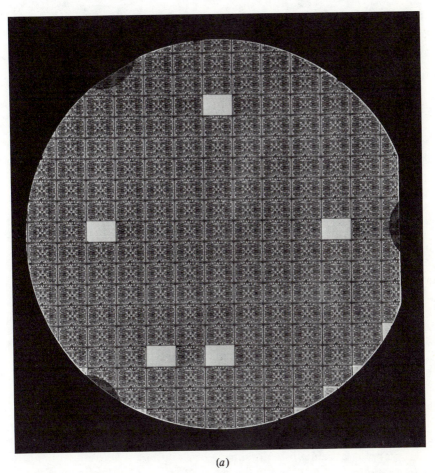

(*a*)

Figure 7.21   A wafer after the metallization step. (*a*) The entire wafer, showing an array of circuits. (*b*) [P. 290] A magnified view, showing the appearance of a single circuit on the wafer. (Courtesy of Siemens Aktiengesellschaft, Munich, Germany)

(*b*)

Figure 7.21   (*Continued*)

Encapsulation   The least dramatic, and at the same time the most expensive, part of semiconductor device and circuit technology is the encapsulation of the finished devices. All through the processing many devices or integrated circuits were processed simultaneously on a single wafer. When processing is complete, the devices are tested, and any defective ones are marked.

Then the wafer is diced, that is, the individual chips, each containing either a single integrated circuit or a single device, are separated. The defective units are thrown away, and the good ones are placed in some sort of container, wired, and sealed for protection. The primary function of the encapsulation is to protect the device from both moisture and mechanical abrasion. Traditionally the containers have been the familiar metal and glass headers; however, in recent times plastic headers, which are much cheaper, have been developed.

Integrated circuits often require a large number of external connections, necessitating a package with many leads. The three most common forms are shown in Figure 7.22.

Mask-Making   The large economies in producing circuits in integrated form stem from the fact that thousands of devices are fabricated simultaneously on a wafer of silicon a few inches in diameter. For each type of circuit to be fabricated in this manner, a set of photographic masks must be prepared. The masks contain the appropriate pattern for each step in the process. As we have seen in Figure 7.11, the mask consists only of light areas and dark areas, according to whether photoresist is to remain or to be removed from the corresponding area on the wafer. A sequence of masks, typically five or six in all, is required to fabricate a complete integrated circuit. Furthermore, the patterns on these masks must correspond closely; for example, the emitter of a transistor must be diffused within the base region, and must not overlap into the collector (Figure 7.19).

The usual process for mask-making consists of first drawing the masks several hundred times final size, and then photographically reducing them to the desired dimensions. The final masks contain the layouts of hundreds or thousands of identical circuits, each of which is made up of numerous transistors and other components. For instance, if a circuit takes up a square area 2 mm on a side, then about 500 to 1000 of these circuits can be fabri-

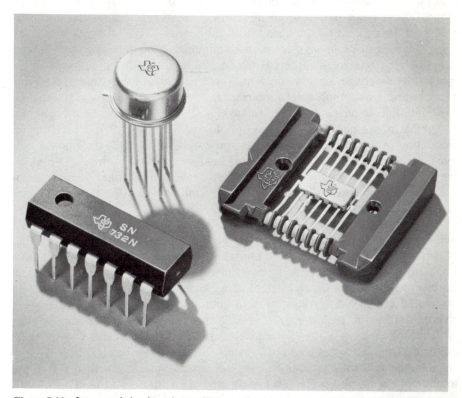

Figure 7.22 Integrated circuit packages. The round package is the traditional transistor header, but with many extra leads to accommodate the larger number of connections to the IC. The other two forms are known as the flat-pack and the dual-in-line package. (Courtesy of Texas Instruments Inc.)

cated on a single wafer. However, the original drawings for each of the masks are not drawn with 500 circuits. The desired pattern, say of the emitter diffusions for the transistors in a single circuit, is first drawn very large, say 1 m across. This drawing is then reduced to final size, say 2 mm across, in an optical system and recorded on a photographic plate. The plate is moved, in a very precise and reproducible fashion, and a second identical image is made beside the first. The plate is again moved, and the third image is made; the process continues until the entire plate is filled with identical images. This process of making a plate with multiple images is aptly named step and repeat. All the masks for a given circuit are produced in identical fashion on the same step-and-repeat camera. In this way, all of the components in the final circuit will be correctly placed with respect to each other.

7.2   Integrated Circuits

Integrated circuits for the most part are constructed from *npn* transistors, diodes, and resistors. The reason these devices are used largely to the exclusion of other types is that these three components are the most versatile, as well as the simplest and the cheapest to fabricate.

It is, of course, always desirable to minimize component size for the area economic reasons given earlier. However, there are several limitations on the minimum component size, arising both from the nature of the photolithographic process and from component requirements themselves. Let us first consider the limitations imposed by the photomasking steps. Because of the wave nature of light, it is not possible to make a photographic image of an object smaller than about a wavelength of light. As shown in Figure 7.11, the photomasks used in IC processing consist simply of light areas and dark areas. The wavelength limitation means that the minimum dimension of either a light area or a dark area on a mask is about 1 μm. This kind of optical limitation on image size is called a resolution limitation.

There are further resolution limitations imposed by the photographic mask, the optical system used for photoreduction and the photoresist used in IC processing. In practice, these limitations degrade the resolution to something closer to 10 μm than to 1 μm. (For example, it is a very difficult optical problem to produce a nearly perfect image over the relatively large area of an integrated circuit.) A further practical limitation on minimum object spacing is imposed by the requirement that each mask used in a process must align with the other masks; for example, the contact to a transistor emitter must contact only the emitter, and not overlap onto the base region.

In the following discussion, we will assume that a practical resolution limitation, taking into account all of the above effects, is 10 μm. This means, for example, that no two diffusions may be made within 10 μm of each other, or that no two sequential steps can be expected to align better than $\pm$10 μm. Thus in Figure 7.20, the spacing between the edge of the emitter and the edge of the base should be at least 10 μm. Further, the emitter must be at least 30 μm wide, in order to be sure that the oxide cut for the contact to the emitter falls entirely within the emitter. The 10 μm figure used here for the resolution limit is typical of routine IC practice; however, it should be realized that the actual value in a given process may be somewhat larger or smaller.

The various components of an integrated circuit are electrically isolated from one another by reverse-biased *pn* junctions, as described in the introduction. We will first examine these isolation junctions, and then the individual components in turn.

Isolation Junctions Figure 7.23 illustrates some of the steps in the formation of isolated integrated circuit components, here a transistor and a resistor. The main steps in the process are (1) the growth of an epitaxial *n*-type

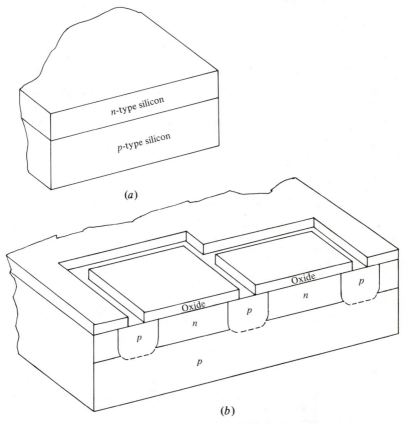

Figure 7.23   The steps in the formation of an integrated resistor and transistor. (*a*) A portion of the wafer after the epitaxial growth of an *n*-type layer on a *p*-type substrate. (*b*) After the isolation diffusion. (*c*) [P. 294] After the base (and resistor) diffusion. (*d*) After the emitter diffusion.

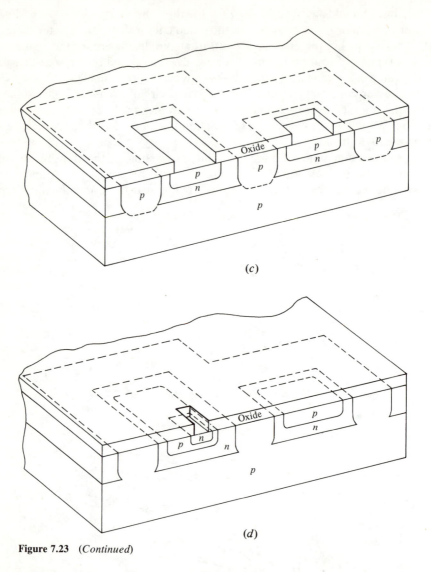

(c)

(d)

Figure 7.23 (*Continued*)

layer on a *p*-type substrate; (2) a *p*-type diffusion to form the isolation junctions; (3) a *p*-type base diffusion that also forms the resistor; and (4) an *n*-type
emitter diffusion. The wafer is shown schematically in Figure 7.23 after each
of these processes. It is understood that oxidation and photolithographic steps
are carried out in each stage of this procedure.

The *isolation diffusion,* shown in Figure 7.23(*b*), transforms the *n*-type
epitaxial layer into a group of *n*-type "islands." When the circuit is complete
and operating, the *p*-type substrate is connected to the most negative potential

in the circuit, assuring that none of the isolation junctions can be forward biased. An extraordinarily large spacing is required for the isolation diffusion. Suppose a 10-μm window is opened, and the diffusion is made to a depth of 12 μm. (This is a minimum "safe" diffusion depth for an epitaxial layer of 10 μm thickness.) Then, as shown in Figure 7.24, the impurity diffuses 12 μm to the side as well, resulting in a total *isolation wall* thickness of 34 μm. In general epitaxial layers are about 5 to 10 μm thick; hence we will adopt 30 μm as an approximate isolation wall thickness. Care must be taken not to diffuse another component, such as a transistor base, within 10 μm of this wall in a subsequent process. Let us now see how the isolation and masking size restrictions affect the design of specific components.

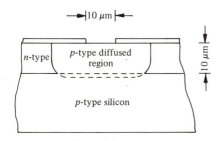

Figure 7.24   The diffusion of an isolation wall. Impurities diffuse in all directions, resulting in an isolation wall about three times as wide as the thickness of the epitaxial layer.

Transistors   We have already used the *npn* transistor as an example in the previous discussion. Two diffusions, one *p*-type and one *n*-type, are required to fabricate it within an *n*-type island. This is the standard process, and it is generally required that all other components be fabricated simultaneously in this process; no extra diffusion steps are allowed.

It is, on the one hand, desirable to make the area of a device as small as possible, to make as many devices as possible in a given silicon wafer. Fortunately, it is also desirable from the point of view of device performance to make transistors as small as possible. It may be shown that the speed of transistor response improves as the size of the transistor decreases. Let us now examine a typical IC transistor, shown in Figure 7.25. The cross-sectional view (*a*) shows half of the complete transistor and island, and the surface view (*b*) shows the complete surface area occupied by the transistor. The transistor turns out to be surprisingly large; let us see how this comes about. It might appear, for example, that it would be possible to diffuse a 10-μm square emitter rather than 30-μm. However, no contact could be made to a 10-μm square emitter without shorting to the base, because of the possible ± 10 μm error in making the oxide cut for the emitter contact. A careful examination of the figure shows that all spacings are held to the minimum allowed value with the exception of the base and collector contact openings. The transistor size could be slightly reduced by reducing the collector and base contacts to 10-μm squares. However, the larger contact openings minimize the collector

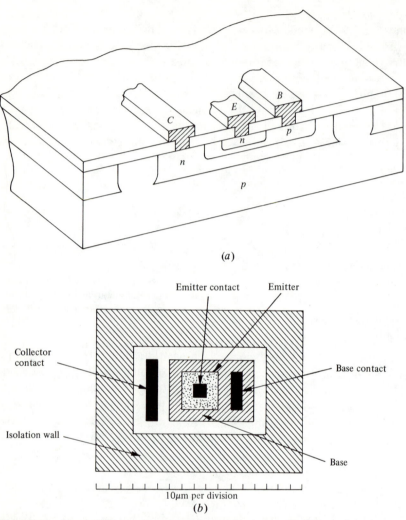

(a)

(b)

10μm per division

Emitter contact   Emitter

Collector contact

Base contact

Isolation wall

Base

Figure 7.25   A typical IC transistor. (a) Cross-sectional view. (b) Surface view with the metal removed. Lateral diffusion of the emitter and base, which are quite shallow, is ignored.

and base series resistance (important for high-frequency operation) at only a small cost in size. Therefore, we will adopt this geometry, which is typical in IC practice.

EXAMPLE 7.4

Find the size of the smallest possible IC transistor if a 10-μm resolution limit must be observed. Use 10-μm square collector and base contacts.

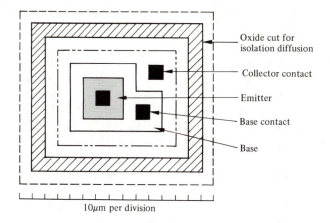

10μm per division

SOLUTION

The transistor geometry may be rearranged into the shape shown in the sketch. All spacings are held to 10 μm. The dimensions of this transistor are 150 μm by 130 μm including the wall, or 90 by 70 μm not including the isolation wall. These numbers compare with 170 by 130 μm or 110 by 70 μm for the transistor of Figure 7.25. ∎

With the spacings used in Figure 7.25, the transistor dimensions turn out to be 110 by 70 μm, which does not include the isolation wall thickness. Realistically, only half of the wall need be included in our estimate of the required total area, because the other half is regarded as belonging to the adjacent device. Thus for a typical wall thickness of 30 μm, the transistor requires an area with dimensions 140 by 100 μm. Although most of the area is taken up by the p-type isolation region and not by the actual transistor, this loss of space is unavoidable.

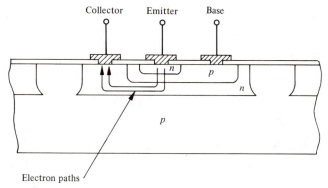

Figure 7.26 Electron flow in the npn transistor of Figure 7.25. The current must flow laterally in the thin n-type collector region, resulting in an excessive series resistance.

It is a simple matter to adjust these figures if the resolution in the processing is different than assumed here. Furthermore, if the basic scheme is altered, for example, if a much thinner *n*-type epitaxial layer is used, then the dimensions may be reduced.

There is a simple modification that is generally made to the basic *npn* transistor shown in Figure 7.25. It can be seen from Figure 7.26 that current flowing between collector and emitter must pass laterally through the *n*-type collector region. Since this *n* region is very thin, an undesired resistance in series with the collector results. This resistance may be minimized by including a highly conductive n^+ layer as a current path under the collector (Figure 7.27). A heavily doped *n*-type layer is diffused into the *p*-type substrate before epitaxial layer growth. During the growth of the epitaxial layer, this n^+ region is "buried" under the future transistor collector; hence, it is called a *buried layer*. The transistor is fabricated in the normal way, and from the surface appears like the transistor without buried layer.

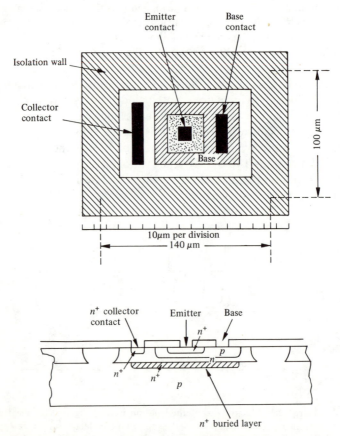

Figure 7.27   Cross section of a typical *npn* integrated circuit transistor with buried layer. The aluminum layer is omitted for clarity.

There is one other small difference between the devices of Figure 7.25 and Figure 7.27. In the latter there is a small n^+ region just under the collector contact. This n^+ region is diffused simultaneously with the emitter, and serves to make a low-resistance connection between the aluminum and the silicon collector region. Figure 7.27 is a realistic integrated circuit transistor, and will be referred to as our "standard" transistor. A photograph of a transistor from the circuit of Figure 7.2 is shown for comparison in Figure 7.28.

The properties of *npn* transistors used in integrated circuits compare favorably with those of discrete transistors. Moreover, integrated transistors have the advantage of similarity; two transistors on the same chip have nearly identical properties (such as β) and the properties remain identical with temperature changes. This close correlation of transistor properties may often be used to advantage in circuit design, as shown in Chapter 11.

Resistors As has been shown in the previous discussion, transistors and resistors are fabricated simultaneously in IC processing. The resistor is made during the transistor base diffusion, and in cross section looks just like the base. However, from the surface they look quite different. The resistor generally follows a meander pattern that allows various resistor values to be achieved. Several resistors may be seen in the photomicrograph of Figure 7.2. A layout of a small resistor, about 5000 Ω, is shown in Figure 7.29.

The doping and the thickness of the resistor are fixed by the transistor base requirements. The resistance is therefore selected by choosing the width and length. As in all resistors, the resistance is proportional to the length and inversely proportional to the area. Since the area equals the thickness times the width, and the thickness is fixed, we may write

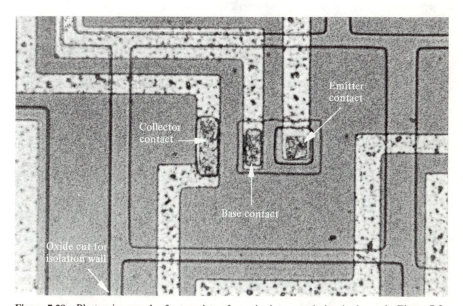

Figure 7.28 Photomicrograph of a transistor from the integrated circuit shown in Figure 7.2.

$$R = R_\square \cdot \frac{L}{W} \tag{7.4}$$

where $R_\square$ is a constant that depends on the thickness and resistivity of the region in which the resistor is formed. The constant $R_\square$ is called the sheet resistivity and is measured in ohms. In a typical IC process, $R_\square$ is about 200 Ω, and by modifying the base diffusion, values of $R_\square$ from 100 to 400 Ω may be obtained.

Assuming that $R_\square$ equals 200 Ω, according to Equation (7.4) a resistor of 2000 Ω would require a geometry in which the length is 10 times the width. The resistor of Figure 7.29 is about 25 times as long as it is wide; hence it has a resistance of about 5000 Ω.[6] Since the resolution limits the minimum width of a resistor to about 10 μm, large resistances can be achieved only by making the resistors very large. From the area economic point of view, the smaller the resistance value the better, because smaller resistors use less chip area and are therefore cheaper. It is interesting to note that the 5000-Ω resistor of Figure 7.29 uses up more chip area than the transistor (Figure 7.28). This apparent high cost of resistors is somewhat ameliorated by the fact that very often several resistors may be placed in the same n-type isolated island, as shown in Figure 7.30. Nonetheless, resistors larger than about 10 kΩ cost more chip area than transistors, and should be avoided in IC design.

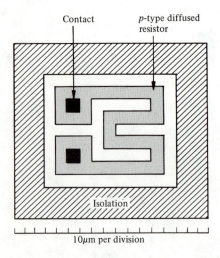

Figure 7.29   An integrated circuit resistor.

---

[6] The calculation is complicated by the corners and by the contacts at the ends. Of the 25 squares in the meander pattern, six are corners, causing the resistance to be somewhat less than 25 times $R_\square$. However, the extra resistance between the contact and the start of the meander pattern partially compensates for this error. In these calculations we are content with a rather rough estimate; variations in $R_\square$ make more accurate computation pointless.

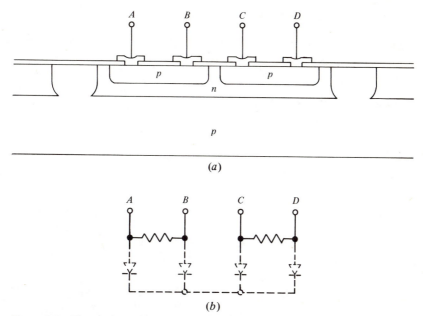

(a)

(b)

Figure 7.30 Two resistors in a common *n* region. (*a*) Cross section. (*b*) The electrical circuit. The circuit paths through the dotted lines all pass through back-to-back *pn* junctions, and therefore may be deleted. The resistors are electrically isolated.

EXAMPLE 7.5

Lay out a 20 kΩ resistor in an approximately square region. Use the minimum area required with the 10-μm spacing rule. Assume $R_{\square} = 200$ Ω.

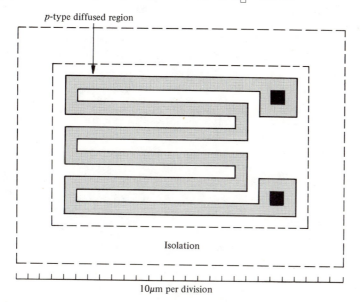

SOLUTION

According to Equation (7.4) the resistor must be approximately 100 times as long as it is wide. One possible arrangement is shown in the sketch. (In this pattern the chain is 97 squares long, and the resistance of the chain would be somewhat less than 20 kΩ if the sheet resistances were exactly 200 kΩ.) It is interesting that three standard transistors would fit inside the isolation wall shown here.   ∎

Integrated circuit resistors differ from ordinary discrete resistors in two important respects: (1) It is impractical to specify IC resistor tolerances to closer than ±10%, and ±20% is preferred. This unavoidable resistance variation stems from the uncertainties in the diffusion process that is used to form the resistor. (2) The IC resistor is temperature sensitive; an increase in resistance of 0.2% per degree centigrade is typical. On the other hand, IC resistors have a characteristic that may often be used to advantage: resistor ratios are determined almost completely by geometry. Thus two identical resistors on the same chip will have nearly identical resistances, even though the resistance value may vary with temperature and from chip to chip. Resistor ratios may be expected to be reproducible within ±1%.

Diodes and Diode Arrays   Diodes are common components in all integrated circuits, but are especially important in some digital integrated circuits. There are numerous ways to make diodes with the same processing steps as are used for transistors. The two most obvious consist of simply using either the transistor emitter-base junction or the base-collector junction as a simple *pn* junction diode, as illustrated in Figure 7.31. These two diodes may be labeled type *BE* or type *BC*, respectively. Actually, there are several possible variations on the type *BE* diode, according to what is connected to the unused *n*-type collector region. Generally, the collector is merely shorted to the base, as suggested in Figure 7.31(*b*). The different types of diodes have somewhat different properties, such as forward voltage drop, series (excess) resistance, switching speed, and, of course, surface area. A major difference between type *BE* and *BC* diodes is that type *BE* diode has a very low reverse breakdown voltage, typically about 6 V. Let us see by means of an example how one particular type might be selected.

If the reader glances ahead to Chapter 8 he will find that many practical logic circuits contain diode **AND** gate arrays of the form given in Chapter 5 (Figure 5.27). The electrical connections are shown in Figure 7.32(*a*), and a possible physical realization is shown in Figure 7.32(*b*). Because the diodes have a common connection, namely, the *p* side, it is possible to save much space by incorporating them within a single isolated island. This saving stems from the need for only a single isolation wall, which, as we have seen in the previous discussion, is the most space-wasting part of any component. In the arrangement shown in Figure 7.32(*b*), the common contact *x* is made along the entire strip of diodes. The total area could be reduced by using a smaller contact, but an objectionable series resistance would result, and the larger contact is preferred.

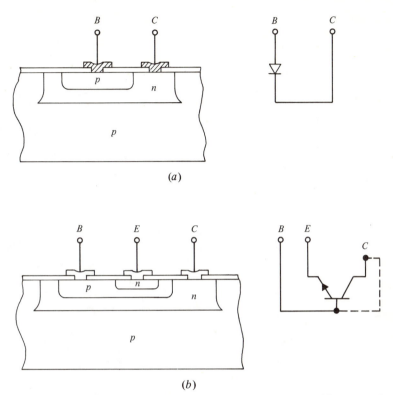

Figure 7.31 Two possible forms of integrated circuit diodes. (*a*) The type *BC* diode. This junction consists of the base-collector junction for an IC transistor process. The emitter is omitted. (*b*) The type *BE* diode. This junction consists of the emitter-base junction for an IC transistor process. Often the collector and base are shorted.

In a sense, the diode array of Figure 7.32(*b*) is really just a big transistor with several emitters. If the base-to-collector short is removed, it can in fact operate as a transistor. Such multi-emitter transistors are useful digital circuit components, as discussed in Chapter 8.

Interconnections and Contacts Just as a minimum space must be allotted for devices within the active semiconductor, the resolution limits also dictate a minimum size and spacing of the aluminum "wiring" on the surface. For simplicity, we assume that the tolerances for the aluminum pattern are the same as for the devices, in the neighborhood of 10 μm. In a relatively small circuit there is generally room on the surface for all the interconnections; however in large circuits there is often difficulty in making all the required interconnections on the surface. It is very commonly the case that two conductors must cross. Just as we find it difficult to draw the circuit diagrams of large circuits without crossovers, it is difficult to interconnect the devices in a large

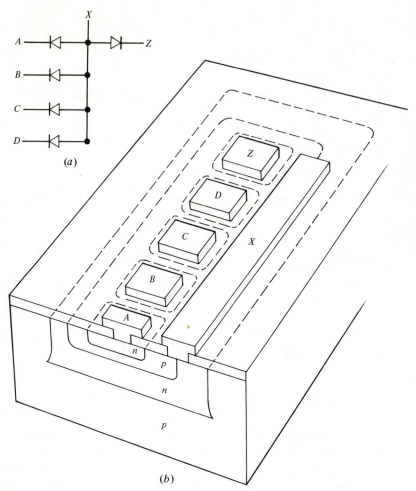

Figure 7.32 An isolated array of five *pn* junction diodes. (*a*) The circuit arrangement.
(*b*) A possible physical arrangement in an integrated circuit.

IC without crossovers. One way of making crossovers is to use multilayer
metallization. An oxide is deposited on top of the completed first layer, holes
are made in the oxide in appropriate places, and a second layer is deposited
and photoetched. This procedure is sometimes used; however it is often avoid-
able by using an ingenious structure known as the diffused cross-under.
Wherever two wires are to cross, a diffused low-value resistor is fabricated
underneath the oxide. As illustrated in Figure 7.33, the resistor is used to connect the "broken" line. The isolation of the cross-under is omitted in this figure,
although it is necessary in many cases.

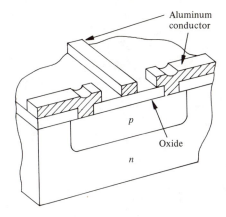

Aluminum conductor

p

Oxide

n

Figure 7.33 The diffused cross-under. Two aluminum wires on the surface can effectively cross over by routing the current under one wire through a low-value resistor.

Special regions must be provided in the aluminization pattern where contacts are made to the external world. These areas are called the *contact pads*. As may be seen from Figure 7.2, an area about 100 μm on a side is provided, to which wires are connected. This relatively large area is necessary to accommodate the large wires that are cold-welded to the aluminum. A schematic view of a bond between an external wire and a bonding pad is shown in Figure 7.34. In small circuits these pads can use up an appreciable portion of the chip area.

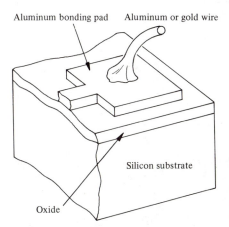

Aluminum bonding pad   Aluminum or gold wire

Silicon substrate

Oxide

Figure 7.34 A contact pad. A large aluminum pad is provided wherever an external wire is to be connected to the circuit.

Special Devices   The majority of integrated circuits are made of the three basic components: *npn* transistors, resistors, and junction diodes. However, it should be realized that there are a number of special devices that may be used if necessary. We briefly examine a few of the more important of these devices that may be fabricated in the standard three-diffusion, single-metallization IC process.

pnp Transistors   In some circuits it is most convenient to have both *npn* and *pnp* transistors available. Let us see how *pnp* transistors can be made during the standard *npn* processing. A so-called substrate *pnp* is shown in Figure 7.35(*a*). A comparison with Figure 7.35(*c*) shows that in this transistor, the emitter is formed by the standard *npn* base diffusion; the base is the normal *npn* collector region and the collector is the *p*-type substrate. Although this transistor functions quite well, it lacks versatility in that the collector is always connected to the most negative voltage source.

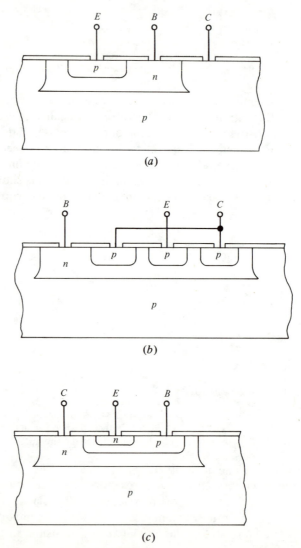

Figure 7.35   *pnp* transistors. (*a*) The substrate *pnp*. (*b*) The lateral *pnp*. (*c*) The standard IC *npn* transistor for comparison.

A second kind of *pnp* transistor, called a lateral *pnp*, is shown in Figure 7.35(*b*). This device makes use of the standard *npn* base diffusion to form both its emitter and collector. The *n*-type epitaxial layer forms the base. This kind of *pnp* transistor is fully isolated, and can be used in any circuit. It suffers, however, from two failings that limit its usefulness to special applications. Both of these limitations stem from its rather wide base region. It takes the injected minority carriers from the emitter a relatively long time (about 10^{-7} sec for a 10-μm wide base) to diffuse from emitter to collector. This delay not only limits the speed of response of the transistor, but it also makes the journey a fatal one for many minority carriers; they recombine before reaching the collector. The result is that beta is rather small, typically in the range of 1 to 10. The lateral *pnp* then is a low-speed, low-gain device.

Pinch Resistors   In some circuit applications where very large resistance is required, and where the precise resistance value is not critical, a special resistor, the so-called pinch resistor, may be used. A cross section of a pinch resistor is shown in Figure 7.36(*a*), and a photomicrograph of a pinch resistor (in the circuit of Figure 7.2) is shown in Figure 7.36(*b*). As may be noted, the

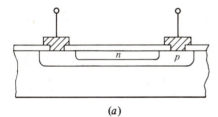

(a)

Figure 7.36   A pinch resistor. (*a*) Schematic cross section. (*b*) A photomicrograph of a pinch resistor from the circuit shown in Figure 7.2. The upper *n*-type layer reduces the thickness of the resistor, making larger resistance values feasible.

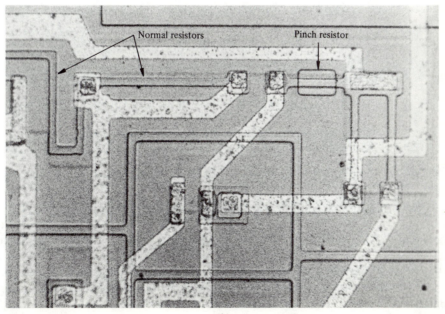

(b)

pinch resistor is much like the standard resistor, except that an n^+ layer diffused into the top reduces the effective thickness. This n^+ layer is introduced in the standard *npn* emitter diffusion step; hence no extra processing steps are required. Resistor values of up to 1 MΩ are feasible if pinch resistors are used. However, because of the small thickness, it is difficult to control the resistance value accurately. Furthermore, pinch resistors are limited to low voltage applications, because the upper *pn* junction typically breaks down at about 6 V.

Capacitors   Many analog circuits require the use of capacitors to "tailor" their frequency response. For example, it is shown in Chapter 12 that a capacitor is necessary to stabilize an operational amplifier against unwanted oscillations. It is possible to build capacitors of good quality into integrated circuits, but as we shall see, the range of practical values is quite restricted. The usual form of an integrated circuit capacitor is shown in Figure 7.37. It is a simple parallel-plate structure, consisting of a silicon dioxide dielectric sandwiched between an aluminum upper plate and a silicon lower plate. The name MOS has been adopted to describe this structure, standing for metal-oxide-semiconductor. The standard parallel-plate formula for the capacitance is valid:

$$C = \frac{A\epsilon}{W} \tag{7.5}$$

where C is the capacitance, A is the area, ϵ is the dielectric constant of the capacitor dielectric (equal to 0.2×10^{-12} F/m for silicon dioxide), and W is the thickness of the dielectric, that is, the spacing of the plates.

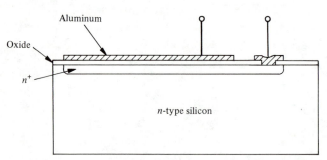

Figure 7.37   A MOS capacitor. This structure is a parallel-plate capacitor formed from a metal-oxide-silicon sandwich.

By controlling A, the capacitance may be made as small as desired. Going in the other direction, the capacitance may be increased by increasing its area or decreasing the oxide thickness. However, there are definite limits to both of these quantities. The area is limited by cost; for example, it would be poor economics to use half the chip for a single capacitor. If a circuit requires a single capacitor, a practical upper limit on its size might be about 10^{-3} cm³, that is, 10 percent of a typical 1-mm² chip. The oxide thickness W cannot be arbitrarily decreased, because there is a maximum electric field that the oxide can stand without breaking down. This maximum allowable field is about 10^7 V/cm. An average application might have a breakdown voltage requirement of 30 V, and this sets a lower thickness limit of 300 Å on the oxide thickness. Allowing a little safety factor, a more reasonable limit might be 500 Å. The largest practical MOS capacitor with $A = 10^{-3}$ cm² and $W = 500$ Å has a capacitance of about 4×10^{-11} F, or 40 pF. It is interesting that the 30 pF capacitor in the circuit of Figure 7.2 takes up about 7% of the total chip area.

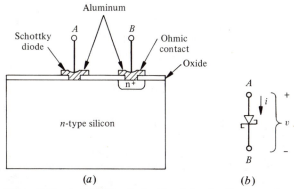

Figure 7.38 A Schottky diode. The Schottky diode consists of a contact between a metal and a lightly doped semiconductor. It rectifies much like a *pn* junction. (*a*) The physical structure. (*b*) The circuit symbol.

Schottky Diodes Another component finding increasing application in integrated circuits is the Schottky diode. This device is merely a special form of a metal-semiconductor contact. As shown in the discussion in Chapter 8, it behaves much like a *pn* junction, except that the forward voltage drop is somewhat lower for the Schottky diode. A typical structure and the corresponding circuit symbol are shown in Figure 7.38. Since no diffusion is required, the device is simple to fabricate, and is small in area. For reasons that become clear in Chapter 8, the Schottky diode is often used in parallel with transistor base-collector junctions, as in the circuit in Figure 7.39. This arrangement is particularly simple to fabricate, because the base contact merely overlaps part of the collector region.

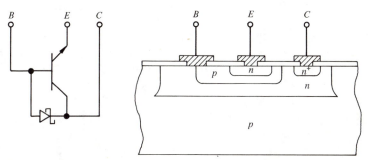

Figure 7.39 A Schottky diode in parallel with the collector-base junction of an *npn* transistor.

Integrated Circuit Layout   Having briefly examined the typical components used in integrated circuits, it is now possible to put these components
together to form a complete integrated circuit. To design an integrated circuit
it is necessary to devise a geometrical arrangement of the standard components. A metallization pattern is also designed, both to interconnect the components and to make external connections.

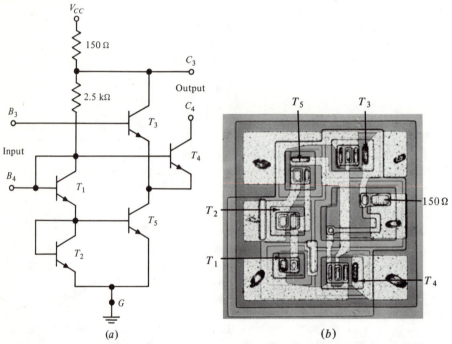

(a)   (b)

Figure 7.40 A simple integrated circuit. (*a*) The circuit diagram. This circuit, the SGS
L103T2, is a high-frequency differential amplifier. (*b*) A photomicrograph of the IC chip. The
transistor, resistor, diodes, and contact pads are easily identified. The chip is 0.5 mm on a side.
(Courtesy of SGS — Societa Generale Semiconductori, Milan, Italy)

An example of a simple amplifier circuit is shown in Figure 7.40(*a*). This
circuit contains two resistors and five transistors, two of which are used as
diodes. The integrated circuit, with the various components identified, is
shown in Figure 7.40(*b*). The size of the components may be compared with
the "standard" components discussed in this section. (The two transistors T_3
and T_4 have double-base contacts for the purpose of obtaining a very low
excess base resistance.) It is interesting that this circuit is contained in a chip
0.5 mm on a side.

EXAMPLE 7.6

The circuit in sketch (*a*) is an elementary flip-flop, and is discussed in Chapter 8 (Figure 8.28). Lay out this circuit using the standard components just derived. Assuming an *n*-type epitaxial layer of thickness 10 μm, a resolution of 10 μm, and a *p*-type (base) diffusion characterized by $R_\square = 200$ Ω, what is the total area of the circuit?

SOLUTION

We wish to fabricate four resistors and two transistors, so it would appear that six isolated *n*-type islands are required. However, as shown in Figure 7.30, all the resistors may be included in a common *n*-type island. Furthermore the resistors may be

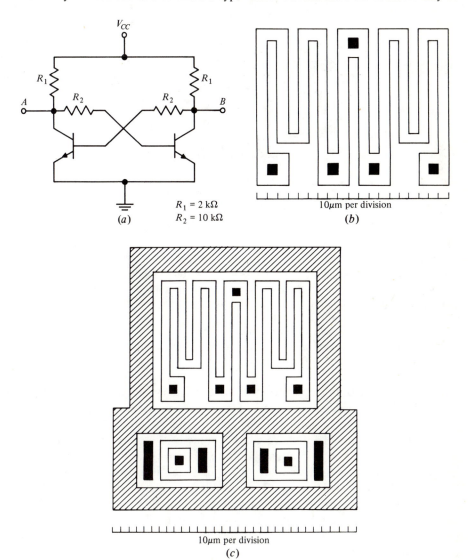

$R_1 = 2$ kΩ
$R_2 = 10$ kΩ

(*a*)

10μm per division

(*b*)

10μm per division

(*c*)

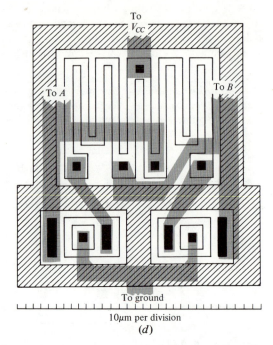

To
V_{CC}

To A

To B

To ground

10μm per division

(*d*)

combined as one resistor with "taps" at the desired connection points. The *p*-type base diffusion that forms the resistors is characterized by $R_\square = 200\ \Omega$, hence a length-to-width ratio of 10 and 50 is required for the 2- and 10-kΩ resistors, respectively. A possible resistor pattern is shown in sketch (*b*). Now the components are arranged in a compact pattern, and an interconnection layout is attempted. For many arrangements, a single layer interconnection pattern is not possible. One fairly compact arrangement for which a single layer interconnect pattern is possible is shown in sketch (*c*). Now the metallization pattern is drawn, again observing the minimum spacing rules [see sketch (*d*)]. The aluminum lines are made wider than the 10-μm minimum. The chip size turns out to be about 320 μm $\times$ 320 μm.   ■

Integrated circuits operate much like ordinary electronic circuits constructed from discrete components. However, the unique properties of IC components can lead to both superior and inferior circuit performance. From the discussion of this chapter, we can list some of the more important factors which should be taken into consideration in IC design: (1) *npn* transistors in integrated circuits are substantially equivalent to discrete transistors in performance; (2) two transistors on the same chip have nearly identical properties; (3) *pnp* transistors with low beta or with collector tied to the most negative supply voltage may be fabricated; (4) resistors of up to 5 or 10 kΩ are equivalent to transistors in cost; (5) large resistors take up a large chip area and should be avoided; (6) IC resistors may be specified to only about ±10%, and the resistance is very temperature sensitive; (7) the resistance ratio of two resistors on the same chip may be accurately specified, and is almost independent of temperature.

Summary

- An integrated circuit is a circuit in which all of the transistors, resistors, and diodes are contained within a single silicon chip. The devices are isolated from one another by junction isolation within the chip, and are interconnected by an aluminum metallization pattern on the surface.

- With planar processing, thousands of devices, or hundreds of integrated circuits, are fabricated simultaneously on a single silicon wafer. The cost of processing a wafer is constant, so the cost of a given device or circuit is proportional to the area it occupies. The cost of packaging a device or circuit can contribute significantly to the total cost.

- The principal steps in integrated circuit technology are mask-making, crystal growth, oxidation, photolithography, diffusion, metallization, and encapsulation.

- The standard integrated circuit components are *npn* transistors, diffused resistors, and diodes. In addition, *pnp* transistors, capacitors, pinch resistors, and Schottky diodes are examples of other components that are occasionally used.

- Diffused resistors of less than 10 kΩ resistance are about equivalent to transistors in cost. The cost of larger resistors is approximately proportional to their resistance.

- Capacitors use up a relatively large amount of chip area, and capacitance values greater than about 40 pF are impractical.

- The parameters of components in integrated circuits cannot be controlled as accurately as in discrete devices. However, identical components, such as two transistors with identical β, are automatically produced. Resistance values cannot be specified to closer than $\pm 10\%$; however, resistance ratios are reproducible to better than $\pm 1\%$.

References

Treatments at approximately the same level as this book:

Camenzind, H. R. *Circuit Design for Integrated Electronics.* Reading, Mass: Addison-Wesley Publishing Company, 1968.

Gray, P. E., and C. L. Searle. *Electronic Principles.* New York: John Wiley & Sons, 1969.

At a more advanced level:

Lynn, D. K., C. S. Meyer, and D. J. Hamilton, ed. *Analysis and Design of Integrated Circuits.* New York: McGraw-Hill, 1967.

Warner, Jr., R. M., and J. N. Fordemwalt, ed. *Integrated Circuits: Design Principles and Fabrication.* New York: McGraw-Hill, 1965.

Grove, A. S. *Physics and Technology of Semiconductor Devices.* New York: John Wiley & Sons, 1967.

Problems

7.1   Suppose the diodes of Figure 7.4 obey the ideal diode equation with $I_S = 10^{-14}$ A. Find and sketch the current flowing from A to B for V_{AB} in the range -10 to $+10$ V.

7.2   Suppose a certain integrated circuit chip has dimensions 1.5×1.5 mm. Assuming that it costs \$40 to process a complete wafer $2\frac{1}{2}$ in. in diameter, but that only the central 2 in. are usable and that the yield of functioning circuits is 10%, what is the basic IC chip cost?

7.3   Evaluate and plot as a function of temperature the constant R in Equation (7.2).

7.4   Suppose the concentration of water molecules near the surface of a growing oxide at 1100°C equals $10^{17}/cm^3$ and that the concentration at the oxide-silicon interface equals zero. What is the value of the diffusion constant K in Equation (7.1)?

7.5   In integrated circuit fabrication it is desired from the point of view of cost to carry out a given process in as short a time as feasible. On the other hand, timed processes, such as oxidation, should not be too hurried (say less than 10 min) or else the timing accuracy becomes poor. Furthermore, there are often other considerations, such as maximum allowable temperature, which restrict the choices in process design. Find suitable time and temperature combinations for producing the following oxide thicknesses: (1) 0.05 μm, (2) 0.3 μm, (3) 1 μm, (4) 2 μm (maximum allowable temperature, 1150°C).

7.6   A wafer has a 1-μm oxide grown, is removed from the furnace, and has "windows" opened (the oxide is removed in certain portions of the wafer). It is replaced in the furnace and a 0.5-μm oxide is grown in the windows. How thick is the oxide over those regions where the original oxide is not removed?

7.7   Sketch a cross section of the wafer and oxide before and after the processes described in Problem 7.6. Is the silicon surface still flat?

7.8   A certain oxidation furnace has a temperature accuracy of ±10°C. Assume that the difference $(C_1 - C_2)$ in Equation (7.1) is proportional to the relative concentration of water vapor in the furnace atmosphere, and that the latter is controlled to ±20%. If the accuracy of the timing is ±1 min, estimate the possible range of oxide thickness in an 1100°C oxidation designed to yield an oxide thickness of 0.2 μm.

7.9   Sketch the concentration (semilog plot) versus depth for an n-type diffusion with surface concentration $10^{20}/cm^3$, and Dt equal to 10^{-8} cm². Find the junction depth if the background p-type impurity concentration is (1) $10^{15}/cm^3$, (2) $10^{16}/cm^3$, (3) $10^{17}/cm^3$. Can you state an *approximate* rule for the junction depth in terms of $(Dt)^{1/2}$, independent of the background concentration?

7.10   It is desired to produce an n^+p junction 4 μm below the silicon surface. Furthermore the n-type surface concentration is to be 10^{21} cm^{-3}. Find the value of Dt necessary, assuming the n-type impurity has a Gaussian profile, and that the background p-type impurity concentration is 10^{18}/cm^3.)

7.11   Two diffusion steps are required in transistor fabrication. Find suitable Dt values to produce (1) a base-collector junction 2 μm below the surface with base surface concentration equal to 10^{19}/cm^3; (2) an emitter-base junction 1.6 μm below the surface with surface concentration 10^{21}/cm^3. The n-type starting wafer is doped 10^{15}/cm^3. (Assume an average base doping of 10^{18}/cm^3.)

7.12   In one typical diffusion process the surface concentration C_0 is proportional to $1/\sqrt{t}$, and the concentration profile given by Equation (7.3) is obeyed. Show that the total number of diffusing impurities is constant under these conditions.

7.13   Find the minimum area of a transistor with two base contacts, such as the device of Figure 7.40. Use 10-μm spacings and a 30-μm isolation wall thickness.

7.14   Compare the area taken up by two transistors: (1) in separate isolation regions, (2) with common collectors. (Include half the isolation wall thickness.)

7.15   Compare the area taken up by (1) a 100-kΩ resistor; (2) two isolated transistors, two 10-kΩ resistors, and one 1-kΩ resistor. (As shown in Chapter 11 these two possibilities correspond to two methods of fabricating a constant current source.) Assume $R_\square = 200$ Ω.

7.16   Estimate the fractional reduction in transistor area possible if the epitaxial layer thickness were to be reduced from 10 to 3 μm.

7.17   Lay out a resistor array of 10 resistors of value 100 Ω; 200 Ω, 400 Ω ... 51.2 kΩ, assuming $R_\square = 200$ Ω. Assume all the resistors have one common contact.

7.18   In a typical forward-biased silicon diode, the voltage drop decreases about 2 mV/$^\circ$C. Assuming a constant current source is available, design a circuit to produce a constant voltage drop of 1.7 V using IC diodes and resistors.

7.19   The capacitor in the circuit shown in Figure 7.2 has a value of 30 pF. Find the oxide thickness.

7.20   It is desired to make an integrated RC filter consisting of a resistor and a capacitor in parallel (Chapter 4), which has an RC time constant of 2×10^{-6} sec. Choose suitable values for R and C with this RC product and lay out the filter.

7.21   What is the minimum area taken up by an integrated, isolated Schottky diode? How does this compare to a type EB diode and to a type BC diode?

7.22   Lay out a lateral pnp transistor. Try to surround the emitter, insofar as possible, by the collector (for the purpose of obtaining a high β).

7.23   Using the specifications given in Example 7.6, lay out a three-input DTL **NAND** gate (Figure 8.8) using type BE diodes.

7.24   Using the specifications given in Example 7.6, lay out a three-input TTL **NAND** gate (Figure 8.23).

7.25   Estimate the value of $R_\square$ in the circuit of Figure 7.40.

7.26   Using the specifications given in Example 7.6, lay out the amplifier stage given in Figure 11.11.

7.27   To estimate integrated circuit cost, let us assume the following: (a) The cost of processing a wafer is C_P. (b) An average wafer has a useful area of A_W cm^2.

(c) The size of an average IC component is K_1L^2, where L is the resolution limit ($L = 10$ μm in the examples used in the text). (d) An extra area A_C must be allowed on each integrated circuit for contacts. (e) The yield, which is the fraction of good circuits, is Y. (f) The cost of a header and encapsulation is H (and only good circuits are encapsulated). (1) Derive a formula for the final cost C_T of an integrated circuit in terms of the number of components N. (2) Simplify the expression and plot C_T versus N assuming $C_P = \$50$, $K_1 = 500$, $L = 10$ μm, $A_W = 20$ cm^2, $A_C = 10^{-2}$ cm^2, $Y = 0.1$, $H = \$1.00$. (ANSWER: \$2.50 for $N = 100$.)

7.28 (1) Assume the yield in Problem 7.27 is a function of the area taken up by the components in the circuit ($N \cdot K_1L^2$); in particular $Y = \exp - \{N K_1L^2/A_Y\}^2$. Find the cost C_T as a function of C_P, A_C, $N K_1L^2$, H, A_Y and A_W. (2) Using the same values given in Problem 7.27 (except Y) and $A_Y = 10^{-1}$, plot C_T versus N. (3) Assume that the cost of a system C_S built from integrated circuits equals the cost of the integrated circuits, plus a handling and interconnection cost C_H per circuit, plus a design cost proportional to the square of the number of circuits (with proportionality constant D). Assume the total system requires M components (M is the number of circuits times the number of components per circuit). Find C_S using the results of (1). (4) Plot C_S versus N assuming $C_H = \$1.00$ and $D = 1 \times 10^{-4}$ (\$), for three values of M: 1000, 10^6, 10^9. What do you conclude about the optimum size of an integrated circuit?

Chapter 8

Introduction to Digital Circuits

The increasing importance of digital electronics has developed with the rapid advances in electronic computers. Computers are constructed from digital circuits, and the increasing demands of the computer industry has spurred the development of a number of flexible digital building blocks. These building blocks, available today as tiny inexpensive integrated circuits, make possible the design of a variety of important digital systems. In this chapter we look inside some typical, practical, digital building blocks. The macroscopic problem, that of designing or analyzing a large electronic system constructed from building blocks, is taken up in Chapter 9.

A look at any semiconductor device manufacturer's catalog shows the availability of a great variety of digital integrated circuits. For purposes of application, the circuits are generally classified into compatible groups, that is, groups of circuits that may be interconnected to form a digital system. Thus the classification is not by function such as **NOR** or **AND**. (For example two **NAND** gates that have entirely different ranges of logical **1** or logical **0** are not compatible – the output of one gate cannot be used to drive the input of another.) The circuits are classified into *families,* such as the diode-transistor logic (or DTL) family considered extensively in this chapter. Generally, all the circuits in a given logic family appear to be similar; they are all constructed from the same components in similar arrangements. Most important, the circuits in a given family are compatible and may be used by a system designer with very little consideration of their details.

The majority of digital circuits are constructed from bipolar transistors, resistors, and diodes. (When it is desirable to incorporate very large circuits into a single integrated circuit, and when somewhat slower operating speeds can be tolerated, a second circuit type becomes important. These so-called MOS circuits are discussed in Chapter 13). We have already seen in Chapter 5 that **AND** and **OR** gates can be constructed from diodes and resistors alone. However, these diode gates have a serious drawback: the signal levels are degraded in passing from input to output. Thus an active element, a transistor, is included in practical circuits. The transistor can be used as an almost ideal electronic switch, in which an output signal at the collector is controlled by an electronic input signal at the base. In particular, the transistor inverter, discussed in Section 8.1, switches the output to low whenever the input is high, and vice versa. The inverter may of course be used as a simple logic inverter, but its importance goes far beyond that; the transistor inverter, when combined with a passive logic array such as a diode logic gate, forms a complete practical logical family of building blocks. This family, the so-called DTL family, is examined in Section 8.2. The emphasis in this section is on the relationship between the external characteristics of the blocks (such as logic levels and switching speed) and the internal circuitry.

In addition to logic circuits, there exists another whole class of digital circuits that are important in all kinds of digital systems. These circuits are distinguished by their ability to retain a history of previous inputs, that is, they have a memory. The purpose of Section 8.3 is to examine a typical circuit of this class and to show its relationship to the more elementary logic blocks.

8.1   The Transistor Inverter

Before analyzing a specific inverter circuit, let us first review the operation of the transistor, with special emphasis on its role as a switch. We first define the properties of an *ideal switch*. An ideal switch is a two-terminal device that is always in one or the other of two states. The switch is either in the "closed" state, in which case its two terminals are shorted together, or in the open state, in which case the terminals are open-circuited. The symbol for the ideal switch, shown in Figure 8.1, derives from the simple mechanical switch. Figure 8.1(*a*) depicts the switch in the closed state and Figure 8.1(*b*) in the open state. The corresponding *I-V* characteristics of the switch are given in Figure 8.1(*c*). In the closed state v is necessarily zero, independent of i, and in the open state i is zero, independent of v.

For a switch to be useful as a component in electronic circuits, it must be possible to control the state of the switch electrically rather than mechanically.

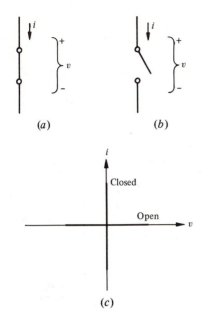

(a)   (b)

(c)

Figure 8.1   The circuit symbol and *I-V* characteristics of an ideal switch. The closed switch (*a*) has an *I-V* characteristic on the current axis, and the open switch (*b*) has a characteristic on the voltage axis.

A current or voltage, applied to a pair of input terminals, controls the state of the switch. The ordinary electromagnetic relay illustrated schematically in Figure 8.2 is an example of an electrically controlled, nearly ideal switch. In the relay, if the input current exceeds some critical current, the field of the electromagnet pulls the switch lever into the closed position. For lower currents, the switch is held open by a spring. The relay has many desirable features as a switch, but it compares badly with transistor switches in terms of speed, cost, and size.

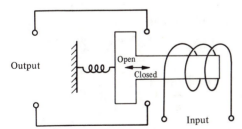

Figure 8.2   The electromagnetic relay. The state of the switch is controlled by the input current.

The Transistor as a Switch   Now let us see how a transistor can function as an electronic switch. In the common emitter configuration, the input appears between the base and emitter and the output between the collector and emitter (Figure 8.3). The *I-V* characteristics at both the input and output terminals of an *npn* transistor in the common-emitter configuration have already been derived in Chapter 6, and examples were shown in Figures 6.14

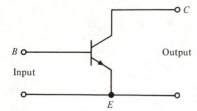

Figure 8.3 The transistor in the common-emitter configuration. The input appears between base and emitter and the output appears between collector and emitter.

and 6.15. The *I-V* characteristics of a typical *npn* silicon switching transistor are shown in Figure 8.4. These *I-V* characteristics are representative of devices in digital integrated circuits and differ slightly from the characteristics given in Chapter 6.[1] For reasons that will become clear, two curves on the output characteristics have been emphasized. In all the transistor circuits that we will be considering, the transistor will be restricted to operate in the region $v_{CE} \geq 0$, $i_C \geq 0$, that is, in the first quadrant of Figure 8.4(*b*). The output curves are drawn only in this region.

The use of a transistor as a switch is suggested by a comparison of Figures 8.4(*b*) and 8.1. If the input current i_B is zero, then the output *I-V* characteristic is that of an open switch. In other words i_C is zero, independent of v_{CE}, for $i_B = 0$. On the other hand, if i_B is large, for example 1 mA in Figure 8.4(*b*), the output *I-V* characteristic is very nearly that of a closed switch; that is, v_{CE} is very nearly zero for all values of i_C, when sufficiently large base current is applied. Thus we see that *the transistor acts as a switch that connects the collector to the emitter for large base currents, and leaves the collector open-circuited for zero base current.*

As is evident from Figure 8.4(*b*), when the base current is zero, the collector current is constrained to be nearly zero, independent of v_{CE}. In the discussion in Chapter 6 we labeled this condition *cutoff*, because essentially no currents flow. If desired, the very minute collector current that does flow may be evaluated by the large-signal transistor equations (see Problem 8.3). Typically, $i_C \ll 10^{-9}$ A in cutoff; hence, to a very good approximation, $i_C = 0$

Table 8.1   The Open Switch

Input	Output
$i_B = 0$	$i_C \cong 0$
$v_{BE} \lesssim 0.3$	

---

[1] Very often the beta of switching transistors is lower than that of general-purpose transistors. The lower value of β comes about from the measures described in Section 6.4, which are taken to increase switching speed.

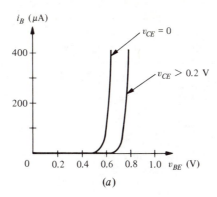

(a)

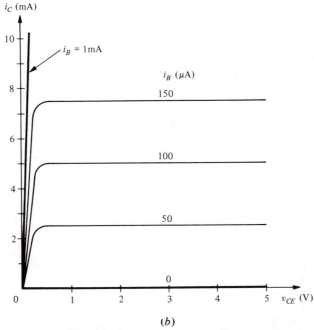

(b)

Figure 8.4 The common emitter *I-V* characteristics of a typical *npn* silicon switching transistor. (*a*) The input characteristics. (*b*) The output characteristics.

for $i_B = 0$. It may be seen from the input characteristics, Figure 8.4(*a*), that if $v_{BE} = 0$ the base current is zero and the transistor is cut off. In fact it requires a forward bias of about 0.5 to 0.7 V to produce any significant base current, so if $v_{BE} < 0.5$ V, the transistor is essentially cut off. We will adopt a very conservative rule: for $v_{BE} < 0.3$ V, the transistor is cut off. Summarizing these results, the transistor appears to be an open switch under the conditions given in Table 8.1.

Over most of the first quadrant of Figure 8.4(*b*) the collector current is proportional to the base current, with proportionality constant β. (We have labeled this region as the active region.) However, there is a region in the first quadrant of Figure 8.4(*b*) where $i_B > i_C/\beta$; this is the saturation region. In fact, the entire curve labeled $i_B = 1$ mA lies in the saturation region. (i_C never exceeds 10 mA in this plot and β is 50.) The transistor operates in the saturation mode when the collector current is limited by the external circuitry to a value less than βi_B.

EXAMPLE 8.1

Find the value of i_1 that is sufficient to saturate the transistor in the circuit in sketch (*a*). The transistor characteristics are given in Figure 8.4.

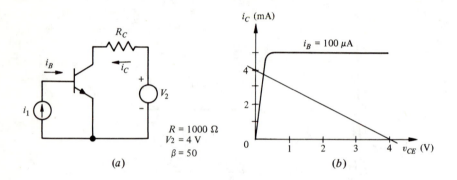

$$R = 1000\ \Omega$$
$$V_2 = 4\ \text{V}$$
$$\beta = 50$$

(*a*)   (*b*)

SOLUTION

From Ohm's law, the collector current is given by

$$i_C = \frac{V_2 - v_{CE}}{R_C}$$

Since only positive voltages are supplied, there is no way the collector could possibly take on a voltage less than zero. Therefore the largest collector current that could possibly flow is given by

$$i_C\ (\text{max}) = \frac{V_2 - (0)}{R_C} = \frac{V_2}{R_C}$$

The base current is i_1; therefore a sufficient condition for saturation is $\beta i_1 > V_2/R_C$. The transistor is saturated if

$$i_1 > \frac{V_2}{\beta R_C} = \frac{4}{(50)\ (1000)} = 80\ \mu\text{A}$$

Let us verify this assertion by finding i_C, for example, under the condition that $i_1 = 100\ \mu$A. We may use the output characteristics and the load-line technique to make a rapid graphical determination of i_C and v_{CE}. The *I-V* characteristic of the load line is given by

$$v_{CE} = V_2 - i_C R_C$$

The transistor output characteristics for $i_B = 100$ μA and the load line are graphed in sketch (b). The approximate graphical solution yields

$$i_C \cong 3.8 \text{ mA}$$
$$v_{CE} \cong 0.2 \text{ V}$$

Clearly, the transistor is in saturation $(\beta i_B > i_C)$, and the collector-to-emitter voltage is close to zero. ∎

In the saturation region the collector-to-emitter voltage is always quite small; in other words, the transistor behaves like a closed switch. Of course the voltage is not exactly zero, and in fact depends on both i_C and i_B. The value of v_{CE} in saturation is such an important parameter that it is given a special name, V_{CESAT}. We will assume that unless otherwise specified $V_{CESAT} < 0.3$ V. It may be seen from Figure 8.4(b) that at constant collector current, the value of V_{CESAT} decreases with increasing base current. The maximum value of V_{CESAT} occurs just at the "corners" of the curves where i_B is only slightly greater than i_C/β. The amount by which i_B exceeds i_C/β is called the *base overdrive*. In a conservative design of a closed transistor switch, a large amount of base overdrive is used to assure that the transistor is well into the saturation region, and thus that the value of V_{CESAT} is minimized.

The conditions under which the transistor behaves like a closed switch are summarized in Table 8.2.

Table 8.2   The Closed Switch

Input	Output
$i_B > i_C/\beta$	$v_{CE} = V_{CESAT} < 0.3$ V
	$i_C < \beta i_B$

It is important to note that a transistor switch never operates in the active region. When the switch is open, $i_B \cong 0$, $i_C \cong 0$, and the transistor is cut off. When the switch is closed $i_B > i_C/\beta$, $v_{CE} \cong 0$, and the transistor is in saturation. (Of course in making a transition between the states, that is, during the moment of switching from one to the other, the transistor may pass through the active operating region.)

The Basic Inverter Circuit   A simple common-emitter transistor switch may function as a logical complementer, that is, as an inverter. Consider the circuit of Figure 8.5. Suppose that this circuit is part of a digital system, so that the input **A** is either logical **0** (with $v_A \cong 0$ V) or logical **1** (with $v_A \cong 5$ V). Let us find the output under each of these two input conditions. If v_A is 0 V, no bias is applied to the emitter-base junction and the transistor is cut off (Table 8.1). Because $i_C = 0$ in cutoff, there is no voltage drop across R_C and $v_F = V_{CC} = 5$ V, corresponding to logical **1**. If v_A is instead 5 V, then the emitter-base junction is forward biased. We know from the input characteristics that in forward bias the voltage drop from base to emitter is in the range of 0.5 to 0.8 V. The base current equals $(v_A - v_{BE})/R_B$ and is therefore in the range of 0.45 to 0.42 mA. Now looking at the collector circuit, it may be seen that an upper limit of the collector current is $V_{CC}/R_C = 5$ mA (which would occur if $v_{CE} = 0$ V). Therefore βi_B is certainly much greater than i_C and the transistor is strongly saturated (Table 8.2). The output voltage v_F equals V_{CESAT} and is close to zero. The output is logical **0**, with an input of logical **1**. We see that the output is always the complement of the input; the circuit is an inverter. When the input is high, the transistor saturates, and the output is low. When the input is low, the transistor is cut off and the output is high.

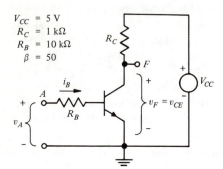

Figure 8.5 A simple transistor inverter circuit. The logical function performed by this circuit is known as **complement.**

There is a commonly used graph for displaying the kind of input-output relationships contained in the statements above. This graph is called a *timing diagram*, and is merely a plot of the output voltage versus time given some graph of the input voltage versus time. An example of a timing diagram is given in Figure 8.6 for the circuit of Figure 8.5. The timing diagram indicates quite clearly that the circuit functions as an inverter. The timing diagram may also be used to indicate the delay between input and output. For example, if the output response is delayed 100 nsec after a change in the input, this delay

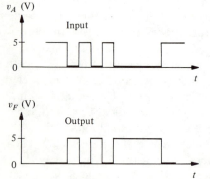

Figure 8.6 A timing diagram illustrating the operation of the circuit of Figure 8.5. When the input is high, the output is low, and vice versa.

is easily indicated on the plot of v_F versus time. Examples will be given in Section 8.2 in connection with practical DTL logic circuits.

The discussion just given in connection with Figure 8.5 is intended only as a simple illustration. It falls short of representing a practical case in two important respects: (1) it was assumed that no current could flow through the output terminal, and (2) it was assumed that a logical input corresponds to a precise input voltage. To be useful, a circuit of this kind must be connected to some load. In general some current will flow through this connection. Furthermore, a range of input voltages must be considered, corresponding to the range of logical **0** and logical **1**. We now consider, therefore, a more general circuit, shown in Figure 8.7. (As is conventionally done in transistor circuit diagrams, for the sake of clarity the collector voltage source V_{CC} is no longer shown explicitly. It is to be assumed, however, that a voltage source of this value is connected between the point marked V_{CC} and ground.) A load of unspecified nature is shown connected to the output terminals. Usually this load consists of the inputs to subsequent logic blocks. The input circuitry is similarly unspecified. We wish to find the conditions that guarantee the proper functioning of this generalized inverter. We consider first the state in which the input is low. According to the discussion above, the transistor should be cut off and the output high. To obtain the output voltage v_C, a node equation is written at the collector:

$$\frac{V_{CC}-v_C}{R_C}+i_L-i_C=0 \tag{8.1}$$

In cutoff, $i_C = 0$ (Table 8.1); thus

$$v_C = V_{CC} + i_L R_C \tag{8.2}$$

As yet it has not been specified whether i_L is positive or negative. The sign and magnitude of i_L depend upon the load circuit. Clearly the output relationship stated in Equation (8.2) puts some restriction on the load; if v_C is to stay within some allowed voltage range, i_L is also constrained.

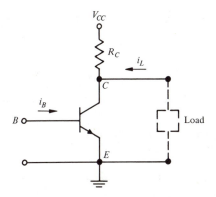

Figure 8.7 The common-emitter transistor inverter circuit. An unspecified load is shown connected at the output terminals.

EXAMPLE 8.2

The circuit below consists of a general inverter connected to a load. Suppose the range of logical **1** is defined to be 4 to 5 V. Determine the restriction, if any, on R_L in order that the output be **1** when the input voltage v_1 is zero.

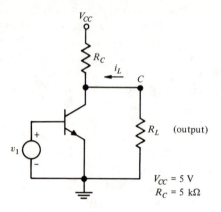

$$V_{CC} = 5 \text{ V}$$
$$R_C = 5 \text{ k}\Omega$$

SOLUTION

Since the circuit is a general inverter with input low, the output relationship, Equation (8.2), may be applied directly. We wish to determine the constraints on R_L to assure that v_C falls in the range of 4 to 5 V. From Equation (8.2)

$$v_C = V_{CC} + i_L R_C$$

If v_C is to fall in the range of 4 to 5 V, then it is necessary that i_L be in the range -0.2 to 0.0 mA. From Ohm's law

$$i_L = -v_C/R_L$$

Thus R_L must be 20 kΩ or greater to assure that v_C does not drop below 4 V.   ■

When the input to the general inverter of Figure 8.7 is high, the transistor should be saturated, and the output low. Table 8.2 gives the input conditions. Again, the load current condition may be derived by writing a node equation at the collector:

$$\frac{V_{CC} - v_C}{R_C} + i_L - i_C = 0 \tag{8.1}$$

If the transistor is in saturation, $v_C = V_{CESAT}$; thus

$$i_C = \frac{V_{CC} - V_{CESAT}}{R_C} + i_L \tag{8.3}$$

In saturation, the input and output currents are related by the condition $\beta i_B > i_C$. Thus the input condition may be stated.

$$\boxed{i_B > \frac{V_{CC} - V_{CESAT}}{\beta R_C} + \frac{i_L}{\beta}} \tag{8.4}$$

The input conditions and the output voltage relationships are summarized in Table 8.3.

Table 8.3 The General Inverter (Figure 8.7)

	Transistor state	Input conditions	Output voltage
Input low (output high)	Cutoff	$i_B \cong 0$ $v_{BE} < 0.3$ V	$v_C = V_{CC} + i_L R_C$
Input high (output low)	Saturation	$i_B > \dfrac{V_{CC} - V_{CESAT}}{\beta R_C} + \dfrac{i_L}{\beta}$	$v_C = V_{CESAT}$

In Section 8.2 we shall use these conditions in analyzing the logic family known as diode-transistor logic.

Worst-Case Design In analyzing circuit operation we have been assuming definite values for the various parameters that appear in the circuit calculations. However, it must be recognized that in designing a circuit, one cannot depend on the value of each component being precisely as specified. In general, even components that are nominally identical vary from one to another, due to production line variations. Moreover, parameters may be expected to vary with changes in temperature, and sometimes with time as devices age. Therefore circuits must be designed so that even if circuit parameters do vary over a certain range, correct operation will still be obtained.

It would appear a formidable task to design a large circuit with many components if a range of values for each component must be considered. However, the procedure known as *worst-case design* makes the computations quite simple. Let us first see, by example, the meaning of worst-case design, and then define the procedure. Suppose for instance, that a transistor switching circuit is being designed and the transistor is to be saturated. The input condition for saturation is $i_B > i_C/\beta$. If we consider variations of the parameter β, the inequality is least likely to be satisfied when β is at the minimum end of its possible range. From the point of view of assuring correct circuit operation (a saturated transistor) the worst possible value of β is the minimum value. The circuit is designed using this minimum value; then if β happens to be larger (it cannot be smaller), the inequality is still satisfied and the circuit functions properly.

Whether the worst case for a parameter is its maximum or minimum value depends upon the particular circuit and the conditions to be fulfilled. Sometimes, in fact, for one design consideration the worst case of a parameter is its maximum value, while from the point of view of another design consideration in the same circuit, the worst case is the minimum value. The general procedure in worst-case analysis is to set up inequalities from the circuit equations, and examine these inequalities to determine the worst cases for each design parameter for each design condition that must be met.

EXAMPLE 8.3

What nominal value should be chosen for R_1 in the circuit below so that the transistor will be saturated under worst-case conditions? The possible ranges of the parameters are given. The resistor used for R_1 may vary $\pm 10\%$ from the nominal value specified (that is, it has a "tolerance" of $\pm 10\%$).

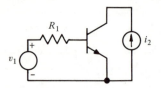

Parameter	Range
v_1	4.3 to 5 V
i_2	30 to 35 mA
β	100 to 1000
R_1	$\pm 10\%$
v_{BE}	0.5 to 0.9 V

SOLUTION

The current source i_2 determines i_C directly. It is possible to solve for i_B by a single loop equation:

$$v_1 - i_B R_1 - v_{BE} = 0$$

or

$$i_B = \frac{v_1 - v_{BE}}{R_1}$$

To insure saturation the base current must exceed i_C/β; thus

$$\frac{v_1 - v_{BE}}{R_1} > \frac{i_2}{\beta}$$

Solving for R_1

$$R_1 < \beta \frac{(v_1 - v_{BE})}{i_2}$$

To do a worst-case design, we must choose a value of R_1 such that the inequality is satisfied for any set of parameter values in the ranges listed. The worst possible value of β, from the point of view of satisfying the inequality, is its *minimum* value. Similarly, the worst-case value of v_1 is its minimum value. In contrast, the inequality is least likely to be satisfied when v_{BE} and i_2 are at their maximum values. Thus the maximum value of R_1 is set by

$$R_1 < 100 \frac{(4.3 - 0.9)}{35 \times 10^{-3}} \, \Omega$$

or

$$R_1 < 9700 \, \Omega$$

We must, however, allow for the possibility that the actual R_1 supplied by the manufacturer may be 10% larger than specified. Thus to guarantee that the actual resistance is less than 9700 Ω, we could specify $R_1 = (0.9) (9700) = 8730 \, \Omega$. (In practice, however, manufacturers do not produce a line of resistors with a continuous range of nominal values; instead they make a series of resistors having a standard set of nominal values separated, approximately, by the tolerance. The standard 10% tolerance resistance values in this range are 6.8, 8.2, and 10 kΩ. Since to be on the safe side R_1 can only be made smaller than the calculated value, the circuit designer could choose a standard 10% resistor, with a nominal value of 8.2 kΩ.   ∎

In the design of digital circuits, the worst-case design philosophy is used to guarantee that every one of the circuits in a digital system will function correctly. (In a system with a million subcircuits, for example, a one-in-a-million chance that a subcircuit will operate incorrectly is intolerable.) In the rest of this book calculations will usually be performed using nominal parameter values. It should be realized, however, that this is being done only for simplicity. The practical circuit designer must make all calculations using worst-case parameter values.

8.2 Diode-Transistor Logic

It was seen in Chapter 5 that combinations of diodes can perform logic operations on binary signals. Diode-transistor logic (DTL) combines transistor inverters with diode arrays similar to those of Chapter 5. A simple, yet practical, diode-transistor logic gate is shown in Figure 8.8. We might expect that the logic function of this gate is **NAND,** from the following argument. The input diode array D_A, D_B,...., performs an **AND** operation. (See Chapter 5, Section 5.5). The remainder of the circuit is an inverter, connected to the **AND** gate at the point X through two diodes D_1 and D_2. The role of these diodes will be brought out in the subsequent discussion, but for the moment we may note that they simply produce a voltage drop between the point X and the base of T_1. (Note that D_1 and D_2 do not interfere with the flow of base current into T_1, which is inward.) An **AND** operation followed by a logical negation constitutes a **NAND;** therefore we may expect that this is a DTL **NAND** gate.

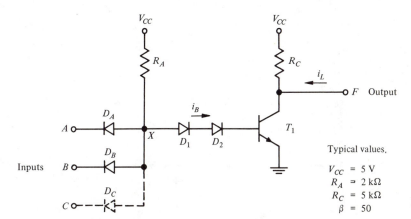

Figure 8.8 A DTL **NAND** gate. The inputs are on the left, and the output is taken at point F. Any number of additional inputs may be added in the manner shown by the dotted line. All voltages are measured with respect to ground.

We shall now analyze the operation of the basic DTL **NAND** gate in some detail. For the purposes of numerical calculations we shall assume the parameter values given in Figure 8.8, which are representative of commercial DTL circuits. Furthermore, the diode *I-V* characteristic and transistor input and output *I-V* characteristics are sketched in Figure 8.9. The possible signal voltage range is 0 to +5 V; let us tentatively assume that the range of 0 to 0.5 V is low and 2.0 to 5 V is high. For simplicity we will deal exclusively with posi-

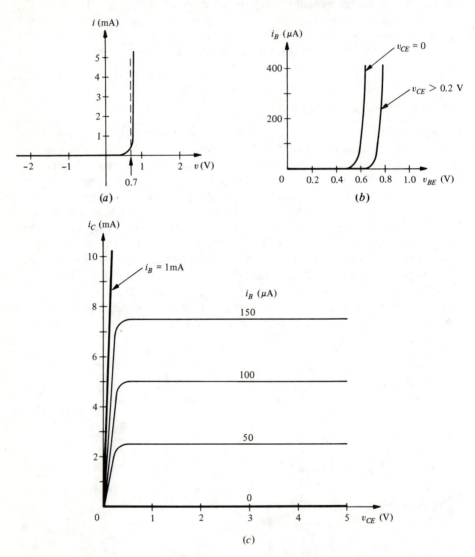

(a)

(b)

(c)

Figure 8.9 The *I-V* characteristics of the devices used in the DTL **NAND** gate. (*a*) The diode characteristic. The approximate characteristic, given by the large-signal diode model, is also shown as the dotted line. (*b*) The transistor input characteristics. (*c*) The transistor output characteristics.

tive logic in this chapter; thus logical **0** corresponds to a voltage in the range of 0 to 0.5 V, and logical **1** to a voltage in the range of 2 to 5 V. These values are summarized in Table 8.4. That these apparently arbitrary choices of voltages are sensible and realistic will become clear from the subsequent discussion.

Table 8.4 DTL Logic Levels

Level	Voltage range	Logical value (Positive logic)
High	2 to 5	1
Low	0 to 0.5	0

The self-consistent analysis method, introduced in the analysis of diode gates, may be used to verify that the circuit of Figure 8.8 indeed functions as a **NAND** gate. The output voltage is computed for a given set of input voltages, making use of the large-signal diode model to simplify the diode characteristics. If any particular current or voltage cannot be readily calculated, a value is assumed and a check is made to see if the assumed value is consistent with the known circuit laws and component *I-V* characteristics.

EXAMPLE 8.4
Calculate the output voltage of the **NAND** gate, Figure 8.8, under the condition $v_A = 0.1$ V, $v_B = 0.1$ V, and $i_L = 0$.

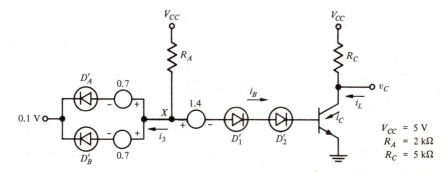

SOLUTION
The circuit is redrawn with the diodes replaced by the large-signal diode model. It is first necessary to compute the transistor-base current; then the output characteristics may be used to find v_C. Since the voltage at node X determines the base current, let us begin by calculating v_X. Writing a nodal equation

$$\frac{5 - v_X}{R_A} = i_B + i_3$$

or $v_X = 5 - (i_B + i_3) R_A$. Because of the perfect rectifiers, both i_B and i_3 can only be zero or positive; therefore v_X is 5 V or less. Suppose $v_X = 5$ V. Then the perfect rectifiers D'_A and

D'_B are forward biased by a voltage $5 - 0.7 - 0.1 = 4.2$ V. This is of course an impossibility, because an infinite current would flow. If the current is to be in a finite range, v_X must not exceed 0.8 V. Let us suppose then that $v_X = 0.8$ V. The perfect rectifiers have exactly 0 V forward bias; hence any current is allowed. A current of magnitude $(5 - v_X)/R_A = 2.2$ mA flows down through R_A, and must flow out through the input diodes (as i_3) or into the transistor base (as i_B). However, a voltage of 0.8 V, when reduced by 1.4 V, is insufficient to forward bias the transistor emitter-base junction. The base current is therefore zero, and the transistor is cut off. In cutoff $i_C = 0$; hence no current flows through collector resistor R_C, and the output voltage v_C equals 5 V.   ∎

It should be pointed out that it is certainly possible to write a set of circuit equations sufficient to solve for v_F, given v_A and v_B. However, the presence of five nonlinear elements makes the solution of these equations quite difficult. Such an accurate analysis is best carried out on a computer. For digital circuit applications, we are primarily interested in the circuit input conditions that result in the desired output, either logical **0** (low) or logical **1** (high). We have already computed the necessary input conditions at the transistor base to assure either saturation (output low) or cutoff (output high). These results, summarized in Table 8.3, may be used, along with an approximate analysis of the rest of the circuit, to rapidly estimate the required circuit input conditions. Again, we make use of the large-signal diode model to eliminate the complications caused by the diode nonlinearities. [A comparison is shown in Figure 8.9(a) of the actual diode characteristic with the characteristic predicted by the large-signal diode model.]

We know from the discussion of the general inverter that the circuit should operate in one of two states, either with transistor T_1 saturated or with T_1 cut off. The input conditions for either of these states are given in Table 8.3. To saturate T_1 the potential v_X must be high enough that a significant base current flows in through diodes D_1 and D_2 and into the base. Conversely, to cut off T_1, v_X must be low enough that essentially no current flows into the base. We may write an equation for v_X in terms of the positive voltage drops across the diodes v_{D1} and v_{D2}

$$v_X = v_{D1} + v_{D2} + v_{BE} \tag{8.5}$$

According to the large-signal diode model, v_{D1} and v_{D2} must each be at least 0.7 V for any current to flow, and from Figure 8.9(b) v_{BE} must also be at least 0.5 V before any significant base current flows. If we adopt the even more conservative estimate that $v_{BE} < 0.3$ V for cutoff (Table 8.3), then we have from Equation (8.5)

$$v_X < 1.7 \text{ V} \quad \text{(cutoff)} \tag{8.6}$$

The condition for saturation (Table 8.3) is not stated in terms of base-to-emitter voltage, but base current. We can relate the base current to the base-to-emitter voltage by Figure 8.9(b). Without knowing the load current i_L, it is impossible to know the exact base current requirement for saturation. However, because of the nature of the transistor input characteristics, it is only necessary to have a rough estimate of i_B to estimate v_{BE}. Therefore we will make a generous estimate of i_L, say 25 mA, and obtain an approximate value for i_B. From Table 8.3 the base current must exceed $i_L/\beta + (V_{CC} - V_{CESAT})/\beta R_C$, or about 0.5 mA, for the circuit values of Figure 8.8. Figure 8.9(b) indicates that a forward bias of approximately 0.8 V is sufficient to insure that $i_B > 0.5$ mA. Summing the two diode drops, and v_{BE}, we obtain an estimate of the voltage v_X to insure that the transistor is saturated.

$$v_X \approx 2.2 \text{ V} \qquad \text{(saturation)} \tag{8.7}$$

The base current requirement is given directly by Table 8.3. Let us now find the relationship between the input voltages and v_X.

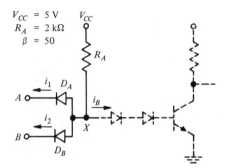

$V_{CC} = 5$ V
$R_A = 2$ kΩ
$\beta = 50$

Figure 8.10 The input portion of the DTL **NAND** gate.

We ignore for a moment the inverter portion of the circuit and concentrate on the input diode array. This part of the circuit, shown in Figure 8.10, functions much like the diode **AND** gate discussed in Chapter 5. If all three currents i_1, i_2, and i_B are zero, then v_X equals V_{CC}, or 5 V. If any of these currents is greater than zero, then v_X is reduced (note that i_1, i_2, and i_B can only be zero or positive because of the diodes). We saw from the previous discussion that the transistor is cut off and the output is high if $v_X < 1.7$ V. In cutoff, the base current is zero; hence i_B (Figure 8.10) equals zero. We wish to find the input conditions that produce $v_X < 1.7$ V. Recalling the operation of the diode **AND** gate, the lesser of the input voltages limits v_X. For example, if $v_A = 0.2$ V, v_X cannot rise much above 0.9 V, otherwise diode D_A would be forward biased by a voltage greater than 0.7 V. (If we idealize the diode characteristic according to the large-signal diode model, then the maximum forward bias is exactly 0.7 V.) Since v_X is limited to 0.7 V plus the lesser of v_A or v_B, v_X will always be less than 1.7 V unless both v_A and v_B equal or exceed 1 V. Clearly, if either input is within the allowed range of logical **0** (0 to 0.5 V), v_X is less than 1.7 V, the transistor is cut off, and the output is high.

EXAMPLE 8.5
Verify that the transistor is cut off for inputs of $v_A = 0.95$ V, $v_B = 0.95$ V (Figure 8.8).

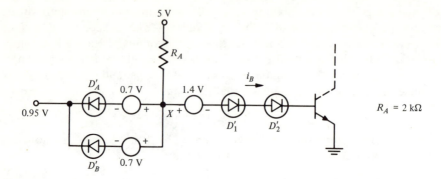

SOLUTION

The input circuitry is redrawn using the large-diode model. We suppose, from the discussion above, that $v_X = 0.95 + 0.7 = 1.65$ V. The current flowing down through R_A flows out through D'_A and D'_B, or into the transistor base, or through both paths. Suppose $i_B > 0$; then D'_1 and D'_2 are forward biased and are effectively short circuits. The forward bias on the emitter-base junction equals $v_X - 1.4 = 0.25$ V. It may be seen from Figure 8.9(b) that this small forward bias is insufficient to produce any significant base current; hence i_B is *not* greater than zero as assumed. The transistor is cut off.   ■

We have determined that either input in the low range guarantees a high output. Let us now find the input condition for a low output. We showed that the transistor will saturate if $v_X \approx 2.2$ V. Furthermore, the input circuitry, Figure 8.10, limits v_X to less than 2.2 V, unless both inputs are greater than $2.2 - 0.7$, or 1.5 V. Let us suppose that $v_A = v_B = 1.55$ V, and determine if the transistor saturates. The circuit is redrawn in Figure 8.11, with the diodes replaced by the large-signal diode model. The voltage at node X cannot rise above 2.25 V, otherwise the perfect rectifier D'_A would conduct an infinite current. However, the transistor limits v_X to an even lower value. Assuming for

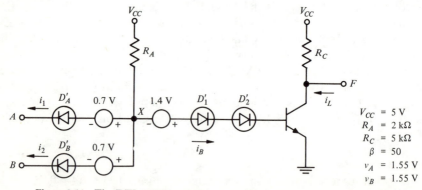

Figure 8.11   The DTL **NAND** gate redrawn using the large-signal diode model.

a moment that diodes D_1' and D_2' are forward biased, the transistor input voltage v_{BE} equals $v_X - 1.4$ V. It may be seen from Figure 8.9(b) that the v_{BE} will not exceed about 0.8 V. Thus v_X cannot exceed 2.2 V. If we assume $v_{BE} \cong 0.8$ V, then $v_X \cong 2.2$ V and a current of magnitude $(V_{CC} - v_X)/R_A$ flows down through R_A and into the transistor base. (Both i_1 and i_2 are zero, because D_A' and D_B' are reverse biased by 0.05 V.) The base current for the values given here is approximately 1.5 mA, which is sufficient to saturate the transistor for even a very large load current, as may be verified from Table 8.3. We have established that the transistor will be saturated if (and only if) both inputs exceed 1.5 V. When the transistor is saturated, the output voltage is V_{CESAT}, which may be estimated from Figure 8.9(c). For example, at a collector current of 8 mA and base current of 1 mA, $v_{CE} \cong 0.18$ V.

It is helpful in understanding the operation of a logic gate to sketch the output voltage as a function of the input voltages. Since the lower of the two inputs determines the output state, we may simply tie one of the inputs, say input B, to +5 V, and sketch v_F versus v_A. This plot, which is known as the gate *transfer characteristic*, is given in Figure 8.12. The transfer characteristic depends upon the load current, and in Figure 8.12 we have assumed $i_L = 0$. Since we have only calculated the characteristic for $v_A < 1$ V and $v_A > 1.5$ V, we show the characteristic with a broken line in the intermediate range. The characteristic in this range is most simply computed numerically. The transfer characteristic shows that, contrary to the behavior of the diode **AND** gate, the **NAND** gate restores the signal levels rather than degrading them. For example, even if the input is near the upper limit of the logical **0** range (0.5 V), the output falls within allowed range of logical **1**. The function of the two diodes D_1 and D_2 (Figure 8.8) now becomes clear; they are included to raise the voltage at node X required for saturating the transistor. If the diodes were not included, a voltage at node X of 0.7 V would saturate the transistor, corresponding to an input voltage of 0 V.

We see from Figure 8.12 that the DTL **NAND** gate, with all inputs but one tied to +5 V, acts as a simple inverter. When it is desired to show the output as a function of more than one input, it is common procedure to use the timing diagram. It should be noted that, like the basic diode **AND** gate,

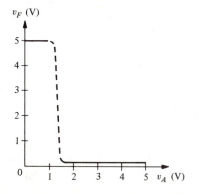

Figure 8.12 The transfer characteristics of a DTL **NAND** gate. In obtaining this characteristic, all inputs but input **A** are tied to +5 V, and the load current is assumed 0.

any number of inputs may be added to the DTL **NAND** gate. For example, a four-input DTL **NAND** gate is shown in Figure 8.13(*a*). The timing diagram for the circuit of Figure 8.13(*a*) is given in Figure 8.13(*b*). The timing diagram indicates the **NAND** function; the output is high unless all four inputs are simultaneously high, in which case it is low.

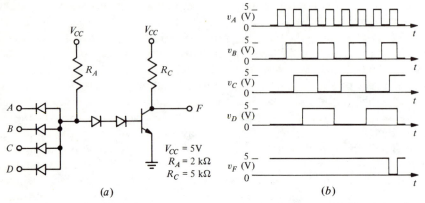

(*a*)   (*b*)

Figure 8.13   A four-input DTL **NAND** gate. (*a*) The circuit diagram. (*b*) The timing diagram. The output is low only when all the inputs are high.

Input and Output Considerations   Up to this point we have analyzed the DTL **NAND** gate in the abstract, without regard to circuits connected to the input and output terminals. In general, in the synthesis of some desired logic function, a number of basic gates are used, and the load on a given gate consists of inputs to one or more subsequent gates. Figure 8.14 shows an arbitrary example in which the output **F** is some desired function of the inputs **A, B, C,** and **D.** A number of gates may load the output of a given gate, and a number of inputs may drive the given gate. For example, gate 2 has two inputs and two outputs. The characteristics of the load are therefore the input characteristics of the gate (or gates) that is (are) connected to the output of a given gate. Because of the presence of the diodes in the input circuitry, the load current can only be positive. Figure 8.15 illustrates, for example, the circuitry of gates 3 and 5 of Figure 8.14. Gate 3 is said to *drive* gate 5, and gate 5 is said to *load* gate 3. When the output of gate 3 is high, diode D_{A5} is either reverse biased or zero biased (v_{A5} equals 5 V, and v_X is at most 5 V). Consequently, the load current for gate 3 is zero. On the other hand, when the output of gate 3 is low, v_{A5} is close to ground, and v_X is close to 0.7 V. A current of $(V_{CC} - v_X)/R_{A5} \cong 2.2$ mA flows downward through R_{A5}, and out through the input circuitry. Depending on the value of the second input to gate 5, this entire current may flow out through D_{A5}. The load current for gate 3 is therefore zero when the output is high, and may be as much as 2.2 mA when the output is low. If the load for a given gate consists of the inputs of several gates, the maximum load current is 2.2 mA times the number of loading gates. For example, in Figure 8.14, gate 2 must drive both

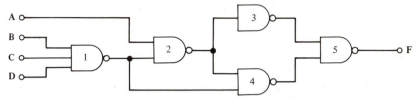

Figure 8.14 A logic circuit constructed from five **NAND** gates. The loads for gates 1, 2, 3, and 4 are the inputs to subsequent **NAND** gates.

the inputs to gates 3 and 4; therefore the load current for gate 2 may be as high as 4.4 mA. There is a limit, imposed by the transistor saturation condition, on the maximum load current. Accordingly, there is a limit to the number of gates that may be connected to the output of a given gate. The term used to describe the driving capability of a gate is the *fan-out;* a gate has a maximum fan-out of 10 if it can drive 10 equivalent gates.

We can estimate the fan-out capability of the basic DTL **NAND** circuit of Figure 8.8: Since a load current flows only for logical **0** output, we concern ourselves only with this state. Table 8.3 gives the transistor saturation requirements:

$$i_B > \frac{1}{\beta}\left(i_L + \frac{V_{CC} - V_{CESAT}}{R_C}\right) \tag{8.8}$$

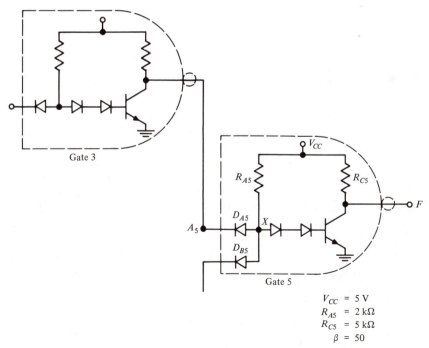

$$
\begin{aligned}
V_{CC} &= 5 \text{ V} \\
R_{A5} &= 2 \text{ k}\Omega \\
R_{C5} &= 5 \text{ k}\Omega \\
\beta &= 50
\end{aligned}
$$

Figure 8.15 Diagram of gates 3 and 5 of Figure 8.14, showing the internal circuitry.

For the output to be low, all the inputs must be high and accordingly carry no current. The circuit may therefore be simplified as shown in Figure 8.16. Also included in Figure 8.16 is an extra resistor R_E, called the base pull-down resistor. As noted in the following section on speed considerations, this resistor is included in most practical DTL circuits to speed up transistor turn-off. Let us now compute the maximum load current i_L that can flow into the circuit of Figure 8.16, without causing the transistor to become unsaturated. The potential of point X, which is determined by the voltage drop across two forward-biased diodes, plus the transistor base-to-emitter voltage drop, is approximately 2.2 V. Thus the current i_A, which equals $(V_{CC} - v_X)/R_A$, is about 1.4 mA. If R_E is omitted, this is the base current. However, if R_E is included, a current of v_{BE}/R_E, or about 0.8 mA, flows down through R_E. In this case the base current is 0.6 mA. Solving Equation (8.8) for i_L, we find that $i_L < 69$ mA if R_E is omitted, or $i_L < 29$ mA if R_E is included. The corresponding fan-out limitations are obtained by dividing the load current by 2.2 mA. Thus the maximum fan-out of the DTL NAND gate is determined to be 31 if R_E is omitted or 13 if R_E is included. In practice this number is reduced by worst-case parameter considerations as well as by considerations of switching speed. Nonetheless, one of the useful features of DTL logic is its large fan-out capability.

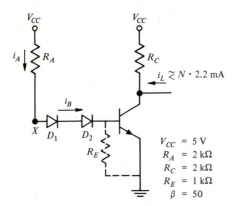

V_{CC} = 5 V
R_A = 2 kΩ
R_C = 2 kΩ
R_E = 1 kΩ
β = 50

Figure 8.16 A DTL NAND gate, with all inputs high. The input diodes carry no current and are omitted. The load consists of N identical NAND gates. (The resistor R_E is often included in practical circuits.)

Let us consider the question of how many inputs may be connected to a DTL NAND gate, that is, what is the maximum *fan-in*. Consider the input to gate 5 in Figure 8.15. If both inputs are 1, then D_{A5} and D_{B5} are reverse biased and essentially zero input current flows. The addition of more inputs does not alter the situation; if all inputs are 1, all input diodes are reverse biased and no current flows. Now suppose one or more of the inputs is logical 0. The voltage v_X is "pulled down" to v_{CESAT} (the voltage of a low input) plus ~0.7 V for a total of about 1 V. The addition of more inputs, having logical values of either 0 or 1, can only lower this voltage, improving circuit operation. Thus the fan-in is essentially unlimited. Of course, in reality there are some limitations imposed by the small diode reverse-biased leakage currents, as well as by considerations of switching speed.

Speed Considerations   When a change in an input of a logic gate occurs, the output does not, in general, respond immediately. Suppose, as shown in Figure 8.17(a) a two-input DTL gate has the gates tied together, and is connected to a pulse generator. The input waveform, and the corresponding out-

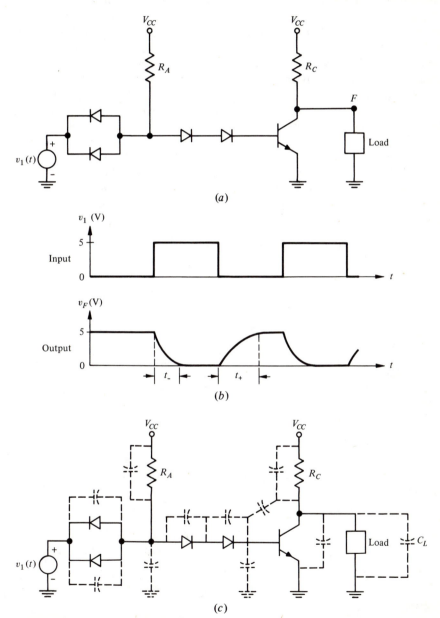

Figure 8.17   (a) A DTL **NAND** gate driven by a pulse generator. (b) The approximate input and output waveforms. (c) The **NAND** circuit redrawn, showing some of the parasitic capacitances.

put waveform, are given in the timing diagram, Figure 8.17(*b*). Changes in output voltage follow changes in the input voltage only after some time delay, labeled t_+ and t_- in the timing diagram. These delays are called the *turn-off delay* and the *turn-on delay,* respectively. They arise both from delays associated with the circuitry and from inherent delays in the transistor response. Associated with every element in the circuit is a parasitic capacitance, as suggested in Figure 8.17(*c*). Furthermore any cables that may be used to interconnect gates may have a very large capacitance. It is these various capacitances that limit the speed of response of the circuit. The capacitance of the load is particularly important, because it is charged through the resistor R_C during turn-off. (During turn-on, the discharge current flows through the transistor, and may be quite large, resulting in a relatively rapid turn-on.)

EXAMPLE 8.6

The output of the two-input DTL **NAND** gate in sketch (*a*) is low (0.2 V) until $t = 0$, when both inputs are suddenly switched to zero. Compute the output waveform, assuming that the delay arises solely from the effects of C_L.

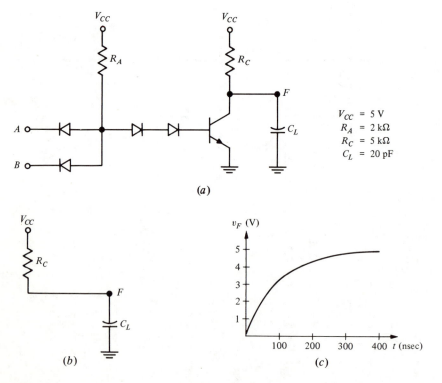

V_{CC} = 5 V
R_A = 2 kΩ
R_C = 5 kΩ
C_L = 20 pF

(*a*)

(*b*)

(*c*)

SOLUTION

At the instant $t = 0$, the transistor is suddenly turned off. According to the assumption above, the transistor is an open circuit, and the circuit may be simplified, as in sketch (*b*). At the instant of switching, $v_F = V_{CESAT} \cong 0.2$ V. Much later, the capacitor will be charged to the voltage V_{CC}. Writing a nodal equation

$$\frac{V_{CC} - v_F}{R_C} = C_L \cdot \frac{d\,v_F}{d\,t}$$

The solution to this equation with the initial condition $v_F = 0.2$ V *is*

$$v_F = 0.2 + (V_{CC} - 0.2)(1 - \exp\{-t/R_C\,C_L\})$$

Sketch (*c*) is the output. ■

The turn-off time, when capacitive loads are present, is limited by the resistor R_C, which charges the capacitance. If R_C is halved, the charging time (proportional to $R_C \cdot C_L$) is approximately halved, and vice versa. One may well ask why a very low value of resistance is not used, minimizing the turn-off time. The answer is found in a consideration of the power dissipation. When the output is low, a current of $(V_{CC} - V_{CESAT})/R_C$, or approximately V_{CC}/R_C, flows from V_{CC} to ground. The power dissipated in R_C thus equals about V_{CC}^2/R_C. If R_C is made very small, not only is a lot of power wasted, but the circuit gets hot. When thousands of such circuits are packed into a small volume, power dissipation becomes a very serious consideration. The selection of R_C thus becomes a compromise between maximizing speed and minimizing power dissipation. This trade-off between power and speed occurs in all forms of integrated circuits. In fact, a parameter called the speed-power product is used as a figure of merit in judging logic gate performance.

While we have primarily focused on delays arising from the charging time of stray capacitances, it should also be recognized that there are, in addition, inherent delays in the switching of the diodes and transistors. Especially important are the minority carrier storage effects discussed in Chapters 5 and 6. Suppose, for example, that the transistor is saturated, and the base current is suddenly reduced to zero. There is a delay called the storage time before the collector current goes to zero. This delay is the time required to rid the transistor of the excess minority carriers that are present under saturation conditions. In an effort to minimize this delay, resistor R_E in Figure 8.16 is generally included. This resistor provides a conducting path for the exit of the carriers in the base when diodes D_1 and D_2 are reverse biased.

Examples of some typical commercial DTL circuits are given in Figure 8.18. The basic circuit is known as the series 930 DTL **NAND** gate, and is produced with minor variations by many semiconductor device manufacturers. It may be noted that the circuit differs in one important respect from the circuit considered in this section. The diode D_1 (Figure 8.8) has been replaced by a transistor. In addition, a resistor is connected from emitter to base of the output transistor. Despite these modifications, the circuit functions substantially equivalently to the simple circuit of Figure 8.8. If the transistor is drawn schematically as an *npn* structure, as shown in Figure 8.19, the similarity of the simple circuit with the series 930 **NAND** gate becomes clear. The function

The DM930, DM936, DM946 and DM962 are a variety of NAND gates with a 6K pull-up resistor. The DM961, DM937, DM949 and DM963 are the 2K pull-up versions of the respective gates. The DM935 is a hex inverter similar to the DM936 with the exception that it has no input diodes.

connection/logic diagrams   schematic diagrams*

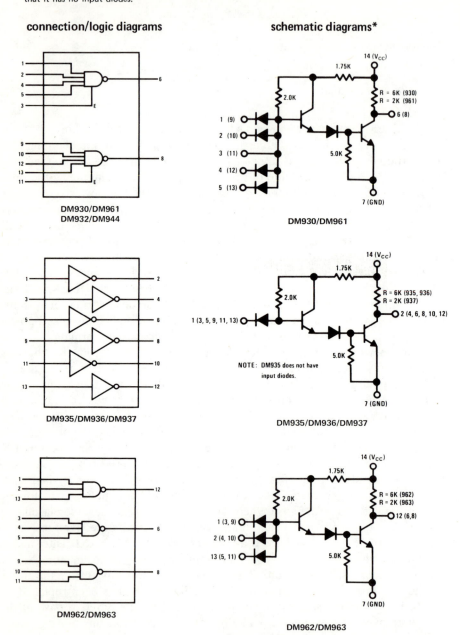

DM930/DM961
DM932/DM944

DM930/DM961

DM935/DM936/DM937

NOTE: DM935 does not have input diodes.

DM935/DM936/DM937

DM962/DM963

DM962/DM963

*Only one circuit element is shown. Pin connections are given in parentheses for other circuit elements.

Figure 8.18 A number of commercial DTL **NAND** gate circuits. These circuits belong to the series 930 DTL family. (Courtesy of National Semiconductor Corp.)

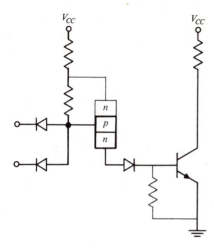

Figure 8.19 The DTL gate of Figure 8.18 redrawn to emphasize the similarity to the basic DTL **NAND** gate of Figure 8.8.

of diode D_1 was to provide a 0.7-V drop in forward bias. The transistor base-emitter junction provides an almost identical voltage drop, hence circuit operation is not impaired. In addition, the transistor provides a current gain of approximately β from base to emitter; thus more base current is supplied to the output transistor in saturation, and a greater load capability results.

The transfer characteristics at a number of temperatures of a typical commercial DTL **NAND** gate (series 930) are shown in Figure 8.20. The curve corresponding to 25°C operation may be compared to the transfer characteristic derived here, Figure 8.12.

Logical Synthesis with NAND Gates We have shown that the relatively simple circuit of Figure 8.8 performs the **NAND** logical operation. Furthermore, the signal levels are not degraded as is the case in simple diode gates. This circuit may be used as a building block from which entire logic systems may be constructed. It is important to realize that any logical function may

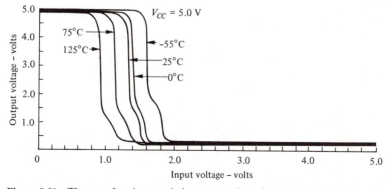

Figure 8.20 The transfer characteristics at a number of temperatures of the series 930 DTL **NAND** gate.

be synthesized with **NAND** gates alone. To illustrate this fact, we shall show here that all of the five basic logical operations can be achieved by a combination of **NAND** gates. It is important to realize that a three-input **NAND** gate can function as a two-input or single **NAND** gate as well. In general, anywhere from 1 to N inputs of an N-input **NAND** gate may be used. Any **NAND** gate input that is high conducts no current, and therefore is effectively removed from the circuit. Thus unused inputs may either be left open or connected to V_{CC}.

First of all a single-input **NAND** gate is just an inverter. This may be verified from the definition of a **NAND** gate (Chapter 3): If the single input is **1** then (since there is only one input) *all* the inputs are **1,** so the output is **0.** If the single input is **0,** the output is **1.** We recognize this behavior as that of an inverter.

Figure 8.21 illustrates a series hookup of two **NAND** gates. The second gate has the effect of just inverting the output of the first, thus the overall

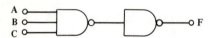

Figure 8.21   A **NAND** gate followed by a single-input **NAND** gate. The overall function is **AND.**

function is that of an **AND** gate. Figure 8.22 illustrates a hookup that acts as an **OR** gate. The **OR** function is most simply verified by constructing the truth table. For example, when $A = 0$ and $B = 0$, the inputs to **NAND** gate 3 are **1** and **1.** The output is **0.** The complete truth table is given in Figure 8.22, and is recognized as the truth table for the **OR** gate. Furthermore, a simple inverter (a single input **NAND** gate) on the output would produce the **NOR** function. Thus we have shown by example that any of the five basic logical functions can be synthesized by a combination of **NAND** gates. This implies that any logical system could be made from **NAND** gates alone. **NAND** gates are not unique in this respect; **NOR** gates can also be used to realize any desired logic function. An example of a logic circuit yielding the **NOR** function is given in Chapter 13.

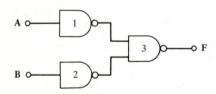

INPUTS		OUTPUT
A	*B*	*F*
0	0	0
0	1	1
1	0	1
1	1	1

Figure 8.22   A combination of three **NAND** gates that result in the **OR** function.

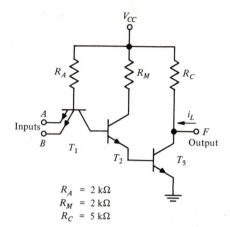

R_A = 2 kΩ
R_M = 2 kΩ
R_C = 5 kΩ

Figure 8.23 A transistor-transistor logic **NAND** gate.

Before leaving the subject of **NAND** gates we wish to consider some circuit modifications that can improve the performance of the logic gate. The first modification is a major one, and results in a whole new family of logic circuits, the so-called transistor-transistor logic family. Yet the modified circuit is so similar in construction and function to the DTL circuit, that it is appropriate to consider it here.

Transistor-Transistor Logic The basic transistor-transistor logic (TTL or T²L) circuit is shown in Figure 8.23. The circuit contains no diodes, but includes a new device, the multiple-emitter transistor (T_1). The operation of the circuit is most simply understood, if the transistor circuit symbols for T_1 and T_2 are replaced by the more physical block symbols, as shown in Figure 8.24(a). The basic DTL circuit (Figure 8.8) is also redrawn with similar symbols in Figure 8.24(b). Both the appearance and the function of the TTL

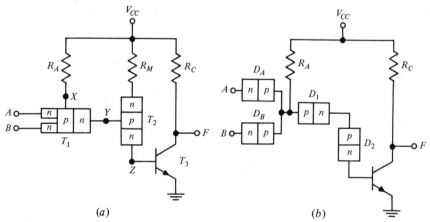

Figure 8.24 The TTL **NAND** gate compared to the DTL gate. (a) The TTL circuit of Figure 8.23 redrawn. (b) The DTL circuit of Figure 8.8 redrawn.

circuit are similar to the DTL circuit. Diodes D_A, D_B, and D_1 in Figure 8.24(b) have an *electrically* common p region. If this p region is made *physically* common, the multiple-emitter transistor T_1 of Figure 8.24(a) results. (Of course the p region must be made very narrow if it is to function as a transistor.) As we will show in the following discussion, transistor T_1 functions substantially equivalently to the diode array D_A, D_B, and D_1.

The other difference in the circuits is the replacement of diode D_2 by transistor T_2. Again T_2 functions substantially equivalently to D_2; its primary function is to produce a 0.7-V voltage drop (from base to emitter). Like the DTL gate, the TTL gate is a **NAND** gate, and it has a transfer characteristic very similar to Figure 8.12.

Suppose all the inputs (there may be more than the two shown in Figure 8.23) are high. Then no current flows out the input terminals; rather a current flows down through resistor R_A and out through the collector of T_1 into the base of T_2. The path through the base-collector junction of T_1, the base-emitter junction of T_2, and the base-emitter junction of T_3 is in the direction of easy current flow for each of these diodes. Furthermore, T_2 produces some current gain, and the current that enters the base of T_3 is larger than the current entering the base of T_2. Under these conditions T_3 is saturated, and the output is low.

Now suppose one or more of the inputs, say input A, is low. In this case the potential of node X [Figure 8.24(a)] cannot be greater than about v_A + 0.7 V because of the forward-biased emitter-base junction of T_1. A very large base current (almost V_{CC}/R_A) flows and T_1 will be saturated. The potential of node Y will be close to the input potential ($v_Y = v_A + V_{CESAT}$). Such a small voltage, at most 0.8 V if $v_A = 0.5$ V, is insufficient to forward bias both the emitter-base junctions of T_2 and T_3; consequently T_3 will be turned off and the output will be high.

EXAMPLE 8.7
    Find the approximate maximum load current i_L of the basic TTL gate (Figure 8.23) when both inputs are high. Assume that the transistors have the characteristics given in Figure 8.9.

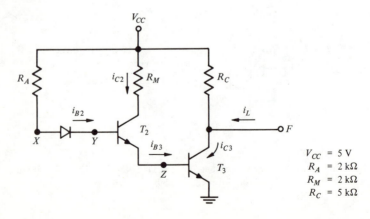

$$V_{CC} = 5 \text{ V}$$
$$R_A = 2 \text{ k}\Omega$$
$$R_M = 2 \text{ k}\Omega$$
$$R_C = 5 \text{ k}\Omega$$

SOLUTION

From the discussion in the text, and from the similar behavior of the DTL gate, we realize that no current flows out a high input. Both inputs are effectively open circuits. As may be best appreciated from Figure 8.24(a), under these conditions transistor T_1 functions merely as a diode (the collector-base diode). The simplified circuit is redrawn as shown above.

A current path exists from V_{CC} to ground via X–Y–Z because a current flowing in this direction forward biases all three pn junctions (B–C of T_1, B–E of T_2, B–E of T_3). Each junction will produce a voltage drop of approximately 0.7 to 0.8 V (Figure 8.9); consequently $v_Z \cong 0.75$ V, $v_Y \cong 1.5$ V, and $v_X \cong 2.25$ V. The base current of transistor T_2 is just the current flowing down through R_A:

$$i_{B2} = \frac{V_{CC} - v_X}{R_A} \cong 1.4 \text{ mA}$$

To find i_{B3} we must first decide if T_2 is saturated. If it is not saturated, a collector current of $\beta\, i_{B2}$ or about 70 mA would flow down through R_M; however this is an impossibility as it would produce a 140-V voltage drop across R_M. Transistor T_2 is saturated, and the collector voltage is given by $v_Z + V_{CESAT}$. Assuming $V_{CESAT} \cong 0.2$ V, the voltage across resistor R_M equals $V_{CC} - (v_Z + V_{CESAT}) \cong 4$ V. Thus i_{C2} equals 4 V/2 kΩ, or about 2 mA. The base current of transistor T_3 is just the current flowing out of the emitter of T_2, which equals $i_{B2} + i_{C2}$.

$$i_{B3} = i_{B2} + i_{C2} = 3.4 \text{ mA}$$

The load current is limited by the requirement that T_3 remain in saturation, which requires

$$i_{C3} < \beta\, i_{B3}$$

or

$$i_{C3} < 170 \text{ mA}$$

The collector current i_{C3} is the sum of the load current and the current flowing down through R_C. The latter equals $(V_{CC} - V_{CESAT})/R_L$, or about 1 mA. Consequently, the load current is limited to less than about 169 mA. ■

Output Circuit Modifications   Both TTL and DTL gates can be improved by a modification of the output circuit. The simple circuits given in Figures 8.8 and 8.23 suffer the same speed limitations when the output transistor is turned off. In both circuits the load, with its parasitic capacitance, is driven only by the resistor R_C. Therefore a second output transistor is often added to drive the load when the output is high. A typical TTL circuit with this modification is shown in Figure 8.25. Despite the apparent added complications, the circuit functions quite similarly to the basic circuit of Figure 8.23. When all inputs are high, a current flows down through R_A, which saturates T_2 and T_3. The voltage at the collector of T_2 is therefore only about 1 V above ground (v_{BE} of T_3 plus V_{CESAT} of T_2). This voltage is insufficient to forward bias the base-emitter junctions of T_4 and T_5; hence they are both cut off (note that the emitter of T_5 is at a potential of about V_{CESAT} because T_3 is saturated). We see that when all inputs are high, T_4 and T_5 are cut off and play no role in circuit operation. The output voltage is V_{CESAT}, as in the simpler circuit of Figure 8.23. Now let us suppose that one or more of the inputs is low. In this case the potential v_X is so low that both T_2 and T_3 are cut off, as in the simpler circuit. With T_2 and T_3 cut off, we may imagine them as removed from the circuit. The circuit parameters are chosen so that if the output voltage is less than V_{CC}, current flowing through R_2 saturates T_4 and T_5. With T_5 saturated, the voltage drop between collector and emitter is very small, and the output (point F) is essentially connected to R_C. Any capacitance connected to the output is charged with time constant $R_C C$, just as in the simpler circuit of Figure 8.23. However, in the circuit of Figure 8.25 it is possible to use a very low value for R_C (because T_5 is off when the output is low and excessive currents do not flow through R_C). Consequently, the $R_C C$ time constant for the circuit of Figure 8.25 is very low, and the output responds quickly to any change in the input. Similar output circuits may be used to improve the performance of DTL circuits.

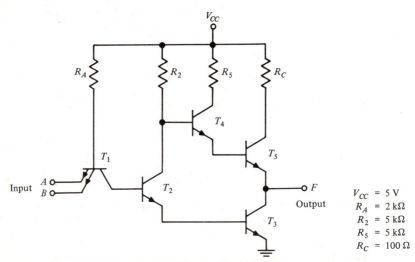

V_{CC} = 5 V
R_A = 2 kΩ
R_2 = 5 kΩ
R_5 = 5 kΩ
R_C = 100 Ω

Figure 8.25   A modified output circuit for the TTL **NAND** gate. A similar output circuit may be used with the DTL gate.

Schottky Diodes   Another important circuit modification that can improve switching speed of both TTL and DTL circuits involves the use of a new, but very simple circuit element called a Schottky diode. The physical form of the Schottky diode and its circuit symbol are shown in Figure 8.26. The device is basically a metal-semiconductor contact. It functions much like a pn junction, in that it passes current much more easily in one direction than the other. The I-V characteristic of a typical Schottky diode is shown in Figure 8.26(c). The significant features of this device are that (1) the voltage drop is lower at a given current than for a pn junction, and (2) the current in forward bias is carried by electrons injected from the semiconductor into the

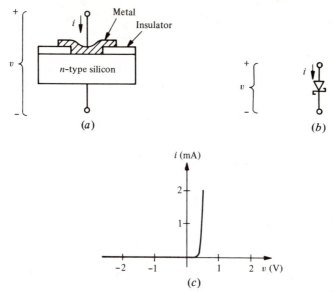

Figure 8.26 The Schottky diode. (*a*) The physical construction of a Schottky diode. It consists of a metal (typically platinum or aluminum) in contact with a lightly doped semiconductor. (*b*) The circuit symbol. (*c*) Typical *I-V* characteristic.

metal. Because of the latter fact, essentially no minority carrier storage effects exist, and the device switches very rapidly. Typically, Schottky diodes are used in parallel with the collector-base junction of transistors, as shown in Figure 8.27. When the transistor is cut off or operating in the active region, the diode SD_1 is reverse biased and carries no current. Hence it does not affect transistor operation. However, when the base current is increased to saturate the transistor, diode SD_1 becomes forward biased. Because it has a lower forward voltage at a given current than the transistor collector-base junction, it prevents a strong forward bias of this junction. The minority carrier storage in the collector and the resulting turn-off delay are eliminated. The Schottky diode is finding increased use in new high-speed digital integrated circuit designs.

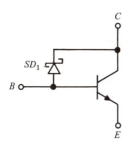

Figure 8.27 A Schottky diode connected in parallel with the collector-base junction of a transistor. The Schottky diode prevents strong forward bias of the collector-base junction in saturation.

8.3   The Flip-Flop

In the previous sections of this chapter we have dealt with logic gates. These are circuits that respond to logical inputs, performing logical operations at the same instant (more or less) that the inputs are applied. We now redirect our attention to a different class of digital circuits: circuits with memory.[2] Whereas logic circuits produce an output (or outputs) that is a function purely of the inputs at hand, circuits with memory produce an output that is a function of both the present inputs and the inputs at some previous time. One can envision the need for memory circuits when operations of a repetitive nature are desired. For example, suppose a number is to be multiplied by another number, and then added to a third. At each stage a record of the previous result must be retained, hence the need for a memory.

The most common and most important digital circuit that has the memory property is known as the *flip-flop*. A flip-flop consists of two electrical switches, each of which may be in an on state or an off state. However, the circuit is constructed so that at any given time one element must be on and the other off. Inputs are provided to reverse the conditions of the two states. Noting that in such a circuit one element "flips" on, while the other "flops" off, the cheerful name flip-flop has been applied. Flip-flops may be realized in any number of ways, in which the switchable elements could be relays, transistors, or various other devices. Here we shall be concerned only with transistor flip-flops that offer both highest performance and lowest cost.

A basic circuit for a flip-flop is shown in Figure 8.28(a). It consists of two symmetrical transistor circuits that are cross-coupled by the resistors R_{B1} and R_{B2}. First let us verify that in the normal state of affairs one transistor is on, the other off. Assume, for example, that T_1 is on, that is, saturated. We shall follow around the circuit and verify that this situation is possible. If T_1 is saturated (this must later be verified), then v_{C1} equals V_{CESAT}, which is nearly zero. In that case T_2 will be cut off, because there is no source of base current for T_2. Practically no base current flows if $v_{BE} < 0.5$ V, and certainly the base current is negligible for base-to-emitter voltage of 0.1 or 0.2 V. If T_2 is cut off, there is no collector current flowing into T_2. Since no current flows through any of the terminals of transistor T_2, it is essentially out of the circuit, as has been suggested in Figure 8.28(b). We may verify that T_1 is saturated by analyzing this circuit. The total resistance in series with the base $(R_{B1} + R_{C2})$ limits the base current to $(V_{CC} - v_{BE})/(R_{C2} + R_{B1})$ or $(5 - 0.8)/(6000) \cong 0.7$ mA. The largest collector current that could possibly flow would have magnitude $V_{CC}/R_{C1} = 5$ mA. Since $i_{C1} \ll \beta\, i_{B1}$, the base current is adequate to saturate T_1.

We have assumed that T_1 is saturated, found that T_2 is cut off, reduced the circuit of Figure 8.28(a) to Figure 8.28(b), and verified that indeed it is

---

[2] Circuits with memory belong to the class of circuits called sequential. The logic circuits already considered belong to the class called combinational.

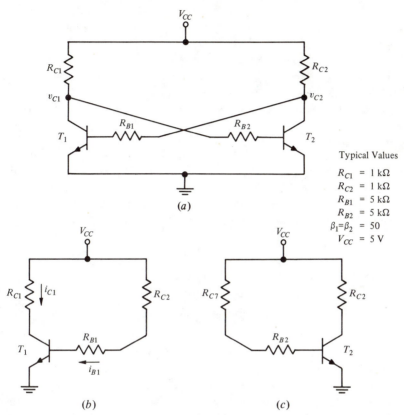

Typical Values

R_{C1} = 1 kΩ
R_{C2} = 1 kΩ
R_{B1} = 5 kΩ
R_{B2} = 5 kΩ
$\beta_1 = \beta_2$ = 50
V_{CC} = 5 V

Figure 8.28 The flip-flop. (*a*) A basic flip-flop circuit. (*b*) The circuit of (*a*) redrawn for the case in which T_1 is saturated and T_2 cut off. (*c*) The circuit redrawn for the case in which T_2 is saturated and T_1 cut off.

possible for the circuit to be in such a state. Let us call this state **1.** The symmetry of the circuit tells us that this state is not unique, the opposite state in which T_1 is off and T_2 on is clearly also possible. We call this state **0,** and the corresponding reduced circuit is given in Figure 8.28(*c*).

It is not yet proven that some other state is not also possible. For example, could it happen that both transistors would be cut off? Both collector currents would have to be zero in that case and the only possible current paths from V_{CC} to ground would be R_{C2} and R_{B1} or through R_{C1} and R_{B2}. There would have to be base currents of magnitude $(V_{CC} - v_{BE})/(R_C + R_B)$, that is, 0.7 mA. But such base currents are not consistent with the transistors being cut off. We conclude that a state in which both transistors are cut off is not self-consistent and could not exist. It is also conceivable that some kind of intermediate state might exist that would indeed be self-consistent, such as each transistor being only partially saturated. Such conditions may indeed be found, but can be shown to be unstable, like a cone balanced on its point. That is, if a flip-flop in such an intermediate state were slightly perturbed, it would immediately fall into states **1** or **0** described above. In practice such unstable states are not usually encountered.

The circuit of Figure 8.28(*a*) can exist stably in either of its two states, so it qualifies as a flip-flop. As yet, however, the circuit is quite useless; we have no provision for either an output or an input. To see how inputs might be provided, let us consider a more generalized form of the circuit. A comparison of Figure 8.28(*a*) with the simple inverter circuit of Figure 8.5 shows that the flip-flop consists of two cross-coupled inverters. A block diagram is given in Figure 8.29. The circuit states may be very quickly evaluated from this figure. For example, if the output of inverter 1, which is the input to inverter 2, is low, then the output of inverter 2 is high. In this case the input to inverter 2 is high, which implies its output is low as initially assumed. Of course the opposite state, with the output of inverter 1 high, and 2 low, is also stable.

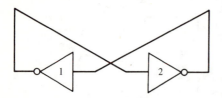

Figure 8.29   Two cross-coupled inverters. This circuit is equivalent to the circuit of Figure 8.28(*a*). It is a flip-flop with no provision for inputs.

Taking the more general view of a flip-flop as consisting of two cross-coupled inverters leads to a more practical form, with provisions for inputs. **NAND** gates can function as inverter circuits, so let us cross-couple two **NAND** gates, as shown in Figure 8.30. In Figure 8.30(*a*) all the inputs but one to each **NAND** gate are tied to V_{CC} (high). Thus the **NAND** gates function as simple inverters, and this circuit is equivalent to the circuit of Figure 8.29. If, however, the extra **NAND** gate inputs are left free, as in Figure 8.30(*b*), they may be used to modify the state of the flip-flop. It is simplest to illustrate the operation of this circuit by a timing diagram. Suppose the inputs A_1 and A_2 vary with

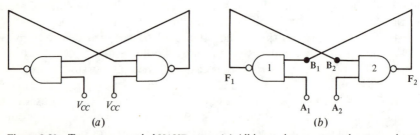

Figure 8.30   Two cross-coupled **NAND** gates. (*a*) All inputs but one to each gate are kept high; thus the **NAND** gates perform as simple inverters. The circuit is equivalent to the circuit of Figure 8.29. (*b*) Inputs A_1 and A_2 may be used to modify the state of the flip-flop.

time according to Figure 8.31(a). Suppose further that at $t = t_0$, $\mathbf{F}_1 = \mathbf{1}$, and $\mathbf{F}_2 = \mathbf{0}$, as shown in Figure 8.31(b). In the interval $t_0 < t < t_1$, the input $\mathbf{A}_1$ and $\mathbf{A}_2$ are both high or $\mathbf{1}$; thus the two **NAND** gates act as simple inverters, and the circuit is stable in the initial state. At time $t = t_1$, $\mathbf{A}_1$ drops to low or logical $\mathbf{0}$. Since input $\mathbf{B}_1$ is already low, this has no effect on the circuit. Again at $t = t_2$, when $\mathbf{A}_1$ returns to logical $\mathbf{1}$ the state of the flip-flop remains unaltered. However, when $\mathbf{A}_2$ is switched to zero at time t_3, the state of the flip-flop is altered. Prior to the instant $t = t_3$, both inputs to **NAND** gate 2 were $\mathbf{1}$. (Both $\mathbf{A}_2 = \mathbf{1}$ and $\mathbf{B}_2 = \mathbf{F}_1 = \mathbf{1}$.) At the instant $t = t_3$, one of the inputs ($\mathbf{A}_2$) is switched to $\mathbf{0}$, consequently the output of gate 2 is switched from $\mathbf{0}$ to $\mathbf{1}$. Suddenly then both inputs to gate 1 are logical $\mathbf{1}$. The output switches to logical $\mathbf{0}$. In this way, a complete change of state of the flip-flop, from $\mathbf{F}_1 = \mathbf{1}$, $\mathbf{F}_2 = \mathbf{0}$ to $\mathbf{F}_1 = \mathbf{0}$, $\mathbf{F}_2 = \mathbf{1}$, is accomplished. When input $\mathbf{A}_2$ returns to $\mathbf{1}$ at $t = t_4$, the circuit remains in the new state. Switching $\mathbf{A}_1$ to $\mathbf{0}$, as at t_5, will return the circuit to its former state, with $\mathbf{F}_1 = \mathbf{1}$, $\mathbf{F}_2 = \mathbf{0}$. The effects of the inputs are summarized in Figure 8.31(c).

It may be noted that the condition in which $\mathbf{A}_1$ and $\mathbf{A}_2$ are simultaneously $\mathbf{0}$ has been ignored. This input condition must be avoided, because the output may become unpredictable. Suppose, for example, $\mathbf{A}_1 = \mathbf{0}$ and $\mathbf{A}_2 = \mathbf{0}$. Both **NAND** gates have output $\mathbf{1}$. Now let $\mathbf{A}_1$ and $\mathbf{A}_2$ return to $\mathbf{1}$. The output state $\mathbf{F}_1 = \mathbf{1}$, $\mathbf{F}_2 = \mathbf{1}$ is not stable, and either one of the two stable states will result. However, it is unpredictable which state will result. Consequently, a circuit

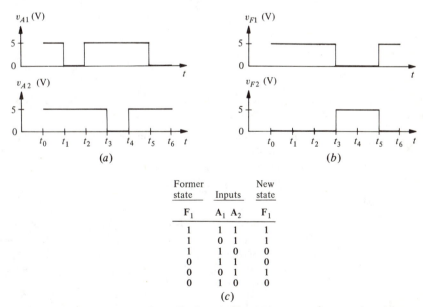

Former state	Inputs	New state
F_1	A_1 A_2	F_1
1	1 1	1
1	0 1	1
1	1 0	0
0	1 1	0
0	0 1	1
0	1 0	0

(c)

Figure 8.31   The response of the flip-flop circuit of Figure 8.30(b) to various inputs. (a) Possible input waveforms. (b) The resulting output waveforms. (c) A summary of the timing diagram presented in (a) and (b).

designer using a flip-flop of the form of Figure 8.30(*b*) must take care that the "forbidden" input condition never occurs.

Flip-flops, like logic circuits, do not change state instantaneously. Thus the timing diagram of Figure 8.31 is idealized. In practice, a delay typically a few nsec to 1 μsec is observed before the output responds to changes in the input. Since flip-flops consist basically of cross-coupled logic gates, the delays are similar to the delays encountered in logic circuits.

The simple flip-flops considered here are intended only to illustrate the nature of these interesting building blocks. Most integrated circuit flip-flops contain many more components, both for the purpose of maximizing the performance and increasing the circuit flexibility of the building block. Some important forms and their circuit applications are considered in Chapter 9.

Summary

- The transistor can function as a nearly ideal controlled switch. The collector-to-emitter characteristic is that of an open switch if the base current is zero, or if the base-emitter voltage is close to zero. The collector-to-emitter characteristic is that of a short circuit for base currents larger than the collector current divided by beta.

- The simple transistor switch functions as an inverter. A high input results in a low output and a low input results in a high output.

- Logic circuits are classified into compatible families. The DTL family is an important commercial family. A DTL **NAND** gate is constructed from a diode **AND** gate, combined with a transistor inverter. Typical logical levels for the DTL family are 0 to 0.5 V for logical **0** and 2 V to 5 V for logical **1** (positive logic). DTL gates feature a high fan-in and fan-out capability.

- The TTL logic family is similar in operation and in logic levels to the DTL family. A TTL **NAND** gate is constructed from a multi-emitter transistor **AND** gate, combined with a transistor inverter.

- All of the basic logic operations may be performed with **NAND** gates alone.

- A flip-flop is an elementary memory circuit that is stable in only two states, labeled **0** and **1**. An elementary flip-flop may be constructed from two cross-coupled inverters. In flip-flop state **0**, the first inverter has output high and the second low. In state **1** the first inverter has output low and the second high.

- A practical flip-flop circuit, with provision for inputs, may be constructed from two cross-coupled **NAND** gates. The state of such a flip-flop may be altered by setting the appropriate input to low.

References

At approximately the same level as this book:

Harris, J. N., P. E. Gray, and C. L. Searle. *Digital Transistor Circuits (SEEC, Vol. 6).* New York: John Wiley & Sons, 1966.

At a more advanced level:

Gray, P. E., and C. L. Searle. *Electronic Principles.* New York: John Wiley & Sons, 1969.

An advanced treatment emphasizing the design of integrated circuits:

Lynn, D. K., C. S. Meyer, and D. J. Hamilton, ed. *Analysis and Design of Integrated Circuits.* New York: McGraw-Hill, 1967.

An advanced treatment of pulse and digital circuits:

Millman, J., and H. Taub. *Pulse, Digital, and Switching Waveforms.* New York: McGraw-Hill, 1965.

Problems

8.1   Calculate v_{CE} in the circuit of Example 8.1 for the following values of i_1: (1) 0, (2) 50 μA, (3) 100 μA, (4) 150 μA. From these numbers make a rough sketch of v_{CE} versus i_1. (ANSWER: (2) 1.5 V.)

8.2   Consider the circuit of Example 8.1 with $i_1 = 100$ μA. Using a graphical technique, find v_{CE} for 10 values of R_C in the range 500 Ω to 2 kΩ. Identify the points in the saturation region and those in the active region. (This problem demonstrates the importance of the collector circuit in determining the operating mode of the device.)

8.3   Evaluate the collector current in cutoff from the large signal equations (Table 6.3). Assume $i_B = 0$, $v_{CE} = 5$ V, and evaluate i_C for the standard transistor of Table 6.4.

8.4   The *I-V* characteristics of a saturated transistor may be computed from the Ebers-Moll equations. (1) Solve Equations (6.1), (6.9), and (6.10) for v_{BC} as a function of i_B and i_C, and for v_{BE} as a function of i_B and i_C. Combine the equations to find v_{CE} as a function of i_B and i_C. [In the saturation mode, both junctions are forward biased, so Equations (6.9) and (6.10) should first be appropriately simplified.] (2) Note that $\beta\, i_B > i_C$ in saturation, and write the equation for v_{CE} as a function of $\beta\, i_B/i_C$. (3) Plot i_C versus v_{CE} over the range $i_C = 0$ to 1 mA for a transistor with $I_{ES} = 10^{-16}$, $I_{CS} = 10^{-14}$, $\alpha = 0.98$, $\alpha_R \cong 0.01$, and assuming $i_B = 0.1$ mA.

8.5   Plot the output voltage of the general inverter (Figure 8.7) as a function of i_L if the input equals 0 V. Assume $R_C = 5$ kΩ, and include the range $i_L = -1$ mA to $i_L = +1$ mA in the plot. Identify the range of i_L for which the output falls in the DTL logical 1 range (2 to 5 V).

8.6   What base current is required to assure that the general inverter (Figure 8.7) is saturated assuming $R_C = 5$ kΩ, $\beta = 50$, $V_{CC} = 5$ V, and $i_L = 23$ mA?

8.7   Suppose the collector resistor of the general inverter with the load resistor shown in Example 8.2 has a value of 5 k$\Omega \pm 10\%$. Find the worst-case minimum value of R_L which is permissible if the output voltage is to remain in the range of 2 to 5 V.

8.8   It is desired to assure that the circuit of Example 8.1 is saturated. Assume that $R_C = 1000\ \Omega \pm 20\%$, $V_2 = 4$ V $\pm 5\%$, and that β lies in the range 30 to 200. Find the value of i_1 sufficient to guarantee saturation under worst-case conditions.

8.9   Find the voltage v_X in Figure 8.8, below which essentially no current flows into the base of T_1. (Ignore the input portion of the circuit consisting of R_A, D_A, and D_B.) Use the large-signal diode model to simplify the circuit.

8.10   Compute the output voltage of the **NAND** gate, Figure 8.8, under the condition $v_A = 0.1$ V, $v_B = 0.2$ V, $i_L = 0$.

8.11   The **NAND** gate, Figure 8.8, has a load resistance of 10 kΩ connected to the output. Find v_F if $v_A = 0.2$ V, $v_B = 0.2$ V.

8.12   Compute the output voltage of the **NAND** gate, Figure 8.8, with $v_A = 5$ V, $v_B = 3$ V, and $i_L = 0$.

8.13   The **NAND** gate, Figure 8.8, is loaded by 10 other **NAND** gates, and must supply a load current i_L of 22 mA when the output is low. Find the required base current.

8.14   Sketch a timing diagram for a four-input **NAND** gate, showing the output for all possible combinations of the four inputs.

8.15   Calculate v_X and i_B in the circuit of Figure 8.8 if $v_A = v_B = 2.5$ V.

8.16   We wish to verify by a worst-case analysis that the **NAND** gate output is high when the input is low. The allowed input range (0 to 0.5 V) is given in Table 8.4. Assume $R_A = 2$ k$\Omega \pm 10\%$, $\beta = 30$ to 90, and $i_L = 0$. Assume further that the forward voltage drops of the diodes are in the range 0.5 to 0.8 V, and that the transistor v_{BE} at any current is given by the value shown in Figure 8.9(b) ± 50 mV. Compute the worst-case value of v_X, the worst-case base current, and the worst-case value of v_F with one or more inputs low.

8.17   Suppose all inputs to the circuit of Figure 8.8 are high. Find the approximate base current, and make a graphical analysis to compute v_F as a function of i_L. [Note: The required curve of i_C versus v_{CE} for the particular base current found here is not given in Figure 8.9(c). Construct the required curve very approximately based on the curves that are given.]

8.18   Suppose the maximum leakage current of a diode in the **NAND** gate is 0.1 μA. Compute the maximum load current of a **NAND** whose output is in the high state which drives 10 other **NAND** gate inputs.

8.19   Find the approximate change in the transistor base current of the DTL **NAND** gate (Figure 8.8) with output low when the pull-down resistor shown in Figure 8.16 is added.

8.20   Compute and plot the output waveform in Example 8.6 for the cases $R_C = 1$ kΩ and $R_C = 500\ \Omega$.

8.21   Compute the approximate power dissipation in the DTL **NAND** gate (the circuit of Figure 8.6) with both inputs high. (1) Let $R_C = 1$ kΩ. (2) Let $R_C = 500\ \Omega$. (3) Using the result of Problem 8.20, show that the speed-power product is approximately constant. (ANSWER: (1) 32 mW.)

8.22   Devise a five-input **OR** gate from **NAND** gates.

8.23   Using only **NAND** gates, construct a four-input **NOR** gate.

8.24   It is possible to construct a **NOR** function by making use of the so-called "wired **OR**" capability of DTL **NAND** gates. An example of **NAND** circuits connected in this manner is given in Figure 8.32. (1) Construct the truth table for this circuit. *Suggestion:* It is helpful to imagine the conditions required for F to be low. (2) Can this circuit function as a two-input **NOR** gate? Which inputs are used, and what is done with the other inputs? (3) Can this circuit be used as a two-input **NAND** gate? (4) Can this circuit be used as an inverter?

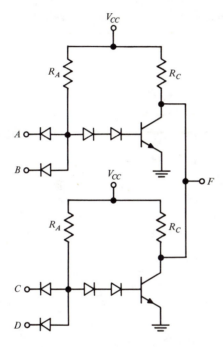

Figure 8.32

8.25   What is the output voltage of the circuit of Figure 8.23 if $v_A = v_B = 0.5$ V, and $i_L = -0.1$ mA? (Assume that the transistor characteristics are those given in Figure 8.9.)

8.26   Suppose the resistor R_M in Figure 8.23 is 4 kΩ. What is the maximum output current if both inputs are high? (Assume that the transistor characteristics are those shown in Figure 8.9.) Compare the results with the result of Example 8.7. (ANSWER: 119 mA.)

8.27   Add a 2-kΩ "base pull-down resistor" to T_2 and T_3 in Figure 8.23. (In each case the resistor is connected from base to emitter.) Compute the maximum output current with both inputs high.

8.28   Redraw the circuit of Figure 8.25 with one or more of the inputs low. (Remove any transistors that are cut off.) What is the approximate output current if the voltage v_F is held at 3 V? (Assume $R_A = R_2 = R_5 = 5$ kΩ, $R_C = 500$ Ω, and let all five transistor characteristics be given by Figure 8.9.)

8.29   Assume that both inputs to the circuit of Figure 8.25 are high. Redraw the circuit, eliminating any transistors that are cut off. Compare with Figure 8.23.

8.30   Suppose that the diodes of D_A and D_B in Figure 8.8 are Schottky diodes. Estimate the effect on the transfer characteristic, Figure 8.12.

8.31   Estimate the values of v_{C1} and v_{C2} in the circuit of Figure 8.28 if T_1 is on and T_2 is off. Use the transistor characteristics given in Figure 8.9.

8.32   A flip-flop circuit is constructed from four **NAND** gates, as shown in Figure 8.33. Start in the state $F_A = 1$, $F_B = 0$, with $A = 0$ and $B = 0$, and vary A and B, one at a time, noting what happens to the output. Use a timing diagram to keep track of the behavior. Is there any combination of A and B that should be forbidden because it leads to an unpredictable output? Summarize the results in a table, as in Figure 8.31.

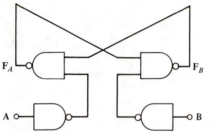

Figure 8.33

8.33   A set of inputs is given in Figure 8.34. Assume input 1 is connected to A_1 and input 2 is connected to A_2 in the circuit of Figure 8.30(b). If F_1 is initially **0**, find F_2 in the time interval shown.

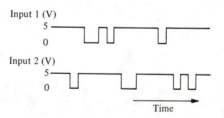

Figure 8.34

Chapter 9

Analysis and Synthesis of Digital Systems

Large digital circuits are generally constructed from sub-circuits or building blocks, such as logic gates and flip-flops. A very large circuit, such as a digital computer, which may contain many thousands of building blocks, is called a *digital system.* The design of a circuit containing many sub-circuits, which are in themselves rather complex, appears to be a profoundly complicated task. Clearly the number of different types of circuits must be restricted, if for no other reason so that the designer can comprehend their operation. In fact a given section of a computer is often constructed entirely from one or two types of subcircuits, such as **NAND** gates.

Understanding the operation of a digital system, even one constructed from identical building blocks becomes a difficult task when the system is very large. Therefore mathematical techniques are used to organize and simplify the analysis, and hence the understanding of circuit operation. These techniques, which are introduced in this chapter, also make possible the design, or *synthesis,* of digital circuits to accomplish a desired function, such as the addition of two numbers. The information contained in this chapter thus serves two purposes: (1) it enables us to write down in some reasonably compact way the function of a given digital circuit consisting of a set of interconnected building blocks, and (2) it enables us to construct from building blocks a circuit accomplishing some desired function. Whereas Chapters 5 and 8 deal with the subcircuit, this chapter deals with the interconnection of subcircuits to form a larger digital system. Little attention is paid in this chapter to the internal details of the subcircuits.

To subdivide further the topics of interest, we break digital circuits into two catagories: combinational (such as the **NAND** logic circuits discussed in Chapter 8) and sequential (such as the flip-flop circuits, introduced in Chapter 8). We shall be concerned with the analysis and synthesis of both kinds of circuit.

In discussing the operation of logic circuits we have found it convenient to use the concept of logical **1** or logical **0** to describe the state of the input or output of a logical block. Referring to Figure 9.1 we say, for example, that the output **F** is **1** if V_F is in the appropriate range, say 4 to 5 V. Similarly, **F** = **0** if V_F is in another range, say 0 to 1 V. In binary logic circuits **F** is allowed only these two values, **0** or **1**; accordingly **F** is known as a *binary variable*. Furthermore, once the logic levels are defined, it is unnecessary to deal with voltages at all; only logical values need be specified.

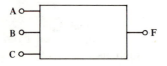

Figure 9.1   A digital block with inputs **A, B, C** and output **F**. The binary variables are **A, B, C,** and **F**.

If the output **F** of the block in Figure 9.1 depends only on the values of the inputs **A, B,** and **C,** then the block is said to be a combinational logic circuit. The value of **F** is said to be a *function* of **A, B,** and **C,** and the function is known as a *logic function* or *logic expression*. In some digital circuits, however, the output at a certain time depends not only on the inputs at that time, but also on what has happened at earlier times. Such circuits are known as *sequential* circuits. A simple example of a sequential circuit is the flip-flop discussed in Chapter 8. The memory, or holding, feature of the circuit is of course quite useful.

The characteristic that causes circuits to behave as sequential is the presence of *feedback* in the circuitry, as shown in Figure 9.2(*b*). A signal path may be identified in this circuit as starting at the output of gate 2, passing through gate 3, and back to the input of gate 2. Since there is necessarily a time delay in the transmission of a signal along this path, the output of gate 2 is a function of its output at some former time. Sequential circuits may therefore have a memory, that is, the ability to retain, or store, digital information. Flip-flops are the simplest practical form of sequential circuit, and are the building blocks for circuits such as counters and registers that make use of the memory function.

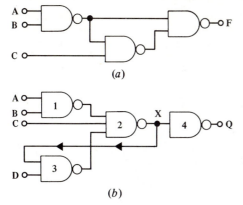

(a)

(b)

Figure 9.2 Comparison of a combinational and a sequential circuit. (*a*) The output **F** of circuit (*a*) is a unique function of the inputs **A**, **B**, **C**; circuit (*a*) is combinational. (*b*) The output **Q** of circuit (*b*) depends on the previous values of the inputs, as well as on **A**, **B**, **C**, and **D**. Circuit (*b*) is sequential.

Before discussing any particular circuits it is useful to develop some formal mathematical background in the manipulation of binary variables. This kind of mathematics, known as *switching algebra,* is the subject of Section 9.1. This discussion may be regarded as an introduction to a general mathematics of binary variables, known as *Boolean algebra.* With this background it is possible to analyze or synthesize combinational logic circuits for the processing of binary signals. Techniques for analysis and synthesis are given in Section 9.2. Sequential circuits are then taken up in Sections 9.3 and 9.4: First the input-output characteristics of flip-flops are established in Section 9.3, and then the analysis and synthesis of a combination of flip-flops are discussed in Section 9.4.

9.1 Switching Algebra

The mathematical relationships between the binary variables in a logic circuit are expressed in switching algebra. The basic operations in switching algebra are **OR, AND,** and **NOR,** and each is given a mathematical symbol. The **OR** operation is represented by a + sign; thus the relationship **A OR B** is written **A + B**. The **AND** operation is represented by a dot, although in practice the dot is usually omitted. Hence **A AND B** is written **A · B** or simply **AB**. A logical complementation is written as a bar over the variable (or over an entire relationship, if it is to be complemented). For example, the complement of **A** is **Ā** and the complement of (**A + B**) is $\overline{(A + B)}$. Any relationship between logical variables is called a logical expression. A logical relationship between a number of variables can be written as an equation. For example, the equation **F = A + B + C**, expresses the statement that **F = 1** whenever **A = 1,** *or* **B = 1,** *or* **C = 1**. Thus if **F** is the name of the output variable and **A, B,** and **C** are the inputs, this equation expresses the action of an **OR** gate. The logical expression for **F** is said to be **A + B + C.** The fundamental logical operations are summarized in Table 9.1.[1]

Table 9.1   The Fundamental Logic Operations

OR	AND	COMPLEMENT
$0 + 0 = 0$	$0 \cdot 0 = 0$	$\bar{0} = 1$
$0 + 1 = 1$	$0 \cdot 1 = 0$	$\bar{1} = 0$
$1 + 0 = 1$	$1 \cdot 0 = 0$	
$1 + 1 = 1$	$1 \cdot 1 = 1$	

Table 9.2

Logical Operation	Mathematical Expression	Definition	Implementation
Multiple Inputs **A, B, C** and Output **F**			
			AND Gate
AND	$F = A\,B\,C$	$F = 1$ when $A = B = C = 1$ $F = 0$ otherwise	A B C → F
			OR Gate
OR	$F = A + B + C$	$F = 0$ when $A = B = C = 0$ $F = 1$ otherwise	A B C → F
			NAND Gate
NAND	$F = (\overline{A\,B\,C})$	$F = 0$ when $A = B = C = 1$ $F = 1$ otherwise	A B C → F
			NOR Gate
NOR	$F = (\overline{A + B + C})$	$F = 1$ when $A = B = C = 0$ $F = 0$ otherwise	A B C → F
Single Input **A** and Output **F**			Inverter
COMPLEMENT	$F = \bar{A}$	$F = 0$ when $A = 1$ $F = 1$ when $A = 0$	A → F

[1]It may at first seem inappropriate that the symbol for **OR** suggests ordinary addition and that for **AND** suggests ordinary multiplication. However, a little reflection shows that this choice is not unreasonable. Suppose C is **1** if **A AND B** are **1.** This relationship, in the notation above, would be written $C = A \cdot B$. This logic equation happens to be entirely consistent with ordinary multiplication. That is, $0 \cdot 0 = 0$, $0 \cdot 1 = 0$, $1 \cdot 0 = 0$, and $1 \cdot 1 = 1$ in ordinary arithmetic. Thus the logical **AND** does have something in common with multiplication. The reader may wish to explore for himself the extent to which the logical **OR** is similar to ordinary addition.

We immediately note that the commutative law of ordinary algebra applies to both the **OR** and **AND** operations. That is, **A • B** is equivalent to **B • A** and **A + B** is equivalent to **B + A**, where **A** and **B** are binary variables. It is also a rule of the algebra that the associative and distributive laws apply. That is, $\mathbf{A + (B+C) = (A+B) + C}$, and $\mathbf{A(B + C) = AB + AC}$. It should be noted in passing that $\overline{\mathbf{A\,B}}$ does *not* equal $\overline{\mathbf{A}}\,\overline{\mathbf{B}}$, nor does $\overline{\mathbf{(A + B)}} = \overline{\mathbf{A}} + \overline{\mathbf{B}}$.

The basic logic gates discussed in Chapter 3 perform single elementary logic operations and thus their input-output relationships may be described by simple logic expressions. These relationships are summarized in Table 9.2, an extension of Table 3.1. Operations performed by combinations of the basic logic gates can be described by more complex logic expressions.

EXAMPLE 9.1

Find the equation relating the output to the input for the following circuit which appeared in Chapter 3.

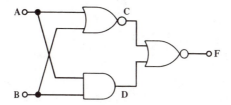

SOLUTION

We first find the intermediate values **C** and **D**. For the **NOR** gate

$$\mathbf{C = \overline{(A + B)}}$$

and for the **AND** gate

$$\mathbf{D = A\,B}$$

The output **NOR** gate relates **F** to **C** and **D**:

$$\mathbf{F = \overline{(C + D)}}$$

The values of **C** and **D** may now be substituted into this expression for **F**:

$$\mathbf{F = \overline{(\,\overline{(A+B)} + AB)}}$$

Example 9.1 demonstrates that one can easily write down the mathematical expression representing a combination of gates. However, several points may be noted that motivate the following discussion. First of all, the expression obtained may be unnecessarily clumsy; it is desirable to simplify it, eliminating especially the brackets and complementation bars. Second, if it is possible to simplify an expression, this implies that an expression is not unique. How do we determine whether or not two expressions are equivalent? What then is the simplest or optimum form of a logic expression? Finally, given a complex expression, how may a circuit be constructed which *implements* the function? The following sections deal with these questions.

Equivalence of Logic Expressions   Two expressions are equivalent if they have the same value for all possible values of their input variables. This may be stated in another form: Two expressions are equivalent if they have identical truth tables. Consider, for example, the expressions $F = AB + A\overline{B}$ and $G = A$. We may show that, in fact, $F = G$ by merely comparing their truth tables. The complete table for F is given in Table 9.3. The truth table for

Table 9.3   Truth Table for $F = AB + A\overline{B}$

A	B	F
0	0	0
0	1	0
1	0	1
1	1	1

G has entries of 1 any time $A = 1$; G is not a function of B. If the values of B are nonetheless considered, and listed in the truth table for G, the table would have the form of Table 9.4. A comparison of Tables 9.3 and 9.4 shows that indeed the expressions are equivalent.

Table 9.4   Truth Table for $G = A$

A	B	G
0	0	0
0	1	0
1	0	1
1	1	1

EXAMPLE 9.2
    Verify that $A + BC$ is equivalent to $A + BC + AB$.

SOLUTION
    We complete a truth table for each expression and compare them. We first compute the values of the individual terms A, BC, and AB, and then write down the truth tables. To construct the complete table, for example, for the expression $A + BC$, we merely set down a 1 whenever A *or* BC equals 1. It so happens that $A = 1$ or $BC = 1$ for

TRUTH TABLE FOR **A, BC, AB**

A	B	C	A	B C	A B
0	0	0	0	0	0
0	0	1	0	0	0
0	1	0	0	0	0
0	1	1	0	1	0
1	0	0	1	0	0
1	0	1	1	0	0
1	1	0	1	0	1
1	1	1	1	1	1

the last five combinations of **A, B,** and **C** considered. The truth table for **A + BC + AB** is constructed similarly; we set down a **1** whenever any of the three terms is **1.**

TRUTH TABLE FOR **A + B C**

A	B	C	A + B C
0	0	0	0
0	0	1	0
0	1	0	0
0	1	1	1
1	0	0	1
1	0	1	1
1	1	0	1
1	1	1	1

TRUTH TABLE FOR **A + B C + A B**

A	B	C	A + B C + A B
0	0	0	0
0	0	1	0
0	1	0	0
0	1	1	1
1	0	0	1
1	0	1	1
1	1	0	1
1	1	1	1

A comparison of the truth tables shows that the expressions are equivalent. ■

In the preceding discussion we saw that an equation in switching algebra is equivalent to a truth table. The value of the algebraic expression is its compactness. (A truth table expressing the relationship **F = A B C D** requires 16 entries because all 16 possible combinations of **A, B, C,** and **D** must be considered.) However not every algebraic expression itself is in the most compact form possible. For example, the expression $F = A + \overline{A}$ may be more compactly written $F = 1$ (because either A is 1, in which case $F = 1$, or A is 0, in which case $\overline{A}$ is 1 and $F = 1$). The search for simpler expressions may be done in various ways. In this book we adopt a simple and powerful visual method of simplifying expressions. The simplification of more complex expressions than are treated here is generally handled on a digital computer.

Before describing the simplification technique, which is the main subject of this section, we wish to point out a theorem of switching algebra that is so important and useful that it deserves special attention. This theorem, known as DeMorgan's theorem, may be stated in two forms:

$$\boxed{\overline{(A\ B)} = \overline{A} + \overline{B}} \tag{9.1}$$

$$\boxed{\overline{(A + B)} = \overline{A}\ \overline{B}} \tag{9.2}$$

EXAMPLE 9.3

Show that the two forms of DeMorgan's theorem are equivalent.

SOLUTION

Complementing both sides of an equation has no effect on its validity ($\overline{1} = \overline{1}$ states that $0 = 0$, and $\overline{0} = \overline{0}$ states that $1 = 1$). Let us complement both sides of Equation (9.1):

$$\overline{\overline{(A\ B)}} = \overline{\overline{A} + \overline{B}}$$

or

$$A\ B = \overline{\overline{A} + \overline{B}}$$

Now if we define new variables $C = \overline{A}$ and $D = \overline{B}$, then $\overline{C} = A$ and $\overline{D} = B$. The last expression may be written as

$$\overline{C}\ \overline{D} = \overline{(C + D)}$$

This is just a statement of Equation (9.2). ∎

The usefulness of DeMorgan's theorem is its ability to remove complementation bars over large expressions. Suppose F is the complement of the expression $\overline{A}\ \overline{B}\ \overline{C}\ \overline{D}$, that is, $F = \overline{(\overline{A}\ \overline{B}\ \overline{C}\ \overline{D})}$. This may be grouped as $F = \overline{((\overline{A}\ \overline{B})\ (\overline{C}\ \overline{D}))}$, and then Equation (9.1) applied. The result is $F = \overline{(\overline{A}\ \overline{B})} + \overline{(\overline{C}\ \overline{D})}$. To each of these terms Equation (9.1) may again be applied; thus $F = A + B + C + D$. We have used the fact that $\overline{(\overline{A})} = A$, and so forth.

EXAMPLE 9.4

Apply DeMorgan's theorem to simplify the expression

$$F = \overline{(\overline{A} + \overline{B} + \overline{C})}$$

SOLUTION

First group any two of the terms, for example, $(\overline{A} + \overline{B})$, and apply Equation (9.2), regarding the group as one term:

$$F = \overline{(\,(\overline{A} + \overline{B}) + \overline{C})} = \overline{(\overline{A} + \overline{B})} \cdot C$$

Now apply Equation (9.2) to the term in parentheses:

$$F = (A\,B) \cdot C = A\,B\,C$$ ∎

We now return to the main topic of simplifying logic expressions. First it is necessary to consider what form of logic expression is desired. Clearly some "simple" form, that is, one containing as few variables and terms as possible, is attractive. Furthermore, as will be shown in the next section, it is useful to cast the logic expression in a special form to simplify the synthesis of actual circuits. The form most desirable for circuit synthesis is the so-called *sum of products* form. In this form the expression is given as a "sum" of terms, each of which is a simple "product" of variables or their complements.[2] Some examples are given in Table 9.5. Note that no parentheses appear in sum of products expressions.

Table 9.5 Some Logic Expressions in the Sum of Products Form

$F = A\overline{B} + AC + \overline{A} \cdot \overline{B} \cdot \overline{C}$	(1)
$F = ABCD$	(2)
$F = A + \overline{B} + \overline{C}$	(3)

In addition to conforming to this sum of products form, it is useful for logic expressions to be as simple as possible. A sum of products expression that has the smallest possible number of terms with each term as small as possible is called a *minimal* expression.

We first suggest a very simple procedure to obtain a sum of products form for an expression. We will then take up the question of minimizing the expression. The procedure for obtaining a sum of products form of an expression is: (1) Make a truth table for the logic expression to be simplified (or circuits to be analyzed), and (2) write down the sum of products form from the truth table. This procedure is best demonstrated by an example.

Suppose that it is desired to find a sum of products form for the expression $F = (A + B)\,(A\,\overline{C})$. According to rule 1 we first form the truth table for

---

[2] As noted previously, these are no ordinary sums or products in the sense of continuous algebra, but are the functions described in Table 9.1.

this expression. To do this we must examine the logical value of the expression for each possible combination of **A, B,** and **C.** For example, when **A, B,** and **C** are all **0,** (**A** + **B**) is **0** and (A$\overline{C}$) is **0.** Therefore **F,** which equals (**A** + **B**) · (**A** $\overline{C}$), is **0.** To consider another possible combination, suppose **A** = **1** and **B** and **C** are **0.** Then (**A** + **B**) is **1** and (**A** $\overline{C}$) is **1.** In this case **F** is **1.** The complete truth table is given in Table 9.6.

Table 9.6   Truth Table for **F** = (**A** + **B**) · (**A** $\overline{C}$)

A	B	C	F
0	0	0	0
0	0	1	0
0	1	0	0
0	1	1	0
1	0	0	1
1	0	1	0
1	1	0	1
1	1	1	0

Now let us see how the sum of products form may be formed from the truth table. We must first recognize that each line in a truth table corresponds in itself to a simple expression. For example, the first line in Table 9.6 represents $\overline{A}\ \overline{B}\ \overline{C}$; that is, $\overline{A}\ \overline{B}\ \overline{C}$ = **1** when **A, B,** and **C** are **0.** Similarly, the last line represents **A B C,** because **A B C** = **1** when **A, B,** and **C** are all **1.** We note that in Table 9.6 **F** = **1** in two cases: (1) **A** = **1, B** = **0, C** = **0,** and (2) **A** = **1, B** = **1, C** = **0.** Case (1) may be expressed as **A** $\overline{B}\ \overline{C}$, because **A** $\overline{B}\ \overline{C}$ = **1** when, and only when, **A** = **1, B** = **0, C** = **0.** Similarly case (2) is expressed as **A B** $\overline{C}$, because **A B** $\overline{C}$ = **1** when **A** = **1, B** = **1, C** = **0.** Since **F** = **1** in either case (1) or case (2), we say that **F** = **A** $\overline{B}\ \overline{C}$ + **A B** $\overline{C}$. This is the desired sum of products expression. (However, it will be shown in Example 9.8 that this expression is not minimal; in fact it may be more simply expressed as **F** = **A** $\overline{C}$.)

The key to forming the sum of products expression from a truth table is the correspondence of each line in a truth table to a simple product of variables expression. This correspondence applies to truth tables with any number of variables. Consider, for example, the truth table for a function of five variables, **A, B, C, D,** and **E.** There must be one line in the truth table for the combination **A** = **0, B** = **1, C** = **1, D** = **0, E** = **0.** This combination expresses the condition $\overline{A}$ **B C** $\overline{D}\ \overline{E}$ = **1,** because the latter expression equals **1** only for this particular combination of variables. Note that the simple expression corresponding to each line in a truth table is readily written by inspection; variables that are **0** have a complement bar in the expression, the others do not.

EXAMPLE 9.5
Find the sum of products form for the expression

$$F = A\ B\ C + B(A + \overline{C}) + \overline{(C + B)}$$

SOLUTION

We first construct the truth table. Since there are three variables, it must have eight entries. The first row will be $A = B = C = 0$. In this case

$$F = 0 \cdot 0 \cdot 0 + 0(0 + 1) + \overline{(0 + 0)}$$
$$F = 0 + 0 + 1$$
$$F = 1$$

The other rows are evaluated similarly, yielding the truth table:

A	B	C	F
0	0	0	1
0	0	1	0
0	1	0	1
0	1	1	0
1	0	0	1
1	0	1	0
1	1	0	1
1	1	1	1

The first line in the truth table corresponds to $\overline{A}\ \overline{B}\ \overline{C}$, because $\overline{A}\ \overline{B}\ \overline{C}$ is 1 when, and only when, A, B, and C are 0. Similarly, the third line corresponds to $\overline{A}$ B $\overline{C}$. The fifth line corresponds to A $\overline{B}$ $\overline{C}$, the seventh line to A B $\overline{C}$, and the last line to A B C. Since the function F is 1 for the conditions expressed by any of these relationships, we may write

$$F = \overline{A}\ \overline{B}\ \overline{C} + \overline{A}\ B\ \overline{C} + A\ \overline{B}\ \overline{C} + A\ B\ \overline{C} + A\ B\ C$$

Each term in this expression corresponds to a line in the truth table above. For example, the second term indicates that $F = 1$ if $\overline{A} = 1$, $B = 1$, and $\overline{C} = 1$. This is precisely the condition expressed by line three of the truth table. This expression will be simplified in Example 9.10. ■

Minimization with Karnaugh Maps   We now consider the simplification of expressions. It is a small step from the sum of products expression to the minimal sum of products expression. By casting the truth table in a slightly different form, the minimal expression may be written by inspection. In particular, it is useful to rearrange the truth table in a two-dimensional form, called a *Karnaugh map*. An example of this form is given in Figure 9.3, in which an ordinary truth table for the expression $F = A\ \overline{B}\ C$ is compared to the Karnaugh map of the same expression. Each square in a Karnaugh map corresponds to some particular combination of A, B, and C; thus each square in the map corresponds to a line in the truth table. The values of A, B, and C are indicated around the boundaries of the map. We see that the lower right-hand square corresponds to the values $A = 1$, $B = 0$, $C = 1$. A 1 is placed in this square signifying that $F = 1$ for this particular combination of A, B, and C. The arrangement of A, B, and C around the boundaries is somewhat arbitrary, for example, A and B or B and C could be reversed (of course the position of the 1 would change in this case). However, the arrangement of the values of

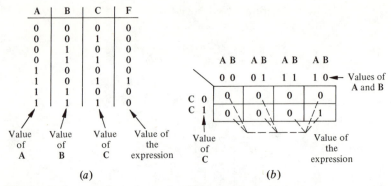

Figure 9.3   A comparison of a truth table and a Karnaugh map. (*a*) A truth table. (*b*) A Karnaugh map. The map is just a two-dimensional rearrangement of the truth table. Each square in the map corresponds to one line in the truth table.

the variables (for example, **00, 01, 11, 10** along the top) is standard, for reasons which will become clear. The Karnaugh map is also not restricted to three variables. Examples will be given shortly of four-variable maps.

We saw earlier that each line in an ordinary truth table corresponds to a simple product of variables term; for example, the line **A = 0, B = 1, C = 1** corresponds to $\overline{A}$ **B C**. Each square in a Karnaugh map similarly corresponds to such a term. All eight possible terms for the three-variable map are enumerated in Figure 9.4. The labeling is also somewhat simplified here in comparison to Figure 9.3(*b*) by not repeating the labels over each value of **A, B,** and **C**. The **A B** appearing at the upper left of Figure 9.4 indicates that the pairs of values along the top are values of the pair **A B**. For example, the upper right-hand square corresponds to the set of values **A = 1, B = 0, C = 0.**

$_A\backslash^B$ C	0 0	0 1	1 1	1 0
0	$\overline{A}\overline{B}\overline{C}$	$\overline{A}B\overline{C}$	$AB\overline{C}$	$A\overline{B}\overline{C}$
1	$\overline{A}\overline{B}C$	$\overline{A}BC$	ABC	$A\overline{B}C$

Figure 9.4   The logical expression corresponding to each square in a Karnaugh map.

The process of finding the Karnaugh map for an expression is called mapping. The Karnaugh map is obtained in precisely the same manner as the truth table. Each possible combination of variables in the expression is examined. A **1** is placed in each square corresponding to each combination for which the expression is **1**. Logical **0**'s may be placed in the other squares, or they may be left blank.

EXAMPLE 9.6
Map the expression **F** = (**A** + **B**) **C.**

SOLUTION
    We must consider each possible combination of **A**, **B**, and **C**. For example, **F** = **0** when **A**, **B**, and **C** are all **0**. Similarly when **A** = **0**, **B** = **0**, and **C** = **1**, then **F** = **0**. In fact, **F** = **1** only when **C** = **1**, and one or both of **A** and **B** equal **1**; that is, for the combination **A,B,C** = **1,1,1**; **A,B,C** = **1,0,1**; and **A,B,C** = **0,1,0**. Therefore, we place a **1** in the three squares of the Karnaugh map corresponding to these latter three combinations of **A**, **B**, and **C**.

C \ AB	0 0	0 1	1 1	1 0
0				
1		1	1	1

In this map we have omitted the zeros. ∎

    A sum of products expression may be written directly from a Karnaugh map, just as from an ordinary truth table. Each square in the map gives rise to one term in the expression.

EXAMPLE 9.7
    Find a sum of products expression for the function that is mapped in Example 9.6.

SOLUTION
    The map is repeated below.

C \ AB	0 0	0 1	1 1	1 0
0	0	0	0	0
1	0	1	1	1

The lower right-hand square indicates that **F** = **1** when **A** = **1**, **B** = **0**, and **C** = **1**; that is, when **A** $\overline{\text{B}}$ **C** = **1**. The next square to the left indicates that **F** = **1** when **A B C** = **1**. The next square to the left indicates that **F** = **1** when $\overline{\text{A}}$ **B C** = **1**. We have then that **F** = **1** when **A** $\overline{\text{B}}$ **C** or **A B C** or $\overline{\text{A}}$ **B C** = **1**; that is,

$$\mathbf{F} = \mathbf{A}\,\overline{\mathbf{B}}\,\mathbf{C} + \mathbf{A}\,\mathbf{B}\,\mathbf{C} + \overline{\mathbf{A}}\,\mathbf{B}\,\mathbf{C}$$ ∎

    We have shown thus far that a Karnaugh map may be used like an ordinary truth table to obtain a sum of products form for an expression. We wish to show now that the Karnaugh map can be used to find the minimal expression, that is, the sum of products expression with the least number of terms. In fact, the Karnaugh map is so arranged that the minimal expression may be recognized by inspection.
    Figure 9.4 shows that the map is so arranged that any two adjacent squares differ only by a complement bar over one of the variables. For example, the two right-most squares indicated in Figure 9.5 correspond to **A** $\overline{\text{B}}$ **C** and **A** $\overline{\text{B}}$ $\overline{\text{C}}$. A map that had **1**'s in only these two squares would be a

^{A}B	0 0	0 1	1 1	1 0
C				
0				$A\overline{B}\overline{C}$
1				$A\overline{B}C$

Figure 9.5 Two adjacent squares in a Karnaugh map. These two squares represent $A\ \overline{B}\ \overline{C} + A\ \overline{B}\ C$, which is more simply written $A\ \overline{B}$.

map of the expression $A\ \overline{B}\ C + A\ \overline{B}\ \overline{C}$. If the common factor $A\ \overline{B}$ is factored out, the expression is written $A\ \overline{B}\ (C + \overline{C})$. But $(C + \overline{C})$ is always **1** (either $C = 0$ and $\overline{C} = 1$ or $C = 1$ and $\overline{C} = 0$). Thus the expression may be more simply written as $A\ \overline{B}$. The two squares indicated in Figure 9.5 taken together correspond to the simple expression $A\ \overline{B}$. Furthermore, this fact may be recognized directly from the map itself; we see from Figure 9.5 that in both of the indicated squares $A = 1$ and $B = 0$. An expression whose map has a **1** in both of these squares has a value of **1** when $A = 1$, $B = 0$, independent of whether $C = 0$ or **1**. Hence the expression may be written $A\ \overline{B}$. Any two adjacent squares on the Karnaugh map correspond to an expression containing only two variables. Some further examples are given in Figure 9.6. In both squares indicated in Figure 9.6(a), $A = 1$ and $C = 0$. The expression has the value of **1** if $A = 1$, $C = 0$, whether $B = 1$ or **0**; hence it may simply be written $A\ \overline{C}$. Similarly, Figure 9.6(b) is the map of an expression that may most simply be written $B\ C$. The right- and left-hand columns of the Karnaugh map should be regarded as being adjacent, as suggested in Figure 9.6(c). The two squares indicated correspond to the condition $B = 0$, $C = 0$; hence the expression mapped here is $\overline{B}\ \overline{C}$.

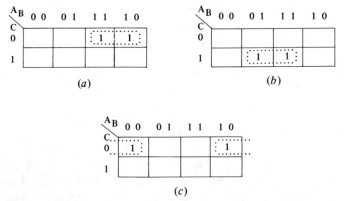

Figure 9.6 Possible adjacent squares in a Karnaugh map. (a) The map of the expression $A\ \overline{C}$. (b) The map of the expression $B\ C$. (c) The map of the expression $\overline{B}\ \overline{C}$.

EXAMPLE 9.8

Find the minimal sum of products form of the expression $F = (A + B)\ (A\ \overline{C})$. (The truth table for this expression is given in Table 9.6.)

SOLUTION

We first find the Karnaugh map of the expression. Since the truth table is already given, we recognize that $F = 1$ for two combinations of A, B, and C: $A = 1$, $B = 0$, $C = 0$ and $A = 1$, $B = 1$, $C = 0$. The map is given below.

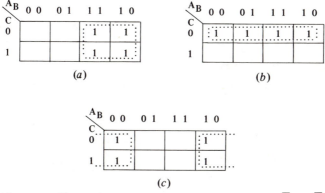

Since the two ones appear in adjacent squares, a single term containing two variables will express the function. These two squares correspond to the condition $A = 1$, $C = 0$; hence the expression is $F = A \overline{C}$. ∎

A single square in a three-variable map corresponds to a single term containing three variables. Two adjacent squares correspond to a single term containing only two variables. Furthermore, four adjacent squares correspond to a single term containing only one variable. For example, the four squares indicated in Figure 9.7(a) correspond to $A = 1$; hence this is a map of the function $F = A$. Similarly, Figure 9.7(b) is a map of the function $F = \overline{C}$. Again the right- and left-hand columns of the map should be regarded as being adjacent; hence Figure 9.7(c) is a map of $F = \overline{B}$.

Figure 9.7 Karnaugh maps of the expression (a) A, (b) $\overline{C}$, (c) $\overline{B}$.

Examples of some various possible groupings are indicated in Figure 9.8. This figure need not be memorized, as any grouping is immediately identified by the values on the perimeter. Any adjacent group of two or four squares may be described by a single term. Furthermore the right-hand column should always be considered as adjacent to the left-hand column, as indicated in Figure 9.8(d) and 9.8(g). If it should happen that all the squares have either 1's or 0's, then the expression is simply $F = 1$ or $F = 0$, respectively. The proce-

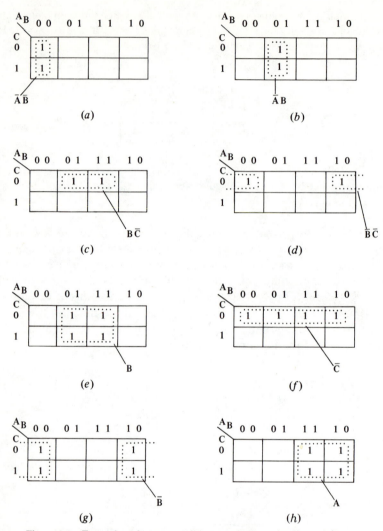

Figure 9.8 Examples of some possible groupings on the Karnaugh map.

dure for finding the minimal sum of products expression may now be stated: (1) Map the function to be minimized. (2) Write the minimal expression directly from the map, grouping the 1's in adjacent squares into groups of two or four. Each group of from one to four squares gives rise to one term in the sum of products expression. Furthermore the groups may overlap.

EXAMPLE 9.9

Map the expression $F = A B C + \overline{A} B + \overline{A} \overline{C} B + A B \overline{C}$ and express F in a simpler form if possible.

SOLUTION

A **1** is placed in each of the squares of the Karnaugh map corresponding to a combination of **A, B,** and **C** for which **F = 1.** For example, **F = 1** when **A = 0, B = 1, C = 0.** When all eight possibilities are examined, the following map is obtained:

```
   A
    B  0 0   0 1   1 1   1 0
   C  ┌─────┬─────┬─────┬─────┐
   0  │     │  1  │  1  │     │
      ├─────┼─────┼─────┼─────┤
   1  │     │  1  │  1  │     │
      └─────┴─────┴─────┴─────┘
```

This arrangement may be recognized as the much simpler expression

$$F = B$$

∎

EXAMPLE 9.10

Rework Example 9.5 using a Karnaugh map; that is, find the sum of products form for the expression

$$F = A\,B\,C + B\,(A + \overline{C}) + \overline{(C + B)}$$

SOLUTION

As in Example 9.5, it is simplest to fill out the map by examining each possible combination. For example, when **A = B = C = 1,** then **F** is given by $F = 1 \cdot 1 \cdot 1 + 1 \cdot (1 + 0) + \overline{(1 + 1)}$. Thus **F = 1** and a **1** is placed in the appropriate square. Examining each possible combination, the map is determined, and is given below.

```
   A
    B  0 0   0 1   1 1   1 0
   C  ┌─────┬─────┬─────┬─────┐
   0  │  1  │  1  │  1  │  1  │
      ├─────┼─────┼─────┼─────┤
   1  │     │     │  1  │     │
      └─────┴─────┴─────┴─────┘
```

Two possible groupings of the 1's are shown below:

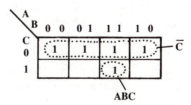

or

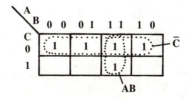

The latter grouping is preferred because the expression is simpler, namely, $\mathbf{F} = \overline{\mathbf{C}} + \mathbf{A}\,\mathbf{B}$ rather than $\overline{\mathbf{C}} + \mathbf{A}\,\mathbf{B}\,\mathbf{C}$. ∎

EXAMPLE 9.11
 Find the minimal sum of products expression for the function mapped in Example 9.7.

SOLUTION
 The map is repeated below:

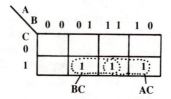

There are again several possible ways of grouping the 1's; however, a simpler expression is always obtained by making the groups as large as possible.

$$\mathbf{F} = \mathbf{A}\,\mathbf{C} + \mathbf{B}\,\mathbf{C}$$ ∎

Examples 9.10 and 9.11 illustrate a common occurrence, the possibility of a variety of expressions. It will be shown shortly that the simplest expression, that is, the one with fewest terms and with fewest variables in each term, leads to the simplest circuit realization with the methods presented in Section 9.2. Thus the minimal expression is preferred.

Karnaugh Maps with Four Variables The minimization technique using Karnaugh maps may be extended in a straightforward manner to functions of four variables. The Karnaugh map is arranged as shown in Figure 9.9. The arrangement is such that common terms are adjacent, as in the three-variable

^{A}B C D	0 0	0 1	1 1	1 0
0 0	$\overline{A}\overline{B}\overline{C}\overline{D}$	$\overline{A}B\overline{C}\overline{D}$	$AB\overline{C}\overline{D}$	$A\overline{B}\overline{C}\overline{D}$
0 1	$\overline{A}\overline{B}\overline{C}D$	$\overline{A}B\overline{C}D$	$AB\overline{C}D$	$A\overline{B}\overline{C}D$
1 1	$\overline{A}\overline{B}CD$	$\overline{A}BCD$	$ABCD$	$A\overline{B}CD$
1 0	$\overline{A}\overline{B}C\overline{D}$	$\overline{A}BC\overline{D}$	$ABC\overline{D}$	$A\overline{B}C\overline{D}$

Figure 9.9 The arrangement of a Karnaugh map for four variables. As in three-variable maps, the common terms are grouped to make the recognition of simplifications straightforward.

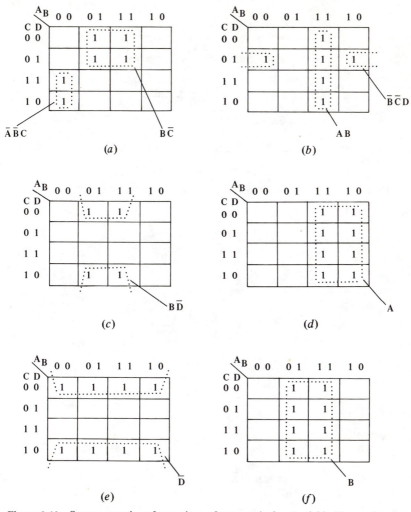

Figure 9.10 Some examples of groupings of squares in four-variable Karnaugh maps.

map. Some examples of combinations are given in Figure 9.10 and examples of their use follow.

EXAMPLE 9.12
Simplify the expression

$$\mathbf{F} = \overline{\mathbf{A}}\,\mathbf{B} + \mathbf{A}\,\mathbf{B}\,\overline{\mathbf{C}} + \mathbf{A}\,\mathbf{B}\,\mathbf{C}\,\overline{\mathbf{D}}$$

SOLUTION
The map is given below.

There are numerous ways to group the terms, one of which is shown above, and yields

$$F = \overline{A}\,B + A\,B\,\overline{C} + B\,C\,\overline{D}$$

A simpler expression results, however, with the grouping shown below.

This grouping yields

$$F = \overline{A}\,B + B\,\overline{C} + B\,\overline{D}$$ ■

EXAMPLE 9.13
    Simplify the expression

$$F = A\,B + \overline{B}\,C + \overline{B}\,D\,C + A\,B\,C\,D$$

SOLUTION
    The map may quickly be found by putting **1**'s in the squares corresponding to each of the terms of the expression. For example, **A B** corresponds to the four squares shown in Figure 9.10(*b*). The complete map is given below.

Three groups of four 1's may be recognized:

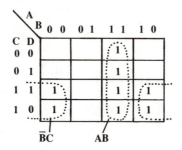

Thus $F = A B + \overline{B} C$.   ∎

9.2 Analysis and Synthesis of Combinational Circuits

The analysis of a combinational circuit has already been demonstrated in Example 9.1. It amounts to merely following through the circuit, setting down the function of each logic block. It is sometimes possible to simplify the expression thus obtained by using DeMorgan's theorem and Karnaugh maps or other techniques.

EXAMPLE 9.14
    Find the logic expression representing the following circuit, and reduce it to the simplest sum of products form.

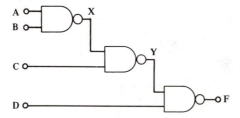

SOLUTION
    All of the blocks in this circuit are **NAND** gates; hence the output of each gate is related to the input by the **NAND** operation. For example, $F = \overline{(D\ Y)}$. At the intermediate point **X**

$$X = \overline{(A\ B)}$$

and at the point **Y**

$$Y = \overline{(C \cdot \overline{(A\ B)}\,)}$$

thus finally at **F**

$$F = \overline{(D \cdot \overline{(C \cdot \overline{(A\ B)})})}$$

It is simplest to remove the complementation bars using DeMorgan's theorem. We proceed first by removing the outermost complement, proceeding to the inner:

$$F = \overline{D} + (C \cdot \overline{(A\ B)})$$

where the complement over the second term disappears because $\overline{(\overline{Z})} = Z$. Applying De-Morgan's theorem to the last complement, we have

$$F = \overline{D} + C \cdot (\overline{A} + \overline{B})$$

To reduce this to the sum of products form, **C** is merely multiplied through $(\overline{A} + \overline{B})$; thus

$$F = \overline{D} + C\,\overline{A} + \overline{B}\,C$$

The reader may wish to verify with a Karnaugh map that this expression is already in its simplest form. ∎

EXAMPLE 9.15
Find the simplest sum of products form for the logic expression representing the circuit.

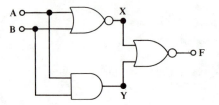

At **X**

$$X = \overline{(A + B)} = \overline{A} \cdot \overline{B}$$

At **Y**

$$Y = A \cdot B$$

Thus

$$F = \overline{(X + Y)}$$

$$= \overline{(\overline{A} \cdot \overline{B} + A\ B)}$$

From DeMorgan's theorem this equals

$$F = \overline{(\overline{A} \cdot \overline{B})} \cdot \overline{(A\ B)}$$

Applying the theorem again we obtain

$$F = (A + B) \cdot (\overline{A} + \overline{B})$$

$$= A\,\overline{A} + B\,\overline{A} + B\,\overline{B} + \overline{B}\,A$$

Clearly $A \bar{A} = B \bar{B} = 0$ (because both a variable and its complement cannot be 1); thus

$$F = B \bar{A} + A \bar{B}$$ ∎

EXAMPLE 9.16

Analyze the following circuit and reduce the expression to the simplest sum of products form.

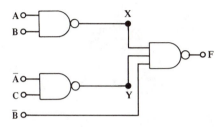

SOLUTION

At X and Y

$$X = \overline{(A\ B)}$$

$$Y = \overline{(\bar{A}\ C)}$$

Now $F = \overline{(\bar{B}\ X\ Y)}$; applying DeMorgan's theorem twice to F

$$F = B + \bar{X} + \bar{Y}$$

and substituting

$$F = B + AB + \bar{A}\ C$$

Mapping this we have

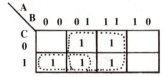

Thus F may be written as

$$F = B + \bar{A}\ C$$ ∎

Synthesis   The process of synthesis, that is, the design of logic circuits, involves several stages. First, the problem or requirement must be stated. Second, the requirement is translated into an algebraic function. Third, the algebraic function is implemented using some combination of logic gates and inverters. Because the main focus of this book is on circuits, we are primarily concerned with the last step.

We shall concentrate on synthesis with **NAND** gates for several reasons. First, as was shown in Chapter 8, and will be demonstrated here, any logic function may be synthesized with **NAND** gates. Second, the very popular DTL and TTL logic families are of the **NAND** form when positive logic is used. Third, other logic families that are of the **NOR** form for positive logic are of the **NAND** form when negative logic is used.

EXAMPLE 9.17
Show that the gate below is a **NOR** gate for positive logic and a **NAND** gate for negative logic.

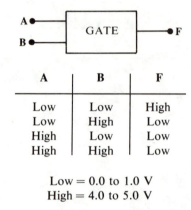

A	B	F
Low	Low	High
Low	High	Low
High	Low	Low
High	High	Low

Low = 0.0 to 1.0 V
High = 4.0 to 5.0 V

SOLUTION
Positive logic means that the high level is assigned logical value **1** and the low level **0.** Thus the truth table becomes

A	B	C
0	0	1
0	1	0
1	0	0
1	1	0

This is the truth table of a **NOR** gate (see Table 9.2).
Negative logic means that the low level is assigned logical value **1** and the high level **0.** Thus the truth table becomes

A	B	F
1	1	0
1	0	1
0	1	1
0	0	1

or, reordering

A	B	F
0	0	1
0	1	1
1	0	1
1	1	0

Either of these tables is recognized as the truth table for the **NAND** gate. ■

To develop a procedure for synthesis with **NAND** gates, we first analyze a simple circuit composed of **NAND** gates, given in Figure 9.11. This is a so-called "two-level" logic circuit, because, in proceeding from input to output,

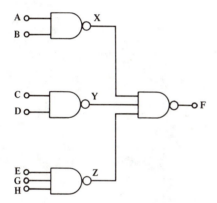

Figure 9.11 A two-level **NAND** gate logic circuit.

all signals must go through two logic gates. For the circuit of Figure 9.11

$$F = \overline{(X\ Y\ Z)} \tag{9.3}$$

where

$$\begin{aligned} X &= \overline{(A\ B)} \\ Y &= \overline{(C\ D)} \\ Z &= \overline{(E\ G\ H)} \end{aligned} \tag{9.4}$$

We wish to remove the complementation bars from each of these expressions. DeMorgan's theorem obviously can be applied to the expressions of only two variables. Furthermore, it may be applied to expressions of three variables as follows: Let $X\ Y = W$; then Equation (9.3) becomes

$$F = \overline{(W\ Z)} = \overline{W} + \overline{Z} \tag{9.5}$$

Substituting for W, and applying DeMorgan's theorem again

$$F = \overline{X} + \overline{Y} + \overline{Z} \tag{9.6}$$

The values of **X**, **Y**, and **Z** may now be inserted directly into Equation (9.6):

$$F = A\,B + C\,D + E\,G\,H \qquad (9.7)$$

A comparison of this result with the circuit of Figure 9.11 suggests immediately a method for synthesis with **NAND** gates. The procedure consists of writing the expression in the sum of products form, and assigning gates as follows: (1) In the first logic level there are as many logic gates as terms in the expression to be synthesized. (2) Each gate corresponds to a single term, and has, as inputs, the variables in the term. (3) The outputs of the first logic level are all inputs to a single **NAND** gate, which is the second logic level. Any expression, no matter how complex, may be implemented in this way. The expression is merely reduced to the sum of products form, and then implemented according to the rules above.

EXAMPLE 9.18
    Implement the following function using **NAND** gates.

$$F = A\,\overline{D}\,\overline{C} + \overline{A}\,\overline{B}\,\overline{C}\,D + \overline{A}\,B$$

SOLUTION
    We first use a Karnaugh map to minimize the expression

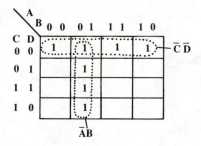

Clearly $F = \overline{A}\,B + \overline{C}\,\overline{D}$. The first term requires a **NAND** gate with two inputs [see sketch (a)], and the second term also requires a two-input **NAND** gate [see sketch

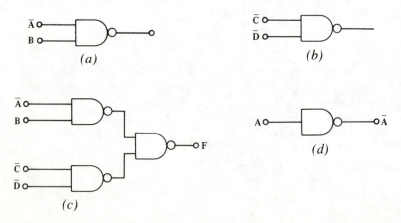

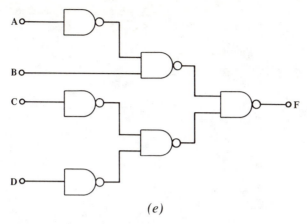

(e)

(b)]. Therefore the entire function is synthesized as in sketch *(c)*. This implemen- tation assumes that $\overline{A}$, $\overline{C}$, and $\overline{D}$ are available as inputs (rather than **A**, **C**, and **D**). If only **A**, **C**, and **D** are available, the complements may be generated by a single input **NAND** gate, which is effectively an inverter [see sketch *(d)*]. The full implementation is given in sketch *(e)*. ∎

Inasmuch as the sum of products form is not a unique way of writing a given expression, the two-level implementation rule is not the only way of synthesizing a given expression. However, it is often the best way, because it leads to minimum propagation delay, that is, to the fastest logic circuit.[3] Furthermore, it is true that the complements of the variables of interest are not always available. However, it is a simple matter to use an inverter to obtain the complement of any variable. To illustrate further the method, as well as to show the usefulness of **NAND** gate synthesis, a few more examples follow.

EXAMPLE 9.19
A half adder is a logic block having the characteristics given in the following truth table:

Inputs		Output
A	**B**	**C**
0	0	0
0	1	1
1	0	1
1	1	0

---

[3] The output of a given logic gate does not reach its final value instantly upon application of the inputs. Instead, there is a delay, called the propagation delay, or delay time, after which the output is at the value prescribed by the inputs. This delay time arises from the finite time constants of the circuits, as discussed in Chapter 8. If there are four levels of logic, it requires four delay times for the effect of any change in the inputs to propagate through to the outputs. The two-level logic circuit therefore leads to minimum delay.

(Note that this characteristic is the same as $A + B$ except for the condition $A = B = 1$.) Synthesize this logic block from **NAND** gates.

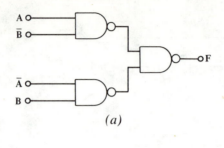

(a)

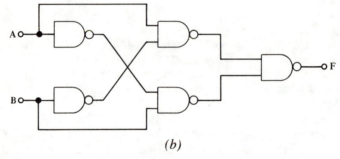

(b)

SOLUTION

It may be recognized from the truth table that

$$F = A \overline{B} + B \overline{A}$$

To synthesize this circuit with **NAND** gates, see sketch (*a*). If, however, the complements $\overline{A}$ and $\overline{B}$ are unavailable, they may be generated, as shown in sketch (*b*). This then is a half adder.   ■

The half adder of Example 9.19 is also called the **EXCLUSIVE OR** and is given a special logic symbol $\oplus$. The function $A \oplus B$ is the same as the function $A + B$, except when A and B both equal **1**. In this case $A \oplus B = 0$, whereas $A + B = 1$. The function is important because it may be used to add binary numbers. However, a complete practical adder circuit must generate a carry signal as well as the sum when two binary numbers are added. (Binary arithmetic is reviewed in Appendix I.) A possible way to synthesize a complete adder circuit, called a *full adder,* is discussed in the next two examples.

EXAMPLE 9.20

In most applications of adders, a carry signal is desired, just as in ordinary addition. Suppose **A** and **B** are two binary numbers to be added. The carry should normally be **0**, unless **A** and **B** are both **1**, in which case the carry signal should be **1**. Synthesize the circuitry to generate a carry signal **C** from **A** and **B**.

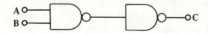

SOLUTION
The truth table for the carry function is

A	B	C
0	0	0
0	1	0
1	0	0
1	1	1

Thus $C = A\,B$. The circuit is thus the simple combination of two **NAND** gates as shown. ∎

EXAMPLE 9.21
The full adder is a circuit that adds two binary numbers, producing both a sum and a carry. For example, in the addition of **A** and **B** shown in the table below, the two values **0** and **1** plus the carry of **1** are added to yield a **0** in column 4. Furthermore, a carry of **1** is transferred into column 5.

Column	6	5	4	3	2	1
A		0	1	0	1	0	1
B		0	0	1	1	0	1

	6	5	4	3	2	1
First sum						0
First carry					1←	
Second sum					1	
Second carry				0←		
Third sum				0		
Third carry			1←			
Fourth sum			0			
Fourth carry		1←				
Fifth sum		0				
Fifth carry	1←					
Sixth sum	1					
Final sum	1	0	0	0	1	0

A full adder is a device that performs the addition of a single column, say column N. A block diagram is given below.

The inputs for column N are $\mathbf{A}_N$ and $\mathbf{B}_N$ as well as the carry from the previous column $\mathbf{C}_{N-1}$. The outputs are the sum $\mathbf{F}_N$ as well as the carry $\mathbf{C}_N$. Design such a full adder with **NAND** gates.

SOLUTION

This box may be considered to contain two circuits, one that generates $\mathbf{F}_N$ and one that generates $\mathbf{C}_N$. It is simplest just to write down a truth table for $\mathbf{F}_N$ and $\mathbf{C}_N$, considering each possibility. Suppose, for example, $\mathbf{A}_N = \mathbf{1}$, $\mathbf{B}_N = \mathbf{1}$, and $\mathbf{C}_{N-1} = \mathbf{0}$; then the sum $\mathbf{F}_N$ must be $\mathbf{0}$ and the carry $\mathbf{C}_N$ must be $\mathbf{1}$. Each possible combination of $\mathbf{A}_N$, $\mathbf{B}_N$, and $\mathbf{C}_{N-1}$ is considered in this way, yielding the combined truth table below:

$\mathbf{A}_N$	$\mathbf{B}_N$	$\mathbf{C}_{N-1}$	$\mathbf{F}_N$	$\mathbf{C}_N$
0	0	0	0	0
0	0	1	1	0
0	1	0	1	0
0	1	1	0	1
1	0	0	1	0
1	0	1	0	1
1	1	0	0	1
1	1	1	1	1

We construct a Karnaugh map separately for $\mathbf{F}_N$ and $\mathbf{C}_N$ from this table.

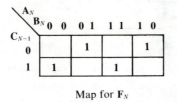

Map for $\mathbf{F}_N$

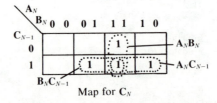

Map for $\mathbf{C}_N$

There are no groupings for $\mathbf{F}_N$; thus

$$\mathbf{F}_N = \overline{\mathbf{A}_N}\overline{\mathbf{B}_N}\mathbf{C}_{N-1} + \overline{\mathbf{A}_N}\mathbf{B}_N\overline{\mathbf{C}_{N-1}} + \mathbf{A}_N\mathbf{B}_N\mathbf{C}_{N-1} + \mathbf{A}_N\overline{\mathbf{B}_N}\overline{\mathbf{C}_{N-1}}$$

The groupings for $\mathbf{C}_N$ are indicated, yielding

$$\mathbf{C}_N = \mathbf{A}_N\mathbf{B}_N + \mathbf{A}_N\mathbf{C}_{N-1} + \mathbf{B}_N\mathbf{C}_{N-1}$$

From these expressions, we immediately construct the circuits as shown in sketch (a). Of course common inputs in these circuits would be connected; these connections are omitted here simply to avoid confusion. Furthermore, if the complements must be provided, the additional circuitry, as in sketch (*b*), is required.

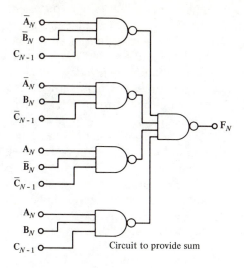

Circuit to provide sum

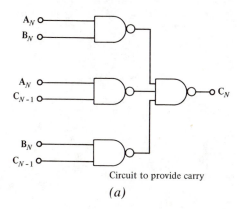

Circuit to provide carry

(*a*)

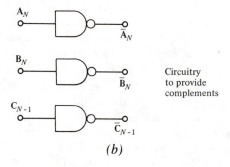

Circuitry
to provide
complements

(*b*)

It is interesting to look at some of the ready-made integrated circuits that provide the functions considered in the examples. The *Fairchild 9304* is an integrated circuit containing two full adders on a single chip. The block diagram is shown in Figure 9.12. Let us examine this figure carefully to understand the meaning of the various parts. First of all, the box around the entire diagram signifies that all the circuitry within the box is contained within the 9304 integrated circuit. Furthermore, two independent circuits are included, labeled FA1 and FA2. The various numbers shown in the terminals are the pin numbers of the integrated circuit. Circuit FA1 is almost an equivalent of the circuit considered in Example 9.21. The inputs are **A** and **B**, and C is the input carry. The output carry is C_0 and **S** is the sum. There are, however, two differences between this circuit and the circuit of Example 9.21: (1) Both the sum (pin 7) and the complement of the sum (pin 6) are provided as outputs in FA1. (2) The output carry is not provided, rather the complement of the output carry is provided (pin 5). The complement is indicated by the small circle at a given terminal; for example, the circle on pin 6 indicates that when the signal corresponding to the sum is high, the output at pin 6 is low.

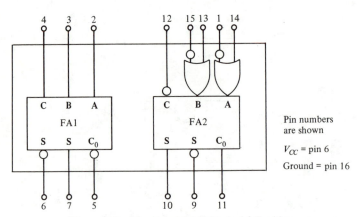

Figure 9.12   The Fairchild 9304 dual full adder.

Circuit FA2 in Figure 9.12 is also a complete full adder, but the input and output arrangement differs from that of FA1. First of all, the carry output, rather than its complement, is given. Again, both the sum and the complement of the sum are provided. In FA2 the complement of the input carry must be provided, as is indicated by the circle on the input carry terminal. Provision is made for either the input numbers or their complements to be added in this circuit. For example, if **A** is to be added to **B**, but only **A** and $\overline{\text{B}}$ are available, then **A** is connected to pin 14 and $\overline{\text{B}}$ connected to pin 15, and the proper sum will appear at pin 10. (Of course, if there is any carry to be taken into account, the complement of the carry must be connected to pin 12.)

9.3   Characteristics of Basic Sequential Circuits

The characteristic property of sequential circuits is that their output depends on both the present input and on the previous history of the inputs. When a change in the values of the inputs occurs, the output may or may not change, according to the values of the inputs and the state of the sequential circuit.

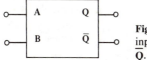

Figure 9.13   Block diagram for a flip-flop. The inputs are **A** and **B** and the outputs are **Q** and $\overline{Q}$.

The most basic sequential circuit is the flip-flop. A flip-flop is a circuit with only two states, thus the output may be either **0** or **1**. A block diagram is shown in Figure 9.13. For convenience, both the output, **Q**, and its complement are generally provided. The output **Q** of the flip-flop depends on the inputs **A** and **B**, as well as on the value of **Q** prior to the last change in **A** or **B**. Thus, upon a change in one or both of the inputs, the new state Q_{NEW} is a function of the old state Q_{OLD}, as well as being a function of the inputs. The function Q_{NEW} (Q_{OLD}, **A, B**) may be expressed in a truth table. For example, a circuit was considered in Chapter 8 that obeys the truth table of Figure 9.14. (This is the circuit of Figure 8.30.) The truth table in Figure 9.14 considers all possible combinations of inputs and previous states of the flip-flop. For example, line 4 indicates that if the previous state of the flip-flop was **1**, that is, Q_{OLD} = **1**, and input **A** is set to **0** while **B** is set to **1**, the output assumes the new value of **0**. The first two lines reflect the fact that the state of the flip-flop is unpredictable if both inputs are set to **0**. This is an "unallowed" condition; the circuit designer using such a flip-flop must make sure that such a combination of inputs never occurs.

Inputs		Old state	New state
A	B	Q_{OLD}	Q_{NEW}
0	0	0	?
0	0	1	?
0	1	0	0
0	1	1	0
1	0	0	1
1	0	1	1
1	1	0	0
1	1	1	1

Figure 9.14   Truth table for the flip-flop of Figure 9.13. This complete table lists all possible combinations of inputs and previous states.

By using an abbreviated notation it is possible to compress the truth table of Figure 9.14 into the compact form of Figure 9.15. This table is made more compact by stating the output in terms of its previous value where necessary. For example, it is seen from the last two lines in Figure 9.14 that if both inputs are **1**, the output remains constant at its previous value. The new table, Figure 9.15, just reflects this statement. Each row in Figure 9.15 summarizes two rows in Figure 9.14. The form of the table given in Figure 9.15 is commonly used to specify commercial flip-flop circuits (although the particular characteristics specified by this table are not typical).

A	B	Q_{NEW}
0	0	?
0	1	0
1	0	1
1	1	Q_{OLD}

Figure 9.15   A compact truth table for the flip-flop of Figure 9.13. This table contains the same information as the longer table of Figure 9.14.

An important question to be answered in connection with this discussion is just when the output changes state (at least in those cases when it does change state). This is a difficult question, involving not only the various time constants of the circuit, but the sequence of inputs as well. Fortunately, there is a modification of the flip-flop that simplifies the operation and description of the timing. In this modification an additional input, called a clock input, is provided that interrupts the ordinary inputs unless the clock input equals **1**. Thus the state of the flip-flop remains constant whenever the clock signal is **0**. However, whenever the clock signal switches to **1**, the flip-flop state may change, according to the prescription of the inputs.[4] A block diagram of such a flip-flop is given in Figure 9.16. The input labeled **CP** is the clock input, often called the *clock pulse input*.

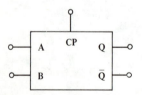

Figure 9.16   A flip-flop with clock input. The inputs **A** and **B** are of the ordinary type, whereas the input **CP** is the clock input. The state of the flip-flop remains constant unless **CP = 1**.

Before examining the characteristics of flip-flops of this type, it is worthwhile to examine briefly the relative merits and applications of flip-flops with and without clock inputs. Circuits that function continuously are called asynchronous, whereas circuits that change state only upon receiving a clock signal are called synchronous. The internal delay in the change of state of a

---

[4] In some commercial flip-flops, the change of state occurs as the clock input goes from **0** to **1**, and in others as the clock input goes from **1** to **0**. However, this is a detail of interest only when a particular circuit is being considered.

synchronous circuit may be calculated or specified, and the rate of receiving clock signals may be set so that all the circuits have time to respond before an inquiry of their output is made. For example, suppose it requires 50×10^{-9} sec for the flip-flop of Figure 9.16 to change state, as illustrated in Figure 9.17. Then the circuit reaches its new state 50 nsec after the rise of the clock pulse. If an inquiry about the new state of the circuit is always made 75 nsec after a clock pulse, then the output surely reflects the new state. The great advantage of synchronous circuits lies in the simplicity of designing systems with them — the designer knows precisely when switching occurs in all the circuits of the system.

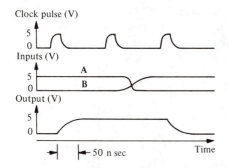

Figure 9.17 The inputs and outputs of a synchronous flip-flop. In this figure 5V corresponds to a **1** and 0V to a **0**. The state of the flip-flop can change only upon receipt of a clock pulse. The conditions for setting the output to **1** in this flip-flop are $A = 1$, $B = 0$, whereas $A = 0$, $B = 1$ sets the output to **0**.

In the following we deal exclusively with flip-flops that have provision for a clock input. Let us first consider what kinds of information are required for the analysis and synthesis of flip-flop circuits, that is, what properties of the flip-flops must be specified to the user.

The S-R Flip-Flop For the analysis of circuits it is only necessary to know what the output state of the flip-flop is at any given time. We will define Q_N as the output at the time clock pulse N is received. The new output, after the receipt of the clock pulse, is Q_{N+1}. Given the previous state Q_N and the inputs, the new output Q_{N+1} must be specified. The truth table is a suitable way of presenting such information, as was shown in connection with Figures 9.14 and 9.15. Whereas the flip-flop of Figure 8.30, which is described by Figures 9.14 and 9.15, is not a standard type, it is closely related to the standard **S-R** flip-flop. If provision for a **CP** input is made, and the complements of **A** and **B** are labeled **R** and **S**, the **S-R** flip-flop, Figure 9.18, results. The **S** and **R** symbols derive from set and reset. The effect of a set input is to set the output to **1**, whereas a reset input resets it to **0**. If both **S** and **R** = **0**, then a **CP** input has no effect. The question mark in the fourth row indicates that the output is unpredictable if both **S** and **R** = **1**. Therefore this condition is to be avoided in circuit operation; that is, whenever **CP** = **1**, either **S** or **R** must equal **0**, or else circuit operation becomes unpredictable.

The J-K Flip-Flop A second flip-flop type, called the **J-K** flip-flop, eliminates this uncertainty problem by specifying that whenever both inputs are **1** the output becomes the complement of its previous value. The block dia-

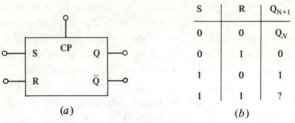

S	R	Q_{N+1}
0	0	Q_N
0	1	0
1	0	1
1	1	?

(a)   (b)

Figure 9.18   The **S-R** flip-flop. (a) Symbol. (b) Truth table.

gram symbol and truth table of a **J-K** flip-flop are given in Figure 9.19. The truth table is identical with the **S-R** flip-flop truth table except in the last row. In addition to the **S-R** and **J-K** flip-flops, there are a number of other types. However, they are less important and may be derived from these types in any case.

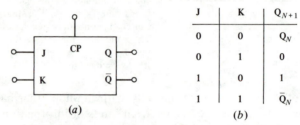

J	K	Q_{N+1}
0	0	Q_N
0	1	0
1	0	1
1	1	$\bar{Q}_N$

(a)   (b)

Figure 9.19   The **J-K** flip-flop. (a) Symbol. (b) Truth table.

The truth table provides an adequate description for flip-flop circuit analysis, because in analysis it is only required that the output be specified in terms of the inputs. However, in synthesis an alternative description is desired; it is specified what the circuit must do, that is, what its outputs should be, and it is the job of the engineer to organize the circuit so that the output is the desired function of the inputs. In other words, the inputs must be programmed to achieve the desired output state. In fact the synthesis operation is often called flip-flop programming. For this purpose a useful description of a flip-flop is one in which the various possible output states are listed, and next to them the inputs required to achieve these states. For example, suppose we want the output of an **S-R** flip-flop to switch from **1** to **0** when **CP = 1.** We know from the truth table that the input **R = 1,** coupled with **S = 0,** will accomplish this result. Thus a line in our synthesis specification table would read as in Table 9.7.

Table 9.7

Q_N	Q_{N+1}	S	R
1	0	0	1

Q_N	Q_{N+1}	S	R
0	0	0	α
0	1	1	0
1	0	0	1
1	1	α	0

Figure 9.20 Excitation table for the **S-R** flip-flop. This table lists the required inputs **S** and **R** to achieve the desired output state $Q_N + 1$. The letter α signifies arbitrary, that is, the value of the particular input is inconsequential.

There are three other possible lines in such a table, and they are given in Figure 9.20. This table is conventionally called an excitation table, because it specifies the input excitation required to put the flip-flop in the desired state. The symbol α appearing in the table means that the value of this particular input is arbitrary. For example, we know from the truth table given in Figure 9.18 that if an **S-R** flip-flop is in state **1** and it is desired to have it remain in this state, it is only necessary that **R** be **0**; the value of **S** is immaterial.

Let us now derive the excitation table for the **J-K** flip-flop. Suppose the flip-flop is in state **1** and it is desired that it switch to state **0**. Clearly **J = 0**, **K = 1** will achieve this as in an **S-R** flip-flop, as may be seen from line 2 of the **J-K** flip-flop truth table, Figure 9.19. However **J = 1**, **K = 1** will also cause the flip-flop to change states, resulting in $Q_{N+1} = \mathbf{0}$, as given in the last line of the **J-K** truth table. Hence the value of **J** is arbitrary in achieving the transition $Q_N = \mathbf{1}$ to $Q_{N+1} = \mathbf{0}$. This forms the third line in the **J-K** excitation table, given in Figure 9.21. The other lines are derived similarly.

Q_N	Q_{N+1}	J	K
0	0	0	α
0	1	1	α
1	0	α	1
1	1	α	0

Figure 9.21 The excitation table for a **J-K** flip-flop.

Although the **S-R** and **J-K** flip-flop truth tables are very similar, the excitation tables do not appear so similar. It will be seen in the next section that the large number of α's in the **J-K** excitation table leads to much simpler circuits using **J-K** flip-flops.

9.4   Analysis and Synthesis of Flip-Flop Circuits

When a complex circuit of flip-flops and logic blocks is to be dealt with, the central problem is bookkeeping. The key to flip-flop circuits is the method of specifying the various combinations of flip-flop states and inputs. Here we shall introduce a simple diagram, called a state diagram, the function of which is to keep track of these states.

Let us see what kind of information should be provided in such a diagram, by considering an example. The block diagram of Figure 9.22(*a*) represents a simple flip-flop circuit. For the moment the exact configuration is not known; it is only specified that the circuit contain two flip-flops, have one input, and have one output. Also shown is a clock input (**CP**). At any given time each of two flip-flops is in a given state. Furthermore, they remain in this state as long as no clock input is received; when a clock input arrives, the states of one or both of the flip-flops may change. The change, if any occurs, is controlled by the state of the input and the previous state of the flip-flops. The output **F** is some logical function of the flip-flop states.

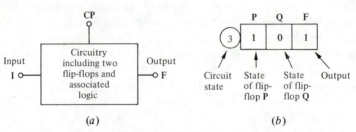

(a)   *(b)*

Figure 9.22   Development of the state diagram. (*a*) A block diagram of a flip-flop circuit constructed from two flip-flops and a number of logic gates. (The actual circuit will be derived in Example 9.26.) The circuit has a single input **I** and a single output **F**. (*b*) A possible partial state diagram showing the flip-flop states **P** and **Q** and the output **F** for one circuit state.

The state diagram is a diagram that indicates the state of each flip-flop in the circuit, as well as the state of the output. A possible form is given in Figure 9.22(*b*).[5] The two flip-flops in the circuit of Figure 9.22(*a*) have been labeled as **P** and **Q**, and their states, as well as the state of the output **F** are indicated. In addition, the particular state of the circuit as a whole is given a label; here state 3 means the state in which flip-flop **P** = **1** and **Q** = **0**. The state label is merely a label and is perfectly arbitrary. The same state might also have been called state *M*, or state Charlie, or whatever. Figure 9.22(*b*) gives information about a single state of the circuit.

Now consider what happens when a clock pulse is received (that is, when the clock input, labeled **CP,** makes a transition from **0** to **1**). The circuit makes a transition into a new state (of course the new state may be identical with the old state). In this circuit there is only one input; hence, upon receipt of a clock pulse the circuit can make a transition into at most two different states, depending on whether the input is **0** or **1**. These two transitions may be indicated on the state diagram by arrows, and the input condition is indicated alongside the arrow, as shown in Figure 9.23. This figure indicates that if the circuit is in state 3 with the input = **1,** then when a clock pulse is

---

[5] Note that the information given in Figure 9.22(*b*) cannot be derived from Figure 9.22(*a*) without knowledge of its contents. Figures 9.22(*b*), 9.23, and 9.24 are only examples describing possible circuit behavior.

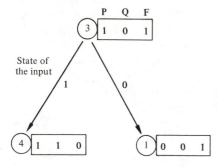

Figure 9.23 Partial state diagram for the circuit of Figure 9.22 (*a*). From circuit state 3, a transition into either state 4 or state 1 occurs, depending on the value of the input at the instant the clock pulse arrives.

received, a transition into a new state, labeled state 4, occurs. In circuit state 4, the flip-flop states are **P = 1, Q = 1** and the output is **0**. If, on the other hand, the input were **0** when the clock pulse arrives, the circuit would make a transition into the new state labeled state 1. In this state, **P = 0, Q = 0,** and **F = 1.**

A complete state diagram shows all the states of the circuit and all possible transitions between the states. Such a diagram is shown in Figure 9.24 for the digital block of Figure 9.22(*a*). Again, we are for the moment concerned with the states of the circuit; the actual circuit itself will be derived in Example 9.26. Since this block contains only two flip-flops, there are only four possible states of the circuit corresponding to the four possible combinations of flip-flop states shown.

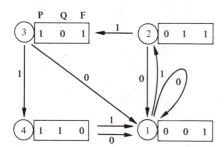

Figure 9.24 A possible complete state diagram for the circuit of Figure 9.22 (**a**)

With the state diagram, the operation of a flip-flop circuit is completely described. Suppose, for example, that the input to the circuit described by Figure 9.24 has the values **0, 1, 1, 0** during the first four clock pulses, and that the initial state is state 1. Then according to the state diagram the following sequence of events occurs: (1) The first clock pulse results in no change — the circuit remains in state 1 because the input is **0.** This is indicated by the fact that the arrow out of state 1 for input **0** curves back toward state 1. (2) The second clock pulse results in a transition into state 2, because the input is **1.** (3) The third clock pulse results in a transition from state 2 into state 3, because the input is **1.** (4) The fourth clock pulse results in a transition from state 3 into state 1. It is shown on the state diagram that state 3 leads to state 1 when the input equals **0.**

This is just a single example of how a state diagram describes in a very compact way the operation of a flip-flop circuit. Of course many variations on this diagram can occur. In the following examples it will be observed that: (1) There may be more flip-flops; in that case there are more states and a larger table is used to represent each state. (2) There may be more than one output, again resulting in a larger table for each state. (3) There may be more inputs (or no inputs at all in the case of a counter). If there are two inputs, then there are four possible combinations of inputs, hence four arrows leading out of each state. (4) There may be fewer states in the state diagram than seem appropriate for the number of flip-flops. For example, a circuit containing two flip-flops may be so arranged that the state in which both flip-flops are **1** never occurs. Another example, which will be treated later, is that of a decimal counter using four flip-flops, which counts 0 to 9, and repeats, thus having only ten states. Such a counter has no inputs; it always changes state when a clock pulse is received. Furthermore, it generally has four outputs to indicate the state of each flip-flop.

EXAMPLE 9.22

A state diagram for a certain flip-flop circuit is given below. The single input is called **I**; the outputs are **F, G,** and **H.** The circuit contains three flip-flops, **P, Q,** and **R.** Here, for convenience, state 1 appears twice on the diagram. (This is simply to avoid drawing a great number of arrows curving back on the uppermost state. The lower state 1 is identical with upper state 1, and transitions out of it are indicated on the upper state.)

If the circuit is initially in a state with **P = 1, Q = 0, R = 1,** specify the values of **P, Q, R, F, G, H,** for the following sequence: during first clock pulse **I = 0;** during second clock pulse **I = 1;** during third clock pulse **I = 0.**

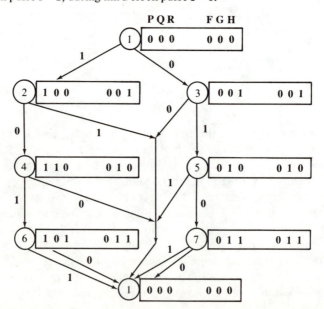

SOLUTION

The initial state is prescribed as **P = 1, Q = 0, R = 1,** that is, state 6 in the diagram. State 6 always leads to state 1, so after the first clock pulse, **P = Q = R = 0,** and the outputs are **F = G = H = 0.** After the second clock pulse **(I = 1)** the state becomes state 2; thus **P = 1, Q = R = 0,** and the outputs are **F = G = 0; H = 1.** After the third clock pulse **(I = 0),** the circuit moves into state 4; thus **P = Q = 1,** and **R = 0.** The outputs are **F = 0, G = 1, H = 0.** ■

The state diagram is the key to both analysis and synthesis of flip-flop circuits. Analysis will be defined as the determination of a state diagram from the circuit. Synthesis involves two steps, writing out the state diagram from the requirements, and the design of the circuit from the state diagram.

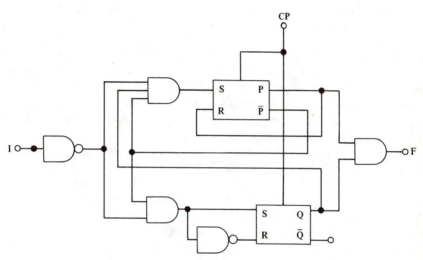

Figure 9.25   A flip-flop circuit to be analyzed.

Analysis of Flip-Flop Circuits   It is easiest to show the analysis procedure with an example. Consider the circuit of Figure 9.25, which has a single input and two flip-flops. The input to the circuit is **I** and the output is **F.** It is desired to find the state diagram of the circuit. The following procedure will be used: (1) Find the logic equations governing the inputs to the flip-flops and the output(s) of the circuit. The equations desired here are the logic equations relating S_P, S_Q, R_P, R_Q, and F to the input **I** and the flip-flop states **P** and **Q.** (2) Start in some initial state, arbitrary unless specified, and examine the transitions for all possible input values.

We begin with step 1. Writing the equations for the inputs to the flip-flops **P** and **Q,** Figure 9.25, we have for **FFP**

$$S_P = \bar{P} \, Q \, \bar{I}$$
$$R_P = P.$$

(9.8)

for **FFQ**

$$S_Q = \overline{P}\,\overline{I}$$

$$R_Q = \overline{(\overline{P}\,\overline{I})}$$

(9.9)

The output **F** is related to the flip-flop states by

$$F = P\,Q$$

(9.10)

Now we proceed with step 2, which involves examining the transitions between the states of the circuit. Because no special state was given we will define state 1 as $P = Q = 0$. From Equation (9.10), we see that in state 1, the output **F** equals **0**. We must then consider what happens when a clock pulse is received. There are two possibilities, corresponding to $I = 0$ or $I = 1$. Suppose $I = 0$. Then, from Equations (9.8) and (9.9), $S_P = 0$, $R_P = 0$, $S_Q = 1$, $R_Q = 0$. According to the **S-R** flip-flop truth table, flip-flop **P** would remain in state **0** and flip-flop **Q** would be set to 1 with the clock input. Let us define this new state ($P = 0$, $Q = 1$) as state 2. Now let us go back to state 1, and determine the effect of an input $I = 1$, rather than **0**. In this case Equations (9.8) and (9.9) yield $S_P = 0$, $R_P = 0$, $S_Q = 0$, $R_Q = 1$. According to the flip-flop truth table both flip-flops would remain in the **0** state under these input conditions. Therefore state 1 is unaltered. The two possibilities just examined are illustrated in the partial state diagram of Figure 9.26(*a*).

We have examined both possible transitions out of state 1, now let us consider the transitions out of state 2. First suppose $I = 0$. Equations (9.8) and (9.9) yield $S_P = 1$, $R_P = 0$, $S_Q = 1$, $R_Q = 0$. This results in no change in flip-flop **Q**; however, flip-flop **P** is set to **1** with the incoming clock pulse.

We define the new state with $P = 1$, $Q = 1$ as state 3. In state 3 $F = 1$. Now suppose that in state 2 $I = 1$ instead of 0. In this case $S_P = 0$, $R_P = 0$, $S_Q = 0$, $R_Q = 1$. Such a combination results in a transition into state 1, because **P** remains **0** and **Q** is reset to **0**. Thus two arrows are drawn from state 2, one to the new state 3 (for $I = 0$) and one to state 1.

We now consider transitions from state 3. Suppose $I = 0$, then $S_P = 0$, $R_P = 1$, $S_Q = 0$, $R_Q = 1$. Both flip-flops are reset to **0**; that is, the circuit again goes to state 1. If, on the other hand, $I = 1$, then $S_P = 0$, $R_P = 1$, $S_Q = 0$, $R_Q = 1$; again both flip-flops are reset to **0**. There are thus two arrows out of state 3, both of which go to state 1.

The almost complete state diagram is shown in Figure 9.26(*b*). It would appear that the diagram is complete because all transitions out of all states shown are considered. However, there are only three circuit states shown, whereas four combinations are possible in a circuit containing two flip-flops. We must consider what happened to the fourth combination. The missing possibility is $P = 1$, $Q = 0$, which, because of the circuitry, is never reached. However, it may happen that under some circumstances (when power is first applied to the circuit, for example) the flip-flops may happen to fall into the condition $P = 1$, $Q = 0$. We should examine this extra state to see what happens when inputs and clock pulses appear. Suppose $I = 0$; then $S_P = 0$, $R_P = 1$, $S_Q = 0$,

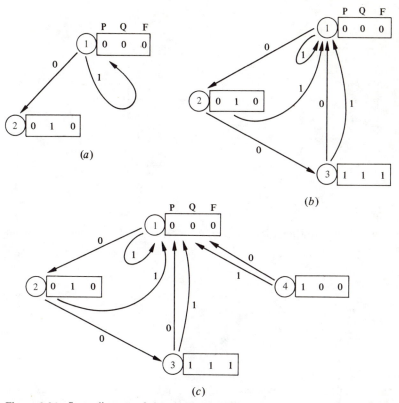

Figure 9.26 State diagram of the circuit of Figure 9.25. (*a*) Partial diagram after examining the transitions out of the first state. (*b*) The complete sequence of possible states for the circuit initially in state 1. (*c*) The state diagram, including the extra state, state 4.

$\mathbf{R}_Q = \mathbf{1}$. In this case the transition to state 1 occurs, as both flip-flops are reset to **0.** If, on the other hand, $\mathbf{I} = \mathbf{1}$ then $\mathbf{S}_P = \mathbf{0}$, $\mathbf{R}_P = \mathbf{1}$, $\mathbf{S}_Q = \mathbf{0}$, and $\mathbf{R}_Q = \mathbf{1}$. Again the circuit moves to state 1. Thus we can say that if, for some reason, the circuit ever gets into the extra state with $\mathbf{P} = \mathbf{1}$, $\mathbf{Q} = \mathbf{0}$, it makes a transition into state 1 with the first clock pulse. From then on, it should not return to the extra state. The complete state diagram is shown in Figure 9.26(*c*). The operation of the circuit can be visualized quickly from this diagram. The output is **1** only in state 3, and state 3 is reached by a sequence of **0** inputs (during two sequential clock pulses). Any time a clock pulse arrives with the input **1**, the circuit is reset to state 1. Thus the circuit is a kind of detector that detects a sequence of two **0**'s on the input line.

EXAMPLE 9.23

Analyze (that is, find the state diagram for) the following circuit. Note that **J-K** flip-flops are used, and that there is no input, other than the clock pulse. The output is **F.** Assume that the initial state is $\mathbf{A} = \mathbf{0}$, $\mathbf{B} = \mathbf{0}$, $\mathbf{C} = \mathbf{0}$.

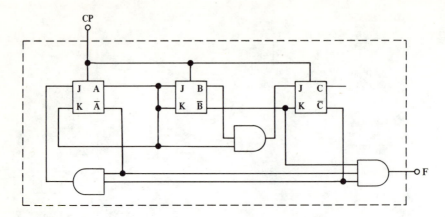

SOLUTION

From the circuit shown, we can immediately write down the following equations:

$$J_A = \overline{A}\,\overline{C}$$
$$K_A = A$$
$$J_B = A$$
$$K_B = A$$
$$J_C = A\,B$$
$$K_C = \overline{B}$$
$$F = \overline{A}\,\overline{B}\,\overline{C}$$

We now generate the state diagram by starting at state 1 ($A = 0$, $B = 0$, $C = 0$) and following through a sequence of clock pulses. This circuit is particularly easy to analyze because there is no input, and thus there is only one transition out of each state. Starting in state 1, we have $A = B = C = 0$. From the equations, $J_A = 1$, $K_A = 0$, $J_B = 0$, $K_B = 0$, $J_C = 0$, $K_C = 1$. From the J-K flip-flop truth table, Figure 9.19, the next state would be $A = 1$, $B = C = 0$. Let us call this state 2. We have identified two states thus far:

A B C

(1) 0 0 0   (2) 1 0 0

In state 2 $J_A = 0$, $K_A = 1$, $J_B = 1$, $K_B = 1$, $J_C = 0$, $K_C = 1$. Following a clock pulse then A goes to 0, B goes to 1, and C remains 0. This is a new state, called state 3.

(3) 0 1 0

In state 3 $J_A = 1$, $K_A = 0$, $J_B = 0$, $K_B = 0$, $J_C = 0$, $K_C = 0$. Thus, following a clock pulse, A goes to 1, and B and C remain unchanged. This is a new state, state 4.

In state 4 $J_A = 0$, $K_A = 1$, $J_B = 1$, $K_B = 1$, $J_C = 1$, and $K_C = 0$. This results in a change of all three flip-flops, into a new state, 5.

In state 5 $J_A = 0$, $K_A = 0$, $J_B = 0$, $K_B = 0$, $J_C = 0$, $K_C = 1$. Following a clock pulse, flip-flop C will be reset to 0, the others remaining 0. We recognize the resulting state as the initial state 1. The complete state diagram may now be drawn (the output F is also evaluated for each of the states):

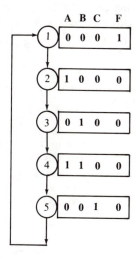

This circuit is seen to be a counter that resets and gives an output after every five clock pulses. ∎

Synthesis — Generating the State Diagram As has been pointed out earlier, the synthesis of flip-flop circuits may be divided into two tasks. First, a state diagram is prepared from the given requirements. Second, the actual circuit is constructed from the state diagram. Since the requirements can be stated in a variety of ways, and there is often a good deal of ambiguity, the procedure for generating the state diagram varies somewhat. However, some general principles may be set down. First of all it is determined how many states are required. The number of states determines the number of flip-flops to be used. For example, four states require two flip-flops and eight states three flip-flops. Furthermore, five states would require three flip-flops. (The reader may wish to prove that if there are M states, N flip-flops are required, where $2^N \geq M$.) Of course, the number of states follows from the stated circuit requirements. For example, a decimal counter that counts up to nine and resets to zero requires ten states. The circuit thus consists of four flip-flops and associated logic. In more complicated specifications the number of states is determined by considering all possible sequences of inputs, as will be shown in the examples.

Once the required number of circuit states and the number of flip-flops have been determined, the flip-flops and the circuit states may be labeled. For example, there may be two flip-flops, labeled **P** and **Q,** and three circuit states, labeled 1, 2, and 3. Specific states for each flip-flop must now be assigned to each circuit state. For example, we may specify that in circuit state 3, **P = 1** and **Q = 1.** Sometimes a natural assignment is apparent, but in general the assignment is arbitrary. Typically, but arbitrarily, circuit state 1 is chosen as the state with all flip-flops set to **0.** Although the state assignment is arbitrary, the resulting circuit may be simpler when some particular assignment is chosen. A circuit designer often examines several assignments, to find the simplest possible circuit.

EXAMPLE 9.24

Generate a state diagram for a circuit that counts clock pulses, producing an output, **F = 1,** after every third pulse.

SOLUTION

In order to count up to 3, the circuit must have three states: 1, 2, and 3. Two flip-flops are required, **P** and **Q.** The output **F** is to be **0,** except in one particular state, say state 3. The circuit must move from states 1 to 2 to 3 and back to 1, and repeat as clock pulses are received. We will (arbitrarily) assign **P = 0, Q = 0** in state 1; **P = 0, Q = 1** in state 2; and **P = 1, Q = 0** in state 3. The state diagram would then appear as follows.

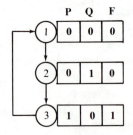

EXAMPLE 9.25

Generate a state diagram for a circuit that examines the input for the sequence **101** during each group of three clock pulses. (In other words, the circuit tests whether the input is **1** during the first clock pulse, **0** during the second, and **1** during the third.) A single output line **F** is to be **0** unless the sequence is observed, in which case it is set to **1.** The circuit is to be reset on the fourth pulse and begin looking for the sequence on the next.

SOLUTION

By considering the possible sequences of inputs, we will determine the number of states. Define the initial state (after a reset) as state 1. From state 1 there must be two possibilities, and two more states, as shown below.

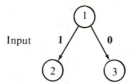

After state 3 it is unimportant whether the input is **0** or **1**; it is only necessary to wait three clock pulses to reset. Let us first examine state 2. After state 2 there are two possibilities; a **0** is the correct code and a **1** incorrect. This leads to two more states:

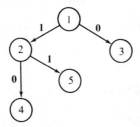

Continuing along the path of the code **101,** that is, through state 4, we define the state achieved with the correct code as state 6, and the alternative after state 4 as state 7:

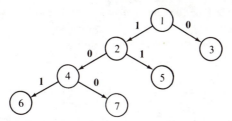

The fourth clock pulse must result in all cases in the circuit reaching state 1. One way of achieving this is given below:

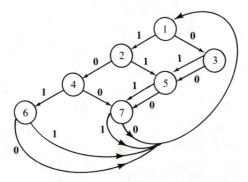

There are seven states; therefore three flip-flops are required. Since no requirements were stated, we can arbitrarily assign the flip-flop states. Let the states be defined as follows: state 1, **000;** state 2, **100;** state 3, **001;** state 4, **110;** state 5, **011;** state 6, **111;** state 7, **010.** The complete state diagram follows.

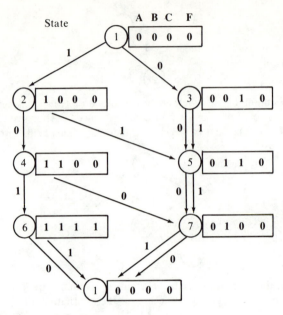

State 1 is merely repeated for convenience, as in Example 9.22.   ■

Synthesis — From State Diagram to Circuit   We have seen in the discussion of analysis that a state diagram may be readily generated if the circuit is known. We wish here to do the opposite — to construct the circuit diagram from the state diagram. We will state the procedure, and illustrate it by applying it to the state diagram given in Figure 9.27. A circuit that obeys this state diagram is to be constructed from **J-K** flip-flops. The first step in the procedure is to construct a kind of truth table, called a *transition table*, which enumerates all possible transitions of the circuit. This table is merely a summary of the state diagram. It lists, under each circuit state, the flip-flop states and the output, as well as the subsequent state, according to the input. An example is given in Figure 9.28. This transition table applies to the state

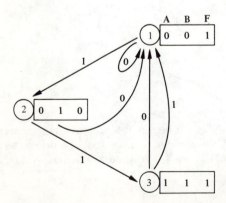

Figure 9.27   State diagram of a flip-flop circuit. The circuit is to be synthesized using **J-K** flip-flops.

diagram of Figure 9.27. It indicates, for example, in line 2, that in state 1 $A = 0$, $B = 0$ and the output $F = 0$. Furthermore, for an input of $I = 1$, a transition into state 2 occurs, and that in state 2 $A = 0$ and $B = 1$.

State N	A	B	F	I	State $N+1$	A_{N+1}	B_{N+1}
1	0	0	0	0	1	0	0
1	0	0	0	1	2	0	1
2	0	1	0	0	1	0	0
2	0	1	0	1	3	1	1
3	1	1	1	0	1	0	0
3	1	1	1	1	1	0	0

Figure 9.28 The transition table for the state diagram of Figure 9.27.

The second step in going from state diagram to circuit diagram involves expanding the transition table to include the flip-flop input conditions which produce the desired flip-flop state changes indicated in the transition table. The expanded transition table is called the *transition-excitation table*. Continuing with the example of Figures 9.27 and 9.28, the transition-excitation table is given in Figure 9.29. For example, line 2 indicates that in state 1, with input 1, the next state is state 2, in which $A = 0$ and $B = 1$. To achieve this state, it is necessary, according to the J-K flip-flop excitation table of Figure 9.21, to set $J_A = 0$ and $J_B = 1$. The values of K_A and K_B are arbitrary. These J and K input conditions are therefore indicated in the transition-excitation table, Figure 9.29.

State N	A	B	F	I	State$_{N+1}$	A_{N+1}	B_{N+1}	J_A	K_A	J_B	K_B
1	0	0	0	0	1	0	0	0	a	0	a
1	0	0	0	1	2	0	1	0	a	1	a
2	0	1	0	0	1	0	0	0	a	a	1
2	0	1	0	1	3	1	1	1	a	a	0
3	1	1	1	0	1	0	0	a	1	a	1
3	1	1	1	1	1	0	0	a	1	a	1

Figure 9.29 The transition-excitation table for the state diagram of Figure 9.27. This table gives the required input conditions to the flip-flops.

The third step in the synthesis procedure is the writing of the input equations. An equation is written for each required input to each flip-flop directly from the transition-excitation table. Any output equations are written similarly. Karnaugh maps should be used to obtain the simplest possible form. In the example being treated, let us write the equation for J_A. The truth

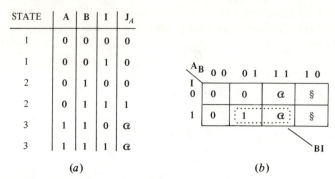

STATE	A	B	I	J_A
1	0	0	0	0
1	0	0	1	0
2	0	1	0	0
2	0	1	1	1
3	1	1	0	⍺
3	1	1	1	⍺

(a)   (b)

Figure 9.30   (a) The truth table for **J**, from Figure 9.29. (b) The Karnaugh map for this truth table. The two squares marked **§** are unspecified by the truth table, and are therefore arbitrary.

table for J_A is extracted (for clarity) from Figure 9.29 and is given in Figure 9.30(a).

To find the simplest expression a Karnaugh map is made, Figure 9.30(b). This map differs somewhat from the previous maps in that the symbols ⍺ and § appear. The ⍺ represents a square that may be either **0** or **1**. The § represents an unspecified state, that is, one not considered in the truth table. The circuit never gets into such states, so these states may also be regarded as arbitrary. From the Karnaugh map, the simplest equation for J_A is $J_A = \mathbf{B\,I}$. As further examples, the equations for $\mathbf{K}_A$, $\mathbf{J}_B$, and $\mathbf{K}_B$ are derived using Karnaugh maps in Figure 9.31. Note that the choice of equations is in some cases arbitrary. For example, $\mathbf{K}_A$ could be written $\mathbf{K}_A = \mathbf{A}$, or $\mathbf{K}_A = \mathbf{B}$, or even $\mathbf{K}_A = \mathbf{1}$. The equations are summarized in Figure 9.32(a).

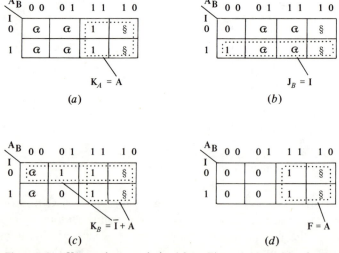

Figure 9.31   Karnaugh maps, derived from Figure 9.29. (a) Map for $\mathbf{K}_A$. (b) Map for $\mathbf{J}_B$. (c) Map for $\mathbf{K}_B$. (d) Map for **F**.

The final step in deriving the circuit consists of drawing the logic circuit required by the equations. In the example considered here, very few components are required. The complete circuit is drawn in Figure 9.32(b). The circuit has been synthesized using **NAND** gates. A few more examples will now be given to illustrate the use of these circuit synthesis principles.

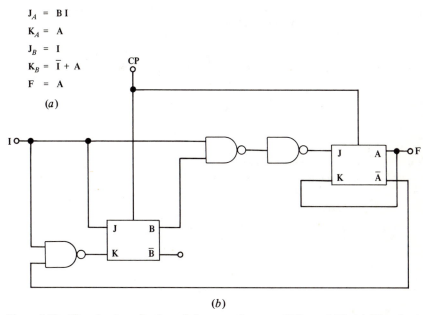

$$J_A = B I$$
$$K_A = A$$
$$J_B = I$$
$$K_B = \bar{I} + A$$
$$F = A$$

(a)

(b)

Figure 9.32 The circuit realization of the state diagram of Figure 9.27. (a) The circuit equations. (b) The circuit realization using **NAND** logic.

EXAMPLE 9.26

Synthesize a circuit that functions according to the state diagram of Figure 9.24. Two **S-R** flip-flops are to be used. Call the output of flip-flop **P, A,** and the output of flip-flop **Q, B.**

SOLUTION

A transition table is first made that contains the present values of **A, B,** and **F,** and the future values of **A** and **B,** for each state and each value of **I.**

State	A	B	F	I	A_{N+1}	B_{N+1}
1	0	0	1	0	0	0
1	0	0	1	1	0	1
2	0	1	1	0	0	0
2	0	1	1	1	1	0
3	1	0	1	0	0	0
3	1	0	1	1	1	1
4	1	1	0	0	0	0
4	1	1	0	1	0	0

The required input conditions are now determined for each transition A_N to A_{N+1} and B_N to B_{N+1}. The excitation table for the S-R flip-flop, Figure 9.20, is used. Thus, for example, state 2 with $I = 0$ requires that A remain 0, while B changes from 1 to 0. Clearly, $S_A = 0$, $R_A = d$, $S_B = 0$, $R_B = 1$ for this line. The reader is asked to derive the values of S and R for the other lines. The resulting transition-excitation table follows.

State	A	B	F	I	A_{N+1}	B_{N+1}	S_A	R_A	S_B	R_B
1	0	0	1	0	0	0	0	d	0	d
1	0	0	1	1	0	1	0	d	1	0
2	0	1	1	0	0	0	0	d	0	1
2	0	1	1	1	1	0	1	0	0	1
3	1	0	1	0	0	0	0	1	0	d
3	1	0	1	1	1	1	d	0	1	0
4	1	1	0	0	0	0	0	1	0	1
4	1	1	0	1	0	0	0	1	0	1

Note that we have not bothered to include the state $N+1$ column (as in Figure 9.29), because in this example we have no need for it. A Karnaugh map is now constructed for each of the variables S_A, R_A, S_B, and F. The logic equation is written from the

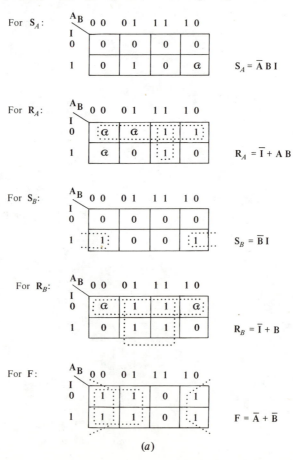

For S_A: $S_A = \overline{A}\,B\,I$

For R_A: $R_A = \overline{I} + A\,B$

For S_B: $S_B = \overline{B}\,I$

For R_B: $R_B = \overline{I} + B$

For F: $F = \overline{A} + \overline{B}$

(a)

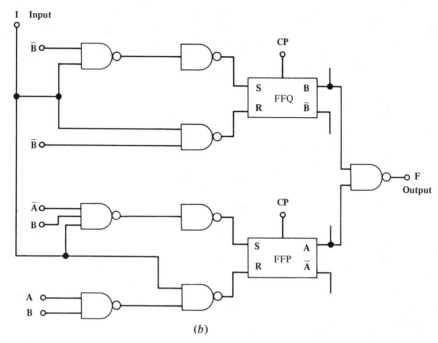

(b)

map [see sketch (a)]. From these equations the circuit may be immediately con-structed. It is fairly complex; hence all of the interconnections are not shown. All the points labeled **A** should be connected; the wires are not shown to avoid confusion. The logic is composed of **NAND** gates, interconnected from the logic equations accord-ing to the methods of Section 9.2. It should be noted that from DeMorgan's theorem, $\overline{A} + \overline{B} = \overline{(A\ B)}$, allowing a single **NAND** gate to generate **F** from **A** and **B** [see sketch (b)].

∎

EXAMPLE 9.27

Design a counter that counts clock pulses up to ten (zero to nine) and resets to zero. The counter should have an output according to the following table, which is recognized as binary coded decimal (Appendix I).

Decimal	T	U	V	W
0	0	0	0	0
1	0	0	0	1
2	0	0	1	0
3	0	0	1	1
4	0	1	0	0
5	0	1	0	1
6	0	1	1	0
7	0	1	1	1
8	1	0	0	0
9	1	0	0	1

Flip-flops of the **J-K** type are to be used in the realization.

SOLUTION

In this circuit with no input other than the clock pulse, we know immediately that ten states are required, one corresponding to each of the values listed. Clearly four flip-flops will be needed. For simplicity, we will label the states 0 to 9, and assign the flip-flop values exactly as given in the table, with $A = T$, $B = U$, $C = V$, $D = W$, and where A, B, C, and D are the flip-flop labels. We may write immediately the transition table:

State	A	B	C	D	A_{N+1}	B_{N+1}	C_{N+1}	D_{N+1}
0	0	0	0	0	0	0	0	1
1	0	0	0	1	0	0	1	0
2	0	0	1	0	0	0	1	1
3	0	0	1	1	0	1	0	0
4	0	1	0	0	0	1	0	1
5	0	1	0	1	0	1	1	0
6	0	1	1	0	0	1	1	1
7	0	1	1	1	1	0	0	0
8	1	0	0	0	1	0	0	1
9	1	0	0	1	0	0	0	0

Using the J-K flip-flop excitation table, the input conditions are determined and the transition-excitation table becomes:

State	A	B	C	D	J_A	K_A	J_B	K_B	J_C	K_C	J_D	K_D
0	0	0	0	0	0	Ⓐ	0	Ⓐ	0	Ⓐ	1	Ⓐ
1	0	0	0	1	0	Ⓐ	0	Ⓐ	1	Ⓐ	Ⓐ	1
2	0	0	1	0	0	Ⓐ	0	Ⓐ	Ⓐ	0	1	Ⓐ
3	0	0	1	1	0	Ⓐ	1	Ⓐ	Ⓐ	1	Ⓐ	1
4	0	1	0	0	0	Ⓐ	Ⓐ	0	0	Ⓐ	1	Ⓐ
5	0	1	0	1	0	Ⓐ	Ⓐ	0	1	Ⓐ	Ⓐ	1
6	0	1	1	0	0	Ⓐ	Ⓐ	0	Ⓐ	0	1	Ⓐ
7	0	1	1	1	1	Ⓐ	Ⓐ	1	Ⓐ	1	Ⓐ	1
8	1	0	0	0	Ⓐ	0	0	Ⓐ	0	Ⓐ	1	Ⓐ
9	1	0	0	1	Ⓐ	1	0	Ⓐ	0	Ⓐ	Ⓐ	1

A few of the Karnaugh maps follow; the remaining are left to the reader as an exercise.

J_A

C D \ A B	0 0	0 1	1 1	1 0
0 0	0	0	§	Ⓐ
0 1	0	0	§	Ⓐ
1 1	0	1	§	§
1 0	0	0	§	§

$J_A = B\,C\,D$

K$_A$

C D \ A B	0 0	0 1	1 1	1 0
0 0	α	α	§	0
0 1	α	α	§	1
1 1	α	α	§	§
1 0	α	α	§	§

$$K_A = D$$

J$_B$

C D \ A B	0 0	0 1	1 1	1 0
0 0	0	α	§	0
0 1	0	α	§	0
1 1	1	α	§	§
1 0	0	α	§	§

$$J_B = C\,D$$

$$K_B = C\,D$$
$$J_C = \overline{A}\,D$$
$$K_C = \overline{A}\,D$$
$$J_D = 1$$
$$K_D = 1$$

Finally the circuit is drawn as shown.

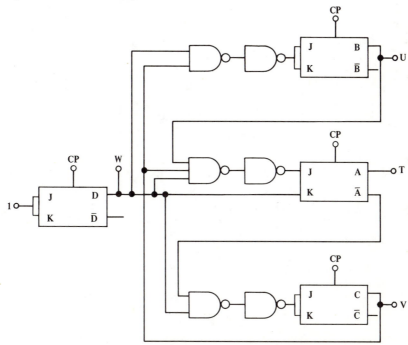

EXAMPLE 9.28

The N-bit binary counter is a device that counts up to 2^N. It then resets and counts again from zero. Design a four-bit binary counter with **J-K** flip-flops.

SOLUTION

A device that counts up to 2^N must have at least 2^N states, hence N flip-flops. Let us label the four flip-flops used in the four-bit counter as **D, C, B,** and **A,** and label the states as given below. Note that **A** has been placed on the right for reasons that will become clear.

D	C	B	A
0	0	0	0
0	0	0	1
0	0	1	0
0	0	1	1
.	.	.	.
.	.	.	.

and so forth.

The transition-excitation table is given below, and includes the **J-K** requirements, as obtained from the excitation tables.

Present State D	C	B	A	Next State D	C	B	A	J_D	K_D	J_C	K_C	J_B	K_B	J_A	K_A
0	0	0	0	0	0	0	1	0	Ⓡ	0	Ⓡ	0	Ⓡ	1	Ⓡ
0	0	0	1	0	0	1	0	0	Ⓡ	0	Ⓡ	1	Ⓡ	Ⓡ	1
0	0	1	0	0	0	1	1	0	Ⓡ	0	Ⓡ	Ⓡ	0	1	Ⓡ
0	0	1	1	0	1	0	0	0	Ⓡ	1	Ⓡ	Ⓡ	1	Ⓡ	1
0	1	0	0	0	1	0	1	0	Ⓡ	Ⓡ	0	0	Ⓡ	1	Ⓡ
0	1	0	1	0	1	1	0	0	Ⓡ	Ⓡ	0	1	Ⓡ	Ⓡ	1
0	1	1	0	0	1	1	1	0	Ⓡ	Ⓡ	0	Ⓡ	0	1	Ⓡ
0	1	1	1	1	0	0	0	1	Ⓡ	Ⓡ	1	Ⓡ	1	Ⓡ	1
1	0	0	0	1	0	0	1	Ⓡ	0	0	Ⓡ	0	Ⓡ	1	Ⓡ
1	0	0	1	1	0	1	0	Ⓡ	0	0	Ⓡ	1	Ⓡ	Ⓡ	1
1	0	1	0	1	0	1	1	Ⓡ	0	0	Ⓡ	Ⓡ	0	1	Ⓡ
1	0	1	1	1	1	0	0	Ⓡ	0	1	Ⓡ	Ⓡ	1	Ⓡ	1
1	1	0	0	1	1	0	1	Ⓡ	0	Ⓡ	0	0	Ⓡ	1	Ⓡ
1	1	0	1	1	1	1	0	Ⓡ	0	Ⓡ	0	1	Ⓡ	Ⓡ	1
1	1	1	0	1	1	1	1	Ⓡ	0	Ⓡ	0	Ⓡ	0	1	Ⓡ
1	1	1	1	0	0	0	0	Ⓡ	1	Ⓡ	1	Ⓡ	1	Ⓡ	1

The reader may verify, by means of Karnaugh maps, the following expressions:

$$\begin{aligned} &J_A = 1 \text{ (or } \underline{A}) & &J_C = A\ B \\ &K_A = 1 \text{ (or } \overline{A}) & &K_C = A\ B \\ &J_B = A & &J_C = A\ B\ C \\ &K_B = A & &K_C = A\ B\ C \end{aligned}$$

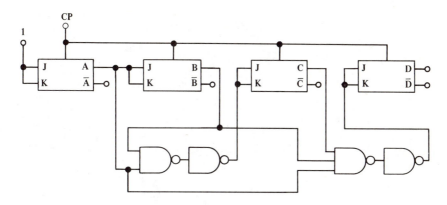

These equations lead to the simple circuit shown, and may easily be generalized to a larger number of bits. (For example, if a fifth bit, **E,** were added, then $\mathbf{J}_E = \mathbf{K}_E = \mathbf{A\ B\ C\ D.}$) ∎

Although the techniques described so far are quite powerful, and allow the rapid design of coding and counter circuits, they are by no means exhaustive. We will conclude this chapter by discussing two important kinds of circuits that require some variations from our standard synthesis techniques.

Shift Registers   A shift register is a digital circuit that is the electronic analog of a bucket brigade. There are a string of cells, each containing a **0** or a **1,** as shown schematically in Figure 9.33. Upon receipt of a clock pulse, the contents of all the cells pass in one direction to the next cell. The first cell receives a **0** or a **1** from the input line and the contents of the first cell pass to the second cell. Finally, the **0** or **1** that was in the last cell passes out. The contents of an N-cell shift register are just the last N values of the input (at the times of the clock pulses). Because each cell stores one binary unit of information, or bit, an N-cell shift register is generally called an N-bit shift register. The shift register is a primitive memory; in an N-bit shift register whatever is put in appears at the output N clock pulses later. Let us now see how a shift register may be constructed from flip-flops.

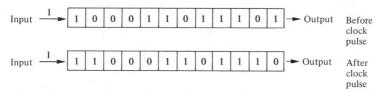

Figure 9.33   The contents of a 12-bit shift register before and after a clock pulse. The digit on the input line (here a **1**) is transferred into the first cell and the contents of each cell are passed one cell to the right.

Each bit in the shift register requires one flip-flop, that is, one flip-flop per cell. Let us assume that **J-K** flip-flops are used. We consider the inputs to two

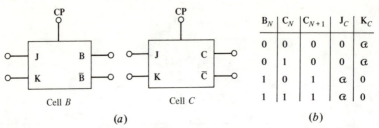

B_N	C_N	C_{N+1}	J_C	K_C
0	0	0	0	α
0	1	0	0	α
1	0	1	α	0
1	1	1	α	0

(a)

(b)

Figure 9.34 (a) Two typical cells in a shift register made from **J-K** flip-flops. (b) The possible combinations of cell contents before and after a clock pulse. The required conditions on **J** and **K** are listed as well.

typical cells, as shown in Figure 9.34(a). Upon receipt of a clock pulse, cell C should contain the former contents of cell B. The various possibilities are summarized in Figure 9.34(b), which also contains the J_C and K_C requirements as obtained from the excitation table. The requirements on J_C and K_C are satisfied if $J_C = B$ and $K_C = \bar{B}$. Thus each cell in the chain is connected directly to the previous cell, and no combinational logic elements are required. The complete circuit of an eight-bit shift register is given in Figure 9.35. Note that the first cell is connected to the input line in the same way as each cell is connected to its predecessor; that is, **J** is connected to the input and **K** is connected to the complement of the input. Some appreciation of the complexity of presently available commercial integrated circuits may be had by considering that 1024-bit shift registers may be purchased. These are made on a single chip of silicon a few square millimeters in area.

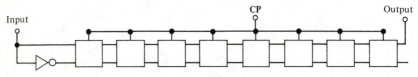

Figure 9.35   An eight-bit shift register made from **J-K** flip-flops.

Binary Ripple Counters   An interesting possibility not considered in our general synthesis technique is the use of the clock pulse input as a data input. Certain integrated circuit flip-flops that do not have critical requirements as to the shape of the clock pulse can be used quite ingeniously in this way. The circuit shown in Figure 9.36 is a good example. It is a so-called binary ripple counter. The operation of the circuit is best conceived with the aid of the timing diagram given in Figure 9.36(b). Each cell divides the pulse repetition rate by two, just as in the binary counter of Example 9.28. The only difference is in the manner in which the carry is transmitted to the other cells. In the present circuit it passes from cell to cell; that is, it must "ripple" through all the cells before reaching the last cell. Any number of cells may be added to the circuit to count to as large a number as desired.

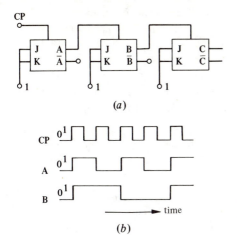

(a)

CP 0 1 ⎍⎍⎍⎍⎍⎍

A 0 1 ⎍⎍⎍⎍

B 0 1 ⎍⎍⎍

→ time

(b)

Figure 9.36   The binary ripple counter. (a) The circuit, constructed from **J-K** flip-flops. (b) The timing diagram.

The shift register and the binary ripple counter are but two of the large number of digital circuits that may be constructed with some ingenuity. In general, both the formal analysis and synthesis techniques and much experience and cleverness go into the design of efficient digital circuits.

Summary

- Digital circuits may be classified into two types: combinational and sequential. The output of a combinational circuit is a logical function of its inputs. The output of a sequential circuit depends upon both the values of the inputs and on the values of the inputs at some previous time.

- Switching algebra is a convenient method of describing the action of a combinational circuit. The output of the circuit is written as a logical function or expression in terms of the logical values of the inputs. The variables in a logical expression are binary; they must always have either the value **0** or **1**.

- The basic logical operations are **AND, OR, NAND, NOR,** and **COMPLEMENT.** The circuits that perform these functions are called **AND** gates, **OR** gates, **NAND** gates, **NOR** gates, and inverters, respectively. Any logical function may be synthesized using **NAND** gates or **NOR** gates alone.

- Truth tables and Karnaugh maps offer a means of displaying the value of a logical expression for the various combinations of the inputs. The Karnaugh map is arranged in a way that facilitates the simplification of logic expressions. The minimal sum of products expression for any logic function of three or four variables is found by mapping the function.

- Logical synthesis with **NAND** gates follows directly from the sum of products expression. Each of the terms in the expression corresponds to a **NAND** gate at the first level of logic, and a single **NAND** gate at the second level of logic combines the outputs of the first level.

- The basic building block for sequential circuits is the flip-flop. Flip-flops are described by their truth tables and excitation tables. The most versatile flip-flop is the standard **J-K** flip-flop.

- The operation of flip-flop circuits is visualized with state diagrams. The state diagram lists all the flip-flop states corresponding to each circuit state. All possible transitions between circuit states are indicated on the diagram.

- A flip-flop circuit is analyzed by finding its state diagram. The synthesis of a flip-flop circuit may be divided into two tasks; the generation of the state diagram and the implementation of the circuit from the state diagram.

- Binary counters, decimal counters, shift registers, and various ripple counters may be synthesized from **J-K** flip-flops and **NAND** gates.

References

Treatments at approximately the same level as this book:

Nashelsky, L. *Digital Computer Theory.* New York: John Wiley & Sons, 1966.

Wickes, W. E. *Logic Design with Integrated Circuits.* New York: John Wiley & Sons, 1968.

At a more advanced level:

Pfister, M. *Logical Design of Digital Computers.* New York: John Wiley & Sons, 1958.

Problems

9.1   Verify that $A C + A \overline{C}$ is equivalent to A by a comparison of truth tables.

9.2   Verify that $A B + A$ is equivalent to A by a comparison of truth tables.

9.3   Verify that $A B C + A B + B$ is equivalent to B by a comparison of truth tables.

9.4   Simplify the expression $F = \overline{(\overline{A} + \overline{B})}$ using DeMorgan's theorem.

9.5   Simplify the expression $F = \overline{(\overline{A} \, \overline{B} \, \overline{C} + \overline{D} \, \overline{E} \, \overline{G})}$ using DeMorgan's theorem.

9.6   Suppose $F = \overline{\overline{A} + \overline{B} + \overline{C} + \overline{(A B)}}$; find F using DeMorgan's theorem to simplify the expression.

9.7   How may the expression $\overline{F} = \overline{A} + \overline{B} + \overline{(\overline{C} \, D)} + \overline{E}$ be more simply written? (ANSWER: $F = ABCDE$.)

9.8   Construct a truth table for the expression $F = A \overline{B} (C + A)$, and write a sum of products expression for F from the truth table.

9.9   Find a sum of products expression for $A = A\overline{B} + \overline{A}B(BA)$ by first constructing a truth table and writing the expression from the truth table.

9.10   Map the following expressions in three variables: (1) $\overline{A}\,\overline{B}\,\overline{C} + A\,B\,C$; (2) $\overline{A}\,\overline{B}\,\overline{C} + A\,\overline{B}\,\overline{C} + \overline{A}\,\overline{B}\,C + \overline{A}\,B\,C$; (3) $A\,B\,\overline{C} + A\,\overline{B}\,\overline{C} + \overline{A}\,B\,C + A\,\overline{B}\,C$; (4) $A\,B + A\,C$; (5) $A\,B + A\,B\,\overline{C} + A\,\overline{B}\,\underline{C}$; (6) $B\,C + A$; (7) $A + B + \overline{A}\,C + A\,C$; (8) $A\,B\,(A\,C)$; (9) $A\,(B + C)$; (10) $\overline{A} + A\,(\overline{B}\,\overline{C})$.

9.11   Write a sum of products expression for each of the maps constructed in Problem 9.10, writing one term in the expression for each **1** in the map. (Do not group the 1's.) (ANSWER: (8) **ABC**.)

9.12   Write the minimal sum of products expression for each of the expressions mapped in Problem 9.10.

9.13   Rearrange the Karnaugh map for three variables so that the values **C** and **B** are given across the top (in place of **A** and **B**), and **A** is given along the side. Remap the expressions represented in Figure 9.6 in the modified arrangement.

9.14   Using the map arrangement suggested in Problem 9.13, remap the three expressions represented in (1) the maps of Figure 9.7; (2) the maps of Figure 9.8.

9.15   Find the minimal sum of products form for the expressions: (1) $(\overline{A} + \overline{B})\,(\overline{A}\,C)$; (2) $(A + C)\,(\overline{B} + C)$; (3) $A\,B\,C + A\,\overline{B}\,\overline{C} + \overline{A}\,\overline{B}\,C$; (4) $\overline{A}\,B\,\overline{C} + \overline{B}\,\overline{C} + \overline{A}\,B\,C$; (5) $(A + B + \overline{C})\,(A + \overline{B} + \overline{C})$; (6) $B\,C + A\,B + A\,\overline{C}$. (ANSWER: (4) $\overline{A}\,\overline{C} + \overline{B}\,\overline{C}$.)

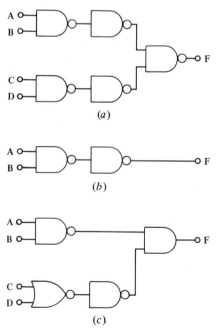

(a)

(b)

(c).   **Figure 9.37**

9.16   Map the following expressions of four variables: (1) $\overline{A}\ \overline{C}\ \overline{D} + \overline{A}\ C\ D + A\ \overline{B}\ D$; (2) $A\ B\ C\ \overline{D} + A\ \overline{B}\ C\ D + A\ \overline{C}\ \overline{D} + B\ \overline{C}\ \overline{D}$; (3) $\overline{A}\ \overline{B}\ C\ \overline{D} + A\ B\ C\ \overline{D} + \overline{B}\ C\ \overline{D} + \overline{A}\ B\ \overline{C}\ D$ $+ \overline{A}\ \overline{B}\ \overline{C}\ D + B\ C\ \overline{D}$; (4) $A\ B(C\ D + \overline{C}\ D) + A\ B\ C\ D + A\ B\ C\ \overline{D}$; (5) $\overline{A}\ \overline{B} + \overline{A}\ B$ $+ \overline{A}\ \overline{B}\ \overline{C}\ D + \overline{A}\ \overline{B}\ \overline{C}\ \overline{D}$; (6) $\overline{A}(D + C) + \overline{A}\ \overline{B}\ \overline{C}\ \overline{D}$.

9.17   Find the minimal sum of products forms for the expressions mapped in Problem 9.16. (ANSWER: (3) $C\ \overline{D} + \overline{A}\ \overline{C}\ D$.)

9.18   Remap the expressions of Figure 9.10 in a Karnaugh map that has **C** and **B** along the top and **A** and **D** along the side.

9.19   Find the logic expressions for the function performed by the circuits of Figures 9.37(a) to 9.37(c).

9.20   Find logic expressions for the function performed by the circuits of Figures 9.38(a) to 9.38(c).

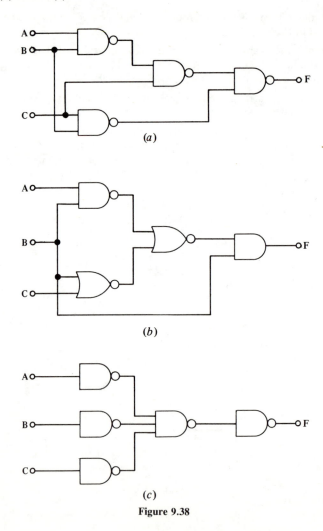

Figure 9.38

9.21 Map the logic functions derived in Problem 9.19, and derive simpler expressions where possible (in the sum of products form). Use DeMorgan's theorem where necessary.

9.22 Put the expressions derived in Problem 9.20 into the sum of products form. Use a Karnaugh map and find the simplest expression representing the function.

9.23 (1) Draw the complete circuit that performs the function $\mathbf{F} = (\mathbf{A} + \mathbf{B})(\mathbf{A} + \mathbf{C})$, using **OR** gates, and **AND** gates. (2) Write the logic equation for **F** from the circuit. (3) Simplify the logic equation using a Karnaugh map. (4) Reconstruct the circuit from the logic equation using **NAND** gates.

9.24 Show how the circuit of Example 9.15 might be constructed from **NAND** gates alone.

9.25 How might the circuit of Figure 9.12 be used as a two-column adder?

9.26 Synthesize the address decoder circuitry of Figure 13.36 using **NAND** gates.

9.27 Implement the functions given in Problem 9.16 using **NAND** gates.

9.28 Implement the functions given in Problem 9.15 using **NAND** gates.

9.29 An important circuit for use in computers is a comparator, which has zero output unless the inputs are identical. Synthesize a circuit that compares $\mathbf{A}_1 \mathbf{A}_2$ (a two-digit binary number) with $\mathbf{B}_1 \mathbf{B}_2$ (another two-digit binary number). Use **NAND** gates for the implementation.

9.30 Many times it is useful to have a circuit that indicates if two binary numbers are equal, and if not, which one is greater. Design a circuit, using **NAND** gates, which has inputs **A** and **B** and three outputs **F, G,** and **H; F = 1** only if $\mathbf{A} > \mathbf{B}$, **G = 1** only if $\mathbf{A} = \mathbf{B}$, and **H = 1** only if $\mathbf{A} < \mathbf{B}$.

9.31 Analyze the flip-flop circuit of Figure 9.39(a). (Find the state diagram.) Describe the function.

9.32 Analyze the flip-flop circuit of Figure 9.39(b). (Find the state diagram.) The single input is always **1** and the outputs are **A, B, C,** and **D.** Describe the function of the circuit.

9.33 Generate a suitable state diagram for a circuit that counts clock pulses and produces an output after every fourth clock pulse.

9.34 Generate a state diagram for a **0 0** detector — a circuit that looks for an input of **0** during two successive clock pulses. The circuit is to be reset on the third clock pulse.

9.35 Implement the circuit that functions according to the state diagram of Figure 9.40(a), (1) using **J-K** flip-flops; and (2) using **S-R** flip-flops. Describe the function.

9.36 Implement the circuit that functions according to the state diagram of Figure 9.40(a), (1) using **J-K** flip-flops; and (2) using **S-R** flip-flops. Describe the function.

9.37 The so-called **D** flip-flop is a flip-flop that responds directly to the single input **D,** independent of the previous state. (Thus if **D** is **1** and a clock pulse is received, the output will be **1.**) Design a **D** flip-flop from a **J-K** flip-flop and any necessary logic gates.

9.38 The so-called **T** flip-flop is a flip-flop that always changes state with the clock

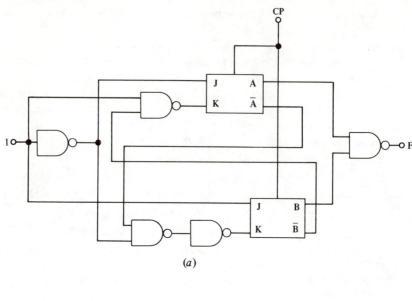

(a)

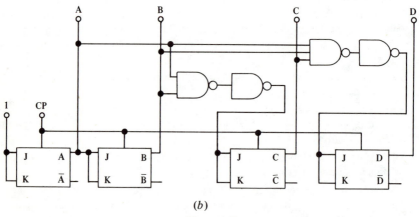

(b)

Figure 9.39

pulse if the input **T** equals **1,** or remains in the same state if **T** equals **0.**   1) Design a **T** flip-flop using a **J-K** flip-flop. (2) Design a **T** flip-flop using **S-R** flip-flops and any necessary logic gates.

9.39   Describe the transitions out of state 1 for the circuit whose state diagram is given in Figure 9.24 if the input equals **1, 0, 0, 1, 1, 1** during six successive clock pulses.

9.40   The input to the circuit of Figure 9.32 is held fixed at **1.** Describe the transitions during a sequence of six clock pulses if the circuit is initially in state 1.

9.41   Using **J-K** flip-flops, synthesize a circuit that functions according to the state diagram of Example 9.24.

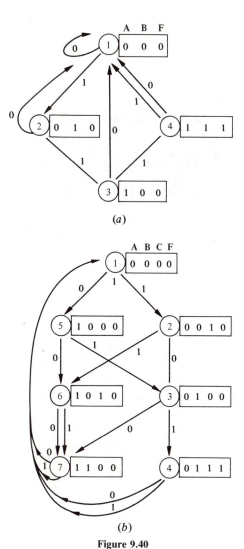

(a)

(b)

Figure 9.40

9.42   Using **J-K** flip-flops, synthesize a circuit that functions according to the state diagram of Example 9.25.

9.43   Design a counter that counts clock pulses up to eight and resets. The output **F** should be **1** *only* after the eighth pulse.

9.44   Design a four-bit shift register using **S-R** flip-flops and **NAND** gates.

9.45   Design a six-bit binary counter (see Example 9.28).

9.46   Design a six-bit binary ripple counter.

Chapter 10 ▬▬▬▬▬▬▬▬▬▬▬▬▬▬▬

Introduction to Amplifiers

In the last two chapters we have considered transistor circuits of the digital type. We shall now consider circuits of the analog type, such as are used in the construction of amplifiers. While in switching circuits one is concerned only with two-state operation, in amplifier circuits continuous ranges of voltages and currents are involved. In an amplifier the objective is usually to make an output voltage or current that, while larger, is linearly proportional to an input voltage or current. In amplifiers, transistors are usually operated in the active mode.

As a first illustration of amplifier principles, let us consider the simple circuit of Figure 10.1. In this circuit a voltage $(V_0 + v_s)$ is applied to the base of the transistor.

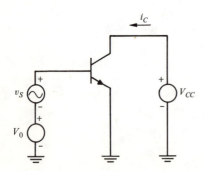

Figure 10.1 An elementary circuit for illustrating the use of a transistor in the active mode.

The time-varying voltage v_S is the input signal and V_0 is a constant voltage whose function will be explained shortly. The collector voltage source V_{CC} causes the collector-base junction to be reverse biased.

Let us first consider the case of $v_S = 0$. The forward bias voltage on the emitter junction is then just equal to V_0. Referring to Figure 10.2(a), which shows the transistor's input characteristic, we find that the base current has the value I_0. Since we are operating the transistor in the active mode, the collector current now has the value $i_C = \beta i_B = \beta I_0$.

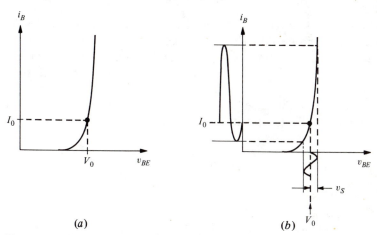

(a)   (b)

Figure 10.2   Input I-V characteristic of the transistor circuit of Figure 10.1. If $v_S = 0$, the base current, determined by V_0, has the value I_0, as shown in (a). If the small-signal voltage v_S is now applied, the base current fluctuates in response to it, as shown in (b).

We now turn on the signal voltage v_S. If $v_S(t)$ is a sinusoid, with amplitude less than V_0, its effect is alternately to increase and decrease the base-to-emitter voltage $v_{BE}(\equiv v_B - v_E)$, as shown in Figure 10.2(b). This causes the base current to vary as shown. The variations in base current give rise to variations of the collector current that are larger by a factor of β. The larger time-varying collector current is sent to a load, and constitutes the output of the amplifier.

From Figure 10.2(b) it now can be seen why the constant voltage V_0 is needed. If V_0 were made zero, v_{BE} would equal v_S. Since the slope of the input characteristic is very nearly zero at $v_{BE} = 0$, the base voltage variations v_S would have almost no effect; the base current would remain zero. If the amplitude of the sinusoidal signal v_S were increased to the order of 0.7 V or so, some base current would indeed appear, at the time v_S was most positive. However, when v_S became negative, the emitter junction would be reverse biased. Thus no base current would ever flow during the negative part of the cycle of v_S. Any time v_S was negative, its value would be unable to influence the collector current; hence those parts of the input signal would not be amplified. The use of V_0 allows us to avoid this problem. From Figure

10.2(*b*) it is seen that through the use of V_0, v_S exerts a controlling effect on i_B (and hence on i_C) at all times.

From the preceding discussion we see that it is desirable, in an amplifier, for the time-varying signals to be accompanied by constant voltages and currents, known as *bias* voltages and currents. For instance, the dc base current (I_0 in the above example) is called the *base bias current,* and the dc voltage V_0 is the *base bias voltage.* The point (V_0, I_0) in Figure 10.2, where operation rests in the absence of signals, is called the *operating point* of the circuit.

Linearization   Figure 10.2(*b*) shows that the changes in the base current are not linearly proportional to the signal voltage v_S. Because the transistor *I-V* characteristic is nonlinear, distortion of the signal waveform occurs. However, we can show that the device approaches linearity more closely as v_S is made smaller.

Let us expand the base current i_B as a Taylor series about the operating point. In general, if $y(x)$ is any non-singular function of x, and $y(x_0) = y_0$, a power series expression of y is

$$y(x) = y_0 + \frac{dy}{dx}\bigg|_{x=x_0} \bullet (x - x_0) + \frac{1}{2!}\frac{d^2y}{dx^2}\bigg|_{x=x_0} \bullet (x - x_0)^2 + \frac{1}{3!}\frac{d^3y}{dx^3}\bigg|_{x=x_0} \bullet (x - x_0)^3 + \cdots \quad (10.1)$$

Here the notation

$$\frac{dy}{dx}\bigg|_{x=x_0}$$

stands for the derivative, evaluated at $x = x_0$. Expression (10.1) for $y(x)$ is in the form of an infinite series. However, if $|x - x_0|$ is a small quantity, its higher powers rapidly become insignificantly small, and one obtains an approximate expression for $y(x)$,

$$y(x) \cong y_0 + \frac{dy}{dx}\bigg|_{x=x_0} \bullet (x - x_0)$$

This approximation becomes more accurate as $|x - x_0|$ becomes smaller, and the neglected higher terms of the series become less significant.

In the case of our transistor circuitry, the role of the function $y(x)$ is taken by the function $i_B(v_{BE})$, shown in Figure 10.2. We know that when $v_{BE} = V_0$, $i_B = I_0$. Expanding in Taylor series about this point, we have the following expression for i_B:

$$i_B = I_0 + \frac{di_B}{dv_{BE}}\bigg|_{v_{BE}=V_0} \bullet (v_{BE} - V_0) + \frac{1}{2!}\frac{d^2i_B}{dv_{BE}^2}\bigg|_{v_{BE}=V_0} \bullet (v_{BE} - V_0)^2 + \cdots \quad (10.2)$$

In our present case we are interested in finding the value of i_B when $v_{BE} = V_0 + v_S$. Making this substitution, we have

$$i_B = I_0 + \left.\frac{di_B}{dv_{BE}}\right|_{v_{BE}=V_0} \bullet (v_S) + \frac{1}{2!}\left.\frac{d^2i_B}{dv_{BE}^2}\right|_{v_{BE}=V_0} \bullet (v_S)^2 + \cdots \tag{10.3}$$

Now if v_S is a small number, an approximate expression for i_B can be found by dropping the quadratic and all higher terms of (10.3):

$$i_B \cong I_0 + \left.\frac{di_B}{dv_{BE}}\right|_{v_{BE}=V_0} \bullet v_S \tag{10.4}$$

Equation (10.4) states that if v_S is made sufficiently small, the variations of the base current around the operating point are, to a good approximation, linearly proportional to the signal voltage v_S. The significance of Equation (10.4) is shown graphically in Figure 10.3. This figure is an expanded version of Figure 10.2(b), showing *small* variations around the operating point. The result, (10.4), basically comes from the fact that even a curved line looks nearly straight if one considers only a small region along its length.

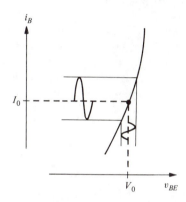

Figure 10.3 Magnified portion of Figure 10.2(b) showing variations in base current due to small v_S. The I-V curve is nearly straight over a small portion of its length. Consequently $(i_B - I_0)$ is more nearly proportional to v_S than in Figure 10.2(b), where v_s is larger.

From Equation (10.4) we may conclude that to construct highly linear circuits, we should restrict their operation to small signals. This procedure is often used when good linearity is desired. However, even when v_S is not extremely small, it is useful to pretend that Equation (10.4) is true. The I-V characteristic of the transistor, after all, is nonlinear; analysis of circuits containing nonlinear elements is a slow numerical procedure. But Equation (10.4) offers a great simplification. It is a linear equation, and solutions for linear equations are easy to find. Thus Equation (10.4) is made the starting point for the technique known as *small-signal analysis,* which is widely used in connection with analog circuits. With small-signal analysis we shall be able to treat transistors as though they were linear circuit elements, constructing linear models for them, just as we did for amplifier blocks in Chapter 3. Of course, Equation (10.4) is not strictly accurate if v_S is not small. Nonetheless, small-signal analysis is such a powerful method that it is most often used anyway, at least for preliminary analysis. Often the approximation is good enough, and further calculations are not necessary.

In a typical amplifier circuit, provision must be made to apply the desired dc voltages and currents to the transistors, so that with signal absent they are biased at the chosen operating point. This part of the circuit will be considered first, in Section 10.1. In Section 10.2 we shall proceed with the subject of small-signal analysis by developing a linear model for the transistor. As was the case in Chapter 3, the model for the transistor will be a collection of linear circuit elements designed to mimic the operation of the transistor. It can be substituted for the transistor in the circuit, so that the techniques of linear circuit analysis can then be used. In Sections 10.3 and 10.4 the modeling technique of the previous section is applied for the analysis of simple amplifier circuits.

The model introduced in Section 10.2 is a simple one, designed to simulate the Ebers-Moll equations in the active mode of operation. As was pointed out in Chapter 6, there exist subtler aspects of transistor behavior that the Ebers-Moll equations do not describe. More refined transistor models can be used, when necessary, to give representation to these subtler effects. Improved transistor models are considered in Section 10.5.

All amplifiers have a maximum frequency at which they will operate properly, and some have a minimum frequency as well. The behavior of an amplifier as a function of signal frequency is known as its *frequency response*. This subject is important because it affects the range of situations in which a given amplifier may be used. A discussion of amplifier frequency response is presented in Section 10.6.

10.1 Biasing Circuits

In each linear amplifier circuit, provision must be made for adjusting the values of the dc currents and voltages, so that under conditions of no signal the circuit rests at the desired operating point. The choice of what point is to be the operating point is a matter of discretion for the circuit designer. Several considerations enter. For one thing, the operating point must be chosen so that the circuit can accommodate the signals of interest. Even though the analysis is based on small deviations from the operating point, the circuit is nonetheless intended for use with signals of finite size. If, for example, we anticipate a base current signal of $10^{-4}\sin(\omega t)$ A, we must first be sure that the operating point base current is *greater* than 10^{-4} A. The base current can never be negative; therefore if it is to decrease by 10^{-4} A from the operating point value, the current at the operating point must be greater than 10^{-4} A. A second consideration in the choice of operating point has to do with power consumption. Large dc bias currents may cause the circuit to consume excessive power. A third consideration in the choice of operating point has to do with the desired operating characteristics of the transistor in the circuit. When we come to the subject of transistor models, it will be seen that the parameters of the model depend on the choice of operating point. These, and perhaps other considerations must be weighed by the circuit designer in choosing an operating point. There is no unique "right answer"; the choice of operating point is an "engineering decision."

Since both time-varying signals and non-time-varying bias quantities are usually present in an amplifier, it is helpful, as well as customary, to use notation which emphasizes the difference between the two. In the following discussion, bias voltages and currents, which are constant in time, will be represented by upper-case symbols. Time-varying voltages and currents will be represented by lower-case symbols.

To illustrate in a simple way how a circuit may be adjusted to its operating point, let us consider the elementary circuit of Figure 10.4. (Other more practical biasing circuits will be developed later in this chapter.) The notation V_{CC} by the terminal at the top indicates, according to convention, that a voltage source is connected between this point and ground, so that the potential at this terminal has the value V_{CC}. In practice this terminal is connected to some source of dc voltage, but to avoid cluttering the diagram, that source is not shown.

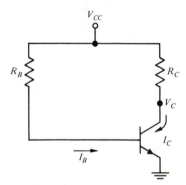

Figure 10.4 A simple biasing circuit for an *npn* transistor. In a discussion of biasing, only dc voltages and currents are considered.

The values of R_B and R_C are selected so as to place the transistor at the desired operating point. Let us assume it has been decided that the dc collector current at the operating point I_C is to have the value 1 mA. The transistor is assumed to have $\beta = 100$ and $V_{CC} = 10$ V. The circuit designer must then choose R_B and R_C so that $I_C = 1$ mA. Because the transistor is to operate in the active mode, the base current is given by

$$I_B = \frac{I_C}{\beta} \tag{10.5}$$

Therefore $I_B = 10^{-5}$ A.

For this amount of base current to flow, the emitter-base junction must be forward biased. As we have seen, this means that there must be a forward-voltage drop across it of approximately 0.7 V. Thus $V_B \cong 0.7$ V. The base current is the same current as that flowing through R_B. But according to Ohm's law, the current through R_B is

$$I_B = \frac{V_{CC} - V_B}{R_B} \tag{10.6}$$

We have already decided that I_B is to have the value 10^{-5} A, and we have adopted 0.7 V as the approximate value of V_B. Thus we have, from (10.6), the result $R_B = (10 - 0.7)/10^{-5} = 9.3 \times 10^5 \ \Omega$, and the proper value for R_B has been found.

Note that the somewhat arbitrary assumption that $V_{BE} = 0.7$ V has little effect on the result just obtained. Inspection of Equation (10.6) reveals that if the assumed value of V_{BE} were changed to 0.75 V, the value obtained for R_B would change by only about 0.5%. This is a desirable characteristic for the circuit to have. The value of V_{BE} is dependent, for instance, on temperature; it is found to decrease about 2 mV for each temperature increase of 1°C. To make circuit operation as much as possible independent of temperature, it is desirable that the operating point not be a strong function of V_{BE}.

Next, the value of R_C will be found. Before doing this, it is necessary to decide what the value of the dc collector voltage V_C should be when the transistor is biased to its operating point. As with I_B, several considerations may bear on this decision. For the present, let us use the following simple line of reasoning. When signals are eventually applied to the circuit, we shall want the collector voltage to vary in response to them. It is desirable that the collector voltage v_C be able to vary as far in the positive direction, in response to signals of one sign, as it can in the negative direction, in response to signals of the opposite sign. However, v_C cannot be allowed to become less than zero, or reverse bias on the collector junction will be lost. It will not be possible for v_C to exceed 10 V either, as this is the largest voltage supplied to the circuit. Therefore the permissible range of v_C is approximately 0 to 10 V. It is reasonable to place the dc operating-point voltage V_C at the middle of this range, so that the excursions around the operating point due to signal may go equally far in either the positive or negative direction. Thus we choose $V_C = V_{CC}/2 = 5$ V.

Now we may proceed to calculate the value of R_C by Ohm's law. It has already been decided that $I_C = 1$ mA. The current I_C flows through R_C; therefore

$$I_C = \frac{V_{CC} - V_C}{R_C} \tag{10.7}$$

Solving, we have $R_C = (10 - 5)/10^{-3} = 5000 \ \Omega$.

EXAMPLE 10.1

Find the dc operating point values of the base current, collector current, and collector voltage for the transistor in the circuit given.

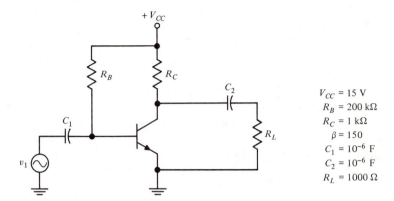

$V_{CC} = 15$ V
$R_B = 200$ kΩ
$R_C = 1$ kΩ
$\beta = 150$
$C_1 = 10^{-6}$ F
$C_2 = 10^{-6}$ F
$R_L = 1000$ Ω

SOLUTION

Although the circuit is fairly complex, at present we need be concerned only with the central part of the circuit. The capacitors C_1 and C_2 are open circuits for dc and therefore v_1 and R_L do not affect the dc operating point.

To find the base current, we write a node equation for the base terminal of the transistor:

$$I_B + \frac{V_B - V_{CC}}{R_B} = 0$$

or

$$I_B = \frac{V_{CC} - V_B}{R_B}$$

If we again adopt, as our standard approximate value, $v_{BE} = 0.7$ V, we have

$$I_B = \frac{15 - 0.7}{2 \times 10^5} = 7.2 \times 10^{-5} \text{ A} = 72 \ \mu\text{A}$$

Since the transistor operates in the active mode, $I_C = \beta I_B$ and thus $I_C = 150 \, (72 \ \mu\text{A})$ = 10.8 mA. To find the dc collector voltage V_C, we write a node equation for the collector terminal:

$$I_C + \frac{V_C - V_{CC}}{R_C} = 0$$

or

$$V_C = V_{CC} - I_C R_C$$
$$= 15 - (10.8 \times 10^{-3})(10^3) = 4.2 \text{ V}$$ ∎

10.2   The Small-Signal Transistor Model

The previous section was concerned with the dc operation, or "biasing," of an amplifier circuit. Now we are ready to deal with its ac behavior. To do this, we shall construct a device model for the transistor.[1] This model is based on a linearized description of the transistor, like that used in obtaining Equation (10.4). Consequently the model is a collection of ideal linear circuit elements. It is intended to be substituted for the transistor in the circuit diagram to facilitate analysis. However, the model is intended to model *only the ac, or signal operation of the transistor.* Our point of view is that analyses of the circuit for dc (biasing) and ac (signals) are separate operations. The ac device model to be considered in this section deals only with the response of the circuit to small signals, and cannot be used to get information about biasing.

Let us consider an *npn* transistor biased in the active region. To model the small-signal behavior of the device, we may first draw upon the input characteristics obtained in Chapter 6. From Equation (6.8) we have

$$i_B \cong \frac{I_{ES}}{(\beta + 1)} \exp \left\{ \frac{q v_{BE}}{kT} \right\} \qquad [npn] \qquad (10.8)$$

[The very small constant term in Equation (6.8) has been dropped.] This equation states the mathematical relationship between i_B and v_{BE}; it is graphed in Figure 10.2. As explained earlier, Equation (10.8) can be linearized by expanding in Taylor series around the operating point:

$$i_B \cong I_B + \frac{di_B}{dv_{BE}}\bigg|_0 \cdot (v_{BE} - V_{BE}) \qquad (10.9)$$

Here the notation $di_B/dv_{BE}|_0$ means that the derivative is to be evaluated at the operating point; I_B and V_{BE} are the base current and base-to-emitter voltage at the operating point. Equation (10.9) is, as previously explained, an approximation, most valid when $(v_{BE} - V_{BE})$ is small.

Let us now introduce an important new notation. We shall define

$$\begin{aligned} i_b &\equiv i_B - I_B \\ v_{be} &\equiv v_{BE} - V_{BE} \end{aligned} \qquad (10.10)$$

The new quantities defined here have small-letter subscripts to indicate that they represent small deviations of current or voltage away from the operating point. We shall refer to them as *small-signal variables.* [Later we shall introduce other small-signal variables; these too will be represented by symbols with small-letter subscripts and will be defined analogously with (10.10).] In terms of the small-signal variables, (10.9) becomes

$$I_B + i_b = I_B + \frac{di_B}{dv_{BE}}\bigg|_0 \cdot v_{be} \qquad (10.11)$$

Subtracting I_B from both sides, we have

---

[1] Some authors refer to this model as "circuit model." However, in Chapter 3 models for entire circuits were considered. To avoid confusion between models of circuits and models of devices, we shall call the model for the transistor a "device model."

$$i_b = \frac{di_B}{dv_{BE}}\bigg|_0 \cdot v_{be} \tag{10.12}$$

This equation expresses a linear proportionality between the small-signal base voltage and the small-signal base current.

The quantity (di_B/dv_{BE}) may next be evaluated by differentiation of Equation (10.8):

$$\frac{di_B}{dv_{BE}} = \frac{I_{ES}}{(\beta+1)} \cdot \frac{q}{kT} \exp\left\{\frac{qv_{BE}}{kT}\right\} \qquad [npn] \qquad (10.13)$$

We evaluate the derivative at the operating point by setting $v_{BE} = V_{BE}$ in (10.13). Substituting (10.8) into (10.13) to eliminate the exponential function, we find

$$\frac{di_B}{dv_{BE}}\bigg|_0 = \frac{qI_B}{kT} \qquad [npn] \qquad (10.14)$$

The value of the derivative, (10.14), may now be substituted into (10.12). This leads to

$$i_b = \frac{qI_B}{kT} v_{be} \qquad [npn] \qquad (10.15)$$

We observe that the quantity (qI_B/kT) must have the units of reciprocal resistance. Accordingly, let us define a resistance r_π by

$$\boxed{r_\pi \equiv \left|\frac{kT}{qI_B}\right|} \tag{10.16}$$

[There is of course no present need for the absolute-value sign in the definition of r_π, as in npn circuits the value of I_B is positive. However, with the absolute-value sign, (10.16) also holds for pnp circuits.] In terms of r_π we have, from (10.13)

$$\boxed{v_{be} = i_b r_\pi} \tag{10.17}$$

Thus we see that as far as the small-signal variables v_{be} and i_b are concerned, the input of the transistor obeys Ohm's law, where the proportionality constant is r_π, given by (10-16). For the device model to exhibit this behavior, the input terminals (base and emitter) must be connected by the resistance r_π. This first stage in the development of the model is illustrated in Figure 10.5.

We now consider the output portion of the model. Since we are dealing with the active mode of operation, $i_C = \beta i_B$. We define the small-signal collector current i_c to be equal to the deviation of the collector current from its

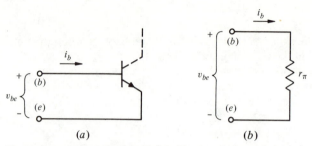

Figure 10.5   Input part of small-signal model of the transistor. The relationship between the small-signal voltage and current, shown in (*a*), is modeled by the connection shown in (*b*).

value at the operating point:

$$i_c \equiv i_C - I_C \qquad (10.18)$$

Then in terms of the small-signal quantities, the equation $i_C = \beta i_B$ becomes

$$I_C + i_c = \beta(I_B + i_b) \qquad (10.19)$$

Subtracting from (10.19) the equality $I_C = \beta I_B$, we have

$$\boxed{i_c = \beta i_b} \qquad (10.20)$$

a linear relationship between the small-signal base and collector currents.

The relationship (10.20) can be modeled by a dependent current source connected to the collector terminal. Thus we can complete the small-signal device model as shown in Figure 10.6. The dependent source imitates the action of the transistor, Equation (10.20), by causing a small-signal current βi_b to flow through the collector terminal.

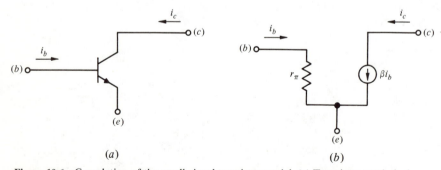

Figure 10.6   Completion of the small-signal transistor model. (*a*) Transistor symbol, showing small-signal base and collector currents. (*b*) The simplified-π model for the transistor, based on Equations (10.17) and (10.20).

The model just obtained is somewhat idealized, but it is sufficiently accurate for almost all calculations, except for those involving questions of frequency response. We shall refer to it as the *simplified-π model*.[2] It has been derived for an *npn* transistor; however, *it is applicable both to* npn *and* pnp *transistors* with no modifications.

Use of the Device Model We have already remarked that the function of the device model is to act as a "stand-in" for the transistor. That is, the transistor is replaced by the model in the circuit diagram, and then the resulting circuit model is analyzed. It should be remembered that the small-signal model we have developed only models the behavior of the small-signal variables. On the other hand, the actual circuit contains not only variable quantities but also dc biasing quantities. Thus the subject of how the model is to be used requires some further discussion.

The function of the small-signal model is to represent the relationships of small-signal quantities to each other. It says nothing about relationships of dc quantities to the small-signal quantities; thus no reference to dc quantities need be made with the model is used. If a certain wire carries a current $i = I_B + i_b$, for small-signal calculations we simply ignore the dc part and say that $i = i_b$. Similarly, a constant voltage is the same as zero voltage, insofar as small-signal calculations are concerned.

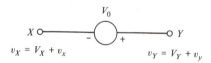

Figure 10.7 Illustrating the rule that in constructing a small-signal model, dc voltage sources are replaced by short circuits.

The latter statement has an important corollary. *When proceeding from the original circuit to the small-signal circuit model, dc voltage sources become short circuits.* The reasoning behind this statement is illustrated by Figure 10.7. Suppose the value of the dc voltage source in the figure is V_0, and let v_X, the voltage at terminal X, consist of a dc part V_X plus a small-signal part v_x. The voltages at terminal Y are similarly defined. The basic property of the voltage source is that $v_X + V_0 = v_Y$, or, equivalently, that

$$V_X + v_x + V_0 = V_Y + v_y \tag{10.21}$$

Now when all signals are turned off, Equation (10.21) reduces to $V_X + V_0 = V_Y$, and since V_X, V_0 and V_Y are non-time-varying quantities, this equality must hold whether the signal is off or on. Subtracting this equality from (10.21), we have $v_x = v_y$. The way to incorporate into a model the statement that two voltages are equal is to connect them by a wire. Thus it has been

---

[2] This name derives from a more general form of this model, known as the hybrid-π model, which has a π-shaped arrangement of components. The hybrid-π model is discussed in Section 10.5.

shown that in the small-signal circuit model, a dc voltage source is replaced by a short circuit.

A similar rule applies to dc current sources. If a branch of the original circuit contains a dc current source, the current through this branch must be constant. But if the current in the branch is constant, the time-varying small-signal current in the branch must be zero. In the model, a branch through which no small-signal can flow is correctly represented by an open circuit. Therefore, *when proceeding from the original circuit to the small-signal circuit model, dc current sources become open circuits.*

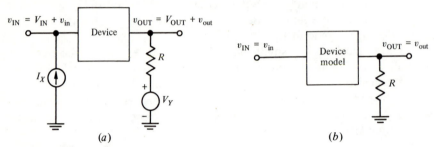

(a)   (b)

Figure 10.8   Comparison of a circuit (*a*) and its small-signal model (*b*). The dc current source I_X appears as an open circuit in the circuit model. The dc voltage source V_Y appears as a short circuit.

The principles of this section are illustrated by Figure 10.8. Figure 10.8(*a*) shows a circuit containing a device, with dc and ac voltages and currents present. Figure 10.8(*b*) shows the small-signal circuit model for the circuit.

For reference, the meanings of the various types of letter symbols are reviewed in Table 10.1.

Table 10.1   Symbols Used in Small-Signal Circuit Analysis

Type Style	*Meaning*	*Example*
Capital letter V or I with any subscript	Non-time-varying quantity	Collector current at operating point I_C
Small letter v or i with capital subscript	Time-varying quantity	Total collector current i_C
Small letter v or i with small subscript	Small-signal quantity	Small-signal collector current i_c

NOTE: The total current (or voltage) equals the dc bias current (or voltage) plus the small-signal current (or voltage). For example

$$i_C = I_C + i_c$$

EXAMPLE 10.2

Obtain the small-signal model for the circuit in sketch (a), using the simplified-π model of the transistor. V_0 is a constant-bias voltage and v_s is a small-signal voltage.

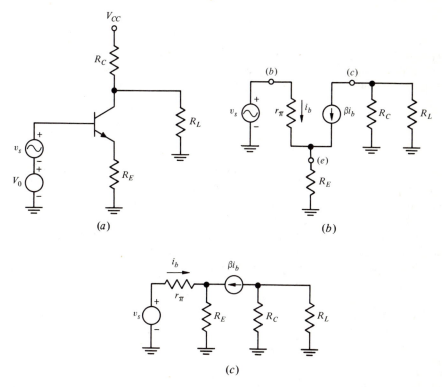

(a)

(b)

(c)

SOLUTION

In constructing the circuit model, we replace the dc source V_0 by a short circuit. There is also a second hidden dc voltage source in the circuit: a source of magnitude V_{CC} is understood to be connected between the terminal marked V_{CC} and ground. Thus when the small-signal circuit model is constructed, the V_{CC} terminal is connected to ground.

The model is substituted for the transistor, with care that the (b) terminal of the model is connected where the base lead of the transistor was, and so forth. After the model has been substituted, the circuit model appears as in sketch (b). Here the locations of the (e), (b), and (c) terminals of the model have been designated to aid in visualizing the process of substitution. Now the circuit may be rearranged, if desired, to tidy it up. One possible arrangement is shown in sketch (c). It is now no longer necessary to keep track of the locations of the (e), (b), and (c) terminals. It is, however, still necessary to show the location of the current i_b, since this current controls the dependent current source. ∎

EXAMPLE 10.3

Suppose that in the circuit of the previous example, the dc base current at the operating point is 10 μA. What is r_π?

SOLUTION

We recall that the value of r_π depends on the base bias current through Equation (10.16):

$$r_\pi = \left(\frac{kT}{q}\right) \cdot \frac{1}{I_B}$$

The question does not specify the temperature; in that case it is reasonable to assume the question means room temperature, 300°K. The value of kT/q at room temperature is 0.026 V. Thus

$$r_\pi = (0.026)\ (10^5) = 2.6 \times 10^3\ \Omega = 2.6\ \text{k}\Omega \qquad \blacksquare$$

10.3 Common-Emitter Amplifier Circuits

We are now ready to use the principles of the two preceding sections to analyze amplifier circuits. Let us consider the simple circuit of Figure 10.9. The parameters assumed for the transistor are given on the right. Resistors R_B and R_C are present in the circuit because of the necessity for biasing; they do, however, affect the ac operation of the circuit, as will be seen. The input signal is applied to the base through the input capacitor C_i. Since the capacitor is an open circuit for dc, the presence of C_i does not disturb the biasing of the transistor. However, we shall assume that the frequency of the signal is high enough that the capacitor can be considered a short circuit, as far as signals are concerned. The same arguments apply to the output capacitor C_o. Capacitors used to transmit signals in and out of circuits while leaving the dc bias unaffected are known as *coupling capacitors*.

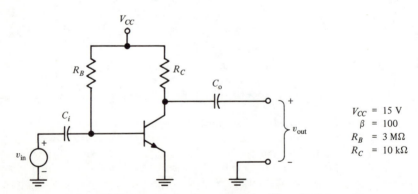

Figure 10.9   A simple transistor amplifier-circuit.

As a first step in the analysis, let us find the dc operating point of the transistor. Since no dc current flows through C_i, the current through R_B is identical to I_B. This current can be found by Ohm's law, if we make our usual assumption that $V_{BE} = 0.7$ V: $I_B = (V_{CC} - V_{BE})/R_B = (15 - 0.7)/3 \times 10^6 = 4.8 \times 10^6$ A $= 4.8\ \mu$A. Assuming for the moment that the transistor is operating in the active mode, we may write $I_C = \beta I_B$. The collector voltage V_C may be found by a node equation. Setting the sum of the currents leaving the collector node to zero, we have $I_C + (V_C - V_{CC})/R_C = 0$. Thus $V_C = V_{CC} - I_C R_C = V_{CC} - \beta I_B R_C = 10.2$ V. We now note that $V_{CB} = V_C - V_B = 10.2 - 0.7 = 9.5$ V. Since the collector junction is therefore strongly reverse biased, our result is consistent with our earlier assumption that operation is in the active mode.

Next we shall calculate the circuit's small-signal output voltage. Substituting the simplified-π model into the circuit, replacing dc voltage sources by short circuits, and, under our assumption that the signal frequency is sufficiently high, replacing the capacitors by short circuits, we obtain the small-signal circuit model of Figure 10.10. The resistance r_π is evaluated by Equation (10.16): $r_\pi = kT/qI_B = 0.026/4.8 \times 10^{-6} = 5400\ \Omega$.

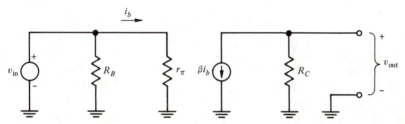

Figure 10.10 Small-signal circuit model for the circuit of Figure 10.9. The simplified-π model has been used.

We observe that the voltage v_{in} appears across r_π. Therefore $i_b = v_{in}/r_\pi$. The small-signal output voltage v_{out} is then given by

$$v_{out} = -\beta i_b R_C = \frac{-\beta R_C}{r_\pi} v_{in} \qquad (10.22)$$

The open-circuit voltage amplification A of the circuit may be defined as v_{out}/v_{in}. Therefore

$$A \equiv v_{out}/v_{in} = \frac{-\beta R_C}{r_\pi} \qquad (10.23)$$

Using the values $\beta = 100$, $R_C = 10,000\ \Omega$, and $r_\pi = 5400\ \Omega$, we find that $A = -185$. The negative sign of A simply means that an increase in v_{in} gives rise to a decrease in v_{out}.

The Amplifier as a Building Block   The amplifier we have been discussing can be used as a building block in a larger system. Its function as an analog building block can be more easily understood if the amplifier as a whole is represented by a simplified circuit model. It is true that the circuit of Figure 10.10 is already a circuit model, but a simpler model can be constructed just as a simpler Thévenin equivalent can be found for a complicated two-terminal subcircuit. A model similar to that of Figure 3.8 is what we have in mind. In Chapter 3 the discussion concerned the form and uses of the ampli-

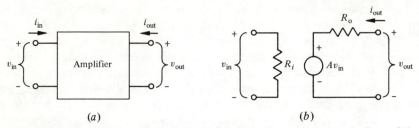

Figure 10.11 The amplifier regarded as a building block. (*a*) Illustrates the form of the block, showing sign conventions for voltages and currents. (*b*) Shows the small-signal model for the block.

fier block, with its parameters A, R_i, and R_o taken as given. Now we are ready to evaluate the parameters of the model.

One minor difference exists between Chapter 3 and the discussion here. The model to be developed here is a *small-signal* model. Strictly speaking, a small-signal model only describes the input and output small-signal variables, and not, for example, large changes of dc voltage. The form of the small-signal amplifier model we shall now develop is shown in Figure 10.11. Analogously with the discussion of Chapter 3, we define the input resistance R_i by the formula

$$R_i = v_{in}/i_{in} \tag{10.24}$$

where the sign conventions for v and i are as shown in Figure 10.11. The output resistance R_o is defined by

$$R_o = -\frac{(v_{out})_{\text{output terminals open-circuited}}}{(i_{out})_{\text{output terminals short-circuited}}} \tag{10.25}$$

The open-circuit voltage amplification A is defined as

$$A = \left(\frac{v_{out}}{v_{in}}\right)_{\text{output terminals open-circuited}} \tag{10.26}$$

We observe that the value arrived at for A in Equation (10.23) is the same A as that defined here. Thus one of the three parameters for the model of the amplifier in Figure 10.9 has already been found.

By inspection of Figure 10.10, we see that R_i is equal to the parallel resistance of the three MΩ biasing resistor and r_π. But $(3\ M\Omega \| r_\pi)$ is very nearly equal to r_π, or 5400 Ω. Thus for the circuit of Figure 10.9

$$R_i \cong r_\pi \tag{10.27}$$

To find the output resistance of this circuit we use Equation (10.25). Referring again to Figure 10.10, the short-circuit output current is seen to be βi_b. The open-circuit output voltage can be found by writing a node equation for the node at the output terminal. This equation is $\beta i_b + v_{out}/R_C = 0$, from which we have $v_{\text{out}} = -\beta i_b R_C$. Then from (10.25), $R_o = -(-\beta i_b R_C)/(\beta i_b) = R_C$. Thus

$$R_o = R_C \qquad (10.28)$$

For this circuit $R_o = R_C = 10 \text{ k}\Omega$. The three parameters for the model of Figure 10.11(b) have now been found. In further work with the amplifier as a part of a larger system, the amplifier can be represented by the model, with its parameters taking on the values we have found.

EXAMPLE 10.4

Calculate the operating point of the circuit in sketch (a). Then find its small-signal open-circuit voltage amplification, input resistance, and output resistance. Use the simplified-π transistor model of Figure 10.6. The capacitors may be regarded as short circuits at the signal frequency.

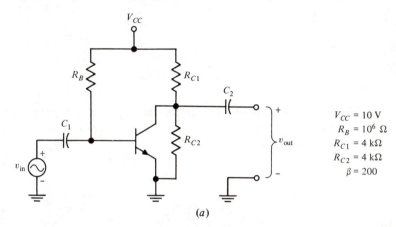

$$V_{CC} = 10 \text{ V}$$
$$R_B = 10^6 \ \Omega$$
$$R_{C1} = 4 \text{ k}\Omega$$
$$R_{C2} = 4 \text{ k}\Omega$$
$$\beta = 200$$

(a)

SOLUTION

First let us find the operating point. The capacitors C_1 and C_2 are regarded as open circuits for this (dc) part of the calculation. Writing a node equation for the node at the base terminal of the transistor

$$I_B + (V_B - V_{CC})/R_B = 0$$

Making our usual assumption that $V_{BE} = 0.7$ V, and noting that in this circuit (although not in general!) $V_B = V_{BE}$, we find

$$I_B = \frac{V_{CC} - 0.7}{R_B} = 9.3 \ \mu\text{A}$$

Since the transistor operates in the active mode, $I_C = \beta I_B$; thus $I_C = (9.3 \times 10^{-6}) \times (200) = 1.86$ mA. The collector voltage may be found by a node equation for the collector node: $I_C + (V_C - V_{CC})/R_{C1} + V_C/R_{C2} = 0$. The solution is

$$V_C = \frac{R_{C2}}{R_{C1} + R_{C2}} V_{CC} - \frac{R_{C1}R_{C2}}{R_{C1} + R_{C2}} I_C$$
$$= 5 \text{ V} - 2000 \ I_C = 1.28 \text{ V}$$

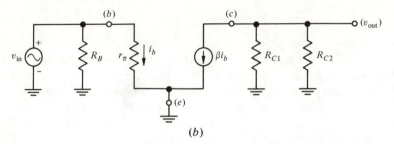

(b)

Now that I_B has been found, r_π is found by $r_\pi = kT/qI_B = 2800\ \Omega$.

After substitution of the transistor model, we obtain the small-signal circuit shown in sketch (b). This circuit is seen to be identical to that of Figure 10.10, except that instead of R_C in Figure 10.10 we now have $(R_{C1}\|R_{C2})$. Thus the calculations proceed exactly as for that circuit, and we find

$$A = -\beta(R_{C1}\|R_{C2})/r_\pi = -140$$

$$R_i = R_B\|r_\pi \cong r_\pi = 2800\ \Omega$$

$$R_o = R_{C1}\|R_{C2} = 2000\ \Omega \qquad\blacksquare$$

An Improved Biasing Scheme for the Common-Emitter Amplifier   Upon examination of the equations governing the biasing of the circuit just considered, we shall see that there is a strong dependence of I_C and V_C on the value of beta. We found that $I_B = (V_{CC} - V_{BE})/R_B$; therefore

$$I_C = (V_{CC} - V_{BE})\beta/R_B \qquad (10.29)$$

In general, for a given transistor type, beta is only specified by the manufacturer within about a factor of 2. (For example, β might be specified to lie in the range 100 to 200.) Thus unless he resorts to preselection of transistors, the circuit designer cannot be sure of I_C for this circuit to better than about a factor of 2. One generally likes the operation of a circuit to be more predictable than this. Thus it is desirable to design a circuit that is less sensitive to variations of β than the simple circuit just analyzed.

Let us consider the circuit of Figure 10.12(a). The capacitors C_i and C_o serve to isolate the biasing circuit, which consists of R_{B1}, R_{B2}, R_E, and R_C. The capacitor C_E is used to "bypass" the resistor R_E at the signal frequency, as will be shown subsequently, but may be regarded as an open circuit now, when biasing is being considered. The biasing circuit is redrawn in Figure 10.12(b). To make the analysis of the biasing circuit somewhat simpler, we can first replace the subcircuit inside the dotted line in Figure 10.12(b) by its Thévenin equivalent. The biasing circuit is redrawn in Figure 10.12(c) with the Thévenin equivalent inserted. The Thévenin resistance R_B' is found to be equal to the parallel combination of R_{B1} and R_{B2}, and the Thévenin voltage V_{BB} is found to be $V_{CC} \cdot R_{B2}/(R_{B1} + R_{B2})$.

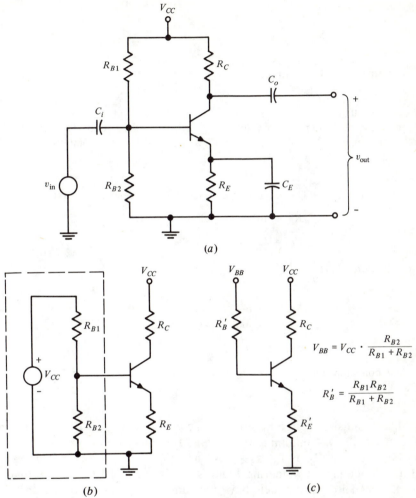

Figure 10.12   An improved biasing scheme for the common-emitter amplifier. (a) The complete circuit; (b) the biasing circuit; (c) the simplified biasing circuit.

We now find the bias point for the circuit of Figure 10.12(c). Starting at V_{BB} and going toward ground, the voltage drops must sum to V_{BB}. Thus

$$I_B R_B' + V_{BE} - I_E R_E = V_{BB} \tag{10.30}$$

We may substitute I_C/β for I_B and $-I_C/\alpha$ for I_E. Alpha may also be written as $\beta/(\beta+1)$. Making these substitutions and solving for I_C, we find that

$$I_C = \frac{(V_{BB} - V_{BE})\beta}{R_E(\beta+1) + R_B'} \tag{10.31}$$

We see that if $(\beta + 1)R_E \gg R_B'$, then Equation (10.31) gives a much reduced β dependence of I_C. In fact, if $(\beta + 1)R_E \gg R_B'$, we have

$$I_C \cong \frac{V_{BB} - V_{BE}}{R_E} \tag{10.32}$$

Thus in this limit the value of I_C is not a function of β at all!

EXAMPLE 10.5
Find the possible range of values for I_C and V_C in the circuit of Figure 10.12(a), if β is in the range 100 to 200. Let $V_{CC} = 15$ V, $R_{B1} = 1$ MΩ, $R_{B2} = 500$ kΩ, $R_C = 10$ kΩ, $R_E = 10$ kΩ.

SOLUTION
The values of the Thévenin source V_{BB} and R_B' are determined first:

$$R_B' = 10^6 \| (0.5 \times 10^6) = 3.3 \times 10^5 \; \Omega$$

$$V_{BB} = 15 \times 0.5/1.5 = 5 \text{ V}$$

From Equation (10.31)

$$I_C = \frac{(5 - 0.7)\beta}{(\beta + 1) \times 10^4 + 0.33 \times 10^6}$$

If $\beta = 100$, $I_C = 0.32$ mA

If $\beta = 200$, $I_C \cong 0.37$ mA

The collector voltage is given by $V_C = V_{CC} - I_C R_C$. Therefore it is between 11.8 and 11.3 V. ∎

The small-signal circuit model for the circuit of Figure 10.12(a) is given in Figure 10.13(a). Combining the parallel resistors and assuming that R_E is short-circuited by C_E as far as signals are concerned, the circuit of Figure 10.13(b) is obtained. Comparing Figures 10.13(b) and 10.10, we see that the two small-signal circuit models have the same form, in spite of the improvements made in the biasing. Thus the calculations previously made for A, R_i, and R_o apply to this improved circuit as well.

The Maximum Voltage Gain of a Common-Emitter Amplifier It is useful to estimate the maximum voltage gain obtainable from a common-emitter amplifier such as we have been considering, for example, that of Figure 10.9. The open-circuit voltage gain was found in Equation (10.23) to be equal to $-\beta R_C/r_\pi$. We can eliminate r_π from (10.23) by means of Equation (10.16), and replace I_B by I_C/β. The absolute value of the open-circuit voltage gain is then found to be

$$|A| = \frac{I_C R_C}{(kT/q)} \tag{10.33}$$

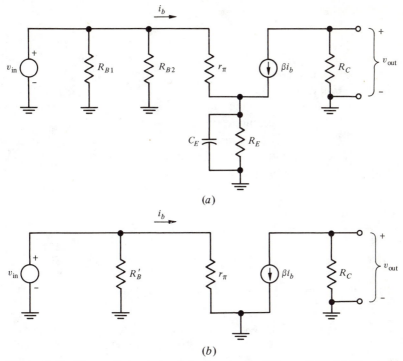

Figure 10.13   The small-signal circuit for the circuit of Figure 10.12. (*a*) The circuit model without simplification. (*b*) The simplified model obtained by combining R_{B1} and R_{B2}.

At first glance one might think it possible to make $|A|$ as large as desired by increasing I_C and/or R_C. However, it is necessary to keep the transistor operating in the active mode. We have already seen that $V_C = V_{CC} - I_C R_C$. Certainly V_C must be greater than zero, or reverse bias on the collector junction will be lost. Thus we must require that $I_C R_C < V_{CC}$. The voltage gain of a single common-emitter amplifier stage therefore is limited in accordance with the expression

$$|A_{\max}| < \frac{V_{CC}}{(kT/q)} \tag{10.34}$$

If V_{CC} is 10 V and kT/q has its room temperature value of 0.026 V, the maximum voltage gain is found to be ~400. Interestingly, this result is independent of β.

The value calculated in Equation (10.34) is an upper limit for the amplification, and not necessarily a typical value. Note that in Example 10.4, where V_{CC} was 10 V, we found $|A| = 140$, while the maximum possible with that value of V_{CC} is 400.

10.4   Multistage Common-Emitter Amplifiers

As an example of a simple analog system, let us consider a *multistage* amplifier. A multistage amplifier consists of several single-transistor amplifiers, called *stages,* connected one after another. Multistage amplifiers are used when more amplification is needed than can be obtained with a single stage. By using the simple model for the amplifier block developed in Figure 10.11, it is possible to quickly determine the overall gain for any number of stages. It will be shown that because of interaction of stages with those that precede and follow, the overall gain of several stages is in general less than the product of the open-circuit amplifications of the individual stages.

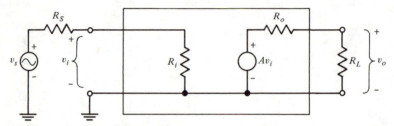

Figure 10.14   A very simple system consisting of an amplifier block, signal source, and load. The source is represented by its Thévenin equivalent (v_s, R_S) and the load is represented by the resistance R_L.

Let us first compute the voltage gain of a single stage, when its output is connected to a load resistance R_L, and it is driven by a source with finite Thévenin resistance R_S, as shown in Figure 10.14. We shall compute v_o/v_i, rather than v_o/v_s, because this quantity will be needed later when we consider the multistage amplifier. The output v_o from the voltage-divider formula equals

$$v_o = A v_i \frac{R_L}{R_L + R_o} \tag{10.35}$$

Therefore the single-stage gain, when load R_L is present, is given by

$$\frac{v_o}{v_i} = \frac{A R_L}{R_L + R_o} \tag{10.36}$$

Now let us consider several amplifier stages connected head-to-tail, as shown in Figure 10.15(*a*). Although this circuit may at first appear formidable, its operation can be easily understood by thinking of it as an assemblage of amplifier blocks. Each amplifier may be replaced by its small-signal circuit model, given in Figure 10.11. The simpler circuit of Figure 10.15(*b*) then results.

Now we can proceed to compute the overall gain of the circuit. Just what is meant by calculating the "overall gain" is not in itself clear; what

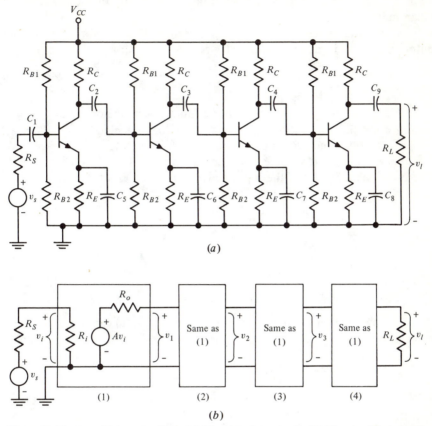

Figure 10.15 A multistage amplifier. (*a*) The complete circuit. (*b*) The simplified circuit obtained by using small-signal models to represent the amplifier blocks.

one calculates depends on what one needs to know. In this case we shall calculate the quantity v_l/v_s. In Figure 10.15(*b*) we can identify three kinds of stages: input, intermediate, and output. We compute the gain of each kind of stage separately.

Any intermediate stage has as its load R_i, the input resistance of the next stage. Therefore we can use Equation (10.36) with $R_L = R_i$. The intermediate stage gain A_{it} is

$$A_{it} = A \cdot \frac{R_i}{R_i + R_o} \tag{10.37}$$

We have already analyzed the individual amplifier stage (Figure 10.11) and have found that $A = -\beta R_C/r_\pi$, $R_i \cong r_\pi$, $R_o \cong R_C$. Thus

$$A_{it} \cong -\frac{\beta R_C}{r_\pi + R_C} \tag{10.38}$$

Clearly the maximum possible gain of an intermediate stage is β. In practice A_{it} is usually a good deal less than β, as r_π is usually of the same order of magnitude as R_C.

The output stage has as its load resistance R_L. Therefore the gain of the output stage A_{op} is given by $A_{op} = A \cdot R_L/(R_L + R_o)$, or

$$A_{op} = \frac{-\beta R_C}{r_\pi} \cdot \frac{R_L}{R_L + R_C} \tag{10.39}$$

The input stage has as its load the input resistance of the next stage. A complication occurs in this stage because we have decided to calculate v_l/v_s. The source resistance R_S and the input resistance R_i form a voltage divider that reduces the input voltage to the input stage by the fraction $R_i/(R_i + R_S)$. Therefore the gain A_{ip} of the input stage is the product of this factor and the stage gain (10.36), with $R_L = R_i$.

$$\begin{aligned} A_{ip} &= \frac{R_i}{R_i + R_S} \cdot A \cdot \frac{R_i}{R_i + R_o} \\ &= \frac{r_\pi}{r_\pi + R_S} \cdot \left(-\frac{\beta R_C}{r_\pi}\right) \cdot \frac{r_\pi}{r_\pi + R_C} \\ &= -\frac{\beta R_C r_\pi}{(r_\pi + R_S)(r_\pi + R_C)} \end{aligned} \tag{10.40}$$

The overall gain of the amplifier of Figure 10.15(b) can now be found. In terms of the intermediate voltages v_1, v_2, and v_3 indicated on the diagram

$$\begin{aligned} \frac{v_l}{v_s} &= \frac{v_l}{v_3} \cdot \frac{v_3}{v_2} \cdot \frac{v_2}{v_1} \cdot \frac{v_1}{v_s} \\ &= A_{op} \cdot A_{it} \cdot A_{it} \cdot A_{ip} \\ &= (\beta R_C)^4 \frac{R_L}{(r_\pi + R_C)^3 (r_\pi + R_S)(R_L + R_C)} \end{aligned} \tag{10.41}$$

As an example, we may take $R_C = R_L = r_\pi = R_S$. Then the result would be $v_l/v_s = \beta^4/32$. If $\beta = 100$, the overall gain for the four stages is about 3×10^6, or 130 db.

10.5   Improved Transistor Models

The function of the small-signal transistor model is to simulate the action of the real transistor in the circuit. However, it is impossible for any model to imitate the action of the transistor exactly, under all conditions. In general, one can always add to the complexity of a model to make it reflect additional aspects of the real transistor's behavior. No one transistor model is best for all circuit problems. One should use a model that is adequately refined to allow whatever problem is at hand to be solved. To use a model that is more refined than necessary just adds to the labor of computation.

The simplified-π model that has been used until now is an unrefined, elementary model. It is nonetheless quite useful, primarily for gaining a quick understanding of how a circuit works. Unfortunately, there are no general rules to indicate when this simple model is adequate. The best procedure is to begin with a more refined model; if the refinements later prove to play no important role in the circuit analysis, one can always revert to the simpler model.

One improvement in the model that is sometimes worthwhile has to do with the *Early effect* (or *basewidth-modulation effect*) already mentioned in Section 6.4. The Early effect gives rise to an increase of $|I_C|$ when the collector-junction reverse bias is increased. This effect, which is caused physically, by a variation in the effective width of the transistor's base when V_{CB} is changed, is not described by the Ebers-Moll equations. Its existence can be seen, however, by comparing the actual transistor characteristics, Figure 6.17, with the idealized output characteristics as predicted by the Ebers-Moll equations, Figure 6.15. The slight increase of I_C with increasing V_{CE} in Figure 6.17 is evidence of the Early effect.

It is evident on inspection of the simplified-π model (Figure 10.6) that it does not simulate the Early effect. In the simplified-π model the value of i_c depends entirely on i_b, and not at all on v_c. However, it is easy to modify the simplified-π model so that the action of the Early effect is simulated. Such an improved model is shown in Figure 10.16. Here the resistor r_o has been added to the simplified-π model. The effect of r_o is to cause an additional collector current v_c/r_o to flow. This current of course increases as v_c increases. Thus r_o gives representation in the model to the Early effect. The value of r_o is rather large, typically in the range 20,000 to 200,000 Ω.

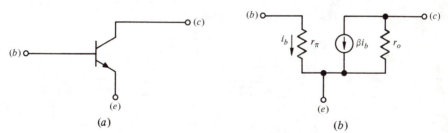

Figure 10.16 Modification of the transistor model to represent the Early effect. (*a*) Transistor; (*b*) improved model obtained by adding r_o to the simplified-π model.

EXAMPLE 10.6

Determine to what extent the voltage amplification of the circuit in sketch (*a*) is influenced by the Early effect. Assume that the capacitors are short circuits at the signal frequency.

SOLUTION

We construct the small-signal circuit model for the circuit, using the transistor model of Figure 10.16(*b*). The circuit model obtained is given in sketch (*b*). Analyzing as we have done earlier [for instance as in Equation (10.22)], we find that

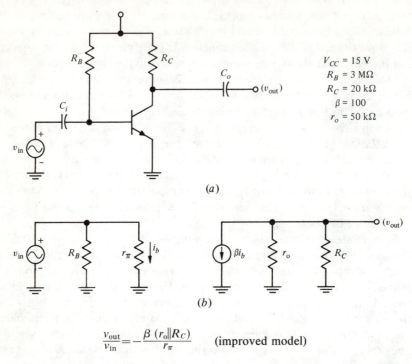

(a)

(b)

$$\frac{v_{\text{out}}}{v_{\text{in}}} = -\frac{\beta\,(r_o\|R_C)}{r_\pi} \qquad \text{(improved model)}$$

If the Early effect were neglected, the simplified-π model would be used to represent the transistor. The circuit model would be the same as above, except that r_o would be missing. The gain calculated in that case would be

$$\frac{v_{\text{out}}}{v_{\text{in}}} = -\frac{\beta\,R_C}{r_\pi} \qquad \text{(simplified-}\pi\text{ model)}$$

In our present case $R_C = 20$ kΩ and $(R_C\|r_o) = 14.3$ kΩ. Thus if one neglects the Early effect by using the simplified-π model for the calculation, one overestimates the voltage gain by about 40%.

In addition to the Early effect, there are many other aspects of transistor behavior not described by the Ebers-Moll equations. These further details have been extensively studied, and a model has been evolved that simulates all

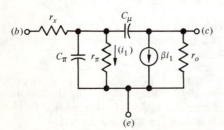

(e)

Figure 10.17 The hybrid-π transistor model.

of the most important transistor phenomena. This model, known as the *hybrid-π model*, is shown in Figure 10.17.

The most apparent difference between the hybrid-π model and the other models discussed up to now is that two capacitances are present in the hybrid-π model.[3] For dc or low-frequency signals these capacitances act as open circuits; that is, as though they were not present. This is why we have been able to neglect them until now; essentially we have been assuming that the frequency is low. Aside from the capacitances, the resemblance of the hybrid-π model to the simplified-π model is apparent. In fact, if one begins with the hybrid-π model and neglects r_x, r_o, C_π, and C_μ, the simplified-π model is obtained. The following values for the newly introduced elements may be regarded as typical: $r_x \sim 50\ \Omega$, C_π (also called C_{be}) ~ 100 pF, C_μ (also called C_{bc}) ~ 5 pF. (The abbreviation pF stands for "picofarad." One pF $= 10^{-12}$ F.) The values of the hybrid-π parameters are usually given in the manufacturer's specifications for the transistor.

The hybrid-π model is the most refined model in common use for analyzing transistor circuits. Nonetheless, the reader should remember that no model is a perfect representation of its original. In particular, even the hybrid-π model fails to be accurate when the frequency is sufficiently high. The range 100 to 500 MHz is typically the limit of its validity.

The hybrid-π model, with its capacitances, is most useful in connection with the subject of amplifier frequency response, which will be considered in the following section.

10.6 Frequency Response of Amplifier Circuits

All amplifiers exhibit variations of performance as the signal frequency is changed. Invariably there is a maximum frequency above which amplification does not occur, and depending on the design of the circuit, there may also be a lower frequency limit below which amplification disappears. The low-frequency limit, if one exists, arises from the action of capacitors intentionally placed in the circuit. The high-frequency limit, however, usually arises not from the intentional capacitors of the circuit, but rather from the limitations of the transistors themselves. These limitations stem from physical effects within the transistor, the same physical effects that give rise to the capacitances in the hybrid-π model.

In general, one is concerned with calculating the *frequency response* of an amplifier, which may be defined as the functional dependence of output

---

[3] One might think of these capacitances as arising from the proximity of emitter and base, and of base and collector. Actually, their physical origins are somewhat more complex. The new resistance r_x of the hybrid-π model comes from the resistance of the very thin base material, which besides being thin is lightly doped and hence a poor conductor.

amplitude and phase upon frequency, for all frequencies. However, it is often necessary only to know the *passband* of the amplifier, which may be defined as the range of frequencies over which maximum amplification is obtained. The passband is bounded by the *upper cutoff frequency* and by the *lower cutoff frequency* if one exists. Let us define the cutoff frequencies as the frequencies where the voltage amplification drops below its maximum value by a factor of $1/\sqrt{2}$.[4] Figure 10.18 shows a typical amplifier frequency response. It will be noted that with this definition of passband, the amplifier may possess considerable gain at frequencies outside the passband, although the gain will be lower outside the passband than within it. Other definitions of passband, of course, are possible.

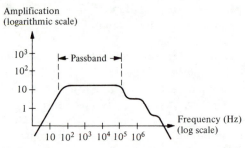

Figure 10.18 Frequency response of a typical amplifier.

We shall now discuss the estimation of amplifier frequency response. First we shall consider the effects of capacitances intentionally placed in the circuit. Then we shall go on to discuss the high-frequency limitations imposed by the transistors themselves.

Effects of Circuit Capacitances It is quite feasible to construct amplifiers which have no low-frequency limit for operation. Such amplifiers are said to be *dc-coupled,* because the stages are coupled together for all frequencies, down to zero, or dc. On the other hand, circuits sometimes contain coupling capacitors, which couple stages together for ac while isolating them for dc. Since at low frequencies these capacitors are unable to transmit signals, circuits in which they are present are limited in their low-frequency response.

As an example, let us consider the simple circuit of Figure 10.19(a), which, after insertion of the simplified-π model of the transistor becomes that of Figure 10.19(b). In this circuit the sinusoidal signal source is coupled to the transistor base through the capacitor C. The capacitor is useful because it prevents the dc transistor base-bias voltage from being affected by possible dc conduction through the source v_s. However, as the signal frequency is reduced, this capacitor begins to act as an open circuit for the signal, preventing the

---

[4] This definition leads to what may be termed the "3-db passband." This passband lies between the frequencies where the amplification is 3 db (that is, $1/\sqrt{2}$) below its maximum value.

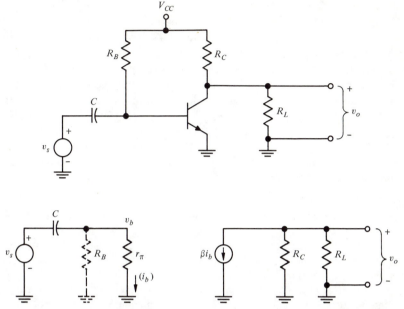

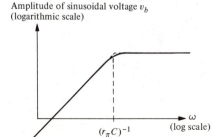

Figure 10.19 (a) A simple common-emitter amplifier circuit with limited low-frequency response. (b) Small-signal circuit obtained by substitution of the simplified-π transistor model. For simplicity, R_B is assumed large compared with r_π so that R_B may be neglected.

signal from reaching the transistor base. The frequency at which this occurs is the lower cutoff frequency of the amplifier.

The input part of Figure 10.19(b) has the same form as the RC circuit discussed in Chapter 2 in connection with Figure 2.16. (Note that the output part of the circuit, containing the dependent source, does not influence the input part.) In Chapter 2 it was shown that the voltage across the resistance of the circuit has the form given in Figure 2.17. In the present case r_π takes the place of R. The amplitude of the sinusoidal voltage appearing across r_π, which we shall call v_b, is accordingly of the same form as Figure 2.17, as shown in Figure 10.20. Inspecting the circuit, Figure 10.19(b), again, we see that $i_b = v_b/r_\pi$, and that the output voltage v_o is given by $v_o = -\beta i_b(R_C\|R_L) =$

Amplitude of sinusoidal voltage v_b
(logarithmic scale)

Figure 10.20 Amplitude of the sinusoidal voltage v_b as a function of signal frequency for the circuit of Figure 10.18. This result is derived by comparison with Figures 2.16 and 2.17.

$-\beta(R_C\|R_L)v_b/r_\pi$. Thus v_o is proportional to v_b, and a graph of v_o versus ω would have the same shape as that of Figure 10.20.

Consistent with our expectation, we see that at low frequencies the output voltage is reduced. The break frequency, at which the output is 3 db below its maximum, is given by $(r_\pi C)^{-1}$. As remarked in Chapter 2, the break frequency is simply the reciprocal of the circuit time constant $(r_\pi C)$.

For simple circuits containing only one capacitor, the frequency response can often be deduced on sight, using physical reasoning and the results obtained in Chapter 2.

EXAMPLE 10.7

Deduce the frequency response of the amplifier in sketch (a), assuming that C is the only important capacitance in the circuit.

SOLUTION

First we construct the small-signal circuit model, using the simplified-π model for the transistor. The small-signal circuit is as in sketch (b). Again for simplicity let us assume that $R_B \gg r_\pi$, so that R_B can be neglected. We may then replace the parallel combination of the signal source i_s and the resistance r_π by a Thévenin equivalent. When this has been done, the input part of the circuit appears as in sketch (c).

From physical reasoning we expect that at low frequencies C will act as an open circuit, and no current will flow through r_t. Thus v_b will equal v_t. At high frequencies we expect that C will act as a short circuit and v_b will approach zero. In fact, through the

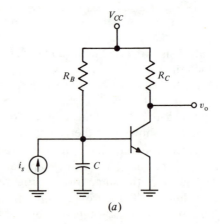

(a)

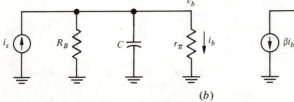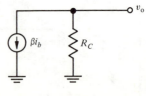

(b)

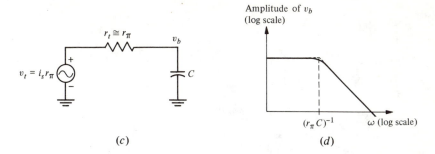

(c)   (d)

use of the Thévenin equivalent we have arrived at a circuit identical to that of Figure 2.14. Comparing, we can immediately conclude that a graph of the amplitude of v_b versus ω has the same form as Figure 2.15 [see sketch (d)].

Looking again at the complete circuit model, we see that $v_o = -\beta i_b R_C = -\beta R_C v_b / r_\pi$. Thus the output v_o is proportional to v_b. Our conclusion is that this amplifier has limited high-frequency response. Its 3-db passband lies between the frequencies zero and $(r_\pi C)^{-1}$ ∎

Use of Phasor Analysis   The phasor techniques of Chapter 4 are extremely useful for the study of amplifier frequency response. Many circuits do not yield to simple intuitive methods, either because they cannot be reduced to a familiar form or because they contain more than one capacitance or inductance. However, even the more complicated circuits can be analyzed with phasor techniques.

EXAMPLE 10.8
Determine the frequency response of the circuit given in sketch (a).

SOLUTION
We begin, as usual, by constructing the small-signal circuit model, using the simplified-π model for the transistor [see sketch (b)]. As before we have assumed

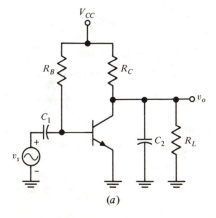

(a)

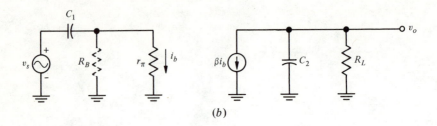

(b)

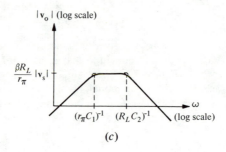

(c)

that R_B is large and may be neglected. We may now write a phasor loop equation for the loop $v_s - C_1 - r_\pi - v_s$:

$$\mathbf{v}_s - \frac{\mathbf{i}_b}{j\omega C_1} - \mathbf{i}_b\, r_\pi = 0$$

Solving, we have

$$\mathbf{i}_b = \frac{j\omega C_1 \mathbf{v}_s}{1 + j\omega C_1 r_\pi}$$

Referring to the output part of the circuit, we write a node equation for $\mathbf{v}_o$:

$$\beta\mathbf{i}_b + j\omega C_2\, \mathbf{v}_o + \frac{\mathbf{v}_o}{R_L} = 0$$

Solving, we have

$$\mathbf{v}_o = -\frac{\beta\mathbf{i}_b\, R_L}{1 + j\omega C_2\, R_L}$$

Substituting the result previously obtained for $\mathbf{i}_b$, we have

$$\mathbf{v}_o = -\frac{\beta R_L}{1 + j\omega C_2 R_L} \cdot \frac{j\omega C_1 \mathbf{v}_s}{1 + j\omega C_1 r_\pi}$$

In finding the passband, we are interested in calculating the absolute value $|\mathbf{v}_o|$ (which is the amplitude of the sinusoid v_o) as a function of frequency. The absolute value of $\mathbf{v}_o$ is

$$|\mathbf{v}_o| = \frac{\omega\beta R_L C_1\, |\mathbf{v}_s|}{\sqrt{1 + \omega^2(C_2 R_L)^2}\,\sqrt{1 + \omega^2(C_1 r_\pi)^2}}$$

Inspecting this result we see that $|v_o|$ approaches zero as ω approaches zero, and that $|v_o|$ also approaches zero as ω approaches infinity. Thus circuit is limited in both its low-frequency response and its high-frequency response. If we assume that $(C_2R_L)^{-1} \gg (C_1r_\pi)^{-1}$, a graph of $|v_o|$ versus ω appears as in sketch (c). We see that in this case the 3-db passband lies between the frequencies $(r_\pi C_1)^{-1}$ and $(R_L C_2)^{-1}$.

Looking back at the small-signal circuit model, we see that the general form of the frequency response curve could have been predicted by physical reasoning. At low frequencies C_1 becomes an open circuit, causing i_b, and hence v_o, to approach zero. Thus C_1 is responsible for the low-frequency limit. At high frequencies C_2 approaches a short circuit. The current βi_b is then diverted away from R_L by C_2; since the current flowing through R_L approaches zero, the voltage across R_L, which is v_o, must also approach zero. Thus C_2 is responsible for the high-frequency limit. ■

EXAMPLE 10.9

Determine the frequency response of the circuit in sketch (a). Assume that C_2 is very large, so that it may be regarded as a short circuit for all frequencies of interest.

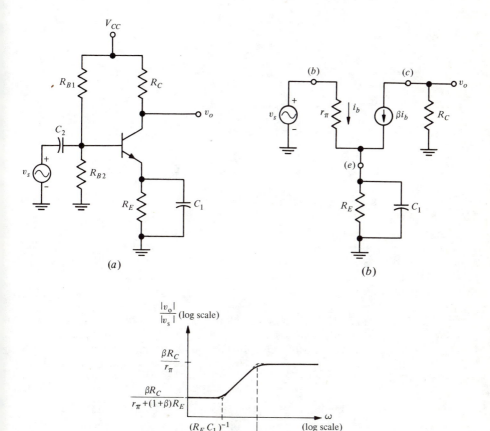

(a)

(b)

(c)

SOLUTION

This circuit is the same as that of Figure 10.12. The capacitor C_1 serves as a "bypass" for the emitter resistor R_E. Previously we analyzed the circuit under a high-frequency assumption: C_1 was regarded as a short circuit for signals. However, when the signal frequency approaches zero, C_1 will act more like an open circuit, and circuit operation will be changed in a way we shall discover.

As usual we shall begin by constructing the small-signal circuit model. Again we shall assume, for convenience, that the base-biasing resistors R_{B1} and R_{B2} are large enough to be neglected, and the coupling capacitor C_2 is regarded as a short circuit, as per instructions. The circuit model is then as in sketch (b).

Although this circuit contains only one capacitor, it is tricky, and cannot easily be reduced to a form that is familiar from Chapter 2. Thus it is best to use phasor analysis. Let us write a node equation for the node at the emitter, marked (e). We shall call the phasor voltage at that point $\mathbf{v}_e$. The node equation is

$$\frac{\mathbf{v}_s - \mathbf{v}_e}{r_\pi}(1 + \beta) - \mathbf{v}_e\left(\frac{1}{R_E} + j\omega C_1\right) = 0$$

Solving for $\mathbf{v}_e$, we have

$$\mathbf{v}_e = \mathbf{v}_s \frac{(1 + \beta)R_E}{(1 + \beta)R_E + r_\pi + j\omega C_1 r_\pi R_E}$$

To obtain $\mathbf{v}_o$, we observe that $\mathbf{v}_o = -\beta \mathbf{i}_b R_C$, and that $\mathbf{i}_b = (\mathbf{v}_s - \mathbf{v}_e)/r_\pi$. Thus

$$\mathbf{v}_o = -\beta R_C \mathbf{i}_b = -\frac{\beta R_C}{r_\pi}(\mathbf{v}_s - \mathbf{v}_e)$$

$$= -\beta R_C \mathbf{v}_s \bullet \frac{1 + j\omega C_1 R_E}{[r_\pi + (1 + \beta)R_E] + j\omega C_1 r_\pi R_E}$$

Inspecting this result, we see that in the limit $\omega \to \infty$, $\mathbf{v}_o \to \mathbf{v}_s \bullet (-\beta R_C/r_\pi)$. This is in agreement with the result previously obtained for this circuit under the assumption that C_1 acted as a short circuit. However, as $\omega \to 0$, $\mathbf{v}_o \to \mathbf{v}_s \bullet [-\beta R_C/\{r_\pi + (1 + \beta)R_E\}]$, a smaller value. Thus we see why the bypass capacitor is used in the circuit. If it were absent, we would have the same situation as exists at low frequencies, where the bypassing action of the capacitor is ineffective. In this case we have found that less amplification is obtained than when the capacitor acts as a short circuit to bypass R_E.

Sketch (c) is a graph of $|\mathbf{v}_o|$ versus ω. Note that in this more complicated circuit, the lower limit of the bandpass is not given simply by $(R_E C_1)^{-1}$, but by the more complicated expression indicated on the graph. The simple technique of estimating RC time constants can be used only after the circuit has been reduced to a known simple form as in Figure 10.19(b) or Example 10.8.   ∎

High-Frequency Limitations of Transistors   In the foregoing paragraphs we have considered the effects of circuit capacitors on amplifier frequency response. Now we shall consider limitations imposed on frequency response by the transistors themselves. If there are no capacitors in the circuit that act to limit high-frequency response, a high-frequency limit will always be imposed by physical effects internal to the transistors. It must be understood that at high frequencies, the simplified-π model no longer correctly describes transistor operation.

Manufacturer's data sheets usually contain information intended to indicate the high-frequency performance of transistors. Unfortunately, the value of a circuit's upper cutoff frequency depends on the form of the circuit as well as on the transistor itself. Therefore upper cutoff frequency is not really a quantity that can be specified by the transistor manufacturer. It is true that on a transistor specification sheet, one may find a quantity called "high-frequency short-circuit current gain," with the symbol "$|h_{fe}|$".[5] For example, a data sheet may say "at $f = 100$ MHz, $|h_{fe}| = 3$." This seems to imply that at 100 MHz, the voltage gain of an amplifier would drop to $3/\beta$ of its low-frequency value. However, such an estimate would often be in error, because the cutoff frequency depends on the circuit as well as on the transistor. For example, in the case of a common-emitter circuit (see Figure 10.9) it can be shown that an estimate based only on $|h_{fe}|$ is accurate only in the limit $\beta(R_C \| R_L)/r_\pi \ll 1$. This inequality is seldom satisfied in practice; when it is not, the upper cutoff frequency is actually lower than one would estimate from $|h_{fe}|$. Another transistor specification that is sometimes given is the "beta-cutoff frequency," f_β, defined as the frequency at which the short-circuit current gain β is reduced by a factor $1/\sqrt{2}$. As with $|h_{fe}|$, there is in general no way to state the upper cutoff frequency of a circuit just from a knowledge of f_β.

The best method, in general, for determining the high-frequency cutoff of a circuit is to use a model more refined than the simplified-π model. The hybrid-π model, introduced in Section 10.5 and shown in Figure 10.17, is most often used for this purpose. It contains several additional circuit elements, and is intended to simulate the action of the transistor at high signal frequencies better than the simplified-π model can do. The parameters of the hybrid-π model are specified in transistor makers' data sheets.

As an example of the use of the hybrid-π model, we shall determine the frequency response of a simple common-emitter amplifier. Let us consider the circuit of Figure 10.21(a). The small-signal model of this circuit, obtained using the hybrid-π transistor model, is then as shown in Figure 10.21(b). We have assumed that $r_o \gg R_C$ and that $R_B \gg r_\pi$, and accordingly have neglected r_o and R_B. The series combination of r_s and r_x has been designated as r_s'.

We can now proceed to analyze Figure 10.21(b) by writing phasor node equations for node x, where the voltage will be called v_x, and the output node. Those equations are

$$\frac{\mathbf{v}_x - \mathbf{v}_s}{r_s'} + \frac{\mathbf{v}_x}{r_\pi} + j\omega C_\pi \mathbf{v}_x + (\mathbf{v}_x - \mathbf{v}_o)j\omega C_\mu = 0 \tag{10.42}$$

---

[5] Manufacturer's data sheets usually use the symbol h_{fe} to stand for $\partial i_C/\partial i_b$ and h_{FE} to stand for I_C/I_B. In this book we have chosen to neglect the rather unimportant difference between the two, and refer to both quantities as β. See Section 6.4.

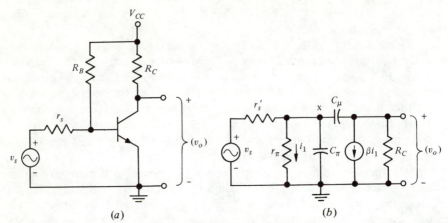

Figure 10.21 Circuit for considering the high-frequency response of a common-emitter amplifier. (*a*) Common-emitter circuit. (*b*) Small-signal circuit obtained with the hybrid-π model. We have defined $r'_s = r_s + r_x$, and assumed that $R_C \ll r_o$, so that r_o may be neglected. We have also assumed that $R_B \gg r_\pi$, so that R_B can be neglected.

and

$$(\mathbf{v}_o - \mathbf{v}_x)j\omega C_\mu + \frac{\beta \mathbf{v}_x}{r_\pi} + \frac{\mathbf{v}_o}{R_C} = 0 \tag{10.43}$$

Solving (10.42) and (10.43) simultaneously, we obtain the formidable result

$$\mathbf{v}_o = \frac{-(\beta - j\omega C_\mu r_\pi) R_C \mathbf{v}_s}{r_\pi + r'_s - \omega^2 C_\pi C_\mu R_C \, r_\pi r'_s + j\omega\left\{ r_\pi r'_s\left[C_\pi + C_\mu + C_\mu R_C\left(\frac{1}{r'_s} + \frac{\beta}{r_\pi}\right)\right]\right\}} \tag{10.44}$$

Equation (10.44) contains the desired information concerning frequency response; one could graph $|\mathbf{v}_o/\mathbf{v}_s|$ as a function of ω and find the upper cutoff frequency. However, it has been found that this complicated result can be simplified if some assumptions are made. Let us assume that $\omega C_\mu r_\pi \ll \beta$. This will allow us to neglect the second term in the numerator compared to the first. Let us also assume that $\omega C_\mu R_C \ll 1$. This will allow us to neglect the term in the denominator containing ω^2. (The assumption makes this term negligible compared with $j\omega r_\pi r'_s C_\mu$.) Later we shall check the validity of the two assumptions. As usual, we can also use $(\beta + 1) \simeq \beta$. The simplified form obtained through these assumptions is

$$\mathbf{v}_o \cong -\beta R_C \, \mathbf{v}_s \cdot \frac{1}{r_\pi + r'_s + j\omega r_\pi r'_s \, C_T} \tag{10.45}$$

where by definition

$$C_T \equiv C_\pi + C_\mu\left(1 + \frac{R_C}{r'_s} + \frac{\beta R_C}{r_\pi}\right) \tag{10.46}$$

It is now very easy to find the upper cutoff frequency. The absolute value of v_o obeys the following proportionality:

$$|v_o| \propto \frac{1}{\sqrt{(r_\pi + r_s)^2 + (\omega r_\pi r_s' C_T)^2}} \tag{10.47}$$

When ω increases enough for the second term in the radical to equal the first, $|v_o|$ will have $1/\sqrt{2}$ of its value when $\omega = 0$. Thus the upper 3-db cutoff frequency ω_{uco} is given by

$$(r_\pi + r_s) = \omega_{uco} \, r_\pi r_s' C_T \tag{10.48}$$

from which we have

$$\omega_{uco} = \frac{r_\pi + r_s'}{r_\pi r_s' C_T} = \frac{1}{(r_\pi \| r_s') \, C_T} \tag{10.49}$$

As typical values, let us take $\beta = 100$, $r_\pi = 2500 \ \Omega$, $r_s' = 50 \ \Omega$, $R_C = 5000 \ \Omega$, $C_\pi = 100$ pF, $C_\mu = 5$ pF. Then from (10.46), $C_T = 1.6 \times 10^{-9}$ F and $\omega_{uco} = 1.25 \times 10^7$ radians/sec. In terms of ordinary frequency $f_{uco} = \omega_{uco}/2\pi = 2 \times 10^6$ Hz $= 2$ MHz. (This frequency is just above the top of the am radio broadcast band.) We may now verify that the two assumptions made earlier were valid. The first, that $\omega \ll \beta/r_\pi C_\mu$, required that $\omega \ll 8 \times 10^9$; the second, that $\omega \ll (R_C C_\mu)^{-1}$, requires that $\omega \ll 4 \times 10^7$. Both assumptions are satisfied up to the cutoff frequency calculated in (10.49), so that the result may be taken as valid. Above $\omega = 4 \times 10^7$, one of the assumptions fails, and (10.45) will not be accurate. However, since by that time $\omega > \omega_{uco}$, the amplifier gain has already dropped more than 3 db below its passband value.

Let us look again at Figure 10.21(b). Suppose that C_μ were not present. Then the Thévenin resistance of the subcircuit in parallel with C_π would be $(r_s' \| r_\pi)$, and we would expect the upper cutoff frequency to be $[(r_s' \| r_\pi) C_\pi]^{-1}$. This is almost the result obtained in (10.49), but there is an important difference: C_T appears in (10.49) instead of C_π. It is a fact that C_π is always much larger than C_μ, but if the reader will substitute the typical values given above into (10.46) he will see that the contribution of C_μ to C_T is much larger than that of C_π. In other words, the frequency response of the circuit is dominated by C_μ, even though it is the smaller of the two capacitances. Physically this can be explained as follows. If a certain time-varying voltage appears at x in Figure 10.21(b), it causes a current to flow through C_π. However, the voltage at x is amplified and causes a much larger v_o to appear at the output terminal. Therefore a very large time-varying voltage $(v_o - v_x)$ appears across C_μ. This large voltage causes a large current to flow through C_μ, larger than the current through C_π. Thus C_μ is more significant in the circuit's operation than C_π. This effect, whereby the importance of a small capacitance is magnified because it is connected between input and output, is known as the *Miller effect*.

EXAMPLE 10.10

Let us consider the circuit of Figure 10.21(a) with $r_s \gg r_\pi$. Evaluate the upper cutoff frequency f_{uco} for the case $\beta R_C/r_\pi \ll 1$. Compare with the result for the case $\beta R_C/r_\pi \gg 1$. The assumptions made in the preceding discussion may be used.

SOLUTION

With $r_s \gg r_\pi$ we have, from (10.49), $\omega_{uco} = (r_\pi C_T)^{-1}$. For the case $\beta R_C/r_\pi \ll 1$, Equation (10.46) reduces to $C_T = C_\pi + C_\mu$. Thus for this case $\omega_{uco} = [r_\pi (C_\pi + C_\mu)]^{-1}$ and $f_{uco} = 1/2\pi r_\pi (C_\pi + C_\mu)$.

The upper cutoff frequency just obtained is the quantity called the "beta cutoff frequency," f_β, in data sheets. It does not represent a typical voltage amplifier because $\beta R_C/r_\pi \ll 1$ is unlikely to be satisfied for a typical circuit.

The situation to be expected in a typical amplifier is $\beta R_C/r_\pi \gg 1$. In this case we find

$$\omega_{uco} \cong \frac{1}{r_\pi [C_\pi + \beta R_C C_\mu/r_\pi]}$$

Thus the beta cutoff frequency overestimates the cutoff frequency for the other, more common case by the factor

$$\frac{C_\pi + \beta R_C C_\mu/r_\pi}{C_\pi + C_\mu} \cong 1 + \frac{\beta R_C C_\mu}{r_\pi C_\pi} \qquad \blacksquare$$

Summary

- Amplifiers are circuits intended to produce an output voltage or current that is larger than, but proportional to, an input voltage or current.

- When used in amplifiers, transistors are operated in the active mode.

- In amplifier circuits, time-varying signal currents are usually superimposed upon dc biasing currents. The biasing currents serve to place the transistor, in the absence of signals, at a desired point of its I-V characteristics. This point is called the operating point.

- Small variations of voltage and current about the operating point are known as small-signal voltages and currents. The relationships between small-signal variables can be shown to be linear.

- A transistor model is a collection of ideal linear circuit elements designed to imitate the relationships between the transistor small-signal variables. It has the same number of terminals as the transistor.

- To analyze an amplifier circuit, the transistor model may be substituted for the transistor in the circuit. This produces a circuit model containing only ideal linear circuit elements, which can then be analyzed using conventional techniques. The circuit model is a small-signal model; that is, it represents relationships between small-signal variables. When constructing the circuit model, dc voltage sources in the original circuit are replaced by short circuits, and dc current sources are replaced by open circuits.

- Amplifier circuits are conveniently treated as building blocks when analyzing larger systems. The amplifier block may be represented by a simple small-signal model. The parameters of this model are the input resistance R_i, the output resistance R_o, and the open-circuit voltage amplification A. The values of these parameters are found by analysis of the circuit inside the block.

- A multistage amplifier is a system obtained by connecting several amplifier blocks in sequence. The individual amplifier blocks are called stages.

- Transistor models of any degree of refinement are possible. A simple model, which we call the simplified-π model, is adequate for most calculations, provided the frequency is not too high. The most detailed transistor model in common use is known as the hybrid-π model.

- The subject of amplifier frequency response has to do with the behavior of an amplifier as a function of signal frequency. All amplifiers have an upper frequency limit for satisfactory operation, and some have a lower frequency limit as well. Limits on the frequency response of an amplifier are imposed either by capacitances in the circuit or by effects internal to the transistors. Circuit capacitances can act to limit either low-frequency or high-frequency response. If high-frequency response is not limited by circuit capacitances, it will be limited by the transistors themselves.

- The action of transistors at high frequencies is best analyzed through the use of the hybrid-π transistor model. Phasor techniques are a powerful tool in the analysis of frequency response.

References

References on approximately the level of this book include:

Durling, A. E. *An Introduction to Electrical Engineering.* New York: The Macmillan Company, 1969.

Brophy, J. J. *Basic Electronics for Scientists.* New York: McGraw-Hill, 1966.

Somewhat more advanced treatments are found in:

Pederson, D. O., J. J. Struder, and J. R. Whinnery. *Introduction to Electronic Systems, Circuits, and Devices.* New York: McGraw-Hill, 1966.

Gray, P. E., D. DeWitt, A. R. Boothroyd, and J. F. Gibbons. *Physical Electronics and Circuit Models of Transistors.* Semiconductor Electronics Education Committee Series, Vol. 2. New York: John Wiley & Sons, 1964.

Searle, C. L., A. R. Boothroyd, E. J. Angelo, Jr., P. E. Gray, and D. O. Pederson. *Elementary Circuit Properties of Transistors.* Semiconductor Electronics Education Committee Series, Vol. 3. New York: John Wiley & Sons, 1964.

Advanced treatments include:

Angelo, E. J., Jr. *Electronics: BJT's, FET's, and Microcircuits.* New York: McGraw-Hill, 1969.

Gray, P. E., and C. L. Searle. *Electronic Principles: Physics, Models and Circuits.* New York: John Wiley & Sons, 1969.

Problems

10.1   Suppose that the *I-V* relationship for a circuit element is $i_E = 10^{-3} v_E + 2 \times 10^{-4} v_E^2$, where i_E is in amperes and v_E in volts. Let v_E be changed from 1.0 to 1.1 V. What is the change in i_E?

10.2   For the case of Problem 10.1, let $i_E = I_E + i_e$, $v_E = V_E + v_e$, where I_E and V_E are constants and i_e and v_e are small-signal deviations about the operating point. Show that for small deviations, $i_e \cong K v_e$. Evaluate the constant K, (1) for $V_E = 1$ V, (2) for $V_E = 10$ V.

10.3   For a certain *pn* junction diode, the saturation current I_S is 10^{-14} A. Calculate the incremental forward resistance of the diode if the operating point is at $I = 2$ mA. [The incremental forward resistance, by definition, is $(di/dv)^{-1}$]. Construct a small-signal model of the diode. Does this small-signal model give any information about the relationship of *total* voltage to *total* current?

10.4   For a transistor operating in the active mode, (1) write an expression for the total emitter current in terms of the base current; (2) write an expression for the small-signal emitter current in terms of the small-signal base current.

$$\text{ANSWER: (1) } i_E = -\frac{i_B}{1-\alpha}; \text{ (2) } i_e = -\frac{i_b}{1-\alpha}$$

10.5   Find the relationship between small-signal emitter current and small-signal base current, as predicted by the simplified-π model. Show that this relationship agrees with the result of Problem 10.4 (2).

10.6   Construct a small-signal model for the network shown in Figure 10.22. Assume the capacitor is a short circuit at the signal frequency.

10.7   In the circuit of Figure 10.23, $V_{CC} = 14$ V, $\beta = 100$. It is desired to bias the transistor to $I_B = 30$ μA, $V_C = 6$ V. Find R_B and R_C.

10.8   In the circuit of Figure 10.24, let $I_B = 10$ μA, $V_{CC} = 10$ V, $\beta = 100$. What should R_C be, so that $V_C = 5$ V?

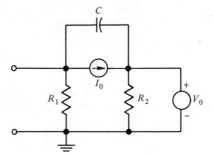

Figure 10.22

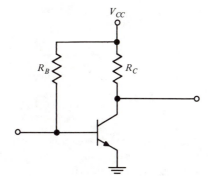

Figure 10.23

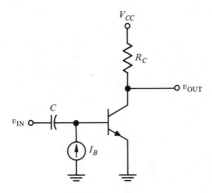

Figure 10.24

10.9   Construct a small-signal model for the circuit shown in Figure 10.24. Use the simplified-π model for the transistor. Assume that the capacitor is a short circuit at the signal frequency.

10.10   Calculate the small-signal voltage gain of the circuit of Figure 10.24. Use the simplified-π model. Assume that capacitors are short circuits for signals.

10.11   Find the small-signal input and output resistances of the circuit shown in Figure 10.24. Use the simplified-π model. Assume that capacitors are short circuits for signals.

10.12   Design a circuit similar to that of Figure 10.24 in which a *pnp* transistor is used. Be sure that the polarities of bias voltages and currents are correct. Use the simplified-π model. Assume that capacitors are short circuits for signals.

10.13   Construct a small-signal model for the *pnp* circuit obtained in Problem 10.12. Use the simplified-π model. Assume that capacitors are short circuits for signals.

10.14   Show that the simplified-π model may be represented in the form shown in Figure 10.25. Evaluate the constant g_m in terms of the parameters β and r_π. What are the units of g_m? This quantity is called the *transconductance*.

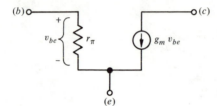

(b)

v_{be} r_π $g_m\, v_{be}$

(c)

(e)

Figure 10.25

10.15   In the circuit of Figure 10.23 let $V_{CC} = 10$ V. Find R_B and R_C, so that at the operating point $I_C = 1$ mA, $V_C = 5$ V, $\beta = 50$.

10.16   The circuit of Figure 10.26 is known as a *common-base* amplifier.
   (1) Explain why the term "common-base" is used.
   (2) Construct the small-signal circuit model, using the simplified-π model for the transistor.
   (3) Calculate the small-signal voltage amplification, using the circuit model of part (2).

10.17   Find the input resistance and output resistance for the common-base amplifier of Figure 10.26. In calculating the input resistance, assume that the output terminals are open-circuited. When finding the output resistance, assume that the input is being driven by an ideal voltage source.

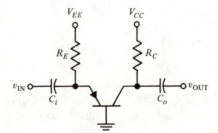

V_{EE} V_{CC}

R_E R_C

v_{IN} C_i C_o v_{OUT}

Figure 10.26

10.18   For the circuit of Figure 10.26, (1) find the input resistance when a load resistance R_L is connected between the output terminal and ground. Does R_i depend on the value of R_L? (2) Find the output resistance, when the input is being driven by an ideal voltage source in series with a source resistance R_S. Does R_o depend on the value of R_S? (3) Repeat part (1), using the transistor model of Figure 10.16.

10.19   Consider the circuit of Figure 10.27.
  (1) Show that the small-signal amplification cannot be found using the simpli-fied-π model.
  (2) Find the small-signal amplification using the more refined model of Figure 10.16, in which the basewidth modulation effect is included.

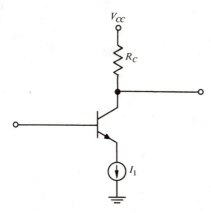

V_{CC}

R_C

I_1

Figure 10.27

10.20   Consider the biasing circuit of Figure 10.12. In this problem we shall obtain a quantitative estimate of the stability of I_C with respect to changes in β. Let us define $(dI_C/I_C) \,/\, (d\beta/\beta)$ as the *desensitivity parameter*. Find the value of the desensitivity parameter if $V_{CC} = 15$ V, $R_E = 1000$ Ω, $R_{B1} = 10$ kΩ, $R_{B2} = 5$ kΩ, $\beta = 100$, $R_C = 1500$ Ω. What is V_C?

10.21   Consider a circuit like that of Figure 10.12, except without R_{B2} (that is, $R_{B2} = \infty$). Let $V_{CC} = 15$ V, $R_E = 1000$ Ω, $R_{B1} = 10$ kΩ, $R_C = 1500$ Ω, $\beta = 100$.
  (1) Estimate V_C.
  (2) If R_{B2} were eliminated (that is, set equal to ∞), what would happen to the operating point?

10.22   Estimate the lower cutoff frequency for the circuit of Figure 10.24. Assume that $I_B = 10$ μA, $V_{CC} = 10$ V, $\beta = 100$. The v_{IN} terminal is driven by an ideal voltage source. Let $C = .01$ μF.

10.23   Estimate the lower cutoff frequency of the common-base circuit of Figure 10.26. Assume that no external load is connected to the output terminal, and that the input is driven by an ideal voltage source. *Suggestion:* Note that since no current flows through it, C_o has no effect. Find the RC time constant asso-ciated with C_i by obtaining the Thévenin equivalent of the subcircuit in parallel with C_i.

10.24   Solve Problem 10.23 using phasor analysis.

10.25   In the circuit of Figure 10.26, assume that a load resistance R_L is connected between the output terminal and ground. The input is driven by an ideal sinu-soidal voltage source. Use phasor analysis to obtain an expression for v_{OUT}. Assume that the frequencies of interest are low, so that the simplified-π model is accurate.

10.26   Using the results obtained in the text with the hybrid-π model, estimate the upper cutoff frequency for the circuit of Figure 10.24. Assume that $I_B = 15\ \mu\text{A}$, $V_{CC} = 10$ V, $\beta = 100$, $R_C = 3$ kΩ, $r_o = 50$ kΩ, $C_\pi = 100$ pF, $C_\mu = 5$ pF, $r_s = 1000\ \Omega$, $r_x = 50\ \Omega$.

10.27   The circuit of Figure 10.26 is driven by an ideal sinusoidal voltage source, and a load R_L is connected between the output terminal and ground. Use the hybrid-π model and phasor analysis to calculate $v_{\text{OUT}}/v_{\text{IN}}$ as a function of ω at high frequencies. Assume that C_i and C_o can be regarded as short circuits at the high frequencies of interest. Graph your result. At what frequency is the amplification reduced 3 db below its value in the passband?

Chapter 11 ▅▅▅▅▅▅▅▅▅▅▅▅▅

Design Techniques for Linear Circuits

In the previous chapter some basic features of amplifier design were considered. In this chapter we shall deal in greater depth with the subject of amplifier design.

There are many different kinds of amplifiers, intended for various purposes. The design of an amplifier depends, of course, on the use for which it is intended. Thus before proceeding to specifics, let us see what considerations will influence the choice of design.

Since the amplifier is in most cases destined for use as a building-block in a system, the parameters that describe the block are usually of importance. If the amplifier is of the general type discussed in Chapter 3, the block is characterized by its input resistance, output resistance, and open-circuit voltage amplification. In general, these parameters will have to have appropriate values, in order for the block to be suited to its eventual use. For example, in many cases a low output resistance is desirable. This is because a low output resistance allows more output current to be supplied to a load, without loss of output voltage; in other words, it allows more power to be transferred to the load. The input resistance may also be an important consideration. For instance, in many cases a large input resistance is needed. This is because if the input resistance is high, little current enters the input terminals and little power is required from the signal source. In addition to the three parameters, R_i, R_o, and A, the usefulness of the block is also affected by its electrical power consumption and of course its cost.

Linearity   Another consideration in amplifier design has to do with *linearity*. An amplifier is said to be *linear* if input voltage results in a linearly proportional change in the output voltage. In other words, an amplifier is said to be highly linear if dv_{OUT}/dv_{IN} is a constant. A graph v_{OUT} versus v_{IN} for an ideally linear amplifier is shown in Figure 11.1(*a*). Often one pretends, for simplicity in systems analysis, that an amplifier is ideally linear; this assumption is inherent, for instance, in all the amplifier models of Chapter 3, because dv_{OUT}/dv_{IN} is assumed to be equal to the constant A. Perfect linearity would indeed be very desirable in most applications. But perfect linearity can never be achieved; some deviations from linearity exist in any real amplifier. On the other hand, there are also more uses than one might expect from amplifiers that are not particularly linear. For instance, in an op-amp some nonlinearity is tolerable.

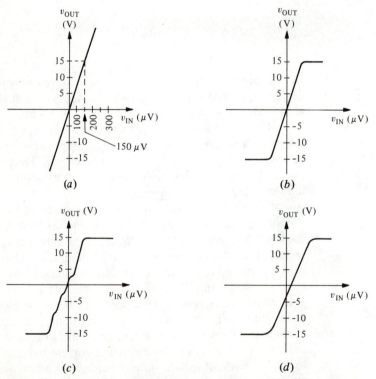

Figure 11.1   Input-output characteristics of amplifiers. (*a*) Idealized characteristic of a perfect amplifier. *A* is assumed to equal 10^5. Note expanded horizontal scale. (*b*) Modified input-out characteristics showing saturation of output voltage. A supply voltage of ± 15 V is assumed. (*c*) Characteristic showing nonlinearity in operating range of amplifier. (*d*) Characteristic of a linear amplifier with imbalance present.

One kind of nonlinearity that is always present arises from *saturation*. The output voltage produced by an amplifier cannot, as a rule, exceed the power supply voltage. Thus if the input voltage is sufficiently increased, the output voltage eventually reaches a maximum value and levels off. This is what is meant by the term "saturation." The effect is depicted graphically in Figure 11.1(*b*). Saturation will not influence operation of the amplifier, or cause distortion, if the signals are kept sufficiently small. For comparison, a more severe kind of nonlinearity is shown in Figure 11.1(*c*). If not compensated for in some way, this nonlinearity will cause distortion of the input signal.

Balance   One might expect that in a perfect amplifier, v_{OUT} would be exactly equal to zero when v_{IN} was zero. However, practical amplifiers, especially dc-coupled ones, often fall short of perfection in this respect as well. An amplifier in which $v_{OUT} \neq 0$ when $v_{IN} = 0$ is said to suffer from *imbalance* or *offset*. A case of offset is shown in Figure 11.1(*d*).

It should be noted that, except for the inevitable saturation, the behavior shown in Figure 11.1(*d*) is linear, because, aside from saturation, dv_{OUT}/dv_{IN} is a constant. Moreover, offset will not cause distortion of an input waveform, but will only add a constant to it. Nonetheless, in some applications, offset may be a matter of concern. For example, if offset is present in the amplifying circuits of an electronic voltmeter, it causes a non-zero reading on the voltmeter dial, even when nothing is connected to the voltmeter input terminals.

Suitability for Construction   The choice of a circuit design is determined, not only by the desired operating characteristics, but also by the manner of construction that is intended. Until a few years ago, every circuit was constructed of *discrete* components. That is, the resistors, capacitors, transistors, and so forth were separate physical components that had to be connected with wires to make the circuit. Recently, however, *monolithic*, or *integrated*, circuits have come into prominence. "IC's," as they are known, are ensembles of components all built into a single chip of silicon crystal. A circuit meant to be constructed in IC form is subject to special design requirements. This subject has already been considered in Chapter 7. However, it may be useful to review the more important considerations of IC design.

Transistors: As a rule it is possible only to build good-quality bipolar transistors of one type into an IC. That is, if transistors with current gains (β) in the usual range are desired, all must be either *pnp* or *npn*. Usually they are *npn* for constructional reasons. It is then possible to include a *pnp* transistor in the circuit, but (except in exceptional circumstances) its performance will be poor.

Resistances: Resistors of high quality generally must be confined to the range 10 to 100,000 Ω; values outside this range are uneconomical or (if made by alternative techniques) difficult to control in value. The cost of a typical IC resistor is usually proportional to its resistance. The cost of a 10,000-Ω resistance is about the same as that of a transistor. Thus situations arise where it is economical to substitute one or more transistors for a single large resistor, when the transistors can be made to serve the same function.

Capacitors: Values greater than 10 pF require so much space as to be undesirable in integrated circuits, and values greater than 200 pF are almost entirely impractical. Thus a major objective of IC design is the elimination of capacitors.

Inductors: There is no good way to build inductors into integrated circuits. Thus any circuit intended for IC realization cannot contain inductors.

Tolerances and component reproducibility: Due to technical complexities in IC construction, control of component values and transistor parameters may be somewhat poorer than with discrete components. However, there is a compensating benefit in that adjacent components in an IC can be made nearly identical to one another (even though the *absolute* values of their parameters cannot be specified precisely). Furthermore, being close together, adjacent components of the IC will be at the same temperature, and so will remain identical even as the temperature of the whole IC changes. This behavior, known as *tracking,* turns out to be very useful in stabilizing integrated circuits against changes in temperature.

The design restrictions on integrated circuits sometimes make these circuits more complicated than the corresponding discrete circuit design would be. However, because of the way IC's can be mass produced, the more complex IC circuit may still be cheaper to make. For example, one amplifier, which originally contained six discrete transistors as well as other individual circuit elements, was redesigned as an IC. Because of the special design considerations of integrated circuits, the redesigned IC version contained 15 transistors instead of the original six; nonetheless the sale price of the IC was less than one-fifth of what the old discrete version used to cost.

It is not possible, in this one chapter, to discuss designs for all the different kinds of amplifiers. Thus we have chosen to deal particularly with those circuits that are used in operational amplifiers. This choice seems appropriate because of the great importance of op-amps in current practice, and also because the circuits found in op-amps are representative of amplifiers generally. The specific design requirements on op-amps include (1) high input resistance — greater than 10,000 Ω; (2) low output resistance — less than 500 Ω; (3) large voltage amplification — greater than 10^4. A certain amount of nonlinearity and offset can be tolerated; these deficiencies can be compensated for through the use of suitable external circuits, as will be explained in Chapter 12.

Some of the circuits to be considered in this chapter are conventional and have been used in discrete circuits for many years. But circuits that suit the needs of integrated construction, and take advantage of its strong points, also are included.

To achieve the various design requirements, a complete amplifier must usually be a multistage circuit. The most convenient way to an understanding of the complete circuit is to discuss its operation one stage at a time. In Section 11.1 we shall discuss the first, or input, stage of the amplifier. The problem of obtaining a high input resistance must be considered here. Moreover, the first stage of an op-amp is usually of the kind known as a *differential amplifier*, which generates an amplified signal proportional to the difference of two input signals. Hence we shall concentrate on differential amplifiers in Section 11.1. Section 11.2 will be concerned with the last, or output, stage; here the problems have to do with reducing the output resistance and power consumption. In Section 11.3 we shall consider how the first, last, and intermediate stages can be connected together. Then in Section 11.4, the stages are assembled, and complete amplifier circuits are presented and discussed.

11.1 Differential Amplifiers

The function of a differential amplifier stage is to provide an output v_{OUT} of the form

$$v_{OUT} = C + D(v_{(+)} - v_{(-)}) \tag{11.1}$$

where $v_{(+)}$ and $v_{(-)}$ are two input voltages to the stage. From the form of Equation (11.1), it is seen that the circuit must treat the two inputs on exactly an equal basis, except that one contributes to the output voltage with a positive sign and the other with a negative sign. A circuit that gave an output of the form $v_{OUT} = C + Dv_{(+)} - Ev_{(-)}$, for example, would not be a differential amplifier in the sense of this discussion. To insure this equal treatment of the two inputs, a symmetrical two-sided design is customarily used. Such a design also helps eliminate changes in output voltage due to changes in temperature.

A typical differential amplifier circuit is shown in Figure 11.2. There are two input terminals, called the "inverting" and "non-inverting" inputs, where

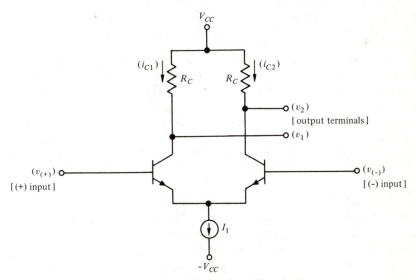

Figure 11.2 A differential amplifier circuit.

the voltages are $v_{(-)}$ and $v_{(+)}$, respectively, and two output terminals, at which the voltages are designated v_1 and v_2, respectively. The two sides of the circuit are constructed to be as nearly identical as possible. (This is easiest to do in an IC, where the two transistors can be constructed very close together. Not only are the transistors then nearly identical, they are also always at nearly the same temperature.) The transistors are biased in the active region, as is usually the case in analog circuits. However, in this circuit a new biasing technique is used; a dc current source applies bias current to the two emitters. Construction of the current source will be considered later in this section.

We note that the dc current from the current source divides between the two transistors into currents i_{C1} and i_{C2}. If $v_{(+)} = v_{(-)}$, the two sides of the circuit are symmetrical, and $i_{C1} = i_{C2} = I_1/2$. Thus $v_1 = v_2 = V_{CC} - I_1R_c/2$, so long as $v_{(+)} = v_{(-)}$. It is clear that any other change that affects both transistors equally, such as a change in temperature, will also preserve the symmetry: the currents still divide evenly and the output voltages are unchanged. On the other hand, if $v_{(+)} \neq v_{(-)}$, the currents will no longer divide evenly. Their sum still must be I_1, however, so we see that if v_1 increases, v_2 must decrease by the same amount.

EXAMPLE 11.1

Can the circuit of Figure 11.2 operate properly if signal sources are connected to the input terminals through series capacitors?

SOLUTION

If the circuit of Figure 11.2 is to operate in the active mode, the dc biasing currents must obey $I_B = I_C/\beta \cong I_1/2\beta$. A capacitor placed in series with the input terminals would block the dc base current from flowing. In that case instead of being forward biased, the emitter-base junction would be biased at $I_B = 0$. This would cause the emitter current to approach zero. The transistor would then not be biased in the active mode and the circuit could not operate properly.

It may be objected that as Figure 11.2 is drawn, with an ideal current source, it is impossible for both I_{E1} and I_{E2} to be zero. However, it should be realized that in practice the current source would not be ideal, but would have, in parallel with it, some finite Norton resistance through which I_1 could be diverted.

Typically, the input stages of operational amplifiers are very much like Figure 11.2. We see that for proper operation, a path for the dc base-bias current must be available through whatever circuitry is connected to the input terminals. This current is usually quite small; for op-amp inputs the range 0.1 to 5 μA is typical. ∎

For small-signal deviations around the bias point, we may analyze the operation of the circuit with the transistor models of Chapter 10. For simplicity, let us use the "simplified-π" model of Figure 10.6. Replacing the transistors by the model, we have the circuit of Figure 11.3. Note that in doing this, dc voltage and current sources have as usual been replaced by short and open circuits, respectively, and incremental voltages and currents, which represent deviations from the operating point, have been denoted by lower-case symbols

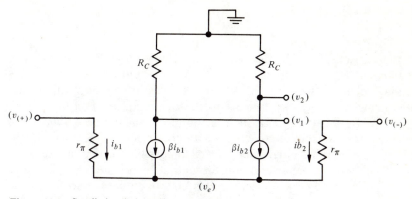

Figure 11.3 Small-signal circuit model of differential amplifier (Figure 11.2), obtained using the simplified-π model of the transistor (Figure 10.6).

with lower-case subscripts. The node at the bottom of the circuit model, corresponding to the point where the two emitters are connected, is defined to be at (small-signal) potential v_e. Assuming that no current flows through the output terminals, the small-signal output voltages are seen to be

$$v_2 = -\beta i_{b2} R_c \tag{11.2}$$

and

$$v_1 = -\beta i_{b1} R_c$$

Furthermore

$$i_{b1} = (v_{(+)} - v_e)/r_\pi$$

and

$$i_{b2} = (v_{(-)} - v_e)/r_\pi \tag{11.3}$$

We now must eliminate v_e from the equations. This may be done by writing a node equation for the emitter node. This equation states that $i_{b1} + \beta i_{b1} + i_{b2} + \beta i_{b2} = 0$. Substituting expressions (11.3) for the base currents, this becomes

$$(1 + \beta)(v_{(+)} - v_e)/r_\pi + (1 + \beta)(v_{(-)} - v_e)/r_\pi = 0 \tag{11.4}$$

which implies that

$$v_e = \frac{(v_{(+)} + v_{(-)})}{2} \tag{11.5}$$

Substituting (11.5) into (11.3) and the resulting expressions for i_{b1} and i_{b2} into (11.2), we find that

$$v_2 = \frac{\beta R_c}{2r_\pi} (v_{(+)} - v_{(-)})$$

and   (11.6)

$$v_1 = -\frac{\beta R_c}{2r_\pi}\,(v_{(+)} - v_{(-)})$$

Both these results have the desired form of Equation (11.1). The voltage amplifications at the two output terminals are equal but opposite in sign. Symmetry between the two sides of the circuit is preserved, one output terminal always going negative with respect to the operating point as the other goes positive by an equal amount. Such an output is said to be *balanced,* and is well-suited for connection to the input terminals of a second differential amplifier stage. On the other hand, it is also possible for one output terminal, say v_1, to be left unconnected, with the output then being regarded as being between the v_2 terminal and ground. This kind of output connection, in which symmetry is no longer maintained, is known as a *single-ended* output. Such a connection must eventually be used in an operational amplifier, after one or two differential stages, since the output of the op-amp as a whole is usually single-ended.

Input Resistance   A differential amplifier stage is usually used as the first, or input, stage in an operational amplifier. When used as an input stage, it should have large input resistance. To calculate the small-signal input resistance, let us look again at Figure 11.3. From this figure we see that $i_{b1} = (v_{(+)} - v_e)/r_\pi$, and $i_{b2} = (v_{(-)} - v_e)/r_\pi$. But from Equation (11.5), $v_e = (v_{(+)} - v_{(-)})/2$. Therefore $i_{b1} = (v_{(+)} - v_{(-)})/2r_\pi$, and $i_{b2} = -(v_{(+)} - v_{(-)})/2r_\pi = -i_{b1}$. Since $i_{b2} = -i_{b1}$, it is possible to make a reasonable definition of an input resistance for the circuit: we can define $R_{in} = (v_{(+)} - v_{(-)})/i_{b1}$. This definition simply states that the current that flows in the (+) input terminal and out the (−) input terminal is equal to $(v_{(+)} - v_{(-)})/R_{in}$. (Thus it is consistent with the op-amp model used in Figure 3.9.) Using the result $i_{b1} = (v_{(+)} - v_{(-)})/2r_\pi$ obtained above, we find that

$$R_{in} = \frac{v_{(+)} - v_{(-)}}{i_{b1}} = 2r_\pi \tag{11.7}$$

An expression for r_π in terms of the operating point dc base current I_{B1} was obtained in Equation (10.16). Using this, and noting that $I_{B1} \cong I_{E1}/\beta \cong I_1/2\beta$ [where I_1 is the value of the dc current source in Figure 11.2], we find that

$$R_{in} = 2 \cdot \frac{kT}{q} \cdot \frac{2\beta}{I_1} = 4\beta \cdot \frac{kT}{qI_1} \tag{11.7A}$$

Thus it is seen that large input resistances may be obtained by using small values of the dc bias current I_1.

EXAMPLE 11.2
      In the differential amplifier of Figure 11.2, find the input resistance if $I_1 = 10^{-4}$ A, $\beta = 100$, at $T = 300°$K.

SOLUTION

Substituting into Equation (11.7A), and noting that at 300°K, $kT/q \simeq 0.026$ V, we find

$$R_{in} = 2 \cdot 0.026 \cdot \frac{200}{10^{-4}} \cong 10^5 \ \Omega \qquad \blacksquare$$

Common-Mode Rejection In an ideal differential amplifier the output voltages would be determined only by the difference between the input voltages. As a practical matter, however, it is instead found that if both input terminals are at exactly the same potential, but the potential of both is varied together, some small variations of the output voltage do occur. A signal v_{CM} applied to both inputs simultaneously, as shown in Figure 11.4(a), is called a *common-mode signal*.

In contrast, a signal applied as a potential difference between the two input terminals, as shown in Figure 11.4(b), is known as a *differential-mode signal, v_{DM}*. In the case of differential-mode signals it is usually assumed that $v_{(+)} = -v_{(-)}$, so that the average value of the two input voltages is zero.

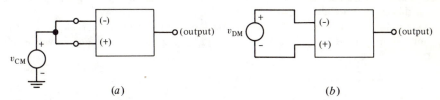

Figure 11.4 Common-mode and differential-mode signals. (a) The signal v_{CM} is applied in the common mode. (b) The signal v_{DM} is applied in the differential mode.

From Equation (11.1), it is seen that an ideal differential amplifier ought to maintain an output voltage of zero, when a purely common-mode signal (that is, $v_{(+)} = v_{(-)}$) is applied to the inputs. The ability of an amplifier to maintain nearly constant output voltage in the presence of common-mode input signals is known as *common-mode rejection*.

A figure of merit known as the *common-mode rejection ratio* is often used as a specification for differential amplifiers. It is defined as follows. Let a common-mode signal v_{CM} be applied. A change in the amplifier output voltage is observed. The same amount of change in the output voltage could be generated by a much smaller voltage v_{DM}, applied in the differential mode as shown in Figure 11.4(b). The common-mode rejection ratio is then defined to be

$$\text{CMRR} \equiv v_{CM}/v_{DM} \qquad (11.8)$$

Since this number is a ratio, it can, like any ratio, be expressed in decibels. High-quality differential amplifiers typically have common-mode rejection ratios in the neighborhood of 30,000 or 90 db.

EXAMPLE 11.3

Consider the differential amplifier of Figure 11.3 with identical transistors. The output will be considered to be v_2 taken single-ended. Calculate the common-mode rejection ratio using (1) the simplified-π transistor model, and (2) the more refined model of Figure 10.16.

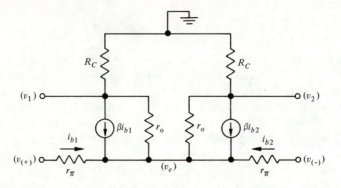

SOLUTION

(1) An analysis using the simplified-π model has already been performed, and the result is given in Equation (11.6). So long as $v_{(+)} = v_{(-)}$, the output voltage v_2 is exactly zero in this approximation, and the CMRR is infinite.

(2) When the more accurate model is used, the new circuit model is as given in the diagram. Writing node equations for v_1, v_2, and v_e (the emitter voltage, shown in the diagram) we have

$$\frac{v_2}{R_C} + \frac{(v_2 - v_e)}{r_o} + \frac{\beta(v_{(-)} - v_e)}{r_\pi} = 0$$

$$\frac{v_1}{R_C} + \frac{(v_1 - v_e)}{r_o} + \frac{\beta(v_{(+)} - v_e)}{r_\pi} = 0$$

$$\frac{(v_e - v_{(+)})(1 + \beta)}{r_\pi} + \frac{(v_e - v_{(-)})(1 + \beta)}{r_\pi}$$

$$+ \frac{(v_e - v_1)}{r_o} + \frac{(v_e - v_2)}{r_o} = 0$$

First let a common-mode signal $v_{(+)} = v_{(-)} = v_{\mathrm{CM}}$ be applied. An easy way to proceed is to note that when a common-mode signal is applied, $v_{(-)} = v_{(+)}$ and, by symmetry, $v_1 = v_2$. Using these relationships in the above three equations and solving, we obtain

$$v_2 = v_{\mathrm{CM}} \cdot \left\{ \frac{1}{(1 + \beta)r_o + r_\pi} \right\} \cdot \left\{ \frac{1}{R_C} + \frac{1}{r_\pi + (1 + \beta)r_o} \right\}^{-1}$$

This result gives the change in output voltage arising from a common-mode signal. No mathematical approximations have been used, and consequently the result is rather complex. Usually, however, one can simplify such a result, because one has in mind the orders of magnitude of the various terms, and can see that some may be neglected. Let us assume the following typical values: $r_\pi = 100$ kΩ, $r_o = 100$ kΩ, $R_C = 5$ kΩ, $\beta = 100$. With these values in mind we see that r_π may be neglected compared with $(1 + \beta)r_o$, and that the last factor becomes simply R_C. Setting $(\beta + 1) \simeq \beta$, we have

$$v_2 \cong v_{\mathrm{CM}} \cdot \frac{R_C}{\beta r_o}$$

Next we must calculate the change in v_2 due to a differential-mode signal $(v_{(+)} - v_{(-)}) = v_{DM}$, with $v_{(+)} = -v_{(-)}$. With this input one finds, from the three node equations, that $v_2 = -v_1$, $v_e = 0$, and that

$$v_2 = \frac{\beta r_o R_C}{2 r_\pi (r_o + R_C)} v_{DM}$$

We wish to find the value of v_{DM} that gives the same v_2 as v_{CM}. Setting equal the expressions for v_2 with common-mode input and differential-mode input, we have

$$\frac{R_C}{\beta r_o} v_{CM} = \frac{\beta r_o R_C}{2 r_\pi (r_o + R_C)} v_{DM}$$

The common-mode rejection ratio is then given by v_{CM}/v_{DM}:

$$\text{CMRR} = \frac{\beta^2 r_o^2}{2 r_\pi (r_o + R_C)}$$

(Notice that if $r_o \to \infty$ we recover the infinite answer given by the simplified-π model.) For the typical values given above, the CMRR is 5000, or 74 db. ∎

Current Sources   In the circuit of Figure 11.2, as well as in other circuits to be described, a dc current source is used. Several methods of constructing such sources have been devised. The simplest approach consists simply of a voltage source in series with a large resistor, as shown in Figure 11.5(a). A voltage source is available in the op-amp, since the power supply terminals are assumed to be connected to ideal voltage sources. With Norton's theorem we find that the circuit is identical, from the point of view of the terminals, to that of Figure 11.5(b). The noninfinite value of the resistance R represents a deviation from the ideal; the larger R is, the closer the circuit is to being an ideal current source. If the desired current I_1 is regarded as fixed, R can be increased but V_0 must be increased proportionately. This method of constructing a current source, though simple, is not a very good one, because to make R large, large voltages (usually unavailable) are required. The method is also poorly suited to IC requirements, as a large-value resistance is needed, and these are difficult to make with integrated circuit techniques.

A better current source is shown in Figure 11.6. The analysis of this circuit is identical to that given in connection with Figure 10.12(b), and the

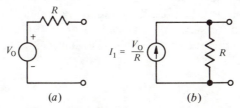

(a)   (b)

Figure 11.5   A voltage source in series with a resistance; (a) is equivalent to a current source with a parallel resistance, as shown in (b).

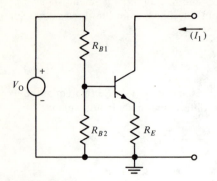

Figure 11.6 Use of transistor collector terminal as a current source.

value of I_1, assuming that $(\beta + 1)R_E \gg (R_{B1}\|R_{B2})$, is given by Equation (10.32):

$$I_1 = \frac{V_{BB} - V_{BE}}{R_E} \tag{11.9}$$

where

$$V_{BB} \equiv V_0 \cdot \frac{R_{B2}}{R_{B1} + R_{B2}}$$

Essentially, this circuit works as a current source because the collector current of a transistor in the active mode equals beta times the base current; it is nearly independent of V_{CE}, provided the collector junction remains reverse biased. Thus the collector terminal of a transistor makes a fairly good current source. The circuit is improved by the presence of R_E because R_E makes I_C independent of β [compare Equation (10.29) with (10.32)], which is important because the value of β is difficult to control in manufacturing. For use in IC's, the circuit of Figure 11.6 has a certain disadvantage. If small I_1 is to be achieved [for example, to increase R_{in} via Equation (11.7)], costly resistances with large values are required.

Often one is interested in the small-signal Norton resistance of a current source (for example, in Problem 11.3). When the circuit of Figure 11.6 is represented by its Norton equivalent, the Norton resistance, which represents deviation from ideal current source behavior, is again not infinite. (It is usually impossible to achieve an infinite value for any circuit quantity.) In this case the Norton resistance turns out to depend on the value of r_0, the resistance representing the Early effect in Figure 10.16, as shown in the following example.

EXAMPLE 11.4

Obtain a Norton equivalent of the current source of Figure 11.6 valid for small-signal deviations of the current around the operating point.

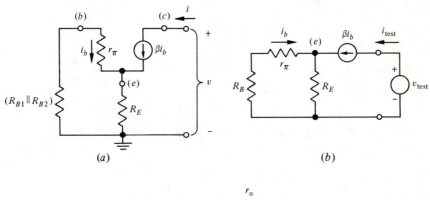

(a)

(b)

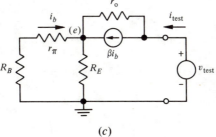

(c)

SOLUTION

Let us substitute the simplified-π model for the transistor. We then obtain the small-signal circuit model given in sketch (a).

It is evident that the short-circuit current for this circuit is zero. This is not surprising, since the circuit does not produce any small signals by itself. Hence the Norton current i_n is zero.

It is not possible to find the Norton resistance by means of $r_n = -v_{oc}/i_{sc}$, since both v_{oc} and i_{sc} are zero. Instead we may proceed by connecting a hypothetical test voltage source v_{test} across the terminals, as explained in Chapter 2. Doing this, we have sketch (b). For convenience we have defined $(R_{B1}\|R_{B2}) \equiv R_B$. Writing a node equation for node (e), we have

$$\frac{v_e(1 + \beta)}{R_B + r_\pi} + \frac{v_e}{R_E} = 0$$

from which we have $v_e = 0$. Since $v_e = 0$, $i_b = 0$, and thus $\beta i_b = 0$ and $i_{test} = 0$. We obtain the result $r_n = v_{test}/i_{test} = \infty$.

This result is not very satisfactory. No circuit is a perfect current source. The problem lies in the inadequacy of the simplified-π transistor model. Let us use the more refined model of Figure 10.16, in which the basewidth-modulation effect is included. With the model, our test circuit is seen in sketch (c).

Now let us again write a node equation for node (e). This time we obtain

$$\frac{v_e(1 + \beta)}{R_B + r_\pi} + \frac{v_e}{R_E} + \frac{v_e - v_{test}}{r_o} = 0$$

Solving for v_e, we have

$$v_e = \frac{v_{\text{test}}}{\dfrac{(1+\beta)r_o}{R_B + r_\pi} + \dfrac{r_o}{R_E} + 1}$$

We now can find i_{test}:

$$i_{\text{test}} = \frac{v_{\text{test}} - v_e}{r_o} - \frac{\beta v_e}{R_B + r_\pi}$$

Substituting the above expression for v_e, we obtain

$$i_{\text{test}} = v_{\text{test}} \cdot \frac{\left[\dfrac{R_E + R_B + r_\pi}{R_E}\right]}{\left[\dfrac{(1+\beta)r_o}{R_B + r_\pi} + \dfrac{r_o}{R_E} + 1\right]\left[R_B + r_\pi\right]}$$

In a typical case $\beta \gg 1$, and $r_o \gg R_E \gg (R_B + r_\pi)$. In that case the result simplifies to

$$i_{\text{test}} \cong v_{\text{test}} \cdot \frac{1}{\beta r_o}$$

The Norton resistance is

$$r_n = v_{\text{test}}/i_{\text{test}} \cong \beta r_o$$

Since $\beta \sim 100$ and $r_o > 10^4$ ohms, the Norton resistance is quite large. Thus the circuit is a good approximation to an ideal current source.   ∎

A current source which is very well-suited for use in IC's, particularly when a small current is needed, is the ingenious circuit of Figure 11.7(a). This circuit makes use of the fact that the two transistors T_1 and T_2, being adjacent in the IC, can be counted upon to have identical characteristics. Although transistor T_1 is "diode-connected," with $V_{CB} = 0$, the collector current I_{C1} still

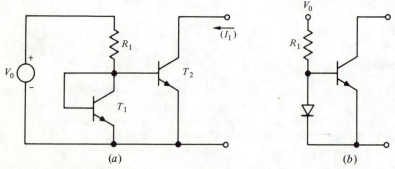

(a)   (b)

Figure 11.7 (a) Two-transistor current source for use in integrated circuits. (b) Shorthand representation.

equals βI_{B1} (see Example 6.8). Using Kirchhoff's current law for the node at the collector of T_1 we have

$$\frac{V_0 - V_{BE}}{R_1} - I_{B1} - I_{C1} - I_{B2} = 0 \qquad (11.10)$$

Because the same base and emitter voltages are applied to the two identical transistors, I_{B1} must equal I_{B2}, and therefore $I_{C1} = I_{C2}$. Using this fact in Equation (11.10), we have

$$I_1 \equiv I_{C2} = \frac{\beta}{\beta + 2} \cdot \frac{(V_0 - V_{BE})}{R_1} \cong \frac{(V_0 - V_{BE})}{R_1} \qquad (11.11)$$

Again, this circuit has the advantage that I_1 is nearly independent of the value of β, plus the additional virtue, for IC applications, that only one resistor is required. (No circuit is perfect; the temperature dependence of V_{BE} and finite small-signal Norton resistance must still be reckoned with.) Because this circuit is widely used, it is sometimes represented in circuit diagrams in the shorthand form shown in Figure 11.7(b). This shorthand is unfortunately rather confusing, since even though T_1 is diode-connected, it is the fact that T_1 and T_2 are identical transistors that makes the circuit work.

11.2 Output Circuits

For the output stage of an amplifier, (as well as in other applications) a low output resistance is often desired. The conventional common-emitter amplifier is not very good from this point of view. From inspection of Figure 10.13(b) it can be seen (by converting the dependent current source and R_C to their Thévenin equivalent) that the common-emitter amplifier of Figure 10.12(a) has output resistance approximately equal to R_C. This resistance generally amounts to several thousand ohms.[1] A circuit that may be more suitable for providing low output resistance is the *emitter follower,* shown in Figure 11.8(a).[2] The corresponding small-signal circuit model (using the simplified-π transistor model) is shown in Figure 11.8(b). Here a small-signal source is represented by v_s and R_s. R_{B1} and R_{B2} are provided for biasing; the resistance of their parallel combination is called R_b'. We shall now calculate the output resistance of the circuit. In doing this it will be found that (unlike

---

[1] A smaller value of R_C could be used. But then the operating-point value of V_C would be nearly V_{CC}, so the allowable "swing" of signal voltage about the operating point would be small. To increase the value of I_C so as to increase $(V_{CC} - V_C)$ to around $V_{CC}/2$ would require excessively large I_C.

[2] This connection is also known as "common-collector" circuit because the collector is connected to ground as far as signals are concerned. Since one input terminal and one output terminal are connected to ground, the collector is connected to the point in the circuit common to both input and output; hence, common collector.

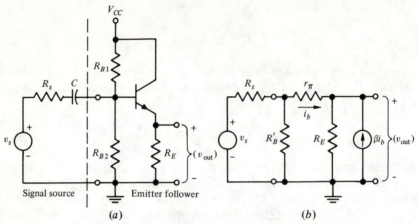

Figure 11.8 (*a*) An emitter-follower circuit. (*b*) Circuit model using simplified-π transistor model. $R'_B = (R_{B1} \mathbin{//} R_{B2})$. Elements v_s and R_s represent a small-signal source; C is a coupling capacitor, assumed to be a short circuit for signals.

the case of the common-emitter circuit) the output resistance is affected by the value of the source resistance connected to the input terminals, that is, R_{out} is affected by R_s.

To simplify matters, let us replace the ensemble v_s, R_s, and R'_B by its Thévenin equivalent (this procedure is identical to Example 2.4). This gives the circuit of Figure 11.9, where

$$v_t = v_s \cdot \frac{R'_B}{(R'_B + R_s)} \;;\qquad R_t = (R_s \| R'_B) \tag{11.12}$$

Writing a node equation for the node connected to the upper output terminal, we have

$$(1 + \beta) \frac{(v_t - v_{\text{out}})}{(R_t + r_\pi)} - \frac{v_{\text{out}}}{R_E} = 0 \tag{11.13}$$

Solving for v_{out} and denoting it as the open-circuit output voltage v_{oc}, we have

$$v_{oc} = \frac{v_t(1 + \beta)}{R_t + r_\pi} \cdot \left(\frac{(1 + \beta)}{R_t + r_\pi} + \frac{1}{R_E} \right)^{-1} \tag{11.14}$$

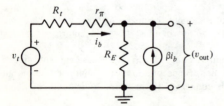

Figure 11.9 Circuit of Figure 11.8(*b*) simplified by use of the Thévenin equivalent.

By inspection we see that

$$i_{sc} = -\frac{v_t}{R_t + r_\pi}(1 + \beta) \qquad (11.15)$$

so the output resistance is given by

$$r_{out} = -\frac{v_{oc}}{i_{sc}} = \frac{1}{\dfrac{(1+\beta)}{(R_t + r_\pi)} + \dfrac{1}{R_E}} \qquad (11.16)$$

In most cases the first term of the denominator dominates because of the large factor $(1 + \beta)$; in such cases

$$r_{out} \cong \frac{R_t + r_\pi}{1 + \beta} \qquad (11.17)$$

EXAMPLE 11.5
  Assume the following typical values: $R_E = 5$ kΩ, $R_B' = 47$ kΩ, $R_s = 4$ kΩ, $\beta = 100$, $r_\pi = 3$ kΩ. Obtain the approximate value of r_{out}.

SOLUTION
  We see that $R_t \equiv R_s \| R_B' \cong R_s$. Moreover

$$\frac{(1+\beta)}{R_t + r_\pi} \gg \frac{1}{R_E}$$

Thus Equation (11.16) simplifies to

$$r_{out} \cong \frac{R_t + r_\pi}{1 + \beta} = 70 \ \Omega \qquad \blacksquare$$

  From Equation (11.14) we may find the open-circuit voltage gain, referred to the signal voltage v_s:

$$A \equiv \frac{v_{oc}}{v_s} \cong \frac{v_t}{v_s} = \frac{R_B'}{R_B' + R_s} \qquad (11.18)$$

where we have again assumed that $(1 + \beta)/(R_t + r_\pi) \gg 1/R_E$. This gain is less than unity; if the source resistance were zero (or if amplification were defined with respect to the voltage at the input terminals, instead of with respect to v_s), the voltage amplification would be (within the accuracy of our approximations) unity. Thus a voltage amplification less than or equal to unity is obtained in an emitter follower stage. Current and power amplification however do occur.

EXAMPLE 11.6
  Find the small-signal input resistance of the emitter follower of Figure 11.8(a), assuming that a load R_L is connected across its output terminals.

SOLUTION

The circuit model using the simplified-π model for the transistor is given in the sketch. To obtain this circuit, a "test" voltage v_t (with zero source resistance) was considered to be connected to the input terminals. When the corresponding "test" current i_t is found, the input resistance may be obtained from $r_{in} = v_t/i_t$. The parallel combination of R_E and R_L has been denoted as R'_E.

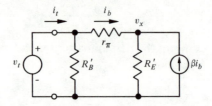

We proceed by writing a node equation for the node whose voltage is indicated as v_x on the diagram:

$$i_b - v_x/R'_E + \beta i_b = 0$$

Noting that $i_b = (v_t - v_x)/r_\pi$, we have

$$\frac{(v_t - v_x)(1 + \beta)}{r_\pi} - \frac{v_x}{R'_E} = 0$$

which on solving gives

$$v_x = \frac{v_t}{1 + \dfrac{r_\pi}{(1 + \beta)R'_E}}$$

We note that from the point of view of the input resistance, R'_B is in parallel with the effective resistance of the ensemble r_π, R'_E, βi_b. Denoting this effective resistance by r_1, we have

$$r_1 = \frac{v_t}{i_b} = \frac{v_t}{(v_t - v_x)/r_\pi} = (1 + \beta)R'_E + r_\pi$$

Thus

$$r_{in} = R'_B \| [(1 + \beta)R'_E + r_\pi]$$

If we take as typical values $\beta = 100$, $R'_E \equiv (R_E \| R_L) = 3$ kΩ, and $r_\pi = 2.5$ kΩ, we find $r_{in} \cong R'_B \| 300$ kΩ. Thus a high input resistance, up to 300 kΩ, can be obtained by using a large R'_B. ∎

One disadvantage of the emitter-follower output stage is that current flows through the transistor at all times, whether or not signal is present. This is wasteful of power. To correct this deficiency, a *push-pull* output stage may be employed, as shown in Figure 11.10. Note that one *pnp* transistor is used along with one *npn*. As was pointed out in Chapter 7, the one time it is possible to construct a *pnp* transistor of good quality in an IC is when its col-

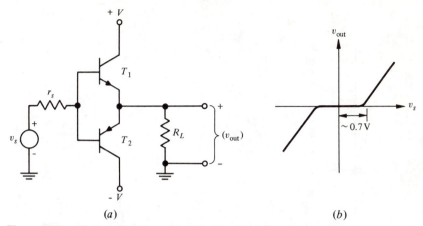

Figure 11.10 (*a*) A push-pull amplifier. In this circuit the output voltage deviates considerably from linear proportionality to the input voltage, as shown in (*b*).

lector can be connected to the point of lowest potential in the circuit, which is the case here. In this circuit, when $v_s = 0$ neither transistor's emitter junction is forward biased, and no current flows in either one. When v_s becomes more than 0.7 V positive, transistor T_1 is turned on and conducts. In this part of the operating cycle the combination of T_1 and R_L acts as an emitter follower, providing low output resistance; since the emitter junction of T_2 is reverse biased under these conditions, it may be regarded as an open circuit. When the sign of v_s reverses, only T_2 conducts and a negative v_{out} is produced. It may be objected that since T_1 and T_2 will be quite different physically, the positive and negative sides of the operation will not be symmetrical. Fortunately, this is not too serious a problem, since, as was shown above, the values of neither β nor r_π have much effect on the operation of an emitter follower. More serious is the "dead space" of the circuit, arising because no conduction occurs until $|v_s|$ exceeds ~ 0.7 V. Some distortion of this kind (known as crossover distortion) can be tolerated, as it can be overcome through the use of feedback (see Chapter 12). However, the situation can be improved by the modification shown in Figure 11.11. The idea of this modified circuit is to apply a small constant dc forward voltage to the emitter junctions of both transistors. This forward voltage is obtained through the addition of two diodes and two current sources, as shown. The values of the current sources (which in practice might simply be large resistances, as in Figure 11.5) are made quite small, so that in the absence of signal, the voltage across each diode, and hence V_{BE}, is small, say 0.5 V. Such small forward voltages are not quite sufficient to "turn on" the transistors, and little current flows in the absence of signal. Now, however, when signal is applied, less signal voltage is required to turn on each transistor than in Figure 11.10, because 0.5 V of forward bias is already present. This modification does not eliminate "dead space" entirely, since a certain amount of signal voltage still is needed to turn

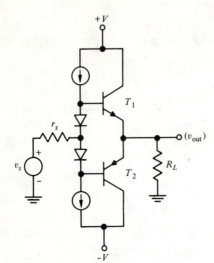

Figure 11.11   Push-pull output stage modified for elimination of "dead space."

either transistor "on"; however, it clearly represents an improvement over Figure 11.10.

Push-pull output stages are widely used because of their high power efficiency, particularly in cases where large amounts of output power are required. It is possible to achieve similar operation using only *npn* transistors, but the circuit is then somewhat more complex.

11.3   Interstage Coupling

In a multistage amplifier the output of one stage must be connected, from the point of view of signals, to the input of the next. However, the dc potential of the output of one stage may not be suitable for the input of the next stage. In a non-integrated circuit the problem is often handled by the method of Figure 10.15, where coupling capacitors were used to connect the stages for signals but isolate them for dc. This "ac-coupled" approach has the serious disadvantage that low-frequency signals or dc voltages cannot be amplified. For IC's, moreover, the use of capacitors is highly undesirable. Another possible approach is *direct coupling,* in which the collector of transistor 1 is connected directly to the base of transistor 2 (Figure 11.12). In this example of direct coupling diode D_1 is used to raise the emitter voltage of the second stage to 0.7 V, so that connecting the collector of the first transistor (at potential 2 × 0.7 = 1.4 V) to the base of the second provides satisfactory biasing for both transistors. If all the transistors are *npn*, as they would be in an IC, this kind of direct coupling causes the stages to be at successively higher dc voltages, so that the operating point voltages at output and input terminals cannot both be zero. In a discrete circuit, however, alternate *npn* and *pnp* stages can be used, so that the voltage level first steps up and then down again.

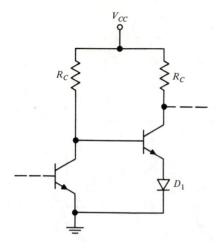

Figure 11.12 Direct coupling of *npn* amplifier stages.

It is possible to obtain low-frequency coupling without stair-step increases in potential from stage to stage by readjusting the dc potential through the use of a circuit called a "level shifter." The principle is shown in Figure 11.13. The constant current I_1 causes a constant-voltage drop across R, so the signal voltage v_s appears at the output, but with a constant dc voltage subtracted. Another way of seeing this is to recall that the dc current source is an open circuit for signals, so (in the absence of a load at the output terminal) there is no signal current through R and thus no signal voltage drop. The ac output resistance (which is the source resistance for the following stage) is increased from r_s to $(r_s + R)$; this may or may not be of significance, depending on the relative sizes of r_s and R, and what the following stage happens to be. The dc, or operating point, value of v_{OUT} is now reduced from V_1 to $V_1 - I_1R$, which may be set to zero by suitable choice of I_1 and R. A circuit example is shown in Figure 11.14. Here a differential amplifier stage is used to drive a single-ended

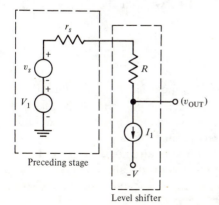

Figure 11.13 Simplified diagram of level shifter, illustrating principle of operation.

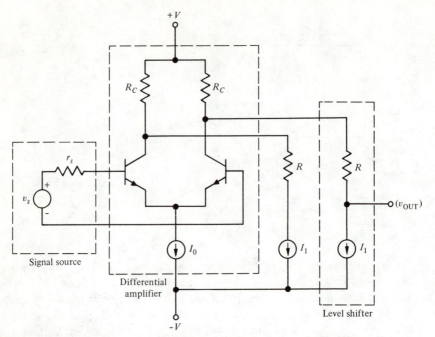

Figure 11.14   An example of a circuit employing a level shifter.

output, and it is desired that the operating point voltage at the v_{OUT} terminal be zero (perhaps so it can be connected to the input of a push-pull output stage). Suppose $V = 6$ V, $R_C = 4$ kΩ, and $I_0 = 2$ mA; then the operating-point voltage at the collectors of the differential amplifier is 2 V. To shift the dc output level down to zero, the product $I_1 R$ must equal 2 V. This could be done by letting $I_1 = 0.2$ mA, $R = 10$ kΩ. The second (I_1, R) combination connected to the other side of the differential amplifier is provided to maintain balance; without it the symmetry of the differential amplifier would be damaged. The small-signal voltage amplification is the same as it would be without the level shifter, although the output resistance is increased from R_C to ($R_C + R$).

11.4   The Complete Amplifier

The example of Figure 11.14 illustrates the fact that complex circuits are most easily understood by imagining them broken up into already familiar sub-circuits. We are now in a position to consider a complete operational amplifier, such as the one shown in Figure 11.15. In spite of the fact that it contains 13 transistors, it can readily be analyzed into circuits already described. The input stage is a differential amplifier, biased by a two-transistor (IC-type) current source. Its outputs are fed as inputs to a second differential amplifier stage.

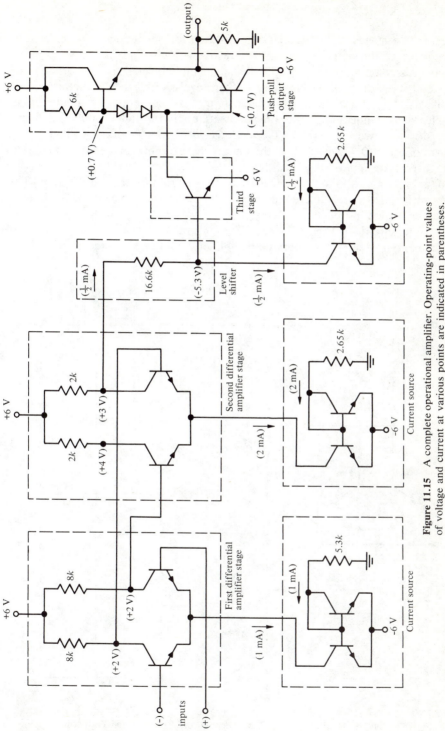

Figure 11.15 A complete operational amplifier. Operating-point values of voltage and current at various points are indicated in parentheses.

Using two stages of differential amplification further reduces the undesirable tendency of the amplifier output to "drift" with changes in temperature and improves common-mode rejection. A single-ended signal from one side of the second-stage differential amplifier is then level-shifted to a value convenient for biasing the third stage, which is a simple common-emitter amplifier whose collector is connected to the +6-V power supply through a 6000-Ω collector resistor and two diodes. The diodes have a constant forward voltage across them of about 0.7 V each, and thus are a short circuit for signals. The output stage is a variant of the push-pull circuit of Figure 11.11. Operating-point voltages and currents are indicated at various points on the diagram. This circuit is suitable for discrete or integrated realization, except for the current sources, which are of the IC type.

The circuit of Figure 11.15, although practical, was chosen for ease of explanation, whereas a commercial amplifier would be expected to be more re-

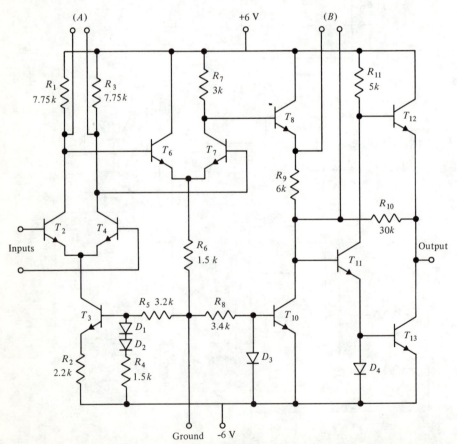

Figure 11.16 The Motorola type MC1530 integrated operational amplifier. Terminal pairs (A) and (B) are provided for connection of external capacitors to adjust frequency response.

fined. A fairly simple commercial op-amp, the Motorola type MC1530, is shown in Figure 11.16. In this circuit the inputs are connected to a differential amplifier composed of transistors T_2 and T_4. Bias current for this stage is provided by current source T_3, which is of the type shown in Figure 11.6, with diodes D_1 and D_2 added for temperature compensation. The output of the first stage is fed to a second differential stage (T_6 and T_7), which is biased by a simple current source using a single resistance (R_6), as explained in connection with Figure 11.5. The output of one side of the differential amplifier goes through an emitter follower (T_8) to a level shifter (R_9, D_3, and T_{10}). The function of T_8 is to provide a path for the level-shifting current through R_9; otherwise this current would have to flow through R_7, which would reduce the voltage at the collector of T_7. The base of T_8 also presents a fairly high ac load resistance for the single-ended output from T_7.

The output stage (T_{12} and T_{13}) is a push-pull stage that uses two *npn*'s, and is consequently more complex than if a *pnp* had been used. Transistor T_{11} is made necessary by this change, acting essentially as a "phase inverter" to invert the sign of the signal, so that for the lower output transistor an *npn* may be used. Resistor R_{10} gives rise to a feedback path from the output back to the input of T_{11}. This stabilizes the output under varying load conditions and improves the linearity of the output stage.

EXAMPLE 11.7

Explain the function of T_{11} in the circuit of Figure 11.16. Ignore the feedback resistor R_{10}.

SOLUTION

Let us assume that when there is no input signal, neither T_{12} nor T_{13} is forward biased. The base of T_{12} is then connected to T_{11} as though the latter were a common-emitter stage, while the base of T_{13} is connected to T_{11} as though the latter were an emitter follower. (The role of R_E in the emitter follower is taken by the incremental forward resistance of D_4 at its operating point.) When a signal causes the base of T_{11} to go up in voltage, T_{11} conducts more current and the voltage drop in D_4 increases. Consequently, the base of T_{13} goes up in voltage and T_{13} conducts, pulling current in through the output terminal. Meanwhile the increase in current through T_{11} has caused the base of T_{12} to go down in voltage, so that T_{12} has remained "turned off." When a negative signal appears at the base of T_{11}, the reverse occurs and only T_{12} conducts, current going from its emitter out through the output terminal. Thus we have push-pull operation of the output stage. T_{11} is known as a "phase inverter" because if a sinusoidal signal is applied to its base, the two output voltages it supplies to T_{12} and T_{13} are 180° out of phase with each other. ∎

Summary

- The design of an amplifier is determined by its intended use. Some important characteristics of a design are input resistance, output resistance, open-circuit voltage amplification, linearity, balance, power consumption, and cost. Circuits intended for integrated circuit realization are subject to special restrictions arising from the way IC's are made.

• A differential amplifier generates a signal proportional to the difference between two input signals. The input stage in an operational amplifier is usually a differential stage. In their most common form differential amplifiers contain two balanced transistors, as nearly alike as possible.

• A signal applied identically to both input terminals of a differential amplifier is called a common-mode signal. Ideally, common-mode signals should not influence the output signal of a differential stage. In practice, they do affect the output slightly. The ability of an amplifier to reject common-mode signals is specified by a figure of merit called the common-mode rejection ratio. A signal applied as a potential difference between the input terminals is called a differential-mode signal.

• Output stages of amplifiers are often required to exhibit low output resistance. The emitter-follower circuit is well-suited to this application. Push-pull stages can be used to reduce the power consumption of the circuit.

• The stages of a multistage amplifier must, of course, be coupled together. When connected directly to one another, so as to couple them for dc as well as signals, they are said to be direct-coupled, or dc-coupled. Alternately, stages may be coupled together by means of coupling capacitors, which isolate the stages for dc; this method however entails loss of low-frequency response. A useful coupling circuit is the level-shifter, which couples stages together at all signal frequencies while establishing a dc potential difference between them.

• A complete operational amplifier is best understood in terms of the action of its individual stages. By considering the stages one by one, even quite complicated circuits can be understood and analyzed.

References

Camenzind, H. R. *Circuit Design for Integrated Electronics*. Reading, Mass.: Addison-Wesley Publishing Co., 1968.

Widlar, R. J. "Design of Monolithic Linear Circuits," Section 10. In *Handbook of Semiconductor Electronics*. Lloyd P. Hunter, ed. 3rd edit. New York: McGraw-Hill, 1970.

Giles, J. N. *Fairchild Semiconductor Linear Integrated Circuits Applications Handbook*. Fairchild Semiconductor Co., 1967.

Problems

11.1   Consider the signal sources shown in Figure 11.17. Let a pair of identical signal sources be connected to the differential amplifier of Figure 11.2, one source to each input. Assume that the value of the current source I_1 in Figure 11.2 is 5 mA, $V_{CC} = 10$ V, and $\beta = 100$. (1) For which of the four sources shown in Figure 11.17 will the transistors of the amplifier be properly biased in the active mode? Take $R_C = 20$ kΩ. (2) Repeat part (1) under the assumption that the current source of Figure 11.2 is not ideal, but actually has a Norton resistance of 10 kΩ. *Suggestion:* For sources (c) and (d), proceed by assuming $V_{BE} = 0.7$ V. By means of a node equation, calculate V_B, and hence I_B and I_C. "Proper biasing" may be interpreted as meaning $I_C > 10^{-4}$ A.

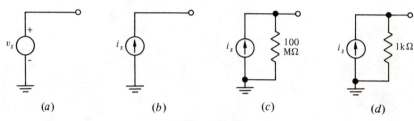

(a)   (b)   (c)   (d)

Figure 11.17

11.2   In the circuit model of Part (2), Example 11.3, verify that $v_e = 0$ when a differential-mode signal is applied, such that $v_{(-)} = -v_{(+)}$.

11.3   In this problem we shall calculate the common-mode rejection ratio of the differential amplifier shown in Figure 11.2, under different assumptions than were used in Example 11.3. Let us now assume that $r_o = \infty$ for the two transistors of the circuit, but that there is, in parallel with source I_1, a Norton resistance R_N. (1) What is the common-mode rejection ratio in this case? (2) Compare the effect on the CMRR of r_o (found in Example 11.3) with the effect of R_N, found in this problem. (3) Suppose R_N is of the same order of magnitude as βr_o (as implied by Example 11.4). Is R_N or r_o likely to be more important in determining the CMRR?

11.4   In Figure 11.4(a), let the input resistance for the common-mode connection be defined as v_{CM} divided by the current which flows through the source v_{CM}. Find the input resistance for signals in the common mode for the circuit of Figure 11.2. Use the circuit model of Example 11.3. How does this result compare with R_i for the differential mode?

11.5   The circuit shown in Figure 11.18 is a differential amplifier with what are known as *Darlington-connected* inputs. The function of T_3 and T_4 is to raise the input resistance of the circuit. Assume that all four transistors are identical. The circuit values are $I_1 = 10^{-4}$A, $R_{C1} = R_{C2} = 50$ kΩ, $\beta = 100$.
(1) What is $r_{\pi 1}$, the value of r_π for T_1? What is $r_{\pi 3}$?
(2) Calculate the voltage gain of the circuit, defined as $A_V = (v_2 - v_1)/(v_{(+)} - v_{(-)})$. Assume that a pure differential-mode signal is applied, so that $v_{(+)} + v_{(-)} = 0$. (*Suggestion:* First show that v_e, the small-signal voltage at the

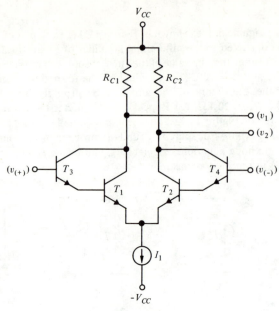

Figure 11.18

emitters of T_1 and T_2, equals zero.) Compare with the amplification of the non-Darlington circuit of Figure 11.2.

(3) Calculate the input resistance under the same assumptions as in Part (2). [The input resistance is defined as in the discussion preceding Equation (11.6).] Compare with the input resistance of the circuit of Figure 11.2.

11.6   In Figure 11.18, T_3 can be regarded as an emitter-follower stage, with the base of T_1 acting as its load. Estimate the differential input resistance of Figure 11.18 using the results of Example 11.6. *Suggestion:* Note that in the previous problem it was found that when $v_{(+)} + v_{(-)} = 0$, the small-signal voltage at the emitters of T_1 and T_2 is zero. This point may thus be regarded as grounded in the small-signal circuit model, so long as only differential signals with $v_{(+)} + v_{(-)} = 0$ are applied to the inputs.

11.7   It is necessary in an IC to construct a dc current source of value 0.1 mA. The power supply voltage available in the circuit is 10 V. The small-signal Norton resistance of the current source must be at least 25 kΩ. The overall current consumption of the circuit should not exceed 0.25 mA, and the value of the source should not depend on β.

Compare the current sources of Figures 11.5, 11.6, and 11.7. Assume that the cost of a resistor is proportional to its value in ohms, and that a 10-kΩ resistor costs as much as a transistor. Which is best for this application?

11.8   In the emitter-follower circuit of Figure 11.8, $R_S = 0$. The output of the circuit is connected through a capacitance to the input of an identical stage. (This capacitance may be regarded as a short circuit for signals.) The second stage has no external load connected to its output terminals.

(1) Find the input resistance as seen by v_s.

(2) Find the output resistance of the second stage.

11.9   Construct a circuit model for the push-pull amplifier of Figure 11.11. Use an incremental (that is, small-signal) model to represent the diodes [see Equation (10.16)]. If the current sources are made small to conserve power, what happens (qualitatively) to the incremental diode resistance? To the input resistance of the circuit?

11.10 Consider the MC 1530 amplifier shown in Figure 11.16. Assuming $\beta = 100$ for all the transistors, estimate the following quantities:

(1) The collector current of T_3.

(2) The input resistance of the amplifier in the differential mode.

(3) The collector voltage of T_7.

(4) The emitter current of T_8. (You may assume that the operating-point voltage at the output terminal is zero.)

11.11 In the circuit of Figure 11.16, assume that the incremental (that is, small-signal) forward resistance of D_4 is 5000 Ω.

(1) Construct a small-signal circuit model for T_{11}, T_{12}, T_{13}, and associated circuit elements.

(2) Imagine an ideal signal-voltage source v_s connected to the base of T_{11}. Find v_{OUT}. Assume that $\beta = 100$ for all the transistors. It is not a straightforward matter to calculate the base currents of T_{11}, T_{12}, and T_{13}. Hence for simplicity assume $r_\pi = 2500 \ \Omega$ for all three transistors.

11.12 How will the frequency response of the amplifier shown in Figure 11.16 be affected if a large capacitance is connected between terminals (A)? Between terminals (B)?

Chapter 12

Operational Amplifiers and Their Applications

Amplifiers are the most important analog building blocks. A particular type of very high-gain amplifier, designed for use as a building block in analog circuits, is called an *operational amplifier*. At first op-amps were used mainly in analog computers to perform various operations on signals, such as addition, subtraction, or integration with respect to time. However, the list of op-amp applications has lengthened greatly. Today the operational amplifier is a nearly universal building block for analog circuits. A great many analog circuits, which until recently were constructed using individual transistors and many other components, are now constructed with an op-amp and just a few additional circuit elements.

The principal factor accounting for the present widespread use of operational amplifiers is their low cost. They are available in integrated circuit form, often for less than $1.00. In addition, integrated op-amps are tiny, in many cases about the same size as a conventional transistor (Figure 12.1). Often it is better to design analog circuits around op-amps than to use individual transistors. The resulting circuit is smaller, and economies are realized in circuit design and construction costs, because the internal circuitry of the op-amp is predesigned and preassembled.

Figure 12.1 Integrated circuit operational amplifiers. Two styles of packaging are shown. The largest diameter of the circular package is about three-eighths of an inch. (Photo courtesy of Burr-Brown Research Corporation.)

The availability of these small, rugged, and inexpensive building blocks greatly simplifies the task of circuit design, and makes practical the design of circuits that were previously either too expensive or too complex to be considered.

In this chapter we shall follow current practice by referring to the basic amplifier block, as in Figures 3.7 to 3.9, as the operational amplifier. This basic amplifier block is then interconnected with other circuit elements to form an operational amplifier circuit.

The op-amp circuit can be analyzed using the methods described in Chapter 3. One substitutes, for the op-amp, a suitable model composed of linear elements. This produces a circuit model that can then be analyzed using conventional techniques. The symbol and model for the op-amp introduced in Chapter 3 are for convenience repeated in Figure 12.2.

In almost all cases, op-amp circuits are of the type known as *feedback circuits*. "Feedback" is said to be present in a system in which some of the output signal is "fed back" to be added to the input. A familiar example serves to illustrate the principle. Let us consider a heating system with and without thermostatic control. A schematic illustration is shown in Figure 12.3. In the system without feedback, Figure 12.3(a), the heat control is set, and the furnace delivers heat at a constant rate to the room. If the outside temperature changes, or the quality of the fuel changes, the room temperature will change. The addition of a simple thermostat and a feedback connection, as shown in Figure 12.3(b), improves the temperature stability of the room considerably. If the wind blows and the temperature of the room starts to drop, the thermostat

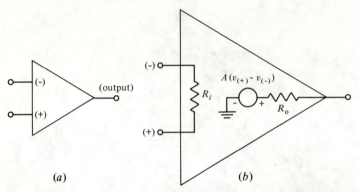

Figure 12.2   The operational amplifier. (*a*) Circuit symbol; (*b*) a suitable model. The symbols $v_{(+)}$ and $v_{(-)}$ stand for the voltages at the "+" (non-inverting) and "−" (inverting) input terminals, respectively, measured with respect to ground.

signals the furnace to increase the heat flow. Similarly, when the sun comes out and the temperature of the room begins to rise, the furnace is instructed, via the feedback path, to reduce the heat flow. This example illustrates several interesting features of a feedback system. (1) The output is determined by only a few simple elements in the system (in this example, the thermostatic control-

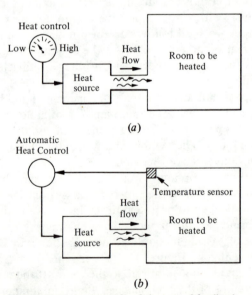

Figure 12.3   An example of the use of feedback. In (*a*) no feedback is used. In (*b*) the action of the heat source is regulated by information "fed back" from the temperature sensor.

ler). (2) The output (here, the temperature of the room) is less sensitive to disturbing variables (here, wind, sun, and so forth) than it would be if feedback were not used. The importance of the other elements in the system is reduced. For instance, note that in Figure 12.3(*a*) such factors as the size of the heat pipe are important, but with feedback, are far less important.

In the example just presented, the feedback is applied in such a way as to decrease the output when the output is too high, or increase the output when the output is too low. This kind of feedback, which exerts a restraining effect on variations of the output, is known as *negative feedback.*

The feedback systems we shall be considering in this chapter exhibit properties similar to those of the foregoing example. We shall find that the output of an op-amp circuit is determined, for the most part, by only a few circuit elements, that the output is insensitive to disturbing influences, and that many useful circuits are negative feedback circuits.

EXAMPLE 12.1

This example is intended to illustrate the insensitivity of feedback circuits to variations of the circuit parameters. A simple feedback circuit using an op-amp is shown in sketch (*a*). Using the op-amp model of Figure 12.2, find the output voltage in terms of the input voltage. Calculate the output voltage under two sets of conditions:

(1) $A = 10^5$, $R_i = 10^4\ \Omega$, $R_L = 10^3\ \Omega$, $R_S = 10^3\ \Omega$
(2) $A = 2 \times 10^5$, $R_i = 3 \times 10^4\ \Omega$, $R_L = 5 \times 10^3\ \Omega$, $R_S = 2 \times 10^3\ \Omega$

To simplify the calculations, assume that the amplifier's output resistance R_o is zero.

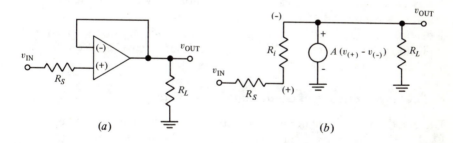

(*a*)   (*b*)

SOLUTION

We first replace the op-amp by its model. This gives the circuit model in sketch (*b*). We wish to solve for v_{OUT} as a function of v_{IN}. We write a node equation at the node labeled (+):

$$\frac{v_{\text{IN}} - v_{(+)}}{R_S} + \frac{v_{\text{OUT}} - v_{(+)}}{R_i} = 0$$

However, v_{OUT} is also related to $v_{(+)}$ and $v_{(-)}$ by the relationship $A\,(v_{(+)} - v_{(-)}) = v_{\text{OUT}}$; furthermore, in this circuit $v_{(-)} = v_{\text{OUT}}$. Thus

$$v_{\text{OUT}} = A\,(v_{(+)} - v_{\text{OUT}})$$

Solving these two equations simultaneously, we obtain

$$v_{OUT} = v_{IN} \cdot \frac{A}{A + 1} \cdot \frac{R_i}{R_i + R_S/(1 + A)}$$

It can now be seen that $v_{OUT} \cong v_{IN}$ for either set of parameters given. Evaluating v_{OUT}/v_{IN} according to this result, we obtain

Case 1:	$v_{OUT}/v_{IN} = 0.999989$	
Case 2:	$v_{OUT}/v_{IN} = 0.999994$	■

Thus we see that in this feedback circuit, the output is very little dependent on the values of R_i, R_L, R_S, and A.

Example 12.1 illustrates the use of feedback with a high-gain amplifier to produce a simple circuit with useful properties. It is the aim of this chapter to present a number of such circuits, and furthermore, to show how a large class of similar circuits may be constructed from op-amps.

As we have seen from the example, the use of feedback in op-amp circuits reduces the importance of the internal details of the op-amp. Consequently, rather simple models can be used to represent the op-amp without serious error. In Section 12.1 we shall discuss op-amp models and techniques for analysis of op-amp circuits. We shall then be ready to consider a number of specific circuits that use op-amps; see Section 12.2.

All the circuits of Section 12.2 are practical circuits, and may be hooked up with an op-amp and a few external components. There are, furthermore, any number of additional useful op-amp circuits that may be found in handbooks or invented by the user. However, not every new circuit will function exactly as planned. For example, op-amp nonlinearity, frequency response, and noise are some factors that should be considered. As a guide to further applications, then, Section 12.3 provides a brief discussion of considerations that bear on design of new op-amp circuits.

12.1   Characteristics of Operational Amplifiers

The standard circuit symbol for an operational amplifier is shown in Figure 12.2(a). By convention, the output voltage is measured with respect to ground; hence only a single output terminal is needed on the circuit symbol. Power supply connections are assumed to be present, but are usually not shown in a circuit diagram.

A convenient model for the operational amplifier block is shown in Figure 12.2(b). The value of the dependent voltage source in the model is determined by the *difference* in potential between the "+" and "−" input terminals. The "−" input terminal is known as the *inverting input,* because if a small signal is applied to it, and the "+" terminal is grounded, a larger signal with the *opposite* sign appears at the output. Similarly, the "+" terminal is called the *non-inverting input.* The constant A is called the *open-circuit voltage amplification,* because when the output terminal is open-circuited, $v_{OUT}/(v_{(+)} - v_{(-)}) = A$. The resistances R_i and R_o are the input and output resistances, respectively. In op-amps the value of A is very large, typically on the order of 10^5. The value of R_i is typically 10 kΩ to 10 MΩ, although in special op-amps it may be as high as 10^{12} Ω. The value of R_o is generally in the range 10 to 100 Ω.

A manufacturer's specification sheet for a typical op-amp is given in Figure 12.4. This amplifier, the so-called "Type 741" is produced by a number of manufacturers, and is a good example of a modern high-performance integrated operational amplifier.

Operational amplifier circuits can be analyzed by substituting the model for the amplifier into the circuit diagram, as explained in Chapter 3. Conventional loop or node analysis is then applied. This technique has already been illustrated in Example 3.7 and again in Example 12.1.

On the other hand, let us again consider Example 12.1. In that example it was found that the values of the op-amp parameters A and R_i had very little effect on the output voltage of an op-amp circuit. This behavior is quite characteristic of practical op-amp circuits in general. So long as A, R_i, and R_o are in the ranges typical for op-amps, their numerical values affect the output voltage very little. This might lead one to think that, since the values of the parameters do not matter, it should be possible to find v_{OUT} by a simpler method in which A, R_i, and R_o are not carried along as "excess baggage" in the calculations. In fact, a very simple and powerful method for finding v_{OUT} does exist. This method, which we shall call the *ideal op-amp technique,* will now be considered.

The Ideal Op-Amp Technique So long as A and R_i are large and R_o is small, the output voltage of an op-amp circuit is usually almost independent of the values of these parameters. The ideal op-amp technique takes advantage of this fact by, in effect, assuming that $A = \infty$, $R_i = \infty$, $R_o = 0$. These assumptions then allow a simplified procedure to be used for finding v_{OUT}. An op-amp with $A = \infty$, $R_i = \infty$, $R_o = 0$ may be referred to as an *ideal op-amp;* hence the name of the technique.

The procedure for using the ideal op-amp technique is based on two approximate rules: (1) *The voltage difference between the two op-amp input terminals,* $v_{(+)} - v_{(-)},$ *equals zero;* (2) *The currents flowing into the two op-amp input terminals equal zero.* Neither of these statements is precisely true. For instance, if statement (1) were strictly correct, one would expect, according to Figure 12.2(*b*), that the op-amp output voltage would always be zero. However, these two statements are almost true, so that if the rest of the circuit is analyzed using these two assumptions, accurate results are usually obtained.

Before proceeding to demonstrate the use of these assumptions, let us discuss their origins. First, in a correctly designed op-amp circuit, the output voltage of the op-amp will always lie within a certain range, defined by the capabilities of the amplifier. As a practical matter, the absolute value of the

μA741C HIGH PERFORMANCE OPERATIONAL AMPLIFIER

FAIRCHILD LINEAR INTEGRATED CIRCUITS

- NO FREQUENCY COMPENSATION REQUIRED
- SHORT-CIRCUIT PROTECTION
- OFFSET VOLTAGE NULL CAPABILITY

- LARGE COMMON-MODE AND DIFFERENTIAL VOLTAGE RANGES
- LOW POWER CONSUMPTION
- NO LATCH UP

GENERAL DESCRIPTION — The μA741C is a high performance monolithic operational amplifier constructed on a single silicon chip, using the Fairchild Planar* epitaxial process. It is intended for a wide range of analog applications. High common mode voltage range and absence of "latch-up" tendencies make the μA741C ideal for use as a voltage follower. The high gain and wide range of operating voltages provide superior performance in integrator, summing amplifier, and general feedback applications. The μA741C is short-circuit protected, has the same pin configuration as the popular μA709 operational amplifier, but requires no external components for frequency compensation. The internal 6dB/octave roll-off insures stability in closed loop applications. For full temperature range operation (–55°C to +125°C) see μA741 data sheet.

ABSOLUTE MAXIMUM RATINGS

Supply Voltage	±18 V
Internal Power Dissipation	500 mW
Differential Input Voltage	±30 V
Input Voltage (Note 1)	±15 V
Storage Temperature Range TO-99	–65°C to +150°C
Dual-In-Line	–55°C to +125°C
Operating Temperature Range	0°C to +70°C
Lead Temperature (Soldering, 60 sec) TO-99	300°C
(Soldering, 10 sec) Dual-In-Line	260°C
Output Short-Circuit Duration (Note 2)	Indefinite

PHYSICAL DIMENSIONS

in accordance with
JEDEC (TO-99) outline

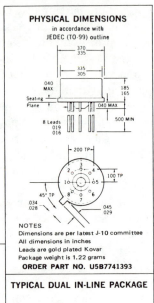

NOTES
Dimensions are per latest J-10 committee
All dimensions in inches
Leads are gold plated Kovar
Package weight is 1.22 grams

ORDER PART NO. U5B7741393

TYPICAL DUAL IN-LINE PACKAGE

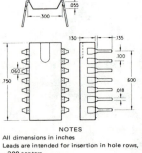

NOTES
All dimensions in inches
Leads are intended for insertion in hole rows,
300 centers

ORDER PART NO. U6E7741393

CONNECTION DIAGRAMS

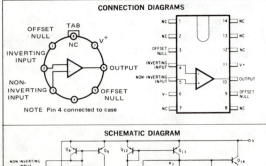

NOTE Pin 4 connected to case

SCHEMATIC DIAGRAM

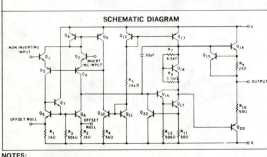

NOTES:
(1) For supply voltages less than ±15 V, the absolute maximum input voltage is equal to the supply voltage.
(2) Short circuit may be to ground or either supply.

*Planar is a patented Fairchild process.

Figure 12.4 Manufacturer's specification sheet for type 741 operational amplifier. (Courtesy Fairchild Semiconductor Company.)

FAIRCHILD LINEAR INTEGRATED CIRCUITS μA741C

ELECTRICAL CHARACTERISTICS ($V_S = \pm 15$ V, $T_A = 25°C$ unless otherwise specified)

PARAMETER	CONDITIONS	MIN.	TYP.	MAX.	UNITS
Input Offset Voltage	$R_S \leq 10$ kΩ		2.0	6.0	mV
Input Offset Current			30	200	nA
Input Bias Current			200	500	nA
Input Resistance		0.3	1.0		MΩ
Large-Signal Voltage Gain	$R_L \geq 2$ kΩ, $V_{out} = \pm 10$ V	20,000	100,000		
Output Voltage Swing	$R_L \geq 10$ kΩ	±12	±14		V
	$R_L \geq 2$ kΩ	±10	±13		V
Input Voltage Range		±12	±13		V
Common Mode Rejection Ratio	$R_S \leq 10$ kΩ	70	90		dB
Supply Voltage Rejection Ratio	$R_S \leq 10$ kΩ		30	150	μV/V
Power Consumption			50	85	mW
Transient Response (unity gain)	$V_{in} = 20$ mV, $R_L = 2$ kΩ				
	$C_L \leq 100$ pF				
Risetime			0.3		μs
Overshoot			5.0		%
Slew Rate (unity gain)	$R_L \geq 2$ kΩ		0.5		V/μs
The following specifications apply for $0°C \leq T_A \leq +70°C$:					
Input Offset Voltage	$R_S \leq 10$ kΩ			7.5	mV
Input Offset Current				300	nA
Input Bias Current				800	nA
Large-Signal Voltage Gain	$R_L \geq 2$ kΩ, $V_{out} = \pm 10$ V	15,000			
Output Voltage Swing	$R_L \geq 2$ kΩ	±10			V

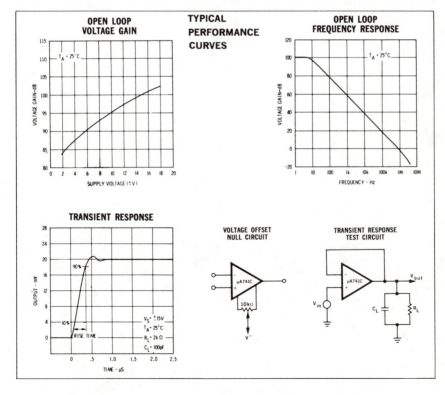

TYPICAL PERFORMANCE CURVES

OPEN LOOP VOLTAGE GAIN

OPEN LOOP FREQUENCY RESPONSE

TRANSIENT RESPONSE

VOLTAGE OFFSET NULL CIRCUIT

TRANSIENT RESPONSE TEST CIRCUIT

output voltage cannot become larger than that of the power supply voltage, that is, $|v_{\text{OUT}}| < V_{\text{SUPPLY}}$. The value of $|v_{\text{OUT}}|$, however, is given by $A|(v_{(+)} - v_{(-)})|$. Thus we have

$$|(v_{(+)} - v_{(-)})| < V_{\text{SUPPLY}}/A \tag{12.1}$$

If A is large, then $|v_{(+)} - v_{(-)}|$ must, according to (12.1) be small; hence the first rule.[1]

The second rule follows from the first, plus the assumption that R_i is large. We see that the input current $(v_{(+)} - v_{(-)})/R_i$ will be a very small quantity; hence the approximation that it is equal to zero.

Let us now demonstrate the use of the ideal op-amp technique, for the circuit of Figure 12.5(a). (This circuit is identical to that of Example 12.1.) We shall calculate v_{OUT}. We observe that $v_{\text{OUT}} = v_{(-)}$. According to rule (1), $v_{(-)} = v_{(+)}$. According to rule (2), there is no current flowing into the (+) input terminal; thus there is no voltage drop in R_S, and $v_{(+)} = v_S$. Thus $v_{\text{OUT}} = v_S$.

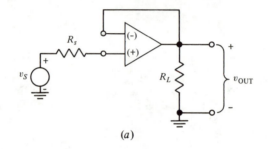

(a)

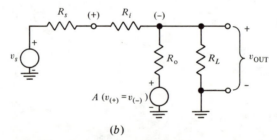

(b)

Figure 12.5 (a) A circuit containing an op-amp, for comparing the ideal op-amp technique with conventional analysis using the op-amp model. (b) Circuit model resulting from substitution of the op-amp model [Figure 12.2(b)] into the circuit of Figure 12.5(a).

---

[1] It is, of course, possible to apply a larger potential difference to the input terminals. In that case, however, the relationship $v_{\text{OUT}} = A(v_{(+)} - v_{(-)})$ would have to fail, since v_{OUT} cannot exceed V_{SUPPLY}. Thus with such large inputs, the op-amp would fail to perform its normal function. In a well-designed op-amp circuit, such large input voltages will not occur.

Note that in this case the approximate technique gives an estimate of the output voltage with no algebraic computation at all!

This result may be checked using the op-amp model and the method of Example 12.1. With the model of Figure 12.2 substituted, the circuit appears as shown in Figure 12.5(b). Writing two node equations for $v_{(+)}$ and $v_{(-)}$ ($= v_{\text{OUT}}$), we have

$$\frac{v_{(+)} - v_S}{R_S} + \frac{v_{(+)} - v_{\text{OUT}}}{R_i} = 0 \tag{12.2}$$

and

$$\frac{v_{\text{OUT}} - v_S}{R_i + R_S} + \frac{v_{\text{OUT}}}{R_L} + \frac{v_{\text{OUT}} - A\,(v_{(+)} - v_{\text{OUT}})}{R_o} = 0 \tag{12.3}$$

Solving these two equations simultaneously, we obtain the result

$$v_{\text{OUT}} = v_S \frac{R_L(R_o + AR_i)}{R_L(R_o + AR_i) + (R_i + R_S)(R_o + R_L)} \tag{12.4}$$

Let us first evaluate this answer for typical values of the circuit parameters. Let us take $R_i = 100 \text{ k}\Omega$, $R_o = 100 \, \Omega$, $A = 10^5$, and $R_S = R_L = 1000 \, \Omega$. Substituting, Equation (12.4) becomes $v_{\text{OUT}} = 0.999989 \; v_S$. Clearly the result obtained by the ideal op-amp technique ($v_{\text{OUT}} = 1.0 \; v_S$) is a very good approximation in this case.

On the other hand, the answer provided by the ideal op-amp technique cannot be correct for the case $R_L = 0$. If R_L approaches zero, it is clear from Figure 12.5 that v_{OUT} must approach zero. The output voltage of any real circuit can, in fact, be reduced toward zero if a sufficiently small load resistance is connected across its output terminals. This effect is known as output *loading*. The ideal op-amp technique fails to be accurate under severe conditions of loading, because under such conditions, the output voltage is no longer independent of the op-amp parameters.

EXAMPLE 12.2

Estimate the output voltage of the following circuit using the ideal op-amp technique.

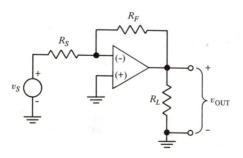

SOLUTION

We write a node equation stating that the sum of all currents entering the $(-)$ input terminal is zero. Using rule (1) of the ideal op-amp technique we assume that $v_{(-)} = 0$, and using rule (2) we assume that no current enters the node from inside the amplifier. Thus the only currents entering this node are those through R_S and R_F, and we have

$$\frac{v_S - (0)}{R_S} + \frac{v_{OUT} - (0)}{R_F} = 0$$

The result is therefore

$$v_{OUT} = -\frac{R_F}{R_S} \bullet v_S \qquad \blacksquare$$

12.2   Operational Amplifier Circuits

In this section several common op-amp circuits are described. In general, the function of each of these circuits is to perform a specific operation on the signal presented to its input. By using these relatively simple circuits, a designer can set up large signal-processing systems quickly and easily.

In Section 3.3 we saw how amplifier circuits can be used as building blocks for use in larger systems. In this section we shall study the properties of op-amp circuits that affect their use as building blocks. The most convenient way to express the properties of the circuits we shall study is to obtain their circuit models. Figure 12.6 shows an amplifier model developed in Chapter 3, suitable for amplifiers with one pair of input terminals and one pair of output terminals. We see that an amplifier represented by this model is fully described when three parameters are given: the input resistance R_i', the output resistance R_o', and the open-circuit voltage amplification A_V'. (The primed symbols distinguish the parameters of the op-amp circuit from those of the op-amp itself, which are R_i, R_o, and A, without the primes.) In feedback amplifier circuits, a complication exists, in that the values of R_i' and R_o' are not purely properties of the amplifier circuit itself, but are affected to some extent by elements connected to the amplifier externally. However, this effect is usually small.

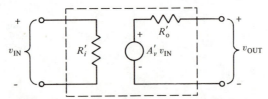

Figure 12.6 A model suitable for an amplifier with a single pair of input terminals and a single pair of output terminals.

As an example, let us consider the op-amp circuit known as the *voltage follower*. This circuit is shown in Figure 12.7. A dashed line has been drawn around the circuit, to indicate that the entire circuit may be regarded as a single building block with one pair of input terminals and one pair of output terminals. If the voltage follower block, as we may now regard it, is then used as part of a still larger system, the system can be analyzed by substituting, for the block, the model of Figure 12.6. Observe the relationship between Figure 12.6 and Figure 12.2. Figure 12.2 shows the model for an op-amp, while Figure 12.6 shows the model for a circuit *containing* an op-amp. The properties of the op-amp block and the op-amp *circuit* block are generally quite different. In this section we shall calculate R_i', R_o', and A_V' for various op-amp circuit blocks.

Since we shall next be concerned with finding the properties of individual op-amp circuits, the significance of the parameters R_i', R_o', and A_V' should be reviewed. The input resistance R_i' is the resistance that would be measured from outside the block, across the input terminals. One way to calculate R_i' is to imagine a "test voltage" v_{TEST} applied to the input terminals, as shown in Figure 12.8(a). If we designate as i_{TEST} the current that flows in response to v_{TEST} in the direction shown, and calculate its value, we then obtain R_i' from $R_i' = v_{TEST}/i_{TEST}$. Since the value of R_i' may be affected by a load resistance R_L connected across the output terminals, the calculation is made with R_L present, as shown. The value of R_i' will then in general be a function of R_L, although the effect of the latter is usually negligible.

Figures 12.8(b) and 12.8(c) show circuits that may be used for calculating R_o'. The latter is equal to the Thévenin resistance of the amplifier block as seen from its output terminals. Thus $R_o' = -v_{OC}/i_{SC}$, where v_{OC} and i_{SC} are, respectively, the open-circuit voltage and short-circuit current at the output terminals, as indicated in Figures 12.8(b) and 12.8(c). In calculating v_{OC} and i_{SC}, a signal source including a source resistance R_S is assumed to be connected at the input terminals, as shown. The value of R_o' that is calculated will in general be a function of R_S, although the effect of R_S (like that of R_L upon R_i') is very often negligible. Last, the open-circuit voltage amplification

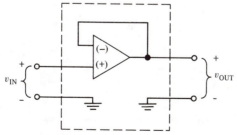

Figure 12.7 The voltage-follower circuit, shown in the form of a building block. This block may be represented by the model shown in Figure 12.6.

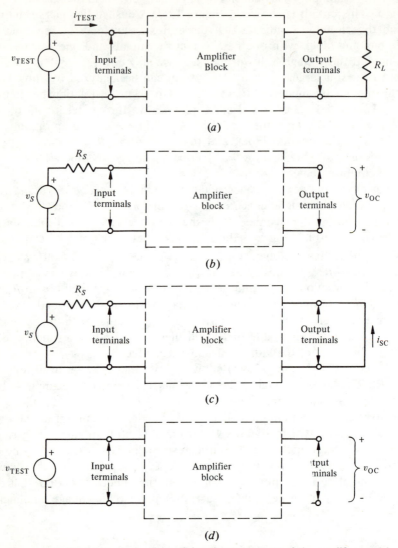

Figure 12.8   Test circuits for determining the parameters of the amplifier model.
(*a*) Circuit for finding R'_i; $R'_i = v_{\text{TEST}}/i_{\text{TEST}}$. (*b*) and (*c*) Circuits for finding R'_0. Values
of v_{OC} and i_{SC} are found using (*b*) and (*c*), respectively; then $R'_0 = -v_{\text{OC}}/i_{\text{SC}}$. (*d*) Circuit
for determining voltage amplification; $A'_v = v_{\text{OC}}/v_{\text{TEST}}$.

is obtained using the test circuit of Figure 12.8(*d*). Here v_{OC} is the output
voltage with no load, when the voltage across the input terminals is v_{TEST};
then $A'_v = v_{\text{OC}}/v_{\text{TEST}}$.

We shall now proceed to discuss the properties of some important op-
amp circuits.

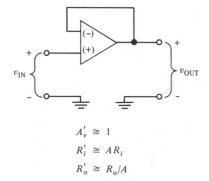

$$A'_v \cong 1$$

$$R'_i \cong AR_i$$

$$R'_o \cong R_o/A$$

Figure 12.9   The voltage-follower circuit.

The Voltage Follower   The voltage follower circuit, which has already been considered in connection with Figure 12.5, is shown in Figure 12.9. This figure shows only the basic circuit itself. To make use of the circuit a source (with its source resistance) would be connected to the input, and a load resistance would be connected to the output (the practical circuit would then have the appearance of Figure 12.5). To find v_{OUT} using the ideal op-amp technique, we note from Figure 12.9 that $v_{OUT} = v_{(-)}$ and that $v_{IN} = v_{(+)}$. But since, according to rule (1), $v_{(-)} = v_{(+)}$, we have that $v_{OUT} = v_{IN}$. Thus we have $A'_v \cong 1$. For comparison, we may use Equation (12.4), obtained by means of the op-amp model. In our present case we are intent on finding the open-circuit voltage gain; hence we must use the limit of (12.4) as $R_L \to \infty$. This limit is

$$A'_v = \frac{R_o + AR_i}{R_o + (A+1)R_i} \tag{12.5}$$

When typical op-amp parameters (for example, $R_i = 100 \text{ k}\Omega$, $R_o = 100 \ \Omega$, $A = 10^5$) are substituted into Equation (13.7), the result is indeed very close to unity.

The values of R'_i and R'_o for the voltage follower cannot be found by the ideal op-amp technique, because R'_i and R'_o, as will be seen shortly, are not independent of the op-amp parameters. (Usually, only the voltages in an op-amp circuit have this convenient property.) We may, however, fall back on the use of the op-amp model. To find R'_i, we use the method of Figure 12.8(a). The voltage follower circuit, Figure 12.9, is inserted for the ampli-fier block in Figure 12.8(a), and the op-amp model of Figure 12.2 is inserted in place of the op-amp itself. The resulting circuit is as shown in Figure 12.10. We may find the value of i_{TEST} by loop analysis. Let us call the two mesh currents i_1 and i_2; then it is clear that $i_{TEST} = i_1$. The two loop equations are

$$v_{TEST} - i_{TEST} \, R_i - R_o(i_{TEST} - i_2) - A(v_{(+)} - v_{(-)}) = 0 \tag{12.6}$$

and

$$A(v_{(+)} - v_{(-)}) - R_o(i_2 - i_{TEST}) - i_2 R_L = 0 \tag{12.7}$$

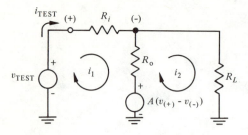

Figure 12.10 Test circuit for determining input resistance of the voltage follower. This circuit is obtained from that of Figure 12.8(a) by substituting the circuit of Figure 12.9 for the amplifier block and then substituting the op-amp model of Figure 12.2 for the operational amplifier.

The quantities $v_{(+)}$ and $v_{(-)}$ may be eliminated by $v_{(+)} = v_{\text{TEST}}$ and $v_{(-)} = v_{\text{TEST}} - i_{\text{TEST}} R_1$. Solving, we find

$$i_{\text{TEST}} = \frac{v_{\text{TEST}}}{R_i(1 + A) + R_o\left[1 - \dfrac{R_o + AR_i}{R_o + R_L}\right]} \qquad (12.8)$$

so that

$$R_i' = v_{\text{TEST}}/i_{\text{TEST}} = R_i \cdot \left[\frac{(A + 1)R_L + R_o}{R_o + R_L}\right] + \frac{R_o R_L}{R_o + R_L} \qquad (12.9)$$

From Equation (12.9) it is readily seen that R_i' is always larger than R_i, and can be very much larger. For example, in the case of $R_L \gg R_o$,

$$R_i' \cong AR_i \qquad \text{(voltage follower)} \qquad (12.10)$$

The voltage follower therefore can have a very large input resistance, typically on the order of 10^{10} Ω. Note that the value of R_i' is dependent on R_i; thus the ideal op-amp technique cannot be used to find R_i'.

A simple physical argument is helpful in understanding why R_i' is so much larger than R_i. The current i_{TEST} is equal to $(v_{(+)} - v_{(-)})/R_i$. If the potential at the $(-)$ terminal were zero, then we would have $i_{\text{TEST}} = v_{(+)}/R_i = v_{\text{TEST}}/R_i$, and R_i' would equal R_i. This is not the case, however; because of the near-unity voltage gain of the circuit, $v_{(-)}(= v_{\text{OUT}})$ is nearly equal to $v_{(+)}(= v_{\text{IN}})$. Thus $v_{(+)} - v_{(-)}$ always has a very small value, and i_{TEST}, which equals $(v_{(+)} - v_{(-)})/R_i$, is much smaller than $v_{(+)}/R_i$. As a consequence of the small value of i_{TEST}, $R_i' = v_{\text{TEST}}/i_{\text{TEST}}$ has a very large value.

We may next calculate R_o' using $R_o' = -v_{\text{OC}}/i_{\text{SC}}$, where v_{OC} is obtained using the circuit of Figure 12.8(b), and i_{SC} is obtained using the circuit of Figure 12.8(c). After substituting the voltage follower circuit into Figures 12.8(b) and 12.8(c), using the op-amp model, and solving for v_{OC} and i_{SC}, we obtain

$$R_o' = \frac{R_o(R_i + R_S)}{R_o + (1 + A)R_i + R_S} \qquad (12.11)$$

We note that in the usual case of $AR_i \gg R_o$, $R_i \gg R_S$. This reduces to the important approximate result

$$R_o' \cong R_o/A \qquad \text{(voltage follower)} \qquad (12.12)$$

Thus the voltage follower typically has an extremely low output resistance, on the order of 10^{-3} Ω! This remarkably low output resistance is caused by the feedback in the circuit.[2]

Since $A_V' \cong 1$ for the voltage follower, it evidently cannot be used to amplify voltages. However, because of its high input resistance and low output resistance the circuit is useful as a separator, or *buffer*, that can be interposed between two parts of the circuit, when it is desired that the parts should not interact. This technique is particularly useful when a low-resistance load is to be driven by a source with high source resistance.

EXAMPLE 12.3

This example demonstrates the use of the voltage follower as a buffer. Let us consider a signal source, represented by a Thévenin equivalent with Thévenin voltage v_S and Thévenin resistance $R_S = 1000$ Ω. Find the voltage v_L that appears across a load resistor $R_L = 10$ Ω; (1) when R_L is connected directly to the source, and (2) when a voltage follower is interposed between the source and R_L. Assume that the properties of the op-amp are $R_i = 10$ kΩ, $R_o = 100$ Ω, and $A = 10^4$.

SOLUTION

In case (1) the situation is as in sketch (*a*). Using the voltage-divider formula, we see immediately that $v_L = R_L v_S/(R_S + R_L) \cong v_S/100$.

In case (2) the circuit is as in sketch (*b*). The voltage follower block may be represented by the amplifier circuit model of Figure 12.6. Substituting the model for the block, we have sketch (*c*). It is now readily seen, by means of the voltage-divider formula, that $v_{\text{IN}} = v_S \cdot R_i'/(R_i' + R_S)$ and that $v_L = A_V' v_{\text{IN}} \cdot R_L/(R_L + R_o')$. Therefore

$$v_L = [A_V' R_L/(R_L + R_o')] \cdot [v_S R_i'/(R_i' + R_S)]$$

We recall that for the voltage follower, $A_V' \cong 1$. If $R_i' \gg R_S$ and $R_L \gg R_o'$ (which will be verified shortly), then $v_L \cong v_S$. In part (1) the result was $v_L = v_S/100$; v_L was reduced by a factor of 100, due to the loading effect of R_L upon the voltage source. Part (2) shows that through the interposition of the buffer stage, the loading effect is removed, and the load voltage is as high as the voltage of the source when the latter is unloaded.

It still remains to be verified that $R_i' \gg R_S$ and $R_L \gg R_o'$ for this particular case. Substituting the given parameter values into Equation (12.9) we find that $R_i' \cong R_i(AR_L/R_o) = 10^3 R_i = 10$ MΩ. Similarly from Equation (12.10) we have $R_o' \cong R_o/A = 10^{-2}$ Ω. Thus indeed we have $R_i' \gg R_S$ and $R_L \gg R_o'$, so that $v_L = v_S$ almost precisely. In this case the voltage follower is a nearly ideal buffer.

---

[2] One should not conclude that since the output resistance is so low, thousands of amperes could be drawn from the output terminals. Independent of its other characteristics, a given op-amp will have a maximum output current, typically 10-50 mA.

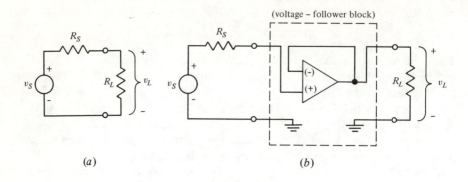

(a)   (b)

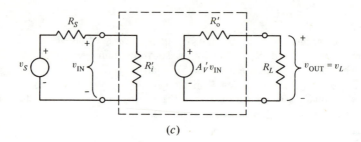

(c)

EXAMPLE 12.4

Find the power amplification of the voltage follower block in part (2) of Example 12.3.

SOLUTION

The power entering the input terminals of the voltage follower block is the product of the voltage across its input terminals times the current flowing into them. The former is $v_{\text{IN}} \cong v_S$ (since $R_i' \gg R_S$) and the latter is $i_{\text{IN}} \cong v_S/R_i'$. Thus

$$P_{\text{IN}} \cong v_S^2/R_i'$$

The power leaving the amplifier is the same as the power dissipated in R_L, which is v_L^2/R_L. Since $v_L \cong v_S$

$$P_{\text{OUT}} = v_S^2/R_L$$

The power amplification is therefore given by

$$A_P = P_{\text{OUT}}/P_{\text{IN}} = R_i'/R_L = 10^7/10 = 10^6$$

Thus we see that, notwithstanding its voltage gain of unity, the power gain of the voltage follower can be very large.

The benefits of the voltage follower circuit, as a result of the use of feedback, are

(1) The voltage gain is nearly independent of the op-amp characteristics (A, R_o, R_i), as well as of the circuit source resistance and load resistance. The circuit is said to be *desensitized* to these parameters.
(2) The input resistance of the circuit block is increased (from R_i to AR_i).
(3) The output resistance of the circuit block is decreased (from R_o to R_o/A).

The general principle is that in feedback circuits gain may be traded for desensitivity and improvement in input and output resistance. Similar effects of feedback will be seen in other circuits in this chapter.

The Non-inverting Amplifier   In cases where voltage gain is desired, the *non-inverting amplifier* of Figure 12.11 may be used. This circuit resembles the voltage follower, except that a voltage divider is inserted in the feedback path. Using the ideal op-amp technique, estimation of A_V' is quite simple. From rule (1) we assume that $v_{(-)} = v_{IN}$; from rule (2) we assume that no current flows from the node between R_1 and R_F into the amplifier terminal. Thus we can write a node equation for $v_{(-)}(= v_{IN})$:

$$\frac{v_{IN}}{R_1} + \frac{v_{IN} - v_{OUT}}{R_F} = 0 \tag{12.13}$$

Solving, we have

$$v_{OUT} = \frac{(R_1 + R_F)}{R_1} \cdot v_{IN} \tag{12.14}$$

so that

$$A_V' = \frac{R_1 + R_F}{R_1} \quad \text{(non-inverting amplifier)} \tag{12.15}$$

The accuracy of this result can be verified by the op-amp model, but since voltages are predicted accurately by the ideal op-amp technique, there is not much to be gained by doing this.

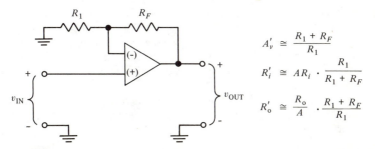

$$A_v' \cong \frac{R_1 + R_F}{R_1}$$

$$R_i' \cong AR_i \cdot \frac{R_1}{R_1 + R_F}$$

$$R_o' \cong \frac{R_o}{A} \cdot \frac{R_1 + R_F}{R_1}$$

Figure 12.11   The non-inverting amplifier. The open-circuit voltage amplification is given, approximately, by $A_v' = (R_1 + R_F)/R_1$.

The input resistance may be calculated by the op-amp model and Figure 12.8(a), just as was done for the voltage follower.

$$R_i' = R_i\left[1 + \frac{AR_1R_L}{R_L(R_1 + R_F) + R_o(R_L + R_1 + R_F)}\right] + R_1\left[\frac{R_F + (R_o\|R_L)}{R_1 + R_F + (R_o\|R_L)}\right]$$

(12.16)

It is easily seen that $R_i' > R_i$. Essentially the same physical explanation applies here as was used to explain the large R_i' of the voltage follower. For the important case of $R_i \to \infty$, $R_o \to 0$, Equation (12.16) implies that

$$R_i' \cong AR_i \cdot \frac{R_1}{R_1 + R_F} \qquad \text{(non-inverting amplifier)} \qquad (12.17)$$

which is smaller than the value of R_i' for the voltage follower by the factor $(R_1 + R_F)/R_1 = A_V'$. The output resistance, obtained using the op-amp model, is given by the formidable expression

$$R_o' = R_o \cdot \frac{R_F[R_1R_F + (R_1 + R_F)(R_i + R_S)]}{(R_o + R_F)[R_1R_F + (R_1 + R_F)(R_i + R_S)] + R_1[AR_iR_F - R_o(R_i + R_S)]}$$

(12.18)

Upon inspection it can be seen that $R_o' < R_o$, and that in the important case $R_i \to \infty$

$$R_o' \cong \frac{R_o}{A} \cdot \frac{R_1 + R_F}{R_1} \qquad \text{(non-inverting amplifier)} \qquad (12.19)$$

Thus this circuit offers nearly the same advantages of high input resistance and low output resistance as the voltage follower.

EXAMPLE 12.5

A multirange electronic voltmeter is to have ranges of 0 to 0.01, 0 to 0.1, and 0 to 1 V. A precision 0- to 1-V panel-mounting voltmeter and an op-amp are to be used. Design a suitable circuit.

SOLUTION

To obtain a voltmeter circuit that covers the range 0 to 0.01 V, we shall need to insert an amplifier with a voltage gain of 100 between the measuring terminals and the 0- to 1-V voltmeter. Then voltages in the range 0 to 0.01 V at the measuring terminals will be converted to voltages in the 0- to 1-V range at the meter. A reading of 0.6 V on the meter, for example, would then be interpreted as a measurement of 0.006 V. To cover the other two ranges, amplifiers with gains of ten and one, respectively, should be used.

Because of its high input resistance, the non-inverting amplifier circuit is well-suited to this application. A possible arrangement is given in the sketch.

Let us arbitrarily choose $R_1 = 1000\ \Omega$. Then for a gain of 100, we need

$$\frac{R_1 + R_{F1}}{R_1} = 100$$

$$R_{F1} = 99\ R_1, = 99,000\ \Omega$$

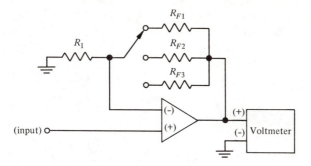

Similarly, for gains of ten and one, we find $R_{F2} = 9000\ \Omega$ and $R_{F3} =$ zero, respectively.

The input resistance of the circuit we have designed will be very high – perhaps 10^9 to $10^{10}\ \Omega$. Thus it will function as an almost ideal voltmeter. ■

The Inverting Amplifier   Another circuit capable of providing voltage amplification is that shown in Figure 12.12. This circuit is known as an *inverting amplifier,* because the output has the opposite sign from the input. The voltage amplification is easily found with the ideal op-amp technique. From rule (1) the voltage at the (−) input terminal is taken to be zero. We write a node equation for this point, assuming, from rule (2), that no current enters the amplifier terminal. This equation is

$$\frac{v_{\text{IN}}}{R_1} + \frac{v_{\text{OUT}}}{R_F} = 0 \tag{12.20}$$

from which we have

$$A_V' = \frac{v_{\text{OUT}}}{v_{\text{IN}}} = -\frac{R_F}{R_1} \tag{12.21}$$

as has already been obtained in Example 12.2.

It should be realized that the Thévenin resistance of a real signal source adds in series with R_1, resulting in a gain change. Thus this circuit will give a voltage amplification that depends on the source resistance, unless the latter is considerably less than R_1.

Unlike the voltage follower and non-inverting amplifier, the input re-

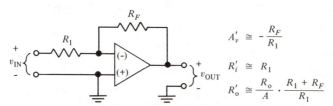

$$A_v' \cong -\frac{R_F}{R_1}$$

$$R_i' \cong R_1$$

$$R_o' \cong \frac{R_o}{A} \cdot \frac{R_1 + R_F}{R_1}$$

Figure 12.12   The inverting amplifier. The open-circuit voltage amplification A_v' is approximately $-(R_F/R_1)$.

sistance of the inverting amplifier can be accurately estimated using the ideal op-amp technique. This is because the input resistance of this circuit is not determined by R_i, but rather by R_1. From rule (1) the voltage at the $(-)$ input terminal of the op-amp is always nearly zero. Thus when v_{TEST} is applied, i_{TEST} is simply v_{TEST}/R_1. Consequently $R_i' \cong R_1$.

The output resistance, calculated using the op-amp model, is

$$R_0' = R_0 \cdot \frac{R_F[R_iR_F + (R_i + R_F)(R_1 + R_S)]}{(R_F + R_0)[R_iR_F + (R_i + R_F)(R_1 + R_S)] + R_i(R_1 + R_S)(AR_F - R_0)} \tag{12.22}$$

Inspection of this result indicates that $R_0' < R_0$. In the limit $R_1 \to \infty$, $A \to \infty$, it becomes

$$R_0' \cong \frac{R_0}{A} \cdot \frac{(R_F + R_1 + R_S)}{(R_1 + R_S)} \tag{12.23}$$

The inverting amplifier is not as useful as a buffer as is the non-inverting amplifier, because of its lower input resistance. It is, however, the basis of the *summing amplifier,* a very useful circuit.

The Summing Amplifier   The circuit known as the summing amplifier is shown in Figure 12.13. This circuit, which is essentially an inverting amplifier with multiple inputs, is capable of adding several input voltages together, each, if desired, with a different scale factor. The operation of the circuit may easily be demonstrated with the ideal op-amp technique. Writing a node equation for the $(-)$ input terminal, we have

$$\frac{v_1}{R_1} + \frac{v_2}{R_2} + \frac{v_3}{R_3} + \frac{v_{\text{OUT}}}{R_F} = 0 \tag{12.24}$$

Solving for v_{OUT}, we have

$$v_{\text{OUT}} = -\frac{R_F}{R_1} \cdot v_1 - \frac{R_F}{R_2} \cdot v_2 - \frac{R_F}{R_3} \cdot v_3 \tag{12.25}$$

Any number of inputs of course may be used. In other respects this circuit resembles the inverting amplifier. However, in the summing amplifier, each input has its own input resistance, equal to the resistance through which it is connected to the op-amp (that is, R_1, R_2, or R_3 in Figure 12.13). The output resistance is identical to the output resistance of the inverting amplifier except that $R_1 + R_S$, in Equation (12.22), must now be replaced by the parallel

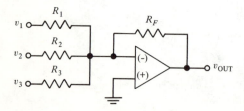

Figure 12.13 The summing amplifier. In the absence of output loading, $v_{\text{OUT}} \cong -(R_F/R_1) v_1 + (R_F/R_2) v_2 + (R_F/R_3) v_3$. (All voltages are measured with respect to ground.)

combination of all the input resistances. That is, if in the circuit of Figure 12.13 the three inputs are fed by sources with Thévenin resistances R_{S1}, R_{S2}, and R_{S3}, then R_1, in Equation (12.22), should be replaced by the parallel combination of $(R_1 + R_{S1})$, $(R_2 + R_{S2})$, and $(R_3 + R_{S3})$.

EXAMPLE 12.6

Two sources of signals may be represented by Thévenin equivalents with Thévenin voltages v_{S1} and v_{S2}, and Thévenin resistances R_{S1} and R_{S2}, both of which are somewhat variable, but in the vicinity of 100 kΩ. It is desired to apply the voltage $v_{S1} + 2v_{S2}$ to a load resistance of 100 Ω. Design a circuit to accomplish this. Available op-amps have $R_i = 100$ kΩ, $R_o = 100$ Ω, $A = 10^5$.

SOLUTION

It is not desirable to connect the voltage sources directly to the inputs of the summing amplifier. The two input resistances of the summing amplifier would be on the order of R_1 and R_2 in Figure 12.13, and in this case would be less than the source resistances R_{S1} and R_{S2}. The two resulting voltage dividers, composed, respectively, of R_1, R_{S1} and R_2, R_{S2}, would cause the input voltages to depend on the source resistances. Sketch (a) illustrates this effect for the case of the v_{S1} input. Since R_{S1} and R_{S2} have been stated to be variable, this effect would make the amplification of the circuit variable, which is highly undesirable.

A better way to handle this problem is to use buffer stages before the summing amplifier. A good circuit would be that in sketch (b). In this circuit the load resist-

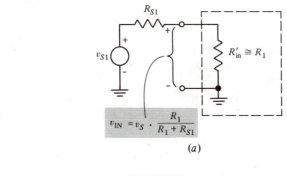

(a)

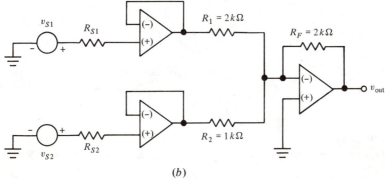

(b)

ances presented to the outputs of the two voltage followers are the two input resistances of the summer, R_1 and R_2, respectively. Since the two R_L's, R_1 and R_2, are much larger than R_o, the input resistances of the two voltage followers are, from Equation (12.9) or (12.10), on the order of 10^{10} Ω. Thus the inputs to the voltage followers load the 10^5-Ω sources practically not at all; that is, for each voltage follower, $v_{IN} = v_S \cdot R_i'/(R_i' + R_S) \cong v_S$. From Equation (12.12) we find that the output resistances of the voltage followers are

$$R_o' \cong \frac{10^2}{10^5} = 10^{-3} \; \Omega$$

This in turn is very low compared with R_1 and R_2, so the summing amplifier does not significantly load the voltage followers. Finally, from Equation (12.23) we can estimate the output resistance of the summing amplifier at approximately 2×10^{-3} Ω. Thus the 100-Ω load does not significantly load the summing amplifier, and the output voltage will be nearly equal to the open-circuit output voltage given by Equation (12.21).   ∎

In addition to the voltage amplifier variations already discussed, op-amp circuits may be devised in which the input or output signals are currents. Two circuits of this type follow.

Current-to-Voltage Converter   The circuit shown in Figure 12.14, which is identical to the inverting amplifier, is also useful for applications requiring a conversion of current to voltage. The output voltage is proportional to the current flowing into the input. To demonstrate this fact we use the ideal op-amp technique. No appreciable current flows into the (−) input, so the source current i_S, which flows out through R_1, must flow in through R_F. Since $v_{(-)} \cong 0$, the output voltage is just the voltage across R_F, which equals $i_S \times R_F$:

$$v_{OUT} = i_S R_F \tag{12.26}$$

The input resistance of the circuit is R_1, as already shown. Since R_1 plays no role in the operation of the converter, we may let $R_1 = 0$, in which case the input resistance $\cong 0$.

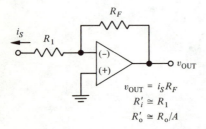

$$v_{OUT} = i_S R_F$$
$$R_i' \cong R_1$$
$$R_o' \cong R_o/A$$

Figure 12.14 Current-to-voltage converter. The output resistance is given by Equation (12.22).

EXAMPLE 12.7
Design a precision electronic ammeter, using an available 0- to 10-V voltmeter

movement with 20,000-Ω resistance. The full-scale reading of the ammeter should be 1 mA.

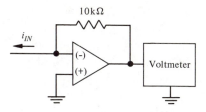

SOLUTION

We may connect the voltmeter to the output of a simple current-to-voltage converter. The voltmeter's full scale voltage, 10 V, should correspond to the maximum current 1 mA. Thus from Equation (12.26)

$$R_F = v_{\text{OUT}}/i_S = \frac{10}{10^{-3}} = 10^4 \ \Omega$$

A suitable circuit is shown in the example sketch. Note that the 20-kΩ resistance of the voltmeter will not load the op-amp output appreciably.   ■

In the case $R_1 = 0$ the input resistance of the current-to-voltage converter is of course not really zero, as suggested by the ideal op-amp technique. When $R_1 \rightarrow 0$, the input resistance is determined by the op-amp parameters. As usual, it may be found with the op-amp model. To solve for the input resistance we again find the ratio $V_{\text{TEST}}/I_{\text{TEST}}$. In this case it is most convenient to imagine a current I_{TEST} flowing into the circuit [as shown in Figure 12.8(a)] and determine V_{TEST}.

Substituting the op-amp model, we obtain the circuit model shown in Figure 12.15. We now write a node equation at node $(-)$:

$$I_{\text{TEST}} - \frac{v_{(-)}}{R_i} + \left(\frac{-Av_{(-)} - v_{(-)}}{R_F + R_o} \right) = 0 \tag{12.27}$$

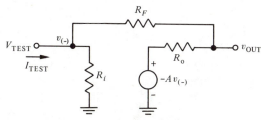

Figure 12.15 Circuit model for finding input resistance of the current-to-voltage converter.

Solving for $v_{(-)}/I_{\text{TEST}}$, we have

$$R_i' = \frac{V_{\text{TEST}}}{I_{\text{TEST}}} = \frac{v_{(-)}}{I_{\text{TEST}}} = \frac{1}{\left(\dfrac{A+1}{R_F + R_o} + \dfrac{1}{R_i}\right)} \tag{12.28}$$

If A is large, the input resistance is approximately $(R_F + R_o)/A$. It is interesting to note that in this circuit the feedback acts to reduce R_i'. (Without feedback, we would have $R_i' = R_i$.) Feedback can either reduce or increase a circuit's input resistance; which it does depends on the form of the circuit.

The output resistance of the current-to-voltage converter is the same as the output resistance of the inverting amplifier and may be found from Equation (12.22). Now, however, the case of greatest interest is $R_S = \infty$, because a current source is generally connected to the input of this circuit. If we give the value of R_o', with $R_S = \infty$, the symbol $R_{o\infty}'$, it is easily shown from (12.22)[3] that

$$R_{o\infty}' \cong R_o \frac{R_i + R_F}{AR_i} \tag{12.29}$$

In most applications $R_i \gg R_F$. When this is so, the output resistance is nearly equal to R_o/A.

Op-amps may also be used to construct current amplifiers or voltage-to-current converters with nearly ideal characteristics.

EXAMPLE 12.8

Show that the magnitude of the output current i_{OUT} in the following circuit is approximately 100 times the input current i_{IN} for any load that satisfies $R_L \ll R_X$. Use the ideal op-amp technique.

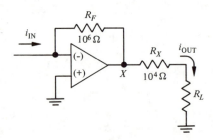

SOLUTION

Because the input current flowing into the ideal op-amp is zero, the current i_{IN} flows through R_F. Further, since $V_{(-)} \approx 0$, $v_X = -i_{\text{IN}} \times R_F$. It is clear that

$$i_{\text{OUT}} = \frac{v_X}{R_X + R_L}$$

---

[3] Typical values of R_o, R_1, and A are assumed.

Thus if $R_L \ll R_X$

$$i_{OUT} \cong -i_{IN} \cdot \frac{R_F}{R_X} = -100\, i_{IN}$$ ∎

We now turn to op-amp circuits that use capacitors as well as resistors in the feedback network. The most important circuit of this type is the integrator circuit, whose output voltage is proportional to the time integral of the input signal.

The Integrator   The basic integrator circuit is given in Figure 12.16. The ideal op-amp technique provides a convenient way to find its output voltage. We write a node equation for the node at the inverting input:

$$C\frac{d}{dt}v_{OUT} + \frac{v_{IN}}{R} = 0 \tag{12.30}$$

Integrating this equation with respect to time, we have

$$v_{OUT} = -\frac{1}{RC}\int_0^t v_{IN}\, dt + V_0 \tag{12.31}$$

where V_0, a constant of integration, is the value of v_{OUT} at time $t = 0$. This constant term in the expression for v_{OUT} arises from charge storage in the capacitor. If the capacitor is uncharged at $t = 0$, then at $t = 0$ the potential difference across C must be zero, and from Figure 12.16 we see that $v_{OUT} = 0$. From (12.31) we then see that if C is initially uncharged, the constant V_0 is zero.

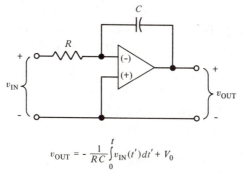

$$v_{OUT} = -\frac{1}{RC}\int_0^t v_{IN}(t')dt' + V_0$$

Figure 12.16   The integrator circuit. This circuit gives an output voltage proportional to the time-integral of the input voltage.

EXAMPLE 12.9

In an electronic camera shutter application, it is desired to generate a signal proportional to the total amount of light that has fallen on a detector during the time it is exposed to the light. (When the output signal reaches some critical voltage, the shutter will be closed.) The detector's open-circuit voltage is given by

$$v_S = \alpha L$$

where L is the light density in photons/(second) and $\alpha = 10^{-15}$ (V) (second)/photon. Furthermore the detector has a Thévenin resistance of 10^6 Ω. Design a circuit that produces an output voltage of -1 V after 10^{11} photons strike the detector.

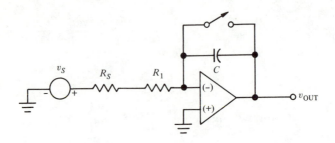

SOLUTION
    The basic circuit is shown in the sketch. We must choose appropriate values for the parameters R and C so that the output design specification is satisfied. Note that a switch is provided to short the capacitor. We shall keep the switch closed until time $t = 0$, so that until this time v_{OUT} equals zero. Then at $t = 0$ we shall open the switch, and v_{OUT} will depart from zero by an amount proportional to $\int v_S\, dt$. Since at $t = 0$, $v_{\mathrm{OUT}} = 0$, we know from Equation (12.31) that $V_0 = 0$.
    From Equation (12.31), we have

$$v_{\mathrm{OUT}} = -\frac{1}{(R_S + R_1)C} \int_0^t v_S\, dt$$

where we have included the source resistance R_S as part of R in Figure 12.16.
    The simplest circuit is obtained if we let $R_1 = 0$; in that case

$$v_{\mathrm{OUT}} = \frac{-\alpha}{R_S C} \int_0^t L\, dt$$

We desire that $v_{\mathrm{OUT}} = -1$ V when $\int_0^t L\, dt = 10^{11}$; thus

$$-1 = -\frac{\alpha}{R_S C} 10^{11}$$

or

$$C = \frac{10^{11}\alpha}{R_S}$$

For the values given here

$$C = \frac{10^{11} \times 10^{-15}}{10^6} = 100 \text{ pF} \qquad \blacksquare$$

Other circuits can be constructed using capacitors and op-amps that perform double integration, differentiation, and various other operations on signals.

EXAMPLE 12.10
Show that the voltage output of the circuit below is proportional to the time derivative of the input voltage.

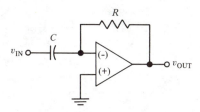

SOLUTION
The operation of this circuit may be demonstrated with the ideal op-amp technique. From rule (1) the voltage at the (−) input terminal is taken to be zero; from rule (2) it is assumed that no current flows into or out of the op-amp input terminal. Writing a node equation for the node between R and C, we have

$$C \frac{d}{dt} v_{IN} + \frac{v_{OUT}}{R} = 0$$

Upon rearrangement, this becomes

$$v_{OUT} = -RC \frac{d}{dt} v_{IN}$$

Note: For reasons having to do with stability and noise, differentiators are less often used than integrators in actual practice. ■

12.3 Design Considerations in Op-Amp Circuits

The previous section dealt with several of the more common op-amp circuits. The ability of an op-amp circuit to perform its intended function depends, however, on several secondary aspects of its behavior. These aspects, for the most part, are not predicted by the op-amp model, which is too simple to give representation to all of an op-amp's properties. Nonetheless, these secondary, or "higher order," effects must be taken into consideration by the designer; otherwise they may interfere with the intended operation of the circuit.

Desensitivity Transistors, op-amps, and other active circuit elements often have properties that vary from unit to unit, or which change with time or temperature. If these variations influence the operation of the circuit as a whole, the circuit will not have dependable properties and will be unreliable. As an example, we recall that the current gain of transistors varies widely, even among transistors of the same type. Thus transistor circuits should be designed so that their operation is affected as little as possible by the value of such parameters. The ability of a feedback circuit to remain unchanged when internal parameters change is called *desensitivity*.

Desensitivity of op-amp circuits with respect to A, the open-circuit voltage amplification of the op-amp, has already been mentioned earlier in this chapter. In production line samples of integrated circuit op-amps, the value of A may vary by a factor of 10 or more from one unit to another. Desensitivity prevents these variations from being translated into variations of the circuit amplification A'_V. We can define a *desensitivity factor* to give a quantitative measure of a circuit's desensitivity with respect to changes in A. Let a change of A, dA, occur. This will cause a smaller change in the circuit's amplification A'_V to occur; we shall call this change dA'_V. What is of interest, really, is the percentage change of A or A'_V that occurs; the percentage changes are determined by the ratios dA/A and dA'_V/A'_V. We may define the desensitivity factor as the ratio of the percentage change in A to the percentage change in A'_V:

$$\text{Desensitivity factor} = \frac{dA/A}{dA'/A'}$$

The larger this factor, of course, the better is the circuit's desensitivity.

EXAMPLE 12.11

For the voltage-follower circuit of Figure 12.5, the ratio v_{OUT}/v_S was given by Equation (12.4). Let us define this ratio as A_L (for "loaded amplification"). Find the desensitivity factor for changes of A_L caused by changes in A. Assume that $R_i = 100 \, \text{k}\Omega$. $R_o = 100 \, \Omega$, $R_S = 10 \, \text{k}\Omega$, $R_L = 10 \, \Omega$, and assume that A is in the vicinity of 10^5.

SOLUTION

With the given values, Equation (12.4) takes the form

$$A_L \cong \frac{10^6 A}{10^6 A + 10^7}$$

Differentiating both sides of this equation, we have

$$dA_L \cong \frac{10^{13}}{(10^6 A + 10^7)^2} \, dA$$

In the vicinity of $A = 10^5$

$$dA_L \cong 10^{-9} \, dA$$

For the voltage-follower circuit, $A_L \cong 1$. Hence the desensitivity factor is given by

$$\text{Desensitivity factor} = \frac{dA/A}{dA_L/A_L} = \frac{dA/10^5}{10^{-9} \, dA/1} = 10^4 \qquad \blacksquare$$

Linearity Desensitivity with respect to A tends to make the op-amp circuit linear, even when the op-amp itself is not. Suppose that for a certain op-amp the graph of v_{OUT} versus $(v_{(+)} - v_{(-)})$ is as shown in Figure 12.17. Obviously for this amplifier v_{OUT} is not linearly proportional to $(v_{(+)} - v_{(-)})$, and if it were used without feedback, severe distortion of the input signal would result. However, if this op-amp is used in a feedback circuit, the output of the entire *circuit* can still be almost linearly proportional to its input. This can be seen as follows. Referring to Figure 12.17, we see that when $|v_{(+)} - v_{(-)}| < 0.5$ mV, $A \simeq 2000$. On the other hand, when $v_{(+)} - v_{(-)}$ reaches 1.0 mV, A [which equals $v_{OUT}/(v_{(+)} - v_{(-)})$] has risen to about 10,000. But as was seen above, the output voltage of the feedback circuit is nearly independent of the value of A. Thus if this op-amp is used in a voltage follower, v_{OUT} will be very nearly equal to $1.0 \times v_{IN}$, notwithstanding the nonlinearity of the op-amp. This is a very useful property of feedback circuits. It makes it possible to construct circuits that are nearly linear using inexpensive simple op-amps, even though the latter may themselves be rather nonlinear. The linearizing action of negative feedback is useful in all kinds of amplifier circuits, whether or not they contain op-amps. For instance, the linearity of the push-pull output stage shown in Figure 11.10 can be greatly improved by providing a negative feedback loop around the stage.

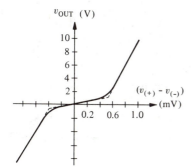

Figure 12.17 Plot of v_{OUT} versus $(v_{(+)} - v_{(-)})$ for a nonlinear op-amp.

Offset From looking at the op-amp model, one might believe that the output voltage of the op-amp would be zero whenever $v_{(+)} - v_{(-)} = 0$. However, in practice real op-amps are always slightly "unbalanced," or asymmetrical, with respect to the two inputs. This leads to the condition known as *voltage offset*, in which the output voltage fails to be zero even when $v_{(+)} - v_{(-)} = 0$.

The amount of voltage offset existing in an op-amp can be specified by stating how large a dc voltage must be applied between the $(+)$ and $(-)$ input terminals in order to make the amplifier's output voltage zero. The required input voltage is called the *input offset voltage*, and is typically in the range 0.5 to 5 mV.

An op-amp with non-zero voltage offset can be represented by the model shown in Figure 12.18(*a*). Here the real op-amp is represented as an offset-

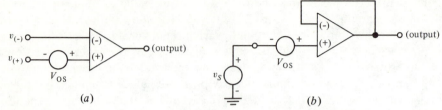

Figure 12.18   Voltage offset in an operational amplifier. (*a*) Model for an amplifier with voltage offset. The model is composed of an offset-free op-amp plus an offset voltage source. (*b*) Circuit model for a voltage follower when the op-amp has voltage offset.

free op-amp combined with an offset voltage source V_{OS}. It is easily seen that when a voltage equal to the input offset voltage V_{OS} is applied between the terminals of this model, the output is reduced to zero, as required by the definition of V_{OS}.

When an op-amp containing voltage offset is used in a circuit, the offset voltage acts to produce a spurious output voltage. For example, consider the voltage follower circuit model shown in Figure 12.18(*b*), in which the op-amp model of 12.18(*a*) has been used. Here it is evident that the effective input voltage being applied to the op-amp is $v_S + V_{OS}$, and the output voltage will accordingly equal $v_S + V_{OS}$, not v_S.

A related effect occurs because small dc currents (which, like the voltage offset, are not indicated by the op-amp model of Figure 12.2) flow through the op-amp input terminals. These currents arise from the biasing of the input transistors of the op-amp, as described in Section 11.1. The average of the two dc currents flowing through the two input terminals is known as the *input bias current,* and is typically in the range 0.5 to 5 μA. The difference between the two dc bias currents, which arises from asymmetry of the amplifier, is known as the *input offset current.* It is typically ten times less than the input bias current.

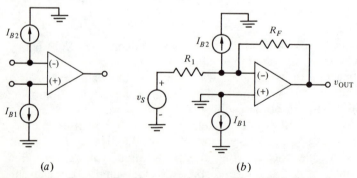

Figure 12.19   Current offset in an operational amplifier. (*a*) Model for the op-amp that gives representation to the dc bias currents that flow through the input terminals. This model consists of two current sources plus an op-amp like that of Figure 12.2 (*b*) Use of this model in an inverting amplifier circuit.

The dc input currents may give rise to additional offset of the output voltage, beyond that caused by V_{OS}. To observe the effects of the dc input currents, the op-amp model shown in Figure 12.19(a) can be used. Here the op-amp is represented as an input current-free op-amp plus two current sources that represent the bias currents that flow into the real op-amp's input terminals. If the op-amp is used in an inverting amplifier circuit, we have the circuit shown in Figure 12.19(b). Let us calculate the output voltage of this circuit arising from I_{B1} and I_{B2}, when $v_S = 0$. Source I_{B1} has no effect, since its current flows directly to ground. However, a current equal to I_{B2} must enter the (−) node through some path. No current flows through R_1; we know this because $v_{(-)} = v_{(+)} = 0$, and we have set $v_S = 0$. Thus there is no voltage across R_1 and hence no current flows through it. We may then conclude that a current equal to I_{B2} enters the (−) node through R_F. Since $v_{(-)} = 0$, we must have $v_{OUT}/R_F = I_{B2}$. Thus the output voltage resulting from the bias currents in this case is

$$\Delta v_{OUT} = I_{B2}R_F \tag{12.32}$$

If $I_{B2} = 1\ \mu A$ and $R_F = 1000\ \Omega$, $\Delta v_{OUT} = 1$ mV.

In applications where precision is required, it may be useful to reduce offset with a special nulling circuit. Such circuits can take on various forms, depending on the rest of the op-amp circuit. A nulling circuit for an inverting amplifier is shown in Figure 12.20(a). The adjustable current source cancels

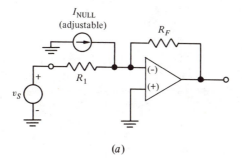

(a)

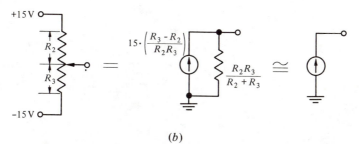

(b)

Figure 12.20 (a) A nulling circuit for canceling out offset in an inverting amplifier. A means of obtaining the adjustable nulling current source is shown in (b).

the effect of I_{B2} by removing the need for any current to flow through R_F. Additional nulling current of either sign can be added to balance out the effects of input voltage offset. A way in which the adjustable current source can be constructed is shown in Figure 12.20(b). If R_2 and R_3 are chosen to be large, the Norton equivalent of the subcircuit shown in Figure 12.20(b) is a current source in parallel with a large Norton resistance. The latter is not desired, but has little effect on the circuit except to reduce the effective value of R_i (the op-amp's input resistance) with which it is in parallel. The arrow in Figure 12.20(b) indicates an adjustable contact on the resistor that can be moved up or down, to adjust R_2 and R_3.[4] In practice, final adjustment of R_2 and R_3 would be made while watching a voltmeter connected to the op-amp output; thus exact balance is obtained by experiment.

EXAMPLE 12.12

An op-amp used in an inverting amplifier has an input offset voltage V_{OS} of approximately 5 mV. It is desired to null the output voltage with the circuit shown in Figure 12.20. Calculate approximate values of R_2 and R_3. Assume that the input bias currents are negligible, and take $R_F = 2$ kΩ, $R_1 = 1$ kΩ.

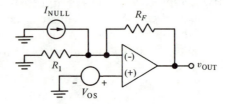

SOLUTION

To see the effect of V_{OS}, let us substitute the model of Figure 12.18(a). The result is given in the sketch . (Since we are now concerned with the output voltage in the absence of input signal, v_S has been set to zero.) Writing a node equation for the ($-$) node, where, according to the ideal op-amp rules, the voltage is V_{OS}, we have

$$I_{\text{NULL}} - \frac{V_{OS}}{R_1} + \frac{v_{\text{OUT}} - V_{OS}}{R_F} = 0$$

We wish to find I_{NULL} when $v_{\text{OUT}} = 0$. Setting v_{OUT} to zero in the above equation, we obtain

$$I_{\text{NULL}} = V_{OS} \left(\frac{1}{R_1} + \frac{1}{R_F} \right) = V_{OS} \cdot \frac{R_1 + R_F}{R_1 R_F}$$

Using the given values for V_{OS}, R_1, and R_F, we have $I_{\text{NULL}} \cong 7.5 \times 10^{-6}$ A.

The value of the Norton equivalent current source in Figure 12.20(b) is to be equal to I_{NULL}. Thus

$$15 \cdot \frac{(R_3 - R_2)}{R_2 R_3} = 7.5 \cdot 10^{-6} \text{ A}$$

---

[4] Such an arrangement is called a *potentiometer*. Potentiometers (or "pots") are widely used as volume controls and so forth.

This equation by itself does not determine the values of R_2 and R_3; there is still an element of choice possible. However, we should make both R_2 and R_3 as large as the above equation will allow, so that the Norton resistance of the combination will be as large as possible. A possible choice would be $R_3 = 10$ MΩ. Solving the above equation for R_2, we find $R_2 = 1.7$ MΩ.

In practice, one might choose for R_2 and R_3 a potentiometer whose total resistance $(R_2 + R_3)$ has a fixed value of, say, 11.7 MΩ. One would then obtain exact null by experiment, adjusting the movable contact on the potentiometer until zero output voltage is observed.   ∎

Although a nulling circuit can be used to set the output voltage to zero at a given temperature, changes in temperature will cause the input offset voltage and bias currents to change, thus spoiling the balance. The resulting change in the dc output voltage with temperature is known as *drift*. To aid in estimating the drift, manufacturers often specify the rate of variation of the offset voltage and bias currents per degree of change in temperature. Some typical values are: for input offset voltage, 5 μV/°C; for input offset current, 4 nA/°C.

Just as imperfections in the op-amp give rise to a spurious dc input voltage, they also give rise to spurious ac input voltages. These spurious signals are known as *noise*, and are present not just in op-amps but in all electronic circuits. Noise, for example, may be represented by the model of Figure 12.18(*a*). Instead of a constant value V_{OS}, the noise voltage source has a constantly changing random value. Noise signals such as this are the source of the "hissing" sound from a phonograph loudspeaker when other signals are absent.

Noise is important because it places a limit on how small an input signal may usefully be applied. It is impossible to distinguish amplifier outputs resulting from signal from other outputs resulting from noise. If input signals are used that are of the same order of magnitude as the noise voltage, the output will be contaminated with noise-caused variations, so that the output signal is unrecognizable and useless. Thus the system designer must take care that his signals exceed the noise level by a safe margin.

The subject of random signals and noise belongs to a branch of electrical engineering known as *communication theory*. This field is large, important, useful, and interesting. Although we shall not explore it further here, it is highly recommended to the student for future study.

Frequency Response of Operational Amplifier Circuits   Up to this point we have been tacitly assuming that the op-amp can always respond instantaneously to input signals. However, real op-amps cannot respond instantaneously. As a consequence, op-amp circuits behave differently for rapidly varying signals than for slowly varying signals.

In practice, the gain of an op-amp A is always a function of the frequency of the signal being amplified. For example, the frequency response of the 741 op-amp is shown in Figure 12.21(*a*). Figure 12.21(*a*) refers to the op-amp alone, without any feedback loop; hence the figure is referred to as a plot of open-loop gain versus frequency. If the bandwidth is defined, as in previous

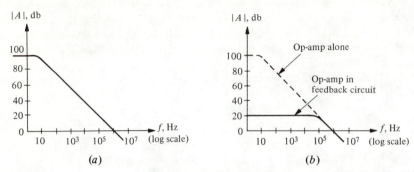

Figure 12.21   (*a*) A plot of open-loop voltage amplification versus frequency for the type 741 operational amplifier. (*b*) Plot of voltage amplification versus frequency for the amplifier of Figure 12.21(*a*), used in a feedback circuit with gain of ten (solid curve). The open-loop response of Figure 12.21(*a*) is repeated for comparison (dashed curve). Note that all scales are logarithmic.

chapters, as being the range of frequencies over which A is within 3 db of its maximum value, then we see that the bandwidth of the 741 amplifier is about 10 Hz. The characteristic displayed in Figure 12.21(*a*) is typical for general purpose op-amps; as will be shown later when we discuss stability, the op-amp internal circuitry is purposely designed to give this characteristic.

It might at first appear that such an op-amp would not be useful for signals with frequency greater than 10 Hz. However, this is not the case. As we have seen, negative feedback makes the op-amp circuit insensitive to the value of A. Consider, for example, the non-inverting amplifier of Figure 12.11. It is a straightforward matter to substitute the op-amp model into this circuit and to calculate its voltage amplification as a function of A and other parameters. [The procedure is like that used to obtain Equation (12.4) for the voltage follower.] After simplifying the result by assuming that R_i is very large and R_o very small, we find that

$$A'_V = \frac{A(R_1 + R_F)}{(A + 1)\, R_1 + R_F} \tag{12.33}$$

Suppose R_1 and R_F are chosen as $R_1 = 1$ kΩ, $R_F = 9$ kΩ, so as to give a gain of ten at low frequencies. We may then plot A'_V as a function of frequency, using $A(f)$ from Figure 12.21(*a*). The result is plotted in Figure 12.21(*b*).

We note from Figure 12.21(*b*) that the bandwidth of the feedback circuit extends approximately to that frequency at which A equals the circuit gain. If the circuit gain were increased by a factor of 10, we see from the figure that the bandwidth would be reduced by a factor of 10. Thus there is a "trade-off" between gain and bandwidth, which may be expressed by stating that the product of gain times bandwidth, known as the *gain-bandwidth product*, is a constant. This is a common situation in amplifier design generally; that is, it is often found that increasing the gain of a given amplifier results in a proportional decrease in bandwidth. Figure 12.21(*a*) shows the gain-bandwidth product of the 741 amplifier to be about 10^6 Hz.

It may be objected that the technique just described for increasing the bandwidth is not helpful if a larger voltage amplification is required. However, one can add additional amplifier stages to obtain the desired amplification. For example, four stages, each with the frequency response shown in Figure 12.21(b), could be used in cascade to provide a gain of 10^4. The bandwidth will still be approximately 10^5 Hz.[5]

EXAMPLE 12.13

Using type 741 op-amps, design a circuit that has a gain of 100 over the frequency range 0 to 10 kHz. The circuit should have a very large input resistance ($> 10^7\ \Omega$).

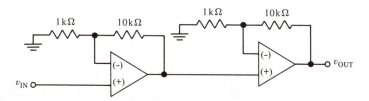

SOLUTION

The required circuit has a gain-bandwidth product of 10^6 Hz. This is just about equal to the gain-bandwidth product of a single 741 amplifier. However, using a single amplifier stage would leave no safety factor, and it is possible that the highest frequencies would be underamplified. This danger can be eliminated by using two amplifier stages, each with a gain of ten. From Figure 12.20(b) we see that the bandwidth of each stage will then be nearly 10^5 Hz. The bandwidth of the two stages in cascade will be 0.64×10^5 Hz, according to footnote 5.

Since it is required that the input resistance of the circuit be high, non-inverting amplifiers are suitable. The required circuit could then appear as in the sketch. ■

Slew Rate The foregoing discussion implies that the frequency response is unaffected by the amplitude of the signal. However for large input signals, op-amps also are bound by another kind of limitation on speed of response, having to do with the rate at which the output voltage of the amplifier can change. The maximum rate of change of the output voltage of which the op-amp is capable is known as its *slew rate*, ρ. In a typical op-amp the slew rate is determined by the internal currents available to charge the various internal capacitances inside the op-amp. Typically the slew rate is in the range 0.5 to 5 V/μsec.

As an example of the kind of limitation imposed on op-amp performance by the slew rate, let us consider a sinusoidal input signal $v_{IN} = V_0 \sin \omega t$. If the voltage amplification of the feedback circuit is A_v', then the output voltage ought

---

[5] The bandwidth of N identical stages in cascade is actually slightly less than the bandwidth of the individual stages. If the bandwidth of a single stage is B, the bandwidth of N identical stages in cascade can be shown to be $B \cdot \sqrt{2^{1/N}-1}$.

to be $A'_V V_0 \sin \omega t$. In this case the rate of change of the output voltage is $A'_V V_0 \omega \cos \omega t$. The largest value reached by this rate of change is $A'_V V_0 \omega$. If $V_0 \omega$, the product of signal amplitude and frequency, is sufficiently small, dv_{OUT}/dt will be less than the slew rate ρ at all times, and normal operation will be obtained. However, if the product of amplitude and frequency of the input signal is too large, so that $V_0 \omega < \rho$ is not satisfied, the output signal of the op-amp will not be able to change fast enough to follow the input signal, and the output will become distorted. The reader should understand that the slew rate limitation is a different kind of restriction than that imposed by frequency response; the former depends on signal amplitude, while the latter does not. In general both must be considered.

EXAMPLE 12.14
   An op-amp used in a voltage-follower circuit has the frequency response shown in Figure 12.21(a). Its slew rate is 0.5 V/μsec. At time $t = 0$ the input voltage is suddenly changed from zero to the value V_1. (1) Estimate the rise time required for the amplifier output to reach its new value, assuming V_1 is small. (2) Determine how large V_1 would have to become before the output begins to be affected by the slew rate limitation.

SOLUTION
   (1) The voltage-follower circuit has a gain of unity, or 0 db. The circuit's bandwidth will extend approximately to the frequency at which A (the open-circuit voltage gain of the op-amp) has decreased to unity. Referring to Figure 12.21(a) we see that the voltage-follower bandwidth is about 10^6 Hz. If we assume that the rolloff of gain above 10^6 Hz is due to a single capacitor somewhere in the circuit, we can estimate, using the reasoning of Chapter 2, that the RC time constant of the circuit is of the order of $(2\pi \times 10^6)^{-1} = 1.6 \times 10^{-7}$ sec. From the discussion of Section 2.3, we expect that the duration of the transient, or rise time, is about equal to the time constant of the circuit, that is, 1.6×10^{-7} sec.

   A graph of the amplifier output versus time under these conditions is, in fact, included in the manufacturer's specifications for the 741 op-amp, and is shown toward the bottom of Figure 12.4. A common definition of rise time is that it is the time required for a signal to increase from 10 to 90% of its ultimate value. The time thus defined is known as the 10 to 90% rise time. From Figure 12.4 we see that the 10 to 90% rise time is about 0.3 μsec, in order-of-magnitude agreement with our estimate.

   (2) Inspecting the step-response curve of Figure 12.4 again, we see that the output voltage rises from 10 to 90% of its final value V_1 in a time on the order of 3×10^{-7} sec. During this rise the average value of dv_{OUT}/dt is about 0.8 $V_1/(3 \times 10^{-7}$ sec) V/sec $= 2.6\ V_1 \times 10^6$ V/sec $= 2.6\ V_1$ V/μsec. The slew rate of the op-amp ρ is $\rho = 0.5$ V/μsec. Thus the value of V_1 at which the slew rate limitation becomes important is given by

$$2.6\ V_1 = \rho = 0.5$$
$$V_1 = 0.19 \text{ V}$$

If V_1 is larger than this, the output voltage will approach its final value ($10\ V_1$) at a rate equal to the maximum slew rate ρ. The time required for the transient in this case will be of the order of V_1/ρ. ∎

Sinusoidal Analysis of Op-Amp Circuits   The phasor techniques introduced in Chapter 4 can be used to study the response of op-amp circuits to
sinusoidal signals. For example, we may calculate a circuit's complex voltage
amplification $\mathbf{A}_V'$. This quantity by definition equals $\mathbf{v}_{OUT}/\mathbf{v}_{IN}$, where $\mathbf{v}_{OUT}$ and $\mathbf{v}_{IN}$
are, respectively, the phasors representing the output and input voltages. Since
the phasors $\mathbf{v}_{OUT}$ and $\mathbf{v}_{IN}$ are complex numbers, $\mathbf{A}_V'$ is also a complex number.
(Complex numbers are represented by boldface symbols.) We can also calcu-
late the input impedance $\mathbf{Z}_{IN}$, defined by $\mathbf{Z}_{IN} = \mathbf{v}_{TEST}/\mathbf{i}_{TEST}$, and the output impedance $\mathbf{Z}_{OUT}$, defined by $\mathbf{Z}_{OUT} = -\mathbf{v}_{OC}/\mathbf{i}_{SC}$. (For the significance of $\mathbf{v}_{TEST}$, $\mathbf{i}_{TEST}$,
$\mathbf{v}_{OC}$, and $\mathbf{i}_{SC}$ refer to Figure 12.8.) The input and output impedances are, of
course, generalizations of the input and output resistances, and play a similar
role.

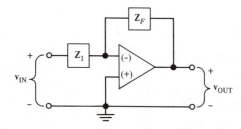

Figure 12.22   A generalized circuit, of which
the integrator and the inverting amplifier are
two special cases.

Let us consider the family of circuits represented by Figure 12.22. This
generalized circuit contains two elements that are designated only by their
impedances, $\mathbf{Z}_1$ and $\mathbf{Z}_F$. [Compare Figure 4.6(*b*).] The generalized circuit can
represent the integrator of Figure 12.16 if we take $\mathbf{Z}_F = 1/j\omega C$, $\mathbf{Z}_1 = R$. However, Figure 12.22 can also represent the inverting amplifier, Figure 12.12,
if we take $\mathbf{Z}_1 = R_1$, $\mathbf{Z}_F = R_F$. Analysis of this circuit can be performed either
by way of the op-amp model or by the ideal op-amp technique. In connection

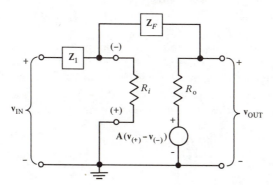

Figure 12.23   Test circuit for finding the complex
voltage gain of the circuit of Figure 12.22, using the
op-amp model.

with the model, it should be understood that the op-amp can introduce phase shifts, and therefore its open-circuit voltage gain should be represented as a complex number $\mathbf{A}$. In general, $\mathbf{A} = \mathbf{A}(\omega)$ is a function of frequency.

Let us use the op-amp model to find the complex voltage amplification $\mathbf{A}'_V$ for Figure 12.22. On substituting the op-amp model, we obtain the circuit model shown in Figure 12.23. We now write two node equations for $\mathbf{v}_{(-)}$ and $\mathbf{v}_{\text{OUT}}$:

$$\frac{\mathbf{v}_{(-)} - \mathbf{v}_{\text{IN}}}{\mathbf{Z}_1} + \frac{\mathbf{v}_{(-)} - \mathbf{v}_{\text{OUT}}}{\mathbf{Z}_F} + \frac{\mathbf{v}_{(-)}}{R_i} = 0 \tag{12.34}$$

and

$$\frac{\mathbf{v}_{\text{OUT}} - \mathbf{v}_{(-)}}{\mathbf{Z}_F} + \frac{\mathbf{v}_{\text{OUT}} - \mathbf{A}(0 - \mathbf{v}_{(-)})}{R_o} = 0 \tag{12.35}$$

Solving, we find that

$$\mathbf{A}'_V = \frac{\mathbf{v}_{\text{OUT}}}{\mathbf{v}_{\text{IN}}} = -\frac{R_i(\mathbf{A}\mathbf{Z}_F - R_o)}{R_i[(\mathbf{A}+1)\,\mathbf{Z}_1 + \mathbf{Z}_F + R_o] + \mathbf{Z}_1\mathbf{Z}_F + R_o\mathbf{Z}_1} \tag{12.36}$$

It is easy to see that in the limit as $R_i \to \infty$, $R_o \to 0$, $\mathbf{A} \to \infty$, (12.36) becomes

$$\mathbf{A}'_V \cong -\frac{\mathbf{Z}_F}{\mathbf{Z}_1} \tag{12.37}$$

This is just the generalization of the former result for the inverting amplifier, Equation (12.21), with $\mathbf{Z}_1$ and $\mathbf{Z}_F$ taking the place of R_1 and R_F. In fact, in all respects analysis of Figure 12.22 is the same as that of the inverting amplifier, Figure 12.12, except for the replacement of R_1 and R_F by $\mathbf{Z}_1$ and $\mathbf{Z}_F$. Consequently we may conclude that just as the input resistance of the inverting amplifier is R_1, the input impedance of Figure 12.22 is $\mathbf{Z}_1$. The output impedance of the circuit may be obtained from Equation (12.22) or (12.23) by replacing R_F by $\mathbf{Z}_F$ and R_1 by $\mathbf{Z}_1$.

When considering questions of frequency response, the assumption $\mathbf{A} \to \infty$ may not be a good one. It will be remembered that the value of $\mathbf{A}(\omega)$ decreases rapidly as frequency increases. In high-frequency calculations one may still use the approximations $R_i \cong \infty$, $R_o \cong 0$, while $\mathbf{A}(\omega)$ remains a finite number. In this approximation the amplification (12.36) becomes

$$\mathbf{A}'_V \cong \frac{\mathbf{A}\mathbf{Z}_F}{(\mathbf{A}+1)\mathbf{Z}_1 + \mathbf{Z}_F} = -\frac{\mathbf{Z}_F}{\mathbf{Z}_1}\left[\frac{1}{1 + \dfrac{1}{\mathbf{A}}\left(\dfrac{\mathbf{Z}_1 + \mathbf{Z}_F}{\mathbf{Z}_1}\right)}\right] \tag{12.38}$$

The second expression given in Equation (12.38) explicitly shows $\mathbf{A}'_V$ as its infinite-gain value, multiplied by a correction factor.

EXAMPLE 12.15

Let the input to the op-amp circuit in sketch (*a*) be sinusoidal, with amplitude 1 mV. Find the amplitude of the output sinusoid as a function of frequency. Assume that the frequency range of interest is low, so that **A** is always very large.

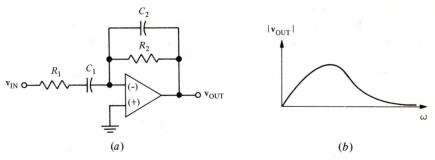

(*a*)   (*b*)

SOLUTION

This circuit is of the general type shown in Figure 12.22. Since **A** is very large at all frequencies of interest, we may use Equation (12.37). The complex voltage gain $\mathbf{A}'_V$ therefore is $-(\mathbf{Z}_F/\mathbf{Z}_1)$. In this case $\mathbf{Z}_F = R_2/(1 + j\omega R_2 C_2)$ and $\mathbf{Z}_1 = (1 + j\omega R_1 C_1)/j\omega C_1$. Thus the complex voltage gain is given by

$$\mathbf{A}'_V = \frac{j\omega R_2 C_1}{1 - \omega^2 (R_1 C_1 R_2 C_2) + j\omega (R_1 C_1 + R_2 C_2)}$$

Accordingly

$$|\mathbf{v}_{OUT}| = |\mathbf{A}'_V \mathbf{v}_{IN}| = |\mathbf{A}'_V| \bullet |\mathbf{v}_{IN}|$$

$$= \sqrt{\mathbf{A}'_V \mathbf{A}'^*_V} \bullet (1 \text{ mV})$$

$$= (1 \text{ mV}) \bullet \frac{(R_2 C_1)\omega}{\sqrt{1 + \omega^2 (R_1^2 C_1^2 + R_2^2 C_2^2) + \omega^4 (R_1 C_1 R_2 C_2)^2}}$$

A graph of $|\mathbf{v}_{OUT}|$ as a function of ω is given in sketch (*b*). An amplifier circuit that has maximum amplification in a certain range of frequencies, and little or none elsewhere is known as a *bandpass amplifier*. Bandpass circuits are sometimes constructed using inductors. (See Example 4.16.) Inductors, however, have size and cost disadvantages. The present circuit shows how similar behavior may be obtained using only resistors, capacitors, and an op-amp. This circuit also has an advantage in that the shape of its transmission curve (the curve of $|\mathbf{v}_{OUT}|$ versus ω) is nearly unaffected by the load to which the output is connected. This would be very difficult to accomplish in a passive circuit.   ■

Stability   When an amplifier performs its function reliably under all normal operating conditions, it is said to be *stable*. The opposite condition, instability, can come about in various ways. Most often, an instability manifests itself by the appearance of an output signal when no input signal is applied. Usually this spurious output signal is sinusoidal in form; in that case *oscillation* is said to occur. Instability in feedback amplifiers is a very common difficulty, and usually must be prevented, as it interferes with normal operation of the circuit.[6]

In general, instabilities arise from feedback of an improper kind. For example, there is usually a time delay, or phase shift, experienced by a signal as it passes through an op-amp. This phase shift typically is nearly zero at the lowest frequencies, but increases at higher frequencies. Since shifting the phase of a sinusoid by 180° is equivalent to multiplying the sinusoid by -1, we see that the negative feedback occurring at low frequencies can reverse in sign as a result of phase shift, and turn into positive feedback at higher frequencies. Positive feedback is very likely to lead to instability. Speaking intuitively, this is because the signal fed back from output to input then tends to reinforce the signal already present at the input (rather than subtracting from it); the reinforced input signal makes the output still larger, which makes the input larger, and so forth; loosely speaking, the circuit "runs away."

In the following discussion we shall develop a simple criterion for the determination of stability in feedback circuits. This criterion is a variant of a general test for stability known as the *Nyquist criterion.*

First it is necessary to express the gain of the feedback circuit $\mathbf{A}_V'$ in a general form. We shall throughout this discussion assume that the op-amp output resistance R_o is zero, and that its input resistance R_i is infinitely large. The open-circuit voltage amplification of the op-amp $\mathbf{A}$ is a complex number; its magnitude and argument are both functions of frequency. $\mathbf{A}$ is not to be regarded as infinitely large in this discussion; in general, its magnitude approaches zero as the frequency increases.

Under these assumptions, it can be shown that the gain $\mathbf{A}_V'$ of a negative feedback circuit can be written in the form

$$\mathbf{A}_V' = \frac{\mathbf{c}(\omega)\,\mathbf{A}(\omega)}{1 + \mathbf{A}(\omega)\,\mathbf{f}(\omega)} \tag{12.39}$$

Here, $\mathbf{c}(\omega)$ is a parameter having to do with the form of the circuit, and $\mathbf{f}(\omega)$ is another parameter called the *feedback coefficient.* For example, for the voltage-follower circuit we see from Equation (12.5) that $\mathbf{c}(\omega) = 1$, $\mathbf{f}(\omega) = 1$. For the inverting amplifier we see from Equation (12.38) that $\mathbf{c}(\omega) = -\mathbf{Z}_F/(\mathbf{Z}_1 + \mathbf{Z}_F)$, $\mathbf{f}(\omega) = \mathbf{Z}_1/(\mathbf{Z}_1 + \mathbf{Z}_F)$.

From Equation (12.39) it is evident that the circuit will be unstable if there is a frequency for which $\mathbf{A}(\omega)\,\mathbf{f}(\omega) = -1$. At such a frequency the circuit gain would be infinite, and an output signal would be produced even in the absence of any input signal. Usually at low frequencies both $\mathbf{A}$ and $\mathbf{f}$ are positive real numbers. However, as the frequency increases, phase shifts occur in both $\mathbf{A}(\omega)$ and $\mathbf{f}(\omega)$. If a phase shift of 180° occurs, the feedback has been changed in sign, from negative to positive. The situation $\mathbf{A}\mathbf{f} = -1$ can then occur.

---

[6] This is not to imply that oscillation is always undesirable. Oscillator circuits are needed in some applications, as a means for generating sinusoidal signals. See Section 4.5.

Actually, although the condition $\mathbf{Af} = -1$ implies oscillation, it is possible for oscillation to occur under more general circumstances. In most cases θ, the argument of the product $\mathbf{A}(\omega)\,\mathbf{f}(\omega)$ (defined by $\theta = \tan^{-1}\,[\mathrm{Im}\,(\mathbf{Af})/\mathrm{Re}(\mathbf{Af})]$) has the value $-180°$ for only a single frequency. For this situation, we can state the Nyquist criterion as follows: *If, at the frequency at which* $\theta = -180°$, $|\mathbf{Af}| \geq 1$, *then the system is unstable.* We have already seen that if $|\mathbf{Af}| = 1$ when $\theta = -180°$, instability occurs. The Nyquist criterion states that if $|\mathbf{Af}|$ is greater than unity when $\theta = -180°$, instability also occurs.[7]

As a first example, let us consider an op-amp in which $\mathbf{A}$ has the dependence on ω shown in Figure 12.24. Here $\mathbf{A}(\omega)$ is shown as a Bode plot, with both $\log |\mathbf{A}|$ and θ_A (the argument of $\mathbf{A}$) graphed versus $\log \omega$. Suppose this op-amp is used in a voltage follower circuit. In this circuit, as we have already mentioned, $\mathbf{f} = 1$. By the laws of multiplication of complex numbers, $\theta = \arg(\mathbf{Af}) = \theta_A + \theta_f$, and in this case $\theta_f = 0$. Hence in this case θ is identical to θ_A, shown in Figure 12.24. We see that $\theta = -180°$ at $\omega = 3 \times 10^7$, and at this frequency $|\mathbf{Af}|$ is about 1000, much greater than unity. Hence the voltage follower circuit will be thoroughly unstable, and will oscillate.

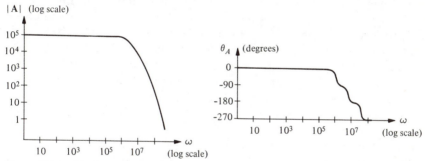

Figure 12.24 Graphs of $|\mathbf{A}|$ and θ_A versus frequency for a typical hypothetical amplifier. A voltage follower constructed using this amplifier will be unstable.

Evidently an op-amp with characteristics like those of Figure 12.24 would not be very useful. To make feedback circuits stable, op-amps usually are designed to have $\mathbf{A}(\omega)$ approximately as shown in Figure 12.25. We see that an op-amp with this characteristic can be used to make a stable voltage follower circuit, because by the time the phase shift reaches $-180°$, $|\mathbf{A}|$ has decreased to about 10^{-1}

An op-amp characteristic like that shown in Figure 12.25 can be obtained by including, somewhere in the op-amp, an RC filter like that shown in Figures 4.4 and 4.5, or the equivalent of such a filter. The simple RC filter gives a rolloff of $|\mathbf{A}|$, above the break frequency, of 20 db/decade, and a phase shift of a safe $90°$ at frequencies above the break frequency. In general, addi-

---

[7] We refer to phase shifts of $-180°$ rather than $+180°$, because in practice θ is usually negative.

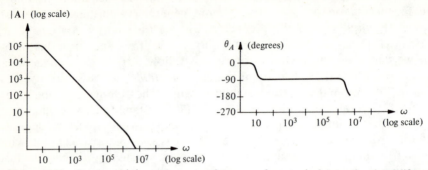

Figure 12.25   Graphs of $|\mathbf{A}|$ and θ_A versus frequency for a typical operational amplifier.

tional phase shifts will occur at higher frequencies, caused by small stray capacitances elsewhere in the op-amp. However, the RC filter should be designed so that at the higher frequencies, where these secondary phase shifts occur, the RC filter has already reduced the value of $|\mathbf{A}|$ to less than unity.

The type 741 op-amp has a characteristic much like that of Figure 12.25, as can be seen from Figure 12.4 or 12.21(a). (Actually, the latter figures only show $|\mathbf{A}|$ versus ω for the 741, and not θ_A. However, it can be proven that if two amplifiers have the same graph of $|\mathbf{A}|$ versus ω, their graphs of θ_A versus ω must also be the same.) In the case of the 741, a capacitor is built right into the IC in order to give it a break frequency at 10 Hz. Some op-amps do not have this capacitor built in, as it is rather large for inclusion in an IC. Instead, such op-amps have terminals, usually designated "external frequency compensation," for external connection of the capacitor.

From the above discussion, it can be seen that the stability of a circuit depends on the Bode plot of the op-amp, and also on the form of the circuit, which determines the feedback coefficient $\mathbf{f}$. Graphs of $|\mathbf{Af}|$ versus ω and of θ ($= \theta_A + \theta_f$) versus ω should be used to check the stability of a given circuit.

EXAMPLE 12.16

The amplifier of Figure 12.24 is used in the circuit of Figure 12.22, the amplification of which was found in Equation (12.38). Let us consider the case in which $\mathbf{Z}_1 = R_1$ and $\mathbf{Z}_F = R_F$ (that is, the inverting amplifier.) Let $R_1 = 1000$ Ω. What is the smallest value R_F can have so that the system will be stable?

SOLUTION

Comparing Equation (12.38) with Equation (12.39), we see that $\mathbf{f}(\omega) = \mathbf{Z}_1/(\mathbf{Z}_1 + \mathbf{Z}_F) = R_1/(R_1 + R_F)$. In this case θ_f, the argument of $\mathbf{f}$, is zero. We must choose R_F so that $|\mathbf{Af}| < 1$ at the frequency that makes θ ($= \theta_A + \theta_f = \theta_A$) equal to $-180°$. From Figure 12.24 we see that this frequency is about 3×10^7 radians/sec. The value of $|\mathbf{A}|$ at this frequency is about 1000. In order to make $|\mathbf{Af}| < 1$, we must have

$$|\mathbf{f}| = \frac{R_1}{R_1 + R_F} < 10^{-3}$$

Thus the smallest value of R_F that will make the circuit stable is $R_F \cong 10^6$ $\Omega = 1$ MΩ.

It is interesting to observe that for voltage amplifications of 1000 or more, the circuit is stable. However, it is not possible to make a stable inverting amplifier circuit with gain less than 1000, using the amplifier of Figure 12.24. ■

Summary

- Operational amplifiers can be used in a variety of simple circuits. Op-amps are inexpensive, compact, and versatile. Often it is more efficient to design a circuit around op-amps than to use individual transistors. Op-amp circuits usually contain negative feedback.

- Op-amp circuits themselves can be regarded as building blocks. These blocks are characterized by their input resistance, output resistance, and open-circuit voltage amplification.

- Some of the more common op-amp circuits are the voltage follower, non-inverting amplifier, inverting amplifier, summing amplifier, current-to-voltage converter, current amplifier, and integrator.

- Feedback circuits have the property known as desensitivity. This property implies that variations in the values of the op-amp parameters have little effect on the output of the circuit.

- Nonlinearity in op-amps usually does not prevent op-amp circuits from being nearly linear, as a result of the desensitivity property.

- Offset is a condition in which the output of an amplifier is displaced from zero by a constant value when the input is zero. A time-varying offset is called a drift. The term noise refers to the small random spurious signals generated by all electronic circuits. Noise places a limit on the smallness of signals that can be used.

- Like all electronic circuits, op-amps have limited frequency response. The passband of an op-amp circuit, however, is usually much larger than that of the op-amp by itself, as a result of the negative feedback of the circuit. In a typical case with a given op-amp, increasing the bandwidth of an op-amp circuit will decrease the voltage gain in the same proportion. Thus for a given op-amp, the product of gain and bandwidth, known as the gain-bandwidth product, is a constant.

- Feedback of an improper kind can lead to instability or oscillation of an op-amp circuit. In general, instability occurs when there is excessive phase shift in the op-amp and feedback loop, so that negative feedback is converted to positive feedback. A mathematical criterion, known as the Nyquist criterion, can be used to determine if a circuit is stable. The frequency response of an op-amp is usually designed to roll off smoothly, at 20 db/decade. This type of frequency response insures stability for the more common op-amp circuits.

References

Handbook of Operational Amplifier Applications. Burr-Brown Research Corp., 1963.

Giles, J. N. *Fairchild Semiconductor Linear Integrated Circuits Applications Handbook.* Fairchild Semiconductor Co., 1967.

RCA Linear Integrated Circuits Handbook. RCA Inc., 1967.

On the subject of stability in feedback circuits, see:

Gray, P. L., and C. R. Searle. *Electronic Principles: Physics, Models, and Circuits.* New York: John Wiley & Sons, 1969.

Angelo, E. J. *Electronics: BJT's, FET's and Microcircuits.* New York: McGraw-Hill, 1969.

Problems

In the following problems, assume type 741 op-amps are used. Typical properties of this op-amp are given in Figure 12.4, except for R_o, which you may take to equal 50 ohms. Use the ideal op-amp technique wherever it is reasonable to do so. In Problems 1–16, assume that the op-amp's open-circuit voltage amplification has its low-frequency value of 10^5.

12.1   Design an op-amp circuit with $R_i' = 1000 \ \Omega$, $A_V' = -10$, $R_o' < 10 \ \Omega$.

12.2   Design an op-amp circuit with $R_i' > 10 \ \mathrm{M}\Omega$, $A_V' = -10$, $R_o' < 10 \ \Omega$.

12.3   Design an op-amp circuit with $R_i' > 10 \ \mathrm{M}\Omega$, $A_V' = +10$, $R_o' <$ ten ohms.

12.4   Two input voltages are $v_1(t)$ and $v_2(t)$. Design an op-amp circuit that will generate the voltage $3v_1(t) - 2v_2(t)$. Its input resistances must exceed 1 kΩ and its output resistance must be less than 10 Ω.

12.5   A certain signal source has a Thévenin voltage $v_S(t)$ and a Thévenin resistance of 1 MΩ. Design an op-amp circuit that produces the voltage $-10v_S$.

12.6   Given an input voltage $v_S(t)$, design a circuit including an op-amp and a switch that produces an output voltage of $-10 \int_0^t v_s(t') \, dt'$. Design your circuit so that it has an output voltage of zero at time $t = 0$.

12.7   Verify Equation (12.16).

12.8   Derive Equation (12.33).

12.9   Consider the inverting amplifier circuit shown in Figure 12.12. A signal source with Thévenin resistance R_S is to be connected to the input as shown. Calculate the output voltage by two different methods:

(1) By combining R_S and R_1 into a single resistance R_1' and using Equation (12.21).

(2) By finding the input resistance R_i' for the inverting amplifier block, and regarding R_S and R_i' as a voltage divider.

Show that the results obtained via (1) and (2) are in agreement.

12.10   Find the open-circuit output voltage of the system shown in Figure 12.26 as a function of the input voltage v_S.

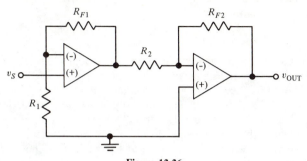

Figure 12.26

12.11   In the circuit of Figure 12.27 find i_L in terms of v_{IN}. What is the function of this circuit?

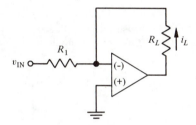

Figure 12.27

12.12   The circuit shown in Figure 12.28 is a voltage regulator. Its function is to supply an output voltage that maintains a constant value in spite of variations of the voltage supplied to it, and in spite of variations of the current drawn from its output terminal. A *reference voltage* is used. The reference voltage source supplies a well-controlled constant voltage but is unable to supply sig-

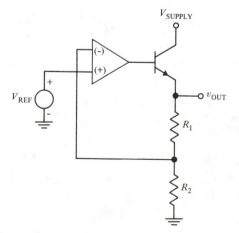

Figure 12.28

nificant current. It could, for example, be a high-quality standard battery cell. Assume that $\beta = 100$.

(1) Verify that the feedback in this circuit is negative.

(2) Find v_{OUT} as a function of v_{REF}.

12.13   It is desired to suspend a steel ball stationary in midair by means of an electromagnet.

   (1) One possible approach would be to set an adjustable current source to a value that is exactly enough to hold the ball suspended, as shown in Figure 12.29(*a*). Comment on the likelihood of success for this approach.

   (2) Suppose that a position detector (something like a miniature radar) is added, as shown in Figure 12.29(*b*). This detector produces a voltage proportional to the instantaneous height of the ball. Design a system in which the signal from the detector is used to stabilize the position of the ball. Assume that the force exerted by the electromagnet is proportional to the current through its windings (which have resistance 10 Ω) and is independent of the height of the ball.

   (3) Can you design your system so as to allow adjustment of the equilibrium height of the ball?

   Systems like that of Figure 12.29(*b*) are in practical use for suspending machinery when it is undesirable to use ordinary bearings. A super-high-speed centrifuge, known as an *ultracentrifuge*, can be made in this way.

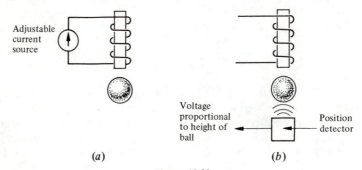

(*a*)   (*b*)

Figure 12.29

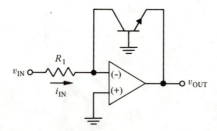

Figure 12.30

12.14   The useful circuit of Figure 12.30 is known as a logarithmic amplifier. Assume (it is quite nearly true) that the transistor obeys $i_C = K \exp\{q v_{BE}/kT\}$, where K is a constant, and that the op-amp is "ideal."

   (1) Find v_{OUT} as a function of v_{IN}.

   (2) Find v_{OUT} as a function of i_{IN}.

12.15 Use phasor analysis to find v_{OUT} as a function of ω, for the circuit of Figure 12.31. Assume that the op-amp is "ideal."

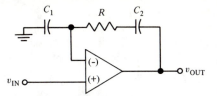

Figure 12.31

12.16 For a certain op-amp, the input bias current is 200 nA, the input offset current is 30 nA, and the input offset voltage is 2 mV. The op-amp is to be used in an inverting amplifier circuit with gain of ten. Design a suitable circuit, including a nulling circuit for cancellation of offset.

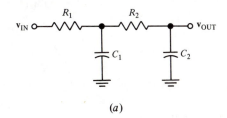

(a)

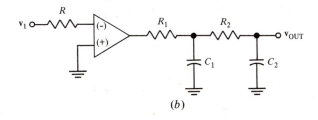

(b)

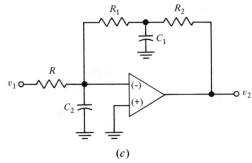

(c)

Figure 12.32

12.17 The purpose of this problem is to illustrate how instability can occur in feedback amplifiers.

(1) Let us first consider the circuit shown in Figure 12.32(a), excited by a sinusoidal v_{IN}. Compute the phase shift of v_{OUT} with respect to v_{IN}, as a function of frequency. What is the limit of the phase shift as $\omega \to 0$? As

$\omega \to \infty$? *Suggestion:* You may be able to simplify the algebra by defining $g_1 = 1/R_1$, $g_2 = 1/R_2$.

(2) Let us assume that the complex open-circuit voltage amplification of an op-amp $\mathbf{A}(\omega)$ is approximately equal to $-jA_0/\omega$ at high frequencies. (That is, at high frequencies the phase shift of the op-amp is $-90°$.) The op-amp is connected as shown in Figure 12.32(b). Assume that R_i for the op-amp is ∞ and $R_o = 0$. Make a sketch of the phase of $\mathbf{v}_{\mathrm{OUT}}$ with respect to $\mathbf{v}_1$. Show that there is a frequency ω_0 at which the phase of $\mathbf{v}_{\mathrm{OUT}}$ is the same as that of $\mathbf{v}_1$.

(3) Let the output now be fed back to the input as shown in Figure 12.32(c). The feedback will now be positive at the frequency ω_0 found in part (2). Show that there is a value of A_0 such that the overall amplification of the circuit $\mathbf{v}_2/\mathbf{v}_1$ is infinite at a frequency ω_1, and that $\omega_1 \to \omega_0$ as $R \to \infty$. The circuit will oscillate at the frequency ω_1. According to the Nyquist criterion, the circuit will in fact be unstable not only for the value of A_0, which makes the gain infinite, but also for any larger value of A_0.

12.18 Repeat Example 12.16, using the op-amp of Figure 12.25 instead of that of Figure 12.24. What is the smallest gain that can be obtained without loss of stability?

12.19 Show that the integrator circuit of Figure 12.16 is stable when the op-amp of Figure 12.25 is used.

12.20 A differentiator circuit can be made by interchanging the resistance and capacitance in Figure 12.16. Comment on the stability of the differentiator when the op-amp of Figure 12.25 is used.

Chapter 13

MOS Transistors and Integrated Circuits

Through the 1960s the bipolar transistor was almost the exclusive active device used in new electronic circuits, whether discrete or integrated. Only in very special applications were other devices employed. However, in recent times another semiconductor device, the MOS transistor, is finding increasing use in one major area of transistor applications, that of digital integrated circuits. MOS transistors, like conventional bipolar transistors, are made in a single crystal of silicon; however, as we shall see, they function in an entirely different manner.

The most important property of an MOS transistor is its simple construction and corresponding small size. As we shall see, only a single impurity diffusion step is required to fabricate MOS integrated circuits, whereas three diffusion steps are required in conventional IC fabrication (Chapter 7). Furthermore, MOS transistors, because they take up less IC area, can be packed more densely. Consequently, more complex circuits can be fabricated in a given piece of silicon using MOS integrated circuits. Extremely complex circuits, such as memories which can store over 1000 bits, are now routinely constructed using MOS technology. Such circuits, containing many thousands of transistors, are called large-scale integrated circuits because they do the job of hundreds of elementary circuit blocks (such as **NAND** gates). Figure 13.1 is a photograph of a 1024-bit shift register (see Section 9.4 or Section 13.2). This circuit contains 1024 cells, each constructed from six MOS transistors and capable of storing one bit of information. (Sixteen columns in the photograph are visible; each has 64 cells.) The entire circuit is constructed on a single chip of silicon approximately 0.1 cm² in area.

Figure 13.1   A microphotograph of a 1024-bit MOS shift register. The complete integrated circuit occupies an area of approximately 0.1 cm². (Courtesy of Intel Corporation.)

MOS circuits also have an important disadvantage in comparison with bipolar circuits. The switching speed of MOS IC's is inherently slower. It should also be noted that reproducible and stable biasing of MOS transistors in large analog integrated circuits has proved difficult. Thus MOS circuits find their major applications in digital systems in which cost and size are more important factors than speed. In such applications as desk calculators, medium-speed computers, medium-speed counters and the like, the use of MOS circuits is rapidly expanding. Large-scale digital integrated circuit applications will be emphasized in this chapter, because it is in such applications that the full potential of MOS devices is realized.

The MOS Transistor   A schematic view of the most common form of MOS transistor is given in Figure 13.2. It consists of an n-type semiconductor into which two p-type regions, labeled *source* and *drain,* have been incorporated. The surface is covered with a nonconducting oxide, and a metal electrode is placed on top of the oxide, covering most of the space between the

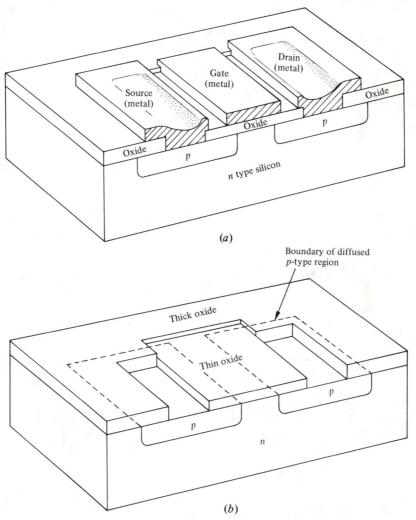

(a)

(b)

Figure 13.2   (a) The cross-section of an MOS transistor. The gate electrode controls the current flow between source and drain. (b) The transistor with metallization removed.

p-type regions. The metal electrode is called a *gate,* because, as will shortly be shown, it controls the flow of current between the *p* regions, that is, between the source and drain.[1]

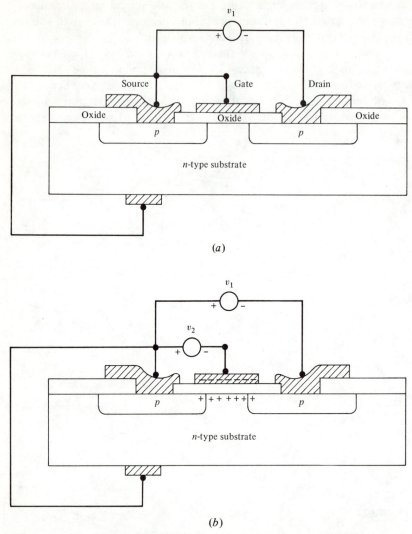

(a)

(b)

Figure 13.3 The MOS transistor with bias applied. (*a*) With no voltage applied to the gate. The drain to substrate pn junction is reverse biased and no current can flow between source and drain. (*b*) With negative voltage applied to the gate. A conducting "channel" of holes is induced by the negative charge on the gate electrode. Current can flow between source and drain via this channel.

----

[1] This use of the word gate is unrelated to and is not to be confused with its use in logic circuits, for example, **NAND** gates. Furthermore, the word source as used here should not be confused with voltage source or current source.

Let us see how this control action in an MOS transistor comes about. Suppose a voltage is applied between source and drain, as shown in Figure 13.3(*a*), and the gate and *n*-type substrate are connected to the source. This voltage, if it is negative as shown, reverse biases the *pn* junction between drain and substrate. As usual for a reverse-biased junction, no current flows, and the resistance between the source and drain appears to be essentially infinite.

Now suppose that the gate is connected to a supply of negative voltage, as shown in Figure 13.3(*b*). The gate and substrate, separated by the oxide, form a parallel-plate capacitor that will charge up with a negative charge on the gate electrode, and a positive charge on the substrate electrode. The negative charge on the gate electrode consists, of course, of a few excess electrons at the surface of the metal. The positive charge on the substrate side of the capacitor consists of both holes attracted there by the negative charge on the metal electrode, and positive donor ions that are left behind when the negative charge on the metal gate drives the electrons away from the semiconductor surface. For small (negative) values of the gate-to-substrate voltage, the charge on the silicon side consists mainly of the positive donor ions. For larger values of applied voltage, an appreciable charge consisting of holes is "induced" under the gate. It is this situation that interests us.

The holes induced in the silicon under the gate constitute a *p*-type layer called a *channel,* which connects the *p*-type drain and source regions. If a sufficient number of holes are present, conduction can take place between source and drain through the channel. Thus the effect of a negative gate voltage is to reduce the resistance between the source and drain. The gate electrode is said to be a field plate. An applied gate voltage results in an electric field in the gate oxide, which in turn induces mobile charges at the surface of the semiconductor. This action is called a *field effect,* and the MOS transistor is a kind of field-effect transistor. The remarkable feature of this device is that a large current can be controlled between source and drain, and no current flows into the control electrode, the gate. Clearly the device is a current amplifier, and the amplification is infinitely large.

The name MOS transistor stems from the configuration: a *m*etal on top of an *o*xide on top of *s*ilicon.[2] Sometimes these devices are constructed from different materials than shown in Figure 13.2; for example, a highly conducting semiconductor layer is often used for the gate material instead of the metallic layer shown. However, such devices are functionally equivalent to the device described here, and will be referred to as MOS devices.

The Circuit Symbol   Two circuit symbols are used commonly for the MOS transistor. The circuit symbols are shown in correspondence to the device in Figure 13.4. The symbol of Figure 13.4(*b*) shows all four connections — source, drain, gate, and substrate — and indicates that the substrate is *n*-type by showing an arrow from the *p*-type channel to the *n*-type substrate. In normal

---

[2] These devices are also known as insulated-gate field-effect transistors.

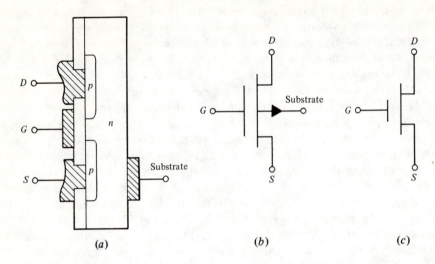

Figure 13.4 The circuit symbol for MOS transistors. (*a*) The structure (schematic). (*b*) The circuit symbol showing all four connections. (*c*) The preferred circuit symbol for an MOS transistor.

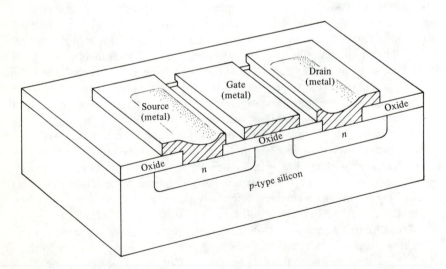

Figure 13.5 An *n* channel MOS transistor. This structure is the analog of the structure of Figure 13.2. A positive gate voltage induces an *n*-type channel between *n*-type source and drain regions.

use the drain-to-substrate and source-to-substrate *pn* junctions are always reverse biased, and no current flows in the substrate wire. To assure this condition, the substrate is usually connected to the most positive voltage in the circuit. The simpler symbol of Figure 13.4(*c*) is preferred for this reason. When using this symbol it is necessary to indicate separately both that the substrate is *n*-type, and to what it is connected.

In the MOS transistor that we have described, the gate voltage controls the flow of current through a *p*-type channel between source and drain. This channel is not present in the absence of a gate-to-substrate voltage, but is enhanced by the presence of negative charges on the metal gate electrode. The device as described is properly called a *p* channel, enhancement mode MOS transistor. We focus on this device because of its practical importance; however, let us briefly examine some possible variations on the structure to see what is possible with the present technology.

An analogous device consists of a *p*-type substrate with *n*-type source and drain regions, as illustrated in Figure 13.5. If a large positive voltage is applied to the gate of this device, electrons are induced into the semiconductor under the gate; that is, an *n*-type channel forms (is enhanced) between source and drain. For this reason the device of Figure 13.5 is called an *n* channel, enhancement mode MOS transistor.

It is also possible to fabricate MOS transistors with a built-in channel. For example, *n*-type impurities could be included in the silicon directly under the gate in the structure of Figure 13.5. Unlike the enhancement mode device, such a device would readily conduct between source and drain, even for zero gate-to-substrate voltage. However, if a large negative gate voltage were applied, electrons would be effectively driven from the channel by the repulsion of the negative charge on the gate electrode. For sufficient negative gate voltage, the *n*-type channel would be completely depleted of carriers; and conduction between source and drain would cease. Such devices are called depletion mode MOS transistors. Enhancement mode MOS transistors are, however, both simpler to construct and have more desirable electronic properties for integrated circuit use than depletion mode devices do. Furthermore the *p* channel device (Figure 13.2) has been almost completely dominant in MOS circuits to date, primarily because a better control of its electrical properties has been achieved. For this reason we use the *p* channel device exclusively in the discussions of this chapter. The simple circuit symbol of Figure 13.4(*c*) will be employed and will be understood to represent the *p* channel device.

To analyze the operation of MOS circuits it is necessary first to establish the electrical characteristics of MOS transistors. We deal with the basic operation of MOS transistors in Section 13.1. Since MOS transistors are used in integrated circuits, both as three-terminal active devices and as two-terminal resistive loads, both modes of operation are discussed. In Section 13.2 we examine the most important MOS digital circuits, including both basic logic circuits and flip-flops. Section 13.3 deals with the technology of MOS integrated circuits.

13.1   Characteristics of MOS Transistors

Let us consider a typical p channel MOS transistor and qualitatively derive the general features of its I-V characteristics. The structure and biasing arrangements are shown in Figure 13.6(a). As in bipolar transistors, all currents are defined as positive inward, and the signs of the voltages are defined by the order of the subscripts.

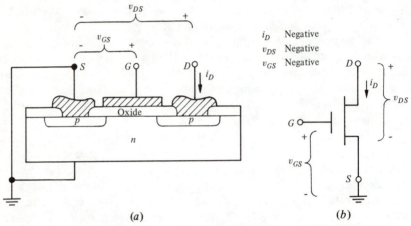

Figure 13.6   The biasing arrangement for the MOS transistor. (a) The MOS structure. All currents are defined as positive inward. (b) The circuit symbol. The gate and drain potentials are measured with respect to the source. Both the gate and drain voltage as well as the drain current are negative.

Because the gate is insulated, the gate current in MOS transistors is zero. Furthermore the n-type substrate is always connected to the most positive voltage in the circuit (here ground); consequently the substrate-to-source and substrate-to-drain junctions cannot become forward biased. The substrate current is also essentially zero. Therefore any current that flows, flows directly from source to drain. The drain-to-source voltage is restricted to negative values; hence the drain current is always negative. Keeping these considerations in mind, it is only necessary to specify the drain current, the drain-to-source voltage, and the gate-to-source voltage as shown in Figure 13.6(b). All three of these terminal variables are negative in the mode of operation of interest here. Furthermore, in the discussion that follows, it is understood, unless specified otherwise, that the source is grounded. Therefore the gate-to-source voltage may be more simply referred to as the gate voltage.

We wish to find the drain current versus drain-to-source voltage I-V

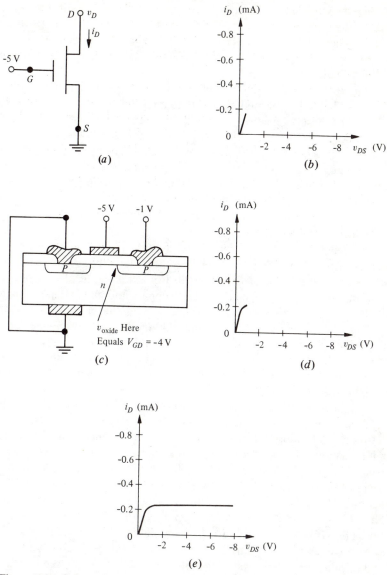

Figure 13.7 Current flow in a p channel MOS transistor. (a) The gate voltage is fixed at $-5\ V$ and the drain current is measured as a function of the drain voltage. (b) For low values of drain voltage, the channel behaves like a resistor. Note that in this and all subsequent plots, i_D and v_{DS} are always negative; the curves have simply been shifted to the first quadrant. (c) The gate-to-channel voltage is reduced near the drain end as the magnitude of the drain voltage is increased. (d) The resistance of the channel increases as the magnitude of the drain voltage increases. (e) At sufficiently high (negative) drain voltage, the drain current saturates.

characteristic of the MOS transistor. Because this *I-V* characteristic is a function of the gate voltage v_{GS}, the graph will be a family of *I-V* curves (i_D versus v_{DS}) with v_{GS} as a parameter. The simplest curve to find is the one for $v_{GS} = 0$. As we have seen, there is no conducting path (channel) between the drain and source when $v_{GS} = 0$; consequently the i_D versus v_{DS} characteristic is a straight line on the voltage axis.

Let us now apply a negative gate voltage v_{GS} of sufficient magnitude that a conducting channel is formed between source and drain. The gate voltage required to just barely produce a conducting channel is called the turn-on voltage. Suppose the turn-on voltage is -3 V, and that we apply -5 V to the gate, as shown in Figure 13.7(a). We now proceed to vary v_{DS} and "measure" i_D. For very small (negative) values of v_{DS}, the conducting channel behaves like a resistor, and the drain current is proportional to the drain-to-source voltage. The characteristic is sketched in this region in Figure 13.7(b). (Note that both v_{DS} and i_D are always negative; for convenience we have shifted the characteristic from the third quadrant.)

As the magnitude of v_{DS} is increased, another phenomenon occurs. The voltage across the gate oxide, the very voltage that induces the channel, is reduced near the drain end of the channel. For example, as shown in Figure 13.7(c), when $v_{DS} = -1$ V, the voltage between gate and channel near the drain end equals -4 V, whereas near the source end of the channel this voltage equals -5 V. The reduction of the magnitude of the gate-to-channel voltage causes a reduction of carriers in the channel, and an increase in the channel "resistance" between source and drain. (The quotes are intended to emphasize the nonlinear character of the resistance.) Consequently, the nonlinear characteristic shown in Figure 13.7(d) is obtained.

The channel resistance approaches infinity as the gate-to-channel voltage approaches the turn-on voltage. In the example here, when v_{DS} equals -2 V, the gate-to-channel voltage equals -3 V near the drain end of the channel. No further increase in $-i_D$ will occur for further increases in $-v_{DS}$. We might at first suspect that the magnitude of the drain current would be reduced to zero for $-v_{DS}$ greater than about 2 V. However we may show that a reduction in the magnitude of i_D is not self-consistent. Suppose i_D were reduced, say to zero. Then there would be no voltage drop along the channel between source and drain. The voltage between gate and channel would be everywhere -5 V, right up to the edge of the drain. Consequently a conducting channel between source and drain would exist, and current would flow from source to drain, in disagreement with our original assumption. The magnitude of the drain current is not reduced; it simply ceases to increase as the drain voltage exceeds -2 V. The complete i_D versus v_{DS} characteristic for $v_{GS} = -5$ V would have the appearance shown in Figure 13.7(e).

We describe once more the origin of this characteristic. As the magnitude of v_{DS} (which is negative) is increased from zero, current flowing from source to drain at first increases linearly with voltage. However, at larger biases, the reduced voltage between gate and channel near the drain end of the channel re-

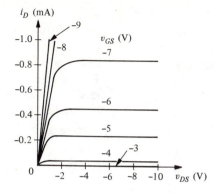

Figure 13.8 The theoretical output characteristics of a p channel MOS transistor. Again, i_D and v_{DS} are negative, but the plot is made in the first quadrant.

duces the channel conductivity, and the current no longer increases linearly with voltage. The rate of increase of current with voltage continues to decrease, until it eventually reaches zero, and the current no longer increases at all. As shown in Figure 13.7(e), the current rises to a maximum value and *saturates*. Beyond a certain voltage the current becomes independent of further increases of $-v_{DS}$.

Up to this point we have kept the gate voltage fixed at -5 V. It is the gate voltage that determines the channel conductivity. The larger the magnitude of the gate-to-channel voltage, the more holes are induced in the channel and the greater the channel conductivity. Thus, if v_{GS} is changed from -5 V to -6 V, the channel conductance is increased, and the magnitude of i_D is increased at

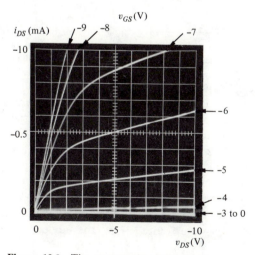

Figure 13.9 The output characteristics i_D versus v_{DS} of a typical p channel MOS transistor intended for digital circuit applications (curve tracer photograph). Horizontal scale: 1 V per large division. Vertical scale: 0.1 mA per large division. These characteristics are redrawn in Figure 13.17.

all values of v_{DS}. A complete set of curves i_D versus v_{DS} for various values of v_{GS} is shown in Figure 13.8. Directing our attention to the region where v_{DS} is close to zero, we see that the MOS transistor behaves like a controlled resistor; the larger the magnitude of the gate-to-source voltage, the lower the resistance. For larger values of v_{DS}, the resistance increases, and current eventually saturates for each value of gate voltage.[3] Figure 13.8 is a display of the output characteristics of a p channel MOS transistor. There is no need to display the input characteristics of this device, because the gate current is always zero.

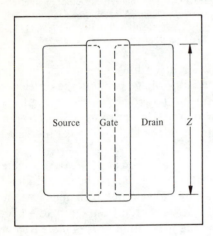

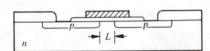

Figure 13.10   Surface view and cross section of an MOS transistor (metal source and drain contacts omitted for clarity). The source to drain current at a given gate and drain voltage is proportional to the channel width Z.

The characteristics of a typical p channel MOS transistor are shown in Figure 13.9. This device is representative of an MOS transistor that might be used in a digital integrated circuit application. The characteristics shown in Figure 13.9 differ from the theoretical behavior (Figure 13.8) in that the drain current does not completely saturate. However, this is of little concern, since we will see that the devices do not operate in this region in digital circuit applications. In general the device will either be strongly turned on (for example, $v_{GS} = -10$ V) or turned off (for example, $v_{GS} = 0$). We can see from Figure 13.9 that when v_{GS} is large (negative) the device behaves much like a resistor, whereas when v_{GS} is between 0 and -4 V, the device behaves like an open circuit.

In the construction of digital MOS circuits we shall have need for MOS transistors with various current capabilities; for example, we may need a de-

---

[3] The terminology is unfortunate. The current saturation of MOS transistors should not be confused with the saturation region in bipolar transistor operation.

vice with source-to-drain resistance of 20 kΩ at -8 V gate voltage, whereas the device of Figure 13.9 has about a 2-kΩ resistance at this gate voltage. It is a simple matter to parallel MOS devices to achieve any current capability desired. The geometry of an MOS device is shown in Figure 13.10. Paralleling devices amount simply to increasing the width of the active device Z. At a given v_{GS} and v_{DS} the source-to-drain current will be doubled for a doubling of Z. Furthermore the current is also controlled by the length of the channel L. Increasing L would have the effect of decreasing the drain current at a given bias. However L plays an important role in determining the speed of MOS transistor response. The shorter L is, the faster carriers travel from source to drain and the faster the device responds to changes in gate voltage. Therefore, in order to have the fastest possible devices, L is generally made as small as possible.

The MOS Transistor as a Circuit Element   In digital circuit applications, the MOS transistor is used as an electronically controlled switch. As may be seen from Figure 13.9, the output characteristic may be switched from one extreme of an open circuit to the other extreme of a low resistance by controlling the gate voltage. Consider the simple voltage-divider circuit shown in Figure 13.11. If the resistance of R_1 is much greater than R_2, say $R_1 = 10\,R_2$, then v_{OUT} is approximately V_1. On the other hand, if R_1 is reduced to a value much smaller than R_2, then v_{OUT} is close to zero. Of course R_1 just represents the resistance from source to drain of an MOS transistor, and the value of R_1 is controlled by the gate voltage. As we shall see in the next section, the circuit of Figure 13.11 is essentially equivalent to the basic MOS inverter.

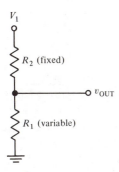

Figure 13.11   A voltage-divider circuit, illustrating the use of the MOS transistor as a digital circuit element. When the resistance of R_1, which represents the MOS transistor, is large, v_{OUT} is close to V_1; however, when R_1 is small, v_{OUT} is close to zero. In p channel MOS circuits, V_1 is negative.

EXAMPLE 13.1

Find the voltage v_{OUT} in the circuit in sketch (a) for two values of v_{IN}, 0 V and -9 V. Assume that the MOS transistor is the device whose characteristics are given in Figure 13.9, and use a graphical technique.

SOLUTION

We shall use the graphical load-line technique introduced in Chapter 2. We plot two graphs, i_D versus v_{OUT}, one for the circuitry above the point D and one for the circuitry below point D. The first graph is obtained from Ohm's law:

$$(-10 - v_{OUT}) = i_D R_D$$

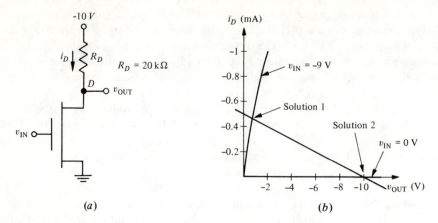

(a)   (b)

The second graph is just the $i_D - v_{DS}$ characteristic for the MOS transistor, plotted for $v_{GS} = v_{IN} = 0$ V and -9 V. The curves are plotted on the same axes in sketch (b). The solutions are read off the graph as follows:

$$v_{IN} = \quad 0 \text{ V}: \quad v_{OUT} = -10 \text{ V}$$
$$v_{IN} = -9 \text{ V}: \quad v_{OUT} = -0.8 \text{ V}$$ ■

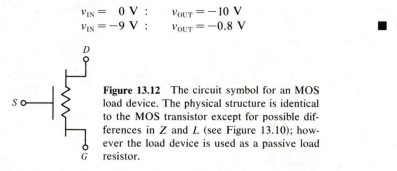

Figure 13.12   The circuit symbol for an MOS load device. The physical structure is identical to the MOS transistor except for possible differences in Z and L (see Figure 13.10); however the load device is used as a passive load resistor.

We wish now to examine another use for the MOS transistor in digital circuits, that of a passive load resistor (in fact, as the resistor R_2 in the circuit just discussed, Figure 13.11). Large valued resistors, 10 to 100 kΩ, are required in MOS integrated circuits. Such large resistors are most conveniently achieved in the smallest space by using a special MOS transistor as a resistor. A symbol for the MOS transistor used as a resistor is shown in Figure 13.12. This device is called the MOS load device.

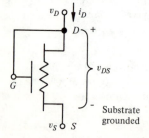

Figure 13.13   The operation of an MOS load device. Typically the gate is connected to the drain, and the substrate is grounded.

It is interesting to compare the *I-V* characteristic of an MOS load device with that of an ordinary resistor. Typically, when an MOS transistor is used as a device, the gate is connected to the drain and the substrate is grounded, as indicated in Figure 13.13. (Furthermore the drain voltage is generally fixed at some large negative value, say -15 V.) The load device *I-V* characteristic is the plot i_D versus v_{DS}, shown for a typical device in Figure 13.14.

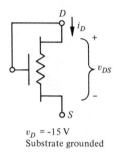

$v_D = -15$ V
Substrate grounded

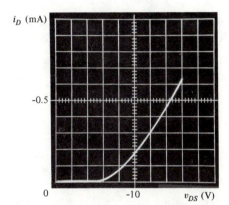

Figure 13.14   The *I-V* characteristic of an MOS load device (curve tracer photograph). Horizontal scale: 2 V per large division. Vertical scale: 0.1 mA per large division. This device has a channel width Z one-tenth that of the MOS transistor whose characteristics are given in Figure 13.9.

The load device characteristic may be understood in terms of the operation of the basic MOS transistor. Note that because drain and source are connected, $v_{DS} = v_{GS}$. For small (negative) values of v_{DS}, v_{GS} is insufficient to cause the device to turn on. However, if the magnitude of v_{DS} is increased, eventually the turn-on voltage is reached, and conduction starts. As may be seen from the graph, the *I-V* characteristic differs from a simple resistor characteristic. For applied voltages less than about 5 V, the resistance is essentially infinite; whereas for voltages in the neighborhood of 15 V, the ratio of V to I is about 30 kΩ. However, the characteristic is quite suitable for digital circuit operation, as will be seen in the next section.

The device characteristic displayed in Figure 13.14 is typical of a load device which might be used in conjunction with the device of Figure 13.9 to produce MOS logic circuits. The device geometry is identical to the device of Figure 13.9, except that the channel width of the latter is ten times as long.

EXAMPLE 13.2

Show that the circuit in sketch (*a*), constructed from a resistor, a voltage source, and a perfect rectifier, is approximately equivalent to the MOS load device.

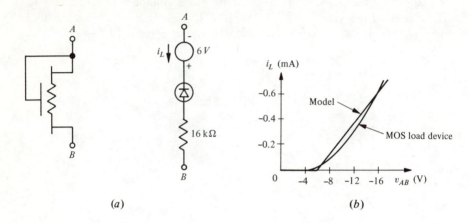

(*a*)   (*b*)

SOLUTION

To show the equivalence of the two circuits we compare their *I-V* characteristics. The characteristics of the circuit on the right are

$$i_L = (v_A + 6\text{ V} - v_B)/16\text{ k}\Omega \qquad (v_B > v_A + 6\text{ V})$$
$$i_L = 0 \qquad (v_B < v_A + 6\text{ V})$$

We plot these on the same graph as the MOS load characteristics [see sketch (*b*)]. We see that the suggested circuit is a reasonable model for the MOS load device.   ■

Gate Protection of MOS Transistors   Because the gate of MOS transistors is insulated, any charge placed on the gate may remain there a considerable time before leaking off. This fact may be taken to advantage in circuit design, as shown in Section 13.2. However, the near-infinite gate resistance also contributes to a well-known problem with MOS transistors, the phenomenon of gate "burnout." A very small charge on the gate of an MOS transistor leads to an appreciable gate voltage because of the small gate capacitance. Typically, the gate-to-substrate capacitance is on the order of a picofarad, so a charge of 10^{-10} C produces a voltage of 100 V ($Q = C\ V$). A voltage of this magnitude will usually rupture the oxide, ruining the device. Furthermore, because of the high resistance, a charge of 10^{-10} C is easily accumulated from static sources. (A classic cause of MOS device failure is sticking the device leads into a foam plastic block, such as used for shipping.) For this reason, any gate leads brought out for external connections are usually protected by a Zener diode. The diode is connected between gate and substrate and is designed to break down (nondestructively) at a voltage above the normal gate operating voltage, but below the oxide breakdown voltage.

13.2 MOS Digital Circuits

The MOS transistor is a very versatile device. Not only can it act as an active device and as a load device, but in the role of an active device it can perform both the **NAND** and **NOR** logic functions. Following the approach of Chapter 8, we shall first discuss the basic inverter circuit, and then examine the various logic circuits that develop naturally from it.

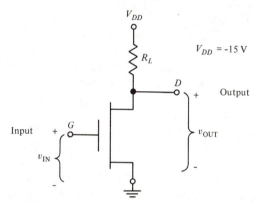

$V_{DD} = -15$ V

Figure 13.15 The basic MOS inverter circuit with resistive load. For a p channel MOS transistor, V_{DD} is typically -15 V.

The MOS Inverter The basic MOS inverter circuit is shown in Figure 13.15. It is simpler than its bipolar equivalent in that no input resistor is needed to limit the current. (The perfectly insulated gate prevents any input current from flowing.) The voltage V_{DD} is a negative supply voltage, typically in the neighborhood of -15 V for p channel MOS circuits. The resistor R_L is generally a MOS load device; however, before considering the complications of a nonlinear load device, we will first analyze the simpler circuit shown here with a linear resistor.

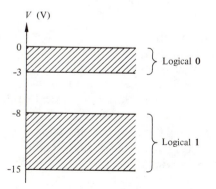

Figure 13.16 Logic assignment for p channel MOS logic. Negative logic is used.

For a p channel MOS transistor with a -4-V turn-on voltage, a reasonable set of logic levels is defined in Figure 13.16. Any voltage in the range of 0 to -3 V is considered in the logical **0** range, and any voltage more negative than -8 V is considered in the range of logical **1**. This assignment is *negative logic,* because the more negative voltage level is assigned as logical **1**. We choose to use negative logic because this assignment is the most common in practice. Furthermore it is easy to remember that the voltage range that includes zero volts corresponds to logical **0**.

We now turn to the operation of the inverter circuit, and wish to find the output voltage for various input voltages. It is simplest to use the graphical load-line technique developed in Chapter 2 and used in several examples, including Example 13.1. We graph the *I-V* characteristics of the load, consisting of the voltage source V_1 and resistor R_D, on the same graph as the device *I-V* characteristics. The composite graph is shown in Figure 13.17. In constructing Figure 13.17, the same device whose characteristics were displayed in Figure 13.9 is used, and the resistor is assumed to be 15 kΩ. The intersections of the load-line and the device characteristics are the solutions for v_{OUT} as a function of v_{IN}. From Figure 13.17 we see that when v_{IN} is in the range of logical **0**, the output is in the range of logical **1** and, conversely, when the input is in the range of logical **1**, the output is in the range of logical **0**. Clearly the circuit is an inverter.

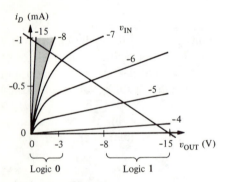

Figure 13.17 Solutions for the input-output relationship of the basic MOS inverter of Figure 13.15. The MOS characteristics are those given in Figure 13.9.

The MOS Inverter with MOS Load   Let us now modify the basic inverter circuit to make it more representative of a real inverter. A practical MOS inverter, as found in a typical MOS integrated circuit, uses an MOS load device instead of the resistor shown in Figure 13.15. The modified circuit is shown in Figure 13.18. In referring to this circuit the lower MOS transistor is called the *driver device,* and the upper MOS transistor is called the *load device.*

We now wish to find the input-output relationship for this circuit. For numerical calculations, we will use Figures 13.9 and 13.14 as representative of a typical driver device and a typical load device, respectively. A graph of v_{OUT} as a function of v_{IN} is known as the *transfer characteristic.* A graphical

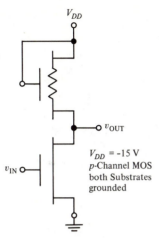

Figure 13.18 The MOS inverter using an MOS load device.

method is the simplest way to solve for this characteristic. We make an imaginary break in the circuit, as shown in Figure 13.19. We compute and plot on the same set of axes: first, the driver device characteristics i_D versus v_D, and second, the load circuit characteristics i_L versus v_L. Of course the break in the circuit is only imaginary; in reality $i_L = i_D$ and $v_L = v_D$. Thus the solution is obtained at the intersection of the two graphs. The graph of i_D versus v_D is obtained directly from Figure 13.9. (Since this graph is a family of curves, there will be a number of solutions.) The graph of i_L versus v_L may be computed from a knowledge of V_{DD} and the load device characteristics, Figure 13.14. The

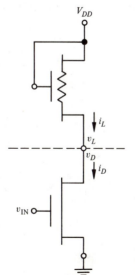

Figure 13.19 The inverter circuit broken into two subcircuits to facilitate a graphical solution.

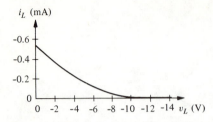

Figure 13.20 The characteristics of the load circuit of Figure 13.19 using the MOS load device of Figure 13.14. The current i_L is the load device drain current, and v_L is V_{DD} minus the load device drain-to-source voltage.

load voltage v_L equals the supply voltage V_{DD} minus the load device drain-to-source voltage. For example, when $i_L = -0.1$ mA, the load device drain-to-source voltage equals -8.5 V, from Figure 13.14. In this case $v_L = -15 - (-8.5 \text{ V})$ or -6.5 V. The complete i_L versus v_L characteristic, obtained in similar fashion, is plotted in Figure 13.20.

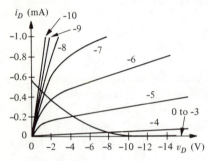

Figure 13.21 Graphical solutions for the inverter circuit of Figure 13.18.

The i_D versus v_D family of curves, and i_L versus v_L, are plotted on the same axes in Figure 13.21. The solutions are indicated for gate voltages in the range of -10 to 0 V. The transfer characteristics of the inverter circuit may now be plotted. The input voltage, Figure 13.18, is the gate voltage parameter of Figure 13.21, and the output voltage is the load or drain voltage of Figure 13.21. The inverter transfer characteristics are shown in Figure 13.22. Again, the logical ranges are indicated, and it may be seen that the circuit indeed functions as an inverter.

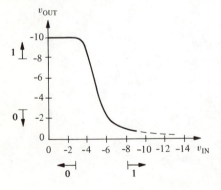

Figure 13.22 The transfer characteristics of the MOS inverter with MOS load, Figure 13.20. The transfer characteristics are obtained from Figure 13.21.

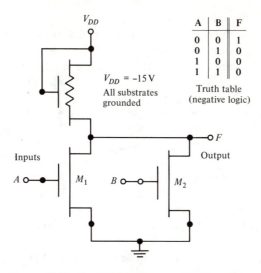

A	B		F
0	0		1
0	1		0
1	0		0
1	1		0

Truth table
(negative logic)

V_{DD} = -15 V
All substrates
grounded

Figure 13.23 An MOS **NOR** gate. If either transistor M_1 or M_2 is turned on, the output is close to zero, that is, in the range of logical **0**. If both inputs are logical **0**, both transistors are turned off and the output is close to V_{DD} (in the range of logical **1**).

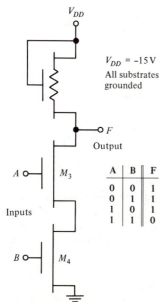

V_{DD} = -15 V
All substrates
grounded

A	B		F
0	0		1
0	1		1
1	0		1
1	1		0

Figure 13.24 An MOS **NAND** gate. Both transistors M_3 and M_4 must be turned on to produce an output in the range of logical **0**.

MOS Logic   It is only a small step from the MOS inverter to MOS logic circuits. Consider the circuit shown in Figure 13.23. Either of the driver devices M_1 or M_2, combined with the load device, acts like an inverter. Turning on either M_1 or M_2 causes the output to be in its most positive state, close to zero volt. In other words the output is logical **0** if either (or both) inputs are logical **1**. This is indicated in the truth table next to the circuit. The output will be most negative only when both inputs are in the logical **0** range, causing M_1 and M_2 to be turned off. The truth table is identical to the truth table for the **NOR** function; hence the circuit is a **NOR** gate. It may be expanded to include more inputs merely by paralleling more transistors. It is equally simply to synthesize the **NAND** function; the circuit is given in Figure 13.24. This circuit operates on the principle that the output is always near the supply voltage V_{DD}, unless both transistors M_3 and M_4 are turned on. To turn on both M_3 and M_4 requires that both gate voltages be at the more negative end of their range. Thus the output is logical **0** only when both **A** and **B** are logical **1**. The various possibilities are again summarized in the table accompanying the circuit. From this table the logical function is identified as the **NAND** function; hence this circuit is a **NAND** gate.

EXAMPLE 13.3

Find the output voltage of the **NOR** gate, Figure 13.23, when both inputs equal -8 V.

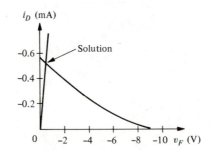

SOLUTION

Assuming that the input devices are identical, the current flowing in the load is just twice the current flowing in each of the input devices. The *I-V* characteristic of the combined input devices may be obtained by doubling the drain current (Figure 13.9) at any given drain-to-source voltage. The characteristic i_D (combined) versus v_{DS} (v_F) for $v_{IN} = v_{GS} = -8$ V is plotted in the sketch, along with the load characteristic. From this graph the output voltage is found to be approximately -0.5 V, well within the range of logical **0**.   ∎

EXAMPLE 13.4

Find the output voltage of the **NAND** gate, Figure 13.24, if both inputs equal -8 V.

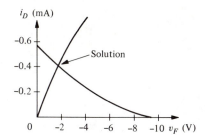

SOLUTION

A composite *I-V* characteristic is made of the two transistors M_3 and M_4, taken together. The two drain-to-source voltages are added at a given drain current. For an input of -8 V, the drain-to-source voltage of M_4 at any drain current is obtained directly from Figure 13.9 (The curve labeled $v_{GS} = -8$ V.) To find the drain-to-source voltage of M_3, it first is necessary to find v_{GS} for this device. The gate voltage is specified to be -8 V and the source voltage the drain-to-source voltage of M_4.

It may be seen from Figures 13.9 or 13.21 that for $v_G = -8$ V, v_{DS} (for M_4) is about 1 V at typical currents; hence v_{GS} (for M_3) is close to -7 V. An exact value may be obtained, given the current, from the *I-V* characteristics. To obtain the composite *I-V* characteristics of the two driver devices, we add the source-to-drain voltages at any given drain current. For example, suppose $i_D = 0.5$ mA, the v_{DS} of M_4 is very close to -1 V. In this case v_{GS} for M_3 is -7 V and, from Figure 13.9, v_{DS} for this device is about -1.4 V[4] The voltage between the drain of M_3 and ground must then equal -2.4 V. The composite *I-V* characteristic obtained in this fashion is sketched, along with the load characteristic. ∎

It should also be realized that **NAND** and **NOR** logic may be combined in a single MOS gate. For example, Figure 13.25 shows a circuit that has two **NAND** inputs and two **NOR** inputs. The output is logical **0** if either M_1 or M_2 is turned on or if M_3 and M_4 are turned on.

EXAMPLE 13.5

Find a logical function that represents the action of the circuit of Figure 13.25.

SOLUTION

The output is **0** if either $\mathbf{A} = \mathbf{1}$ (M_1 turned on), or $\mathbf{B} = \mathbf{1}$ (M_2 turned on), or if both $\mathbf{C}$ and $\mathbf{D}$ equal **1** (M_3 and M_4 turned on). Thus $\mathbf{F}$ is *not* **1** when $\mathbf{A}$ or $\mathbf{B}$ or ($\mathbf{C}$ and $\mathbf{D}$) are **1**. This statement may be written as

$$\overline{\mathbf{F}} = \mathbf{A} + \mathbf{B} + (\mathbf{C}\ \mathbf{D})$$

or

$$\mathbf{F} = \overline{\mathbf{A} + \mathbf{B} + (\mathbf{C}\ \mathbf{D})}$$

This function is seen to be a combination of the **NAND** and **NOR** functions. ∎

---

[4] Strictly speaking the curves of Figure 13.9 cannot be used for device M_3, because these curves are for a device with source and substrate grounded. However, the curves for the same device connected as device M_3 in Figure 13.24.

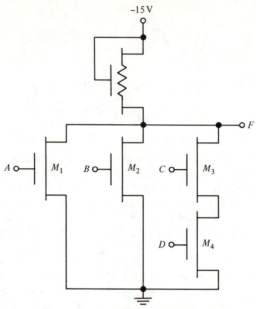

Figure 13.25   A combination of the **NAND** and **NOR** functions in a single logic gate. Inputs **A** and **B** are **NOR** inputs, and inputs **C** and **D** are **NAND** inputs.

Limitations on Logic Gate Expansion   If the MOS transistor were a perfect switch, there would be no dc limitations on the number of devices that could be paralleled to form multiple-input **NOR** gates or placed in series to form multiple-input **NAND** gates. In practice, limitations do exist because the transistor is not ideal. Consider first the **NAND** gate, Figure 13.24. When all the gate voltages are in the logical **0** range, the transistors are all conducting, tending to bring the output voltage close to zero. However, because the devices are not perfect closed switches, there is a finite voltage drop across each one. It may be seen in Figure 13.22 that the voltage drop across one device may be as high as 1 V when the input is in the range of logical **1.** Thus the voltage drop across two may be as high as 2 V, across three about 3 V, and so forth. (See Example 13.4.) If the output is to stay in the range of logical **1,** then no more than three devices may be placed in series in the **NAND** circuit. The maximum fan-in is 3. Of course that number could be modified by altering the device characteristics. In practice, **NAND** gates with more than two inputs are rarely encountered.

The limitations on the fan-in of the **NOR** logic circuit are not nearly as severe. Consider the **NOR** circuit of Figure 13.23 with all inputs in the logical **1** range. Each device is off, and consequently the output is close to the supply voltage V_{DD}. If the devices are properly designed (and fabricated), the source-to-drain resistance is at least 10^8 Ω. In principle, hundreds of input devices could be placed in parallel and the circuit would still function. However, there are very few applications for 100-input **NOR** gates, and a four- or five-input gate is more usual.

Another loading limitation occurs when a large number of inputs are to be driven by the output of a single MOS logic circuit. Figure 13.26 illustrates a system in which a **NOR** logic circuit drives the inputs to two **NAND** logic circuits and a **NOR** logic circuit. (We use the term logic circuit rather than gate temporarily here, to avoid confusion with the gate electrode in an MOS transistor.) The inputs to any MOS logic circuit are connections to the gate electrode of an MOS transistor. Since this electrode is insulated, there is essentially no dc loading and consequently almost no dc limitation on the number of inputs that may be driven by a given output. The fan-out is essentially unlimited. However, each input is a small capacitor and, as more inputs are placed in parallel, the time required to charge or discharge the combined capacitance increases. Thus the switching time limits the fan-out.

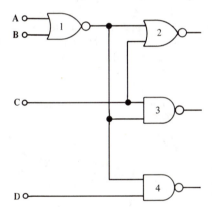

Figure 13.26 A part of a possible digital system to illustrate the loading of one logic gate by the inputs to other logic gates.

EXAMPLE 13.6

Suppose that the gate capacitance of each MOS input transistor is 1 pF. Find the approximate time required for gate 1 of Figure 13.26 to switch the inputs of gates 2, 3, and 4 from logical **0** to logical **1**. Use the model of Example 13.2 to represent the load device.

SOLUTION

Logic gate 1 is a **NOR** gate of the form shown in Figure 13.23. When the output is switched from **0** to **1**, the input transistors M_1 and M_2 are both turned off; hence they may be omitted from the diagram. The load presented by logic gates 2, 3, and 4 may be represented by a 3-pF capacitance. The circuit is given in sketch (*a*). Suppose that the switching occurs at $t = 0$. For simplicity, assume $v_F \approx 0$ at $t = 0$. We wish to find the time required for v_F to reach -8 V. We could write a node equation at F and obtain a differential equation for v_F. However, because of the nonlinear MOS load device, this equation would be difficult to solve. The approximate model for the load

(a)   (b)   (c)

device introduced in Example 13.2 offers a simplification. The circuit is redrawn in sketch (b), replacing the load device with the model. The 6-V voltage source may be combined with the −15-V supply, and the ideal rectifier may be eliminated, since it will never be reverse biased [sketch (c)]. This circuit is a simple RC circuit, familiar from the examples of Chapter 2. The capacitor voltage obeys the equation

$$v_F = -9\,(1 - e^{-t/RC}) = -9\,(1 - e^{-t/(48\times10^{-9})})$$

The output voltage reaches the range of logical 1 when $v_F = -8$ V, which occurs at the time t_s given by

$$-8 = -9\,(1 - e^{-t_s/(48\times10^{-9})})$$

or

$$t_s = 105 \times 10^{-9} \text{ sec}$$

Thus it requires about 100 nsec to switch three inputs from 0 to 1. Thirty inputs would take a factor of 10 longer, approximately 1 μsec. ∎

MOS Flip-Flops It was shown in Chapter 9 that two cross-coupled inverters form a flip-flop. MOS flip-flops are constructed similarly. The basic MOS flip-flop is shown in Figure 13.27. Figure 13.27(a) emphasizes the symmetry of the circuit, whereas the same circuit redrawn in Figure 13.27(b) is more clearly recognized as two inverters. If it is supposed that A = 1, then B = 0 and C = 1. Since C is connected to A, the circuit state is self-consistent; M_1 is on and holds M_2 off, which in turn holds M_1 on. From the basic symmetry of the circuit, we know that the opposite condition, in which A = 0, B = 1, C = 0, is also self-consistent. The circuit is stable in either of two states (either A = 1 or A = 0). We have not considered whether there might be other intermediate stable states. A careful examination of all possibilities, however, shows that for a properly designed circuit only the two states already mentioned are stable.

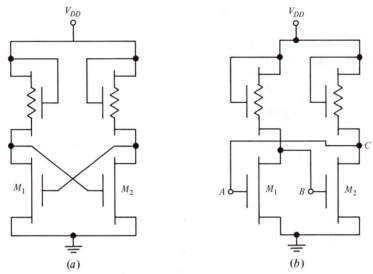

Figure 13.27 The basic MOS flip-flop. (*a*) The flip-flop drawn to emphasize the symmetry of the circuit. (*b*) The flip-flop redrawn to emphasize the interconnection of two inverters.

A practical flip-flop requires provision for inputs to alter the state of the circuit. Suppose two **NOR** gates are cross-coupled as in Figure 13.28. The state of the circuit may be modified by controlling the state of the inputs **R** and **S**. If, for example, **S** is set to **1** and **R** to **0**, $\overline{\mathbf{Q}}$ must be **0**; hence **Q** will be **1**. Conversely, if **R** = **1** and **S** = **0**, then **Q** = **0** and $\overline{\mathbf{Q}}$ = **1**. If **S** and **R** are both **0**, they do not affect the circuit state. This behavior is characteristic of the so-called **S-R** flip-flop (Chapter 9); hence the choice of the input labels. As shown in Chapter 9, it is often desirable to provide for a clock input. The function of the clock input (or "clock pulse input") is to block the **S** and **R** inputs unless the clock input equals **1**. This logic function is realized by "ANDing" the clock input with each of the **S** and **R** inputs, as shown in Figure 13.29(*a*). The corresponding MOS circuit is shown in Figure 13.29(*b*).

Consider the operation of the circuit of Figure 13.29(*b*). The presence of a **1** at input **S** (the "set" input) and a **1** at the **CP** input causes output $\overline{\mathbf{Q}}$ to drop to logical **0**, because M_4 and M_5 form a **NAND** gate, as in Figure 13.24. If $\overline{\mathbf{Q}}$ = **0** M_3 is off, making **Q** = **1**. Because $\overline{\mathbf{Q}}$ = **1**, M_6 is held in the on state. Thus the flip-flop remains in the self-consistent state **Q** = **1**, $\overline{\mathbf{Q}}$ = **0**. A simultaneous input of **1** at **R** and **CP** has precisely the opposite effect, causing **Q** to become **0** and $\overline{\mathbf{Q}}$ to become **1**. Whenever the **CP** input equals **0**, the inputs **S** and **R** have no effect. However, whenever **CP** becomes **1**, that is, whenever a clock pulse is received, the **S** and **R** inputs determine the new state of circuit. The inputs are restricted in that **S** and **R** may never both be **1** when **CP** = **1**; otherwise the state of the circuit is unpredictable. The behavior is summarized in the truth table, Figure 13.29(*c*).

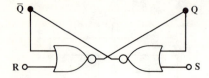

Figure 13.28 An **S-R** flip-flop formed from two **NOR** gates. The outputs **Q** and $\overline{\text{Q}}$ are controlled by the inputs **S** and **R**.

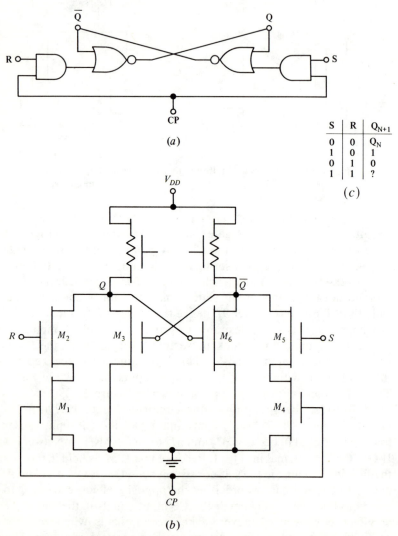

S	R	Q_{N+1}
0	0	Q_N
1	0	1
0	1	0
1	1	?

(c)

Figure 13.29 The clocked **S-R** flip-flop. (a) The logic diagram. (b) The MOS circuit. (c) The truth table. The **CP** input blocks inputs **S** and **R** unless **CP** = 1. The circuit is identical to Figure 13.28 in other respects.

MOS Memories   Fundamentally, a memory, such as might be used in a digital computer, is an array of *cells,* each of which has an *address* and contains one bit of information. A simplified block diagram of a 16-bit memory is given in Figure 13.30. Each cell might be a flip-flop, and the information is stored as the state of the flip-flop. The external connections to the memory are via a *read/write line,* a *data input line,* a *data output line,* and *four address lines.* Suppose it is desired to inquire about the information stored in cell 7.

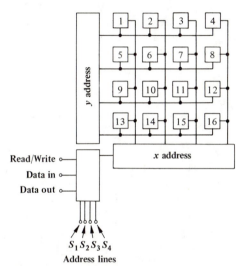

Read/Write
Data in
Data out

$S_1 S_2 S_3 S_4$
Address lines

Figure 13.30   Block diagram of a 16-cell random access memory. Each cell contains a single bit of information. A cell is accessed by giving its address on the address lines. Giving a read instruction on the read/write line causes the contents of the addressed cell to appear on the data output line.

In this case, the address of cell 7 is sent into the memory on the address lines. A code **0111** on lines **S** through S_4 might, for example, represent the address of cell 7. A read signal, which might, for example, be a **1** on the read/write line, instructs the memory to send out on the data output line whatever happens to be in cell 7. A write signal, which might be a **0** on the read/write line, instructs the memory to read into cell 7 whatever is on the data input line.

The kind of memory described here is called a *random access memory,* or RAM, because access, whether for input or output of data, is made with equal facility to any cell in the memory. We wish now to see how a random access memory might be constructed using MOS circuits.

A suitable memory cell for an MOS RAM is the basic flip-flop of Figure 13.27. The flip-flop is stable in either of two states (labeled **0** or **1**) and is composed of only four small devices. However the basic flip-flop of Figure 13.27 has no provision for sending data in or out, so some additional complications are necessary. A typical memory cell with provision for input and output is shown in Figure 13.31(a). The addition of two devices allows data to be taken out from or put into the terminals **A** and **B**. The terminal **C** is called an *address* terminal, and provides a means of connecting or disconnecting the inputs **A** and **B**. (It functions much like the clock input in the **S-R** flip-flop of Figure 13.29.) If **C** = **1**, then the transistors M_1 and M_2 are turned on, and data may flow into and out of the flip-flop. If, on the other hand, **C** = **0,** then M_1 and M_2 are open switches, and the cell is essentially disconnected from the outside world. The circuit of Figure 13.31 is similar to that of Figure 13.29. However, in Figure 13.29 separate terminals must be provided for input to the cell and output from the cell.

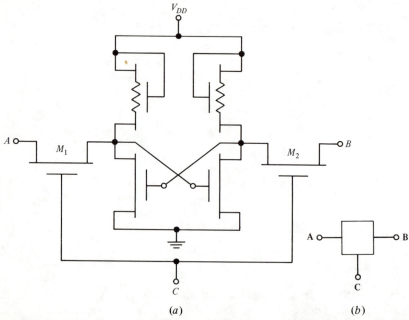

Figure 13.31   The basic memory cell for an MOS memory. (*a*) The circuit diagram. Terminals **A** and **B** are input or output lines, and terminal **C** is an address or control line that connects or disconnects **A** and **B** according to whether **C** is **1** or **0**. (*b*) The block diagram.

Let us see how a simple memory might be constructed from the basic cell of Figure 13.31(*a*). First we adopt a block diagram symbol for the cell, as shown in Figure 13.31(*b*). The functions of **A, B,** and **C** must be kept in mind. Terminal **C** is an address input, and terminals **A** and **B** are data inputs and outputs. Suppose that during a period where **C** = **1, A** is set equal to **0** and **B** equal to **1**. The flip-flop then continues in this state. The address input **C** may be set to **0**; if it is later set to **1**, then the voltage at terminal **A** would continue to be found to be in the logical **0** range, and **B** in the logical **1** range.

A group of memory cells may be arranged in a column as shown in Figure 13.32. The data terminals may all be connected to common data lines because only one of the address lines C_1, C_2, C_3, or C_4 will be allowed to be **1** at a given time. If, for example, $C_2 = 1$, then $C_1 = C_3 = C_4 = 0$, and only cell 2 is "connected" to the common data lines.

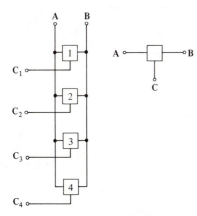

Figure 13.32 Four basic cells combined to form a column of memory cells.

The columns of cells may be combined to form a two-dimensional array, as shown in Figure 13.33. All the address terminals in a given row are connected to a common address line, C_1 for row 1, and so forth. (We have labeled the cells by row and column in this figure.) If, for example, it is desired to read the contents of cell 23, line C_2 is set to **1**, and C_1, C_3, and C_4 set to **0**. Then the cell contents are read on lines A_3, B_3.

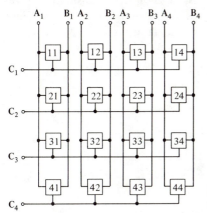

Figure 13.33 A two-dimensional array of memory cells.

Figure 13.33 already represents a practical 16-bit memory, and could be constructed in this form and used. However, it is much simpler to use a memory if it is organized more like Figure 13.30. It should be realized that a practical sized MOS memory may contain as many as 1024 cells, and the system designer who uses such a memory does not want to concern himself with

96 lines (3 × 32); he wants a data input line, a data output line, a read/write instruction line, and sufficient address lines to be able to address any cell on the chip. Any number between 0 and 1024 can be represented by a 10-digit binary code, so 10 address lines are sufficient for a 1024-bit memory. (See Appendix I.) Thus a 1024-bit memory requires about 13 external connections rather than 96.

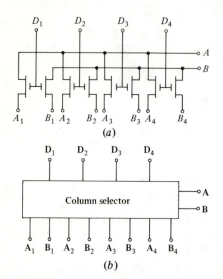

(a)

(b)

Figure 13.34 A circuit used to select the column to which inputs **A** and **B** are connected. (a) The circuit. (b) The block diagram.

The organization of a 16-bit memory into functional blocks is the same as a 1024-bit memory, so we continue the discussion with the smaller system. The data lines of Figure 13.33 are first combined with the circuit of Figure 13.34. Inputs D_1 to D_4 determine which set of data lines A_1, B_1 through A_4,

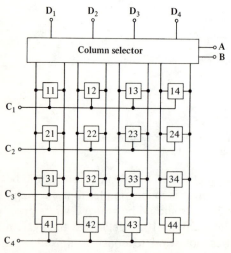

Figure 13.35 The memory cell array with column selector. Input-output terminals **A** and **B** are connected to the **A** and **B** terminals of the cell (MN) if $C_M = 1$, $D_N = 1$. All other **C** and **D** terminals must be **0**.

B_4 is connected to the input-output lines **A, B**. For example, if $D_1 = 1$ and D_2 through $D_4 = 0$, then **A** is connected to A_1 and **B** is connected to B_1.

The memory now appears as in Figure 13.35. Lines C_1 to C_4 and D_1 to D_4 are address lines, and **A** and **B** are the combination input/output lines. To address cell 23, a **1** is placed on lines C_2 and D_3. In this case lines **A** and **B** are connected to the **A** and **B** terminals of cell 23. The address lines may, however, be further condensed, because it requires only four lines to address 16 cells. (A four-digit binary number may represent any decimal number from 0 to 15). Thus an *address decoder* is used. A block diagram and the truth table for a suitable decoder are given in Figure 13.36. It is a simple logical synthesis problem to construct such a decoder from logic gates.

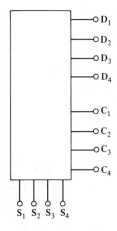

Truth Table For **C** and **D**

S_1	S_2	S_3	S_4	C_1	C_2	C_3	C_4	D_1	D_2	D_3	D_4
0	0	0	0	1	0	0	0	1	0	0	0
0	0	0	1	1	0	0	0	0	1	0	0
0	0	1	0	1	0	0	0	0	0	1	0
0	0	1	1	1	0	0	0	0	0	0	1
0	1	0	0	0	1	0	0	1	0	0	0
0	1	0	1	0	1	0	0	0	1	0	0
0	1	1	0	0	1	0	0	0	0	1	0
0	1	1	1	0	1	0	0	0	0	0	1
1	0	0	0	0	0	1	0	1	0	0	0
1	0	0	1	0	0	1	0	0	1	0	0
1	0	1	0	0	0	1	0	0	0	1	0
1	0	1	1	0	0	1	0	0	0	0	1
1	1	0	0	0	0	0	1	1	0	0	0
1	1	0	1	0	0	0	1	0	1	0	0
1	1	1	0	0	0	0	1	0	0	1	0
1	1	1	1	0	0	0	1	0	0	0	1

Figure 13.36 The block diagram of an address decoder suitable for the 16-cell memory of Figure 13.35. Each of the 16 cells is selected by a binary code, given in the truth table.

With the addition of address decoders, the memory of Figure 13.35 is almost complete. It is only necessary to add some input-output circuitry to send data into or take data from lines **A** and **B**. A block diagram of a possible input-output circuit block appears in Figure 13.37(*a*). This circuit functions according to the truth tables in Figure 13.37(*b*). Again, it is a simple logical synthesis problem to construct such a circuit from **NAND** or **NOR** gates.

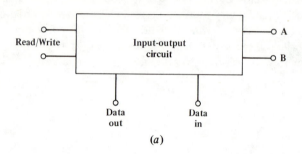

(*a*)

Truth table for input–output circuit

	Read/Write	Data in	A	B	Data out
Write mode $\Big\{$	1	0	0	1	—
	1	1	1	0	—
Read mode $\Big\{$	0	—	1	0	1
	0	—	0	1	0

(*b*)

Figure 13.37 (*a*) The block diagram of an input-output circuit block for the memory of Figure 13.35. (*b*) The truth table. In the input mode (READ/WRITE = 1) the circuit sets A equal to whatever is on line DATA IN, and B to the complement. In the output mode, the DATA OUT line is set to A.

As an example of a practical MOS memory, the circuit schematic of the INTEL 1101 memory is given in Figure 13.38. It closely resembles the basic memory circuit described here, except that it is much larger. This 256-bit memory contains almost 2000 transistors and is constructed in a single chip of silicon. A photomicrograph of the chip is shown in Figure 13.39.

MOS Shift Registers   The shift register consists of a string of memory cells, each of which contains a **0** or **1**. Upon command, every cell passes its contents to its neighbor on the right, and receives the contents of its neighbor on the left. As shown in Chapter 9, shift registers may be simply constructed from flip-flops. However, MOS transistors offer the possibility of an even simpler construction.

An example of a cell of a typical MOS shift register is shown in Figure 13.40(*a*). This circuit is not a flip-flop, and the memory function is performed quite differently. The cell consists of two inverters, coupled by transistor M_2.

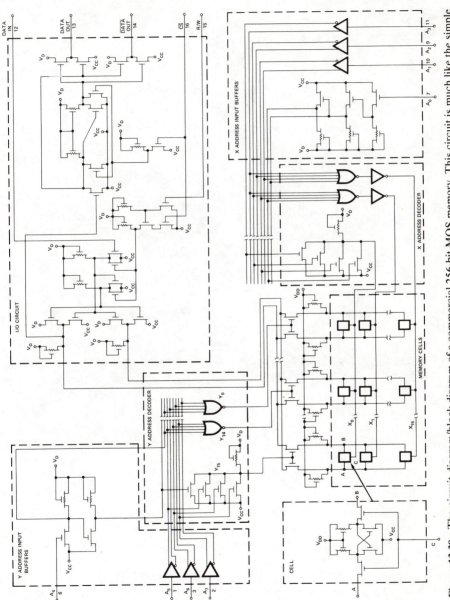

Figure 13.38 The circuit diagram/block diagram of a commercial 256-bit MOS memory. This circuit is much like the simple 16-bit memory of Figures 13.31 to 13.37. (Courtesy of Intel Corporation.)

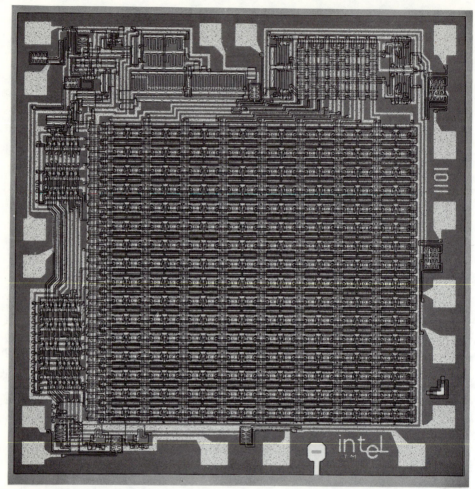

Figure 13.39 A microphotograph of the silicon chip containing the 256-bit memory of Figure 13.38. The chip dimensions are 2.7 mm on a side. (Courtesy of Intel Corporation.)

There are two clock inputs, labeled C_1 and C_2, which alternate, but which are never simultaneously logical **1** as shown in Figure 13.40(*b*). Suppose the input to the cell equals **1**. As soon as a clock pulse is received in line C_1, M_2 and M_3 are turned on, causing the input to M_4 to be the complement of **A**, which is $\overline{1}$ (or **0**) in this case. Now, as shown in Figure 13.40(*b*), at time t_2, C_1 returns to **0**, causing M_2 to become an open switch. It might at first appear that the information about what entered the cell would be lost in the interval between t_2 and t_3. However, the gate of M_4 is a tiny capacitor that is disconnected in this interval and thus retains the charge placed in it. In the example here with **A** = **1** at time t_1, the gate of M_4 is close to zero volts (logical **0**) at

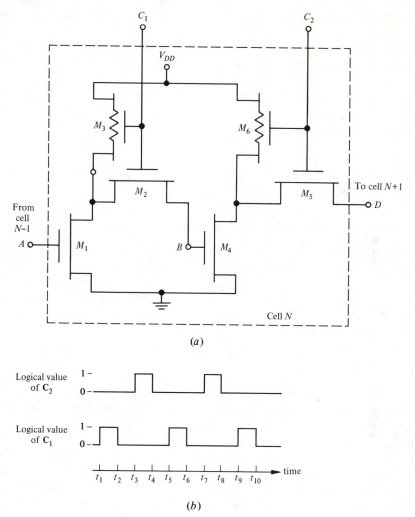

Figure 13.40 An MOS shift register using charge storage. (*a*) The circuit diagram of a typical cell. (*b*) The clock signals. Two clocks with the same frequency, but out of phase, are required.

time t_2 and remains at this voltage as long as M_2 is an open switch. At time t_3, $C_2 = 1$, and M_5 and M_6 become conducting. The output **D** is just the complement of **B**; thus **D** = **1**. The contents of cell N have been passed to cell $N + 1$. It is seen from Figure 13.40(*b*) that at time t_5 the process starts all over again. The time for a complete cycle, or shift, is $t_5 - t_1$, the period of the clock signal.

The charge storage feature of an MOS transistor gate precludes the need for a flip-flop. Of course the gate is not perfectly insulated, so the charge eventually leaks off if no clock pulse arrives to restore the gate charge. In practice, a minimum clock frequency is about 1 kHz and a maximum about 5 to 10 MHz.

Shift registers are very useful for the temporary storage of information in small computers and computer terminals, and represent one of the major applications of MOS transistors. Because of the large sales volume of shift registers, they are quite cheap, at present typically about 1¢ per bit in 100- to 1000-bit registers. A microphotograph was shown in Figure 13.1 of a 1024-bit shift register in a single silicon chip.

EXAMPLE 13.7

Suppose that the input A to cell N equals 0 at time t_1, Figure 13.40. Find the value of D at time t_5.

SOLUTION

If the shift register functions properly, D at time t_5 should equal A at time t_1, because $t_5 - t_1$ is the clock period, or the shift time. Let us demonstrate that the shift of 0 from cell N to cell $N + 1$ takes place as it should. Initially $A = 0$; thus when M_3 and M_2 turn on at $t = t_1$, $\overline{A}$, which equals 1, appears at B. At $t = t_3$, M_2 is turned off and remains off until t_5. Because the charge on the gate of M_4 cannot leak off, B remains 1 in the interval $t_2 < t < t_5$. Thus at time t_3, when M_5 and M_6 are turned on, D is set to $\overline{B}$, or 0. At $t = t_4$, M_5 turns off and D remains 0 until M_5 turns on again at t_7. Clearly D, at $t = t_5$, equals A at $t = t_1$, as claimed.   ■

13.3   MOS Integrated Circuits

It was shown in the previous section that MOS digital circuits may be constructed entirely from interconnected MOS transistors. Because no other components are required, the integrated circuits may be constructed with a minimal number of processing steps. It is only necessary to fabricate an array of MOS transistors on a silicon chip, and provide suitable interconnections.

Let us see what steps are involved in the fabrication. The process starts with an n-type wafer covered with a protective oxide. As shown in Figure 13.41(b), the oxide is removed in a photolithographic step wherever source and drain regions are to be diffused. (The basic oxidation, diffusion, and photolithographic steps are described in Chapter 7). A p-type impurity is diffused into the desired areas, and a thick oxide is regrown over the whole wafer, Figure 13.41(c). It is necessary to have a very thin oxide in the region of the gate for device operation; therefore the oxide is removed in this region and a thin oxide regrown, Figure 13.41(d). Windows are now opened in the oxide wherever it is desired to make contact to the source and drain, and aluminum is evaporated over the entire wafer. The last photolithographic step is the aluminum etching step to define the gate, source and drain contacts, and the interconnections. The final appearance of one MOS inverter on the wafer is shown in Figure 13.41(e).

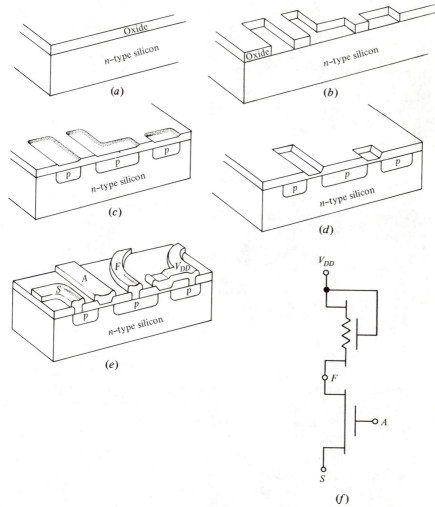

Figure 13.41   The appearance of a silicon wafer at various stages in the fabrication of an MOS integrated circuit. (*a*) The starting wafer, an oxidized *n*-type silicon crystal. (*b*) The oxide cuts for *p*-type diffusion. (*c*) The wafer after diffusion and reoxidation. (*d*) The appearance after a second oxide cut and a thin reoxidation in the regions of the gates. (*e*) The final structure. (*f*) The circuit diagram for the integrated circuit of (*e*).

The extreme simplicity of the processing in comparison with bipolar integrated circuit processing is evident. It derives largely from the fact that the isolation in MOS transistors is automatic. The *n*-type substrate is connected to the most positive voltage in the circuit (ground), and all of the *p* regions in the wafer are isolated by reverse-biased *pn* junctions. Furthermore many of the interconnections between *p* regions are made internally;

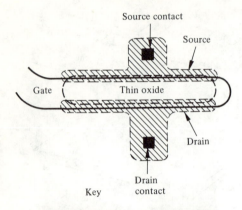

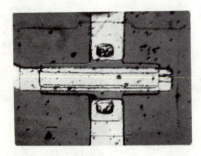

Figure 13.42 Micrograph of the MOS transistor whose *I-V* characteristics are given in Figure 13.9. *Z* is about 100 μm and *L* is about 10 μm. The various oxide cuts and the metallization boundary are shown.

the *p* regions are merely brought into contact rather than interconnected with aluminum on the surface. The connection between the two MOS transistors at the circuit point *F*, Figure 13.41(*f*), is made in this fashion.

In the circuits discussed in this chapter, two types of *p* channel MOS transistors were used; the driver devices and the load devices. For the driver device we have used the characteristics of Figure 13.9. A micrograph of the

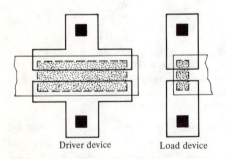

Driver device   Load device

■ Contact regions
□ *p*–type diffused regions
▨ Thin oxide region
□ Gate metal

Figure 13.43 The relative geometry of the MOS driver device and the MOS load device used as examples in this chapter.

device is shown in Figure 13.42. The width Z is approximately 100 μm and the length L equals 10 μm. The load device whose characteristics are given in Figure 13.14 is similar to the driver device except that the channel width Z is reduced by a factor of 10. Thus the relative geometry of the load device and the driver device might appear as in Figure 13.43.

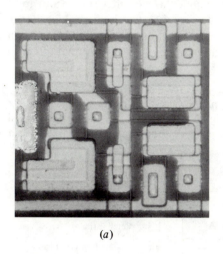

(*a*)

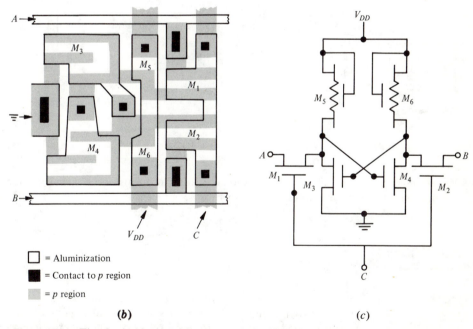

☐ = Aluminization

■ = Contact to *p* region

▨ = *p* region

(*b*)   (*c*)

Figure 13.44   The physical layout of the memory cell of Figure 13.31. (*a*) Photomicrograph of the integrated circuit. (*b*) Schematic drawing showing the location of the various components. (*c*) The circuit diagram.

Figure 13.44 shows the layout of a circuit that has been discussed extensively, the basic memory cell of Figure 13.31. Note that the devices are "folded" to conserve space. Again, the minimum allowable spacing of 10 μm suggested in Chapter 7 is maintained between all boundaries. Figure 13.44 by no means represents the ultimate in compact circuit layout. With a somewhat reduced allowable spacing, this circuit is layed out in an area of 130 μm² in some commercial MOS circuits. The high packing density in MOS integrated circuits may be appreciated when it is realized that a single bipolar transistor often takes up this much area in a conventional integrated circuit.

EXAMPLE 13.8

Lay out a two input **NAND** gate using an active device Z/L of 10 and a load device Z/L of 1, as in Figure 13.43. Find the total area required including contacts and 50 μm square bonding pads. Assume that a 10-μm tolerance must be observed.

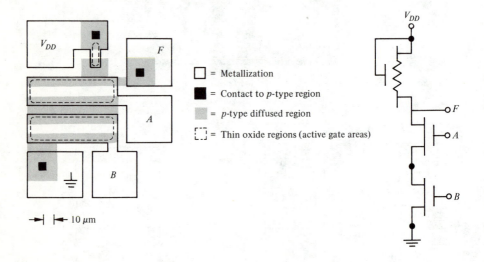

SOLUTION

A possible layout, using a linear geometry for the transistors, is given in the sketch. The total area used is 160×190 $(\mu$m$)^2$ or 3×10^{-4} cm². ■

Summary

- MOS transistors are constructed on the surface of a silicon crystal. The source and drain electrodes are diffused regions of the opposite conductivity type from the substrate. The gate is a metal electrode insulated from the semiconductor surface by a nonconducting oxide. A p channel MOS transistor is made on an n-type silicon substrate.

- The gate voltage controls the current flow between source and drain in a MOS transistor. In a p channel device a negative gate voltage induces a conducting channel of mobile holes that connects the p-type source and drain region. A typical p channel device has a turn-on voltage of -4 V.

- For small drain-to-source voltages, the MOS transistor behaves like a controlled resistor. The drain-to-source resistance decreases as the magnitude of the gate-to-source voltage increases. As the magnitude of the drain-to-source voltage increases, the channel resistance increases, and eventually the drain current saturates.

- The MOS transistor may be used as an electronic switch. If the magnitude of the gate-to-source voltage is less than the magnitude of the turn-on voltage, the source-to-drain characteristic is that of an open switch. For large gate-to-source voltages that greatly exceed the turn-on voltage, the characteristic is that of a low resistance, or closed switch.

- MOS transistors may be used as load resistors in integrated circuits. Typically the gate is connected to the drain and the substrate is grounded. The Z/L ratio of an MOS load device should be less than the Z/L of the driver device.

- The basic MOS inverter for use in integrated circuits consists of an MOS transistor (the driver device) and an MOS load device. A suitable set of logic levels for a device with -4-V turn-on voltage is 0 to -3 V for logical **0** and -8 V or less for logical **1**.

- Both **NAND** and **NOR** gates may be constructed using MOS devices. Mixed functions may also be simply synthesized. The practical limit of **NAND** gate fan-in is 2 or 3; however **NOR** gate fan-in is essentially unlimited.

- Fan-out limitations of MOS logic circuits arise from the switching speed. The combined input capacitance of the gates being driven must be charged by the output of the driving gate. Typical switching speeds for MOS circuits are in the range of 100 to 1000 nsec.

- Flip-flops may be constructed by combining MOS logic circuits in sequential fashion. The basic flip-flop is constructed from two cross coupled inverters. A clocked **R-S** flip-flop may be constructed from eight MOS transistors.

- MOS circuits offer a means of achieving a compact electronic memory. A basic MOS memory cell may be constructed from six transistors, four to produce a flip-flop and two for input-output. The memory cells may be arranged in two-dimensional arrays to yield large memories. Memories storing over 1000 bits on a single chip of silicon are available. MOS shift registers also have a memory function, and are commonly constructed as integrated circuits storing from 100 to 1000 bits.

- MOS integrated circuits are simpler to construct than bipolar integrated circuits. Only a single diffusion step is required, as opposed to three diffusions in a typical bipolar circuit. Furthermore no special isolation of devices is required; all devices in the circuit are automatically isolated by reverse-biased *pn* junctions.

References

At approximately the same level as this book:

Crawford, R. H. *MOSFET in Circuit Design.* New York: McGraw-Hill, 1967.

At a more advanced level:

Wallmark, J. T., and H. Johnson. *Field Effect Transistors.* Englewood Cliffs, N.J.: Prentice-Hall, 1966.

Emphasizing the physical theory of MOS transistors:

Grove, A. S. *Physics and Technology of Semiconductor Devices.* New York: John Wiley & Sons, 1967.

Problems

13.1   Suppose that the circuit of Example 13.1 is used as the basic inverter in an MOS logic system. Define an appropriate voltage range for logical **0** and logical **1**. (Use negative logic.)

13.2   Devise a circuit for a three-input **NOR** gate (negative logic) based on the inverter of Example 13.1. Using a graphical technique, solve for v_{OUT} when all three input voltages equal -8 V.

13.3   Using the "model" for the MOS load device suggested in Example 13.2, find the output voltage of the basic inverter, Figure 13.18, for two values of the input voltage, -4 V and -8 V. (Use the MOS device characteristics of Figure 13.9, and solve for v_{OUT} graphically.) Compare with the exact results given in Figure 13.22.

13.4   Find the output voltage of the circuit of Figure 13.15 for an input voltage of -5 V. Assume that $R_L = 15$ kΩ.

13.5   Find the transfer characteristics, v_{OUT} versus v_{IN}, of the inverter circuit of Figure 13.15.

13.6   Find the *I-V* characteristic of the MOS load circuit (such as given in Figure 13.20) using, however, the model for the load device given in Example 13.2.

13.7   Suppose the channel width Z of the MOS driver device used in Figure 13.18 were halved. Find the new transfer characteristics of the circuit.

13.8   Draw the circuit diagram for a four-input **NOR** gate analogous to Figure 13.23. (1) Construct the truth table. (2) Find the value of v_{OUT} when all inputs equal -3 V. (3) Find the output voltage when all four inputs equal -8 V.

13.9 Find the approximate output voltage of the two-input **NAND** gate (Figure 13.24) when $v_A = -7$ V, $v_B = -7$ V.

13.10 Find the approximate output voltage v_F of the circuit of Figure 13.25 if $v_A = v_B = v_C = v_D = -8$ V. (Use the device characteristics given in Figures 13.9 and 13.14, and a graphical technique.)

13.11 Show that the logical function of the circuit of Figure 13.25 may be expressed as $F = \overline{A}\ \overline{B}(\overline{C} + \overline{D})$. (See Example 13.5.)

13.12 Find a logic circuit constructed from one **NOR** gate and one **NAND** gate that has the same logical function as the circuit of Figure 13.25.

13.13 The MOS transistor, when strongly turned on, behaves much like a resistor. (1) Find a suitable resistance to represent the device of Figure 13.9 when the gate voltage equals -8 V. (2) Using this resistor as an equivalent circuit for the MOS driver device, find the output voltage of the basic inverter, Figure 13.18, when the input voltage equals -8 V. Compare with the exact result (Figure 13.22).

13.14 The input of the basic inverter of Figure 13.18 is switched from 0 to -8 V. Find the approximate time for the output to reach -4 V, assuming that the load consists of a 50-pF capacitor (representing the input capacitance of gates connected to the output). Use the model for the load device given in Example 13.2, and model the driver device as a resistor, as suggested in Problem 13.13.

13.15 A possible set of inputs to the **S-R** flip-flop of Figure 13.29 is given in Figure 13.45. Find the values of **Q** and $\overline{\mathbf{Q}}$ in the time interval t_0 to t_{10}, assuming that $\mathbf{Q} = \mathbf{0}$ at $t = t_0$. (Assume that the flip-flop changes state in a time short compared to the intervals shown, so that switching time is of no concern.)

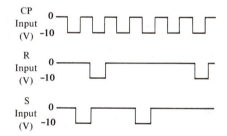

Figure 13.45

13.16 Suppose that the circuit of Figure 13.31 is constructed from four driver devices whose characteristics are given in Figure 13.9, and two load devices whose characteristics are given in Figure 13.14. If $V_{DD} = -15$ V, **C = 1, A = 0, B = 1**, find the voltages v_A and v_B. (Assume that nothing is connected to terminal **A** or **B**.)

13.17 M_1, M_2, M_4, and M_5 of Figure 13.40 are driver devices (Figure 13.9), and M_3 and M_6 are load devices (with the characteristic given in Figure 13.14 when the clock signal is 1). Assume that $v_A = -7$ V at time t_1 and, (1) find v_B at $t = t_2$; (2) find v_D at $t = t_3$. Assume that $v_A = -4$ V at $t = t_1$, and (1) find v_B at $t = t_2$; (2) find v_D at $t = t_3$. (Assume that the gates are perfect insulators.) This problem demonstrates the restoring action of the circuit that makes up for gate leakage. (ANSWER: (2) -10 V.)

13.18   Lay out an MOS transistor with a Z/L ratio of 100 in an approximately square region.

13.19   Lay out an MOS two-input **NOR** gate. Use the device geometry of Figure 13.43.

13.20   Lay out a two-input **NOR** gate using the same Z/L ratio, as in Problem 13.19, but with "folded" driver devices to conserve space.

13.21   The characteristics of two MOS devices made by silicon gate technology are given in Figure 13.46. These devices have a lower turn-on voltage than the basic p channel devices of Figures 13.9 and 13.14. Find the transfer characteristics of an inverter constructed from the devices of Figure 13.45. Assume a power supply voltage of $V_{DD} = -12$ V.

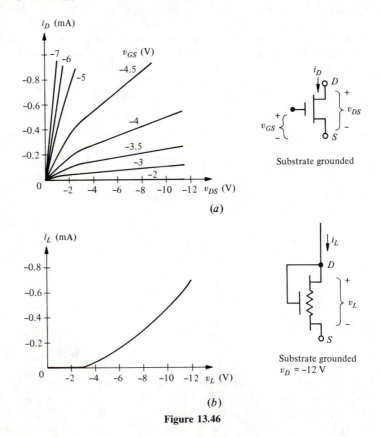

Figure 13.46

13.22   The voltage levels of MOS circuits may be translated by shifting the power supply voltages by some constant amount. Suppose that in the basic inverter, Figure 13.18, all voltages are shifted (such that $V_{DD} = -5$ V, and the source of the driver transistor is connected to $+5$ V). (1) Find the modified transfer characteristics. (2) Find the modified transfer characteristics of the inverter of Problem 13.21, again using the same voltages. (3) Compare with the characteristics of the DTL **NAND** gate (Figure 8.20). Comment on compatibility.

13.23 Find an approximate model for the load device whose characteristics are given in Figure 13.46. The model should be constructed from a resistor, a voltage source, and a perfect rectifier. (See Example 13.2.)

13.24 Use the model derived in Problem 13.23 to replace the load device in the inverter circuit (Figure 13.18). Derive the transfer characteristics, using $V_{DD} = -12$ V, and using the driver device characteristics given in Figure 13.46.

13.25 Lay out a clocked **S-R** flip-flop integrated circuit. Use a Z/L of 1 for the load device and a Z/L of 10 for the other transistors. Use 10-μm spacings, and allow 100-μm square areas for bonding pads.

13.26 Lay out a typical cell in a MOS shift register. Use 10-μm mask spacings, a Z/L of 10 for the driver devices, and a Z/L of 1 for the two load devices.

Appendix I

The Binary
Number System

Most persons are familiar with at least two different number systems, the ordinary decimal number system and the number system of Roman numerals. Here we introduce another system, the binary number system, which is in many ways the simplest of all. In the decimal system, each digit in a number may take on ten different values; hence it is necessary to have ten different symbols, the familiar 0, 1, 2,..., 9. The binary system is simpler in that only two values are allowed and consequently only two symbols are required, 0 and 1. A restriction to two symbols is appropriate for digital systems such as computers, because the memory devices used to store the numbers in general have only two states.

A number system is merely a way of representing a number. Let us review the representation of numbers in the decimal system, and proceed to see how they may be also represented in the binary system. For the moment only integers will be considered. For example, seven objects are represented by the symbol 7 in the decimal number system. When more than nine objects are to be represented, a subtle but very familiar notation is used. If, for example, a class contains two hundred forty-five students, we represent the number of students by the decimal number 245. This notation, because it is so familiar, is "decoded" by us auto-matically. We recognize that the meaning of 245 is $200 + 40 + 5$; that is, each digit in itself represents a number, and the complete number is formed by tak-ing the sum of the values represented by the individual digits. The digit farthest to the right, in this case 5, represents simply five. The next digit to the left, 4 in this example, does not mean simply four, but four times ten, or forty. The next digit stands for two times one hundred, or two hundred. In other words

$$245 = 2 \times 10^2 + 4 \times 10^1 + 5 \times 10^0$$

Similarly, the number 30245 may be written

$$30245 = 3 \times 10^4 + 0 \times 10^3 + 2 \times 10^2 + 4 \times 10^1 + 5 \times 10^0$$

Because each digit in the decimal number system corresponds to a higher power of 10, 10 is called the base of the system.

The *binary number system* is constructed analogously using the base 2 rather than the base 10. In this case there are only two digits available, 0 and 1; hence a number is represented by a string of 0's and 1's. Consider for example the binary number 101. This notation represents

$$1 \times 2^2 + 0 \times 2^1 + 1 \times 2^0$$

In other words, 101 in the binary number system represents five. The binary equivalents of the decimal numbers from 1 to 20 are given in Table I.1.

Table I.1   The First Twenty Binary Numbers

Decimal	Binary
0	0
1	1
2	10
3	11
4	100
5	101
6	110
7	111
8	1000
9	1001
10	1010
11	1011
12	1100
13	1101
14	1110
15	1111
16	10000
17	10001
18	10010
19	10011
20	10100

Note that a given number can have different meanings, according to the base of the number system in which it is written. The number 11, for example, means eleven in the decimal system, or three in the binary system. Generally, it is known from context which number system is being used. Whenever there is a possibility of confusion, the base may be written as a subscript. Thus $11_{(10)}$ means eleven, and $11_{(2)}$ means three.

The method of converting a binary number to a decimal number follows from the meaning of the binary number. The contributions represented by the individual digits are simply summed using decimal notation. For example the binary number 10001 is converted as follows:

$$1 \times 2^4 + 0 \times 2^3 + 0 \times 2^2 + 0 \times 2^1 + 1 \times 2^0$$
$$= 16 + 0 + 0 + 0 + 1$$
$$= 17$$

EXAMPLE I.1

Convert the binary number 11010 into its decimal equivalent.

SOLUTION

The binary number 11010 represents the sum

$$1 \times 2^4 + 1 \times 2^3 + 0 \times 2^2 + 1 \times 2^1 + 0 \times 2^0$$

Converting each of these terms to the decimal equivalent, we obtain

$$16 + 8 + 0 + 2 + 0$$

or the number 26 in decimal notation.   ■

The inverse process, that of converting a decimal number to a binary number, is not so obvious. However, there is a simple method consisting of repeated divisions by two. The method is summarized in Table I.2. The repeated divisions are carried out as in the example of Table I.2, until the quotient equals

Table I.2 An Example of the Conversion of a Decimal Number to Its Binary Equivalent

Decimal number to be converted →	Quotient		Remainder
$245 \div 2$	122	+	1
$122 \div 2$	61	+	0
$61 \div 2$	30	+	1
$30 \div 2$	15	+	0
$15 \div 2$	7	+	1
$7 \div 2$	3	+	1
$3 \div 2$	1	+	1
$1 \div 2$	0	+	1

Binary Result   1 1 1 1 0 1 0 1

$$245_{(10)} = 11110101_{(2)}$$

zero. At each step the remainder is noted, and the binary number is formed from the remainder as shown. No proof will be given here for the validity of this method; however the reader may convince himself of its validity by converting a few numbers from decimal to binary and back to decimal.

EXAMPLE I.2

Convert the decimal number 29 to its binary equivalent. Check that the answer is correct by converting it back to decimal.

SOLUTION

Following the recipe of Table I.2, we obtain

$$
\begin{aligned}
29 \div 2 &= 14 + 1 \\
14 \div 2 &= 7 + 0 \\
7 \div 2 &= 3 + 1 \\
3 \div 2 &= 1 + 1 \\
1 \div 2 &= 0 + 1
\end{aligned}
$$

Binary result   1 1 1 0 1

Thus the binary equivalent of 29 is 11101. To check the answer, we convert back to decimal:

$$1 \times 2^4 + 1 \times 2^3 + 1 \times 2^2 + 0 \times 2^1 + 1 \times 2^0 = 16 + 8 + 4 + 0 + 1 = 29 \qquad \blacksquare$$

All the operations of ordinary arithmetic may also be performed using binary numbers. For example, the addition of binary numbers is quite similar to the addition of decimal numbers, as demonstrated in Table I.3.

Table I.3   An Example of Binary Addition

1	0	1	0	0		
	1	1	0	1		
				1		Sum
			(0)			Carry from preceding sum
			0			Sum (including carry)
		(0)				Carry
		0				Sum
	(1)					Carry
	0					Sum
(1)						Carry
0						Sum
(1)						Carry
1	0	0	0	0	1	Final sum

EXAMPLE I.3
Carry out the same addition as in Table I.3 using the decimal representation, and verify the correctness of the final sum.

SOLUTION
We first convert the two binary numbers that are added in Table I.3 into their decimal equivalent. From Table I.1 they are 20 and 13. Thus the sum is 33. Converting back into the binary representation, we obtain

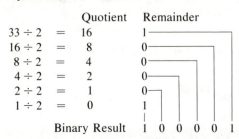

The same answer is obtained by doing the addition in either the decimal or the binary representation.   ∎

As demonstrated in the problems, subtraction of binary numbers may be carried out using complementation and addition, and in general this method is employed in digital computers. Similarly, multiplication is carried out by multiple additions, and division by multiple subtractions.

Codes   A general definition of a code is a string of digits that represents a number. Thus both the binary number system and the decimal number system are examples of codes. The most useful codes for use in digital electronic systems are those in which the values of the digits are restricted to one of the two binary values, 0 or 1. In addition to the binary number system there are other codes that are common in digital technology. The most important is the binary coded decimal, or BCD code. The BCD code is a direct translation of the individual digits of a decimal number into binary numbers. The decimal digits 8 and 9 require four binary digits; consequently each digit in the decimal number gives rise to four digits in the BCD representation. An example is given in Table I.4.

Table I.4   An Example of the BCD Code

Decimal number to be represented	BCD Representation
1 8 0 6	
0001 1000 0000 0110	0001100000000110

The conversion of BCD to decimal and decimal to BCD is elementary. Each decimal digit corresponds to a group of four binary digits. The four-digit binary group is a simple binary representation of the decimal digit, as given, for example, in Table I.1.

EXAMPLE I.4

Convert the decimal number 256 to BCD form.

SOLUTION

We convert each of the digits separately to binary form using Table I.1:

$$2 \quad 5 \quad 6$$
$$\underbrace{0010}\ \underbrace{0101}\ \underbrace{0110}$$

Thus $256_{(10)} = 001001010110$ in BCD form.   ∎

EXAMPLE I.5

Find the decimal equivalent of the representation 011000000001, which is known to be in BCD code.

SOLUTION

We group the coded number into groups of four binary digits, which are then converted to their decimal equivalents.

$$\underbrace{0110}\ \underbrace{0000}\ \underbrace{0001}$$
$$6 \qquad 0 \qquad 1$$

Thus 011000000001 in BCD code represents the decimal number 601.   ∎

Summary

- A number system is a way of representing numbers. The familiar decimal number system has a base of 10, which means that each digit can take on ten different values. The binary number system has a base of 2, and each digit can take on only two values, 0 or 1.

- The standard arithmetic operations may be carried out in the binary number system.

- Binary codes other than the simple binary number system may be used to represent numbers. An important binary code for use in digital computers is the BCD code.

References

Treatments at approximately the same level as this book:

Smith, R. J. *Circuits, Devices, and Systems.* New York: John Wiley & Sons, 1970.

Wickes, W. E. *Logic Design with Integrated Circuits.* New York: John Wiley & Sons, 1968.

A discussion emphasizing computer applications is given in:

Nashelsky, L. *Digital Computer Theory*. New York: John Wiley & Sons, 1966.

Problems

I.1   Convert the following binary numbers into their decimal equivalent: (1) 1001, (2) 1110, (3) 0011, (4) 1110001, (5) 1101011101, (6) 1011110011. (ANSWER: (5) 861.)

I.2   Convert the following decimal numbers into their binary equivalent: (1) 15, (2) 9, (3) 13, (4) 44, (5) 64, (6) 127, (7) 333, (8) 999. (ANSWER: (7) 101001101.)

I.3   Add the numbers in Problem I.1 in the binary representation and check with addition in the decimal representation.

I.4   Convert the following decimal numbers into BCD form: (1) 17, (2) 6, (3) 13, (4) 115, (5) 999, (6) 222, (7) 128. (ANSWER: (6) 001000100010.)

I.5   Convert the following numbers, which are in BCD code, into decimal form: (1) 100001010111, (2) 000100010001, (3) 010101010101, (4) 100110011001, (5) 0001011110000101.

I.6   Convert the following numbers from binary form into BCD code: (1) 11010011001, (2) 110111101, (3) 100011001, (4) 1111001101.

I.7   A simple way of subtracting binary numbers in digital circuits is based on the following rule. To subtract a binary number X from a binary number Y, form the 2's complement of X and *add* it to Y. Drop the last carry. (The 2's complement of a number is obtained by complementing the value of every digit in the number, and adding 1.) Verify the above rule by example, using the following numbers: (1) $X = 10010$, $Y = 10101$; (2) $X = 101$, $Y = 11110$; (3) $X = 11111$, $Y = 11111$; (4) $X = 1011111$, $Y = 1111000$; (5) $X = 1100$, $Y = 1110000$. Check the answers by converting the numbers to their decimal equivalent, and subtracting in the normal fashion.

I.8   Noninteger decimal numbers are defined similarly to integer decimal numbers. For example, the decimal number 245.16 means: $2 \times 10^2 + 4 \times 10^1 + 5 \times 10^0 + 1 \times 10^{-1} + 6 \times 10^{-2}$. Noninteger binary numbers are defined similarly. Find the approximate decimal equivalent of the following binary numbers: (1) 1101.10 (2) 1111.01, (3) 10011.1011, (4) 1111.1111, (5) 1000.0001, (6) 1.00101, (7) 10.111. (ANSWER: (2) 15.25.)

I.9   Just as 2 is the base of the binary number system, and 10 is the base of the decimal number system, eight is the base of the octal number system. (1) Find the decimal representation of the following octal numbers: 3, 7, 4, 17, 26, 45, 100, 33. (2) Find the octal representation of the following decimal numbers: 4, 8, 18, 100, 128.

I.10 The BCD is only one possible "weighted decimal code." In the BCD code the most significant digit is "weighted" a value of eight, the next digit is weighted four, the next two, and the least significant digit is weighted one. For example, 1001 in BCD form equals, in decimal form, $1 \times 8 + 0 \times 4 + 0 \times 2 + 0 \times 1 = 9$. Suppose that the most significant digit is weighted six, the next three, and the last two digits are both weighted one. (This is the so-called 6-3-1-1 code.) Thus, 1011 in 6-3-1-1 code represents the decimal number 8. In the 6-3-1-1 code, as in BCD, each group of four digits represents one decimal digit. (1) Find the decimal equivalent of the following numbers written in 6-3-1-1 code: 1000, 1011, 1100, 10011001, 11000001. (2) Find the 6-3-1-1 code representations of the following decimal numbers: 3, 7, 9, 22, 46, 174, 554, 28, 100.

Appendix II ▬▬▬▬▬▬▬▬▬▬▬▬▬▬▬▬

An Elementary Theory of *pn* Junctions

A qualitative description of *pn* junction operation is given in Chapter 5. This appendix extends this discussion, and derives the *I-V* characteristic of a *pn* junction. We first consider the general relationships for current flow in semiconductors, and then apply these relationships to derive the currents that flow in a *pn* junction.

II.1 Current Flow in Semiconductors

There are two kinds of particles in semiconductors, and a current is associated with any net flow of either particle. The magnitude of the current depends on several factors, including the carrier charge, concentration, distribution, and velocity. The concentration of carriers is considered first.

Carrier Concentrations in Extrinsic Semiconductors
It was shown in Section 5.1 that in an extrinsic semiconductor the majority carrier concentration is determined directly by the doping. However, it is the recombination and generation processes that control the minority carrier concentration. To compute the latter, a particular mechanism for recombination and generation of electrons and holes will be examined. The result to be obtained, namely, that the

product of the concentrations of electrons and holes is a constant, is perfectly general and does not depend on the particular mechanism of recombination and generation.

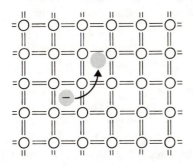

Figure II.1 Recombination of an electron with a hole. A free electron hops into a hole, annihilating both the electron and the hole. This sketch portrays the crystal just before recombination takes place.

A possible process for recombination of electrons and holes is illustrated in Figure II.1. A free electron hops into an empty bonding site, that is, into a hole, resulting in the annihilation of one electron and one hole. The reverse of this process, the breaking away of a bonding electron to produce a free electron and a hole, can also take place. This process, the generation of an electron and a hole, is illustrated in Figure II.2. The rate of recombination of electrons with holes should be proportional to the probability that an electron and a hole encounter each other. The more electrons, the greater this probability; similarly, the greater the concentration of holes, the greater the probability of an encounter of an electron and a hole. Thus we expect that the recombination rate R is proportional to both the electron and hole concentration, that is

$$R = r_1\, n\, p \tag{II.1}$$

where r_1 is some constant. In contrast, the generation rate G does not depend on the electron or hole concentration. (We are assuming that n and p are small compared to the concentration of bonds in the crystal; thus the density of bonds available to be broken by thermal energy is essentially independent of n and p.) Hence the generation rate in thermal equilibrium is given by

$$G = g_{th} \tag{II.2}$$

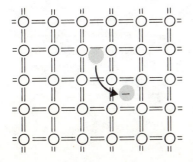

Figure II.2 Generation of an electron and a hole. When a bond is broken, both an electron and a hole are created. This sketch portrays the crystal just after the generation event.

where g_{th} is a constant. Both r_1 and g_{th} are constants in the sense that they do not depend on the electron or hole concentration; however they do depend on other parameters, such as the temperature of the semiconductor.

Since the number of carriers is constant in thermal equilibrium, the rate of generation must equal the rate of recombination; that is

$$g_{th} = r_1 \, n \, p \qquad \qquad (II.3)$$

Thus

$$n \, p = g_{th}/r_1 \qquad \qquad (II.4)$$

Equation (II.4) states the important result that in equilibrium the product of electron concentration and hole concentration is a constant. If, for example, the electron concentration is made very large by adding many donors, the hole concentration is reduced accordingly. (Physically, the high density of electrons makes it more probable that holes will experience recombination, thus reducing the hole population.) Equation (II.4) may be stated in a simpler form by considering the special case of a pure or "intrinsic" semiconductor. If no dopants are present, then the density of electrons, which must equal the density of holes, is given the special symbol n_i. In this case the product of n and p must equal n_i^2. We may now write Equation (II.4) in a simpler fashion:

$$\boxed{n \, p = n_i^2} \qquad \qquad (II.5)$$

In silicon, at room temperature, $n_i = 1.5 \times 10^{10}/\text{cm}^3$, which is very small compared to, for example, the density of electrons in metals ($5 \times 10^{22}/\text{cm}^3$). In fact n_i is so small that in almost all cases of practical interest the semiconductor is deliberately made "extrinsic," that is, doped sufficiently that the majority carrier concentration is fixed by the doping to a value much greater than n_i.

In an extrinsic semiconductor, the majority carrier concentration equals the doping and Equation (II.5) can be used to obtain the minority carrier

Table II.1   Equilibrium Carrier Concentration in Extrinsic Semiconductors

	n-type	*p*-type
Dopant	Group V	Group III
Concentration	N_D	N_A
Majority carrier	Electrons ($n = N_D$)	Holes ($p = N_A$)
Minority carrier	Holes ($p = n_i^2/N_D$)	Electrons ($n = n_i^2/N_A$)

$n_i = 2.5 \times 10^{13}$ for germanium at 300° K
$n_i = 1.5 \times 10^{10}$ for silicon at 300° K

concentration. Suppose we have donor impurities; then $n = N_D$, the donor concentration. From Equation (II.5), $p = n_i^2/n = n_i^2/N_D$. The electrons are majority carriers, holes the minority carriers, and the semiconductor is n-type. If acceptors are added instead of donors, holes are the majority carriers, and the material is p-type. These results are summarized in a convenient form in Table II.1.

EXAMPLE II.1

One part per million (atomic) boron is incorporated in a silicon crystal. What are the electron and hole concentrations at room temperature?

SOLUTION

The majority carrier concentration, that is, the hole concentration, just equals N_A, the boron concentration. Since there are 5×10^{22} silicon atoms/cm³, $N_A = 5 \times 10^{22} \times 10^{-6} = 5 \times 10^{16}$ cm⁻³. The electron concentration can be found from the equation $n = n_i^2/N_A$ (Table II.1). Thus $p = 5 \times 10^{16}$ cm⁻³ and $n = (1.5 \times 10^{10})^2/5 \times 10^{16} = 4.5 \times 10^3$ cm⁻³. ■

If both donors and acceptors are present simultaneously, the net doping equals the difference in concentrations.[1] Figure II.3 illustrates, for example, a region in which four donors and two acceptors are present. Two of the donated electrons have recombined with the two holes given up by the acceptors, leaving a net of two free electrons. It may be stated as a general rule in extrinsic semiconductors that if $N_D > N_A$, $n = N_D - N_A$ and if $N_A < N_D$, $p = N_A - N_D$. In both cases, Equation (II.5) can then be used to calculate the minority carrier concentration.

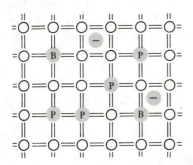

Figure II.3   A semiconductor with four donors and two acceptors present. The density of electrons equals $N_D - N_A$. Note that according to Table II.1, there will be some holes present, but this number is very much smaller than the number of electrons in an n-type sample.

Carrier Concentrations in Weakly Extrinsic Semiconductors   Table II.1 may be used to compute the electron and hole concentrations in extrinsic semiconductors, which are materials in which $|N_D - N_A| \gg n_i$. In this section we derive a general result for n and p in terms of n_i, N_D, and N_A, independent of their relative values. The principle upon which this derivation is based is

---

[1] We assume for the moment that this difference is much larger than n_i (which is usually the case).

charge neutrality. In the homogeneous bulk region of a semiconductor in thermal equilibrium, there must be no net charge in order that there be no electric field, otherwise an electric current would be produced.[2] The possible charges are: electrons, holes, donors (which at room temperature give up one electron each and are therefore positively charged), and acceptors (which at room temperature have each accepted an electron, becoming negative). The charge neutrality condition may be written as

$$n + N_A = p + N_D \tag{II.6}$$

where all the possible negative charges are grouped on the left and the positive charges on the right. We have not yet specified N_A or N_D; they may have any value, according to whatever doping has been put into the crystal.

Equations (II.5) and (II.6) may be solved simultaneously for n and p. If $N_D > N_A$, it is convenient to solve for n by eliminating p in Equation (II.6). The result, from the quadratic formula, is

$$n = \frac{N_D - N_A}{2} + \sqrt{\frac{(N_D - N_A)^2}{4} + n_i^2} \tag{II.7}$$

where the possible minus sign in front of the square root has been dropped because n can never be negative. The value of p may be obtained from Equation (II.5), once n is determined from Equation (II.7).

It is interesting to examine Equation (II.7) in several limiting cases. Suppose, for example, N_A and N_D were both much less than n_i; then the equation reduces to $n = n_i$. This is the case of an intrinsic semiconductor. Suppose on the other hand $N_D - N_A \gg n_i$; then the equation reduces to $n = N_D - N_A$. If, furthermore, $N_A = 0$, then $n = N_D$. This is the usual case of the doped semiconductor, described by Table II.1. In general it is necessary to use Equation (II.7) only if the semiconductor is lightly doped with the magnitude of $(N_D - N_A)$ comparable to n_i. This is indeed a rare situation in silicon.

The Equations of Current Flow   There are two basic mechanisms for current flow in semiconductors, corresponding to two basic mechanisms for the transport of mobile carriers. The first and most familiar arises from the presence of an electric field. Whenever a field is present in a medium with mobile carriers, a drift of the carriers, and an associated current, result. Such a current, called a *drift current,* is proportional to both the carrier concentration and the electric field. In other words the more particles drifting, or the faster they drift, the larger the current. Stating this quantitatively, a drift current density J is defined by

$$J = Q N v \tag{II.8}$$

where Q is the charge of the drifting particle, N is the particle concentration, and v is the average velocity. If Q is given in coulombs, N in particles per

---

[2] In an inhomogeneous region, such as near the interface of a *pn* junction, charge neutrality does not obtain; however this example involves a homogeneous region of a semiconductor.

cubic centimeter, and v in centimeter per second, then J is given in amperes per square centimeter.

Consider a uniformly doped n-type semiconductor, containing many more electrons than holes. Neither the atoms of the crystal, nor the dopant atoms, nor the bonding electrons can move in response to a field. The flow of current in the presence of an electric field may be simply illustrated as in Figure II.4. Only the electrons (free electrons) are shown; everything else is irrelevant to the conduction process. Because of their negative charge, the electrons move in a direction opposite to the electric field. This gives rise to a current in the direction of the field, again because of the negative charge on the electrons.

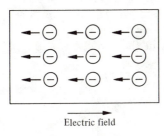

Electric field

Figure II.4 Drift current in an n-type semiconductor. The electrons move in a direction opposite to the field. The current is proportional to both the density of particles and the strength of the electric field.

The picture given in Figure II.4 is somewhat oversimplified. The electrons (as well as the holes) in semiconductors move about continuously, in the same way that the molecules of a gas move about. Even in the absence of any electric field the electrons are in continuous motion. Figure II.5 (*a*) illustrates the motion of one electron. At any temperature above absolute zero, the particle is always in motion because it possesses thermal energy; however it constantly collides with things, causing it to change directions. In thermal equilibrium the particle moves quite randomly; since the net result is no net

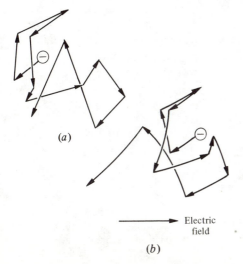

(*a*)

Electric field

(*b*)

Figure II.5 The random motion of an electron. The particle moves because of thermal energy, but changes directions frequently because of collisions. (*a*) In thermal equilibrium the result of all the motion and collisions is no net motion at all. (*b*) In an electric field the path of the electron is bent slightly in the direction of the force. A net motion results.

motion, all of this random motion may be ignored. However, if an electric field is present, the electron is continuously pushed by the field. Consequently each segment of the journey is altered by an extra motion in the direction of the force. The result, shown in Figure II.5(*b*), is a net motion, on the average, in the direction opposite to that of the electric field, that is, in the direction of the force. Again, the random part of the motion results in no net flow and may be ignored. Therefore we can use the simpler picture of Figure II.4, neglecting the random components of the particle motion.

The behavior of holes in a uniformly doped *p*-type semiconductor is analogous. The silicon and dopant atoms cannot move. There are negligibly few free electrons so the only carrier motion that occurs is that of holes. Thus the situation for hole conduction is as illustrated in Figure II.6. The holes behave as particles with positive charge and move on the average in the direction of the electric field. Current thus flows in the direction of the electric field.

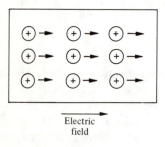

Electric
field

Figure II.6 Drift current in a *p*-type semiconductor. The mobile charged particles are the holes, which move in the direction of the electric field.

In their random thermal motion the electrons and holes in a semiconductor typically collide with various scattering centers in the solid about 10^{10} to 10^{14} times per second. As shown in Figure II.5, for electrons, an applied field slightly distorts the electron paths, resulting in a small average velocity, called the *drift velocity*, in the direction of the applied force. This average velocity is proportional to the applied field.[3] The constant of proportionality is defined as the *mobility* μ. The symbol μ_e refers to the mobility of electrons, while μ_h stands for the mobility of holes. Accordingly, the drift velocity of electrons is given by

$$v_e = -\mu_e E \quad \text{(electrons)} \tag{II.9}$$

where the minus sign appears because the direction of motion is opposite to the direction of the electric field E. The drift velocity of holes is given by

$$v_h = \mu_h E \quad \text{(holes)} \tag{II.10}$$

[3] The reader may wish to convince himself of this proportionality. Assume, for example, that the velocity is set to zero every 10^{-12} sec by a collision, and find the average distance traveled per unit time as a function of the field.

where the velocity is in the same direction as the electric field. Typical values are: $\mu_e = 1300$ cm²V⁻¹sec⁻¹, $\mu_h = 500$ cm²V⁻¹sec⁻¹ in lightly doped silicon. (If the material is heavily doped, numerous collisions with impurities occur and mobility is reduced.) We can now give the current density due to an applied field as

$$J_{e\ \text{drift}} = q\mu_e nE \qquad\qquad\text{(II.11)}$$
$$J_{h\ \text{drift}} = q\mu_h pE \qquad\qquad\text{(II.12)}$$

where q is the absolute value of the electronic charge. In any given piece of semiconductor the total drift current density is just the sum of the hole and electron drift current densities. The *conductivity* σ is defined as the ratio of the total drift current density to the electric field, and from Equations (II.11) and (II.12) is given by

$$\sigma = q\mu_e n + q\mu_h p \qquad\qquad\text{(II.13)}$$

The reciprocal of the conductivity is the *resistivity* ρ. Hence

$$\rho = \frac{1}{q\mu_e n + q\mu_h p} \qquad\qquad\text{(II.14)}$$

(Generally one of the terms in both Equations (II.13) and (II.14) is negligible, because in extrinsic silicon either $n \gg p$ or $p \gg n$.) The resistivity of a semi-conductor is important because it is a very easily measured property that is directly related to the doping. In fact the doping of a commercial sample is usually not specified by the manufacturer; rather the carrier type and resistivity are given.[4] The electron and hole concentrations may then be computed from Equations (II.14) and (II.5).

EXAMPLE II.2

A crystal is specified to be n-type silicon of 15-Ω cm resistivity. Compute the electron and hole concentrations.

SOLUTION

Because the crystal is specified as n-type, it is assumed that $n \gg p$. Thus from Equation (II.14), $\rho \cong 1/q\mu_e n$.

$$n = (q\mu_e\rho)^{-1} = (1.6 \times 10^{-19} \times 1.3 \times 10^3 \times 15)^{-1}$$
$$= 3.1 \times 10^{14}\ \text{cm}^{-3}$$
$$p = \frac{n_i^2}{n} = \frac{(1.5 \times 10^{10})^2}{3.1 \times 10^{14}} = 2 \times 10^6\ \text{cm}^{-3} \qquad\blacksquare$$

The second important mechanism in semiconductors for the transport of particles is diffusion. A diffusion of mobile particles of any kind is the result

---

[4] MKS units are standard in semiconductor work, except that the centimeter rather than the meter is the standard unit of length. Thus the units of resistivity are ohm cm, usually written Ω cm. One Ω cm equals, of course, 0.01 Ω m.

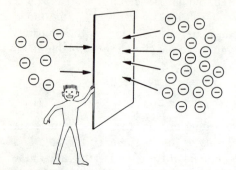

Figure II.7 Small observer standing inside a semiconductor. If there are more electrons on the right side of his hoop than on the left, the average diffusive flow of electrons through it is from right to left.

of a nonuniform concentration. If some molecules of a gas, for example, are introduced into one end of an empty container, they soon distribute so that their concentration is uniform throughout the container. Similarly in the case of electrons and holes in a semiconductor, there is a net flow of particles toward any region of lower concentration. No electric field is necessary to cause this motion. Figure II.7 illustrates the diffusive flow of electrons. A small observer stands inside a semiconductor, holding up a hoop of unit area. Because of random thermal motion, electrons are continuously going through the hoop, from left to right and from right to left. However, there are more particles per unit volume on the right-hand side than on the left; hence the flow from right to left is greater than from left to right. Thus if the particle concentration in a solid increases in, say, the x direction, a flow of particles will move, on the average, in the negative x direction. Associated with the flow of charged particles (here electrons) is a current, called the diffusion current. For electrons, the current is opposite to the flow; hence the current is in the direction of increasing concentration, as illustrated in Figure II.8(a).

The diffusive flow of holes, illustrated in Figure II.8(b), is analogous. With the concentration increasing to the right as shown, holes flow to the left. A current in the same direction as the flow results.

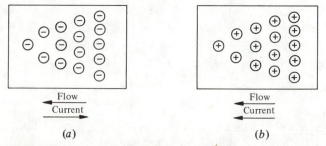

Flow

Current

(a)

Flow

Current

(b)

Figure II.8   The diffusion of electrons and holes. In (a) the electron concentration is higher on the right than on the left; hence there is a net flow to the left, constituting a current toward the right. In (b) the hole concentration is higher on the right than on the left, resulting in a flow of holes, and a current, toward the left.

The current density passing through a certain plane is obtained by multiplying the number of particles that pass through the plane per unit area, per unit time, by their charge. According to the physical arguments just given, it seems reasonable that the average rate of diffusive flow would be proportional to the derivative with respect to position of the particle density. Thus we may expect equations of the following form:

$$J_{e \text{ diffusion}} = qD_e \frac{dn}{dx} \qquad (\text{II.15})$$

$$J_{h \text{ diffusion}} = -qD_h \frac{dp}{dx} \qquad (\text{II.16})$$

These equations define the constants of proportionality D_e and D_h, called the *diffusion coefficients* for electrons and holes, respectively.

The constants D and μ depend on the same physical parameters, primarily temperature and the frequency of particle collisions. For example, if for some reason the collision rate were to increase, we would expect both D and μ to decrease. In fact there exists a mathematical relationship between D and μ, known as the Einstein relation:

$$D_e = \frac{kT}{q} \mu_e \qquad (\text{II.17})$$

$$D_h = \frac{kT}{q} \mu_h \qquad (\text{II.18})$$

The total current in the semiconductor is the sum of drift and diffusion currents, the individual values of which were given in Equations (II.11), (II.12), (II.15), and (II.16). Combining these equations, we have

$$\boxed{J_e = q\mu_e nE + qD_e \frac{dn}{dx}} \qquad (\text{II.19})$$

$$\boxed{J_h = q\mu_h pE - qD_h \frac{dp}{dx}} \qquad (\text{II.20})$$

II.2   *pn* Junction Theory

A very crude picture of *pn* junction operation was given in Chapter 5. This section gives a more detailed physical description of the mechanisms involved in the operation of a *pn* junction, and derives the ideal diode equation. It is easiest first to consider the processes in a junction with no bias applied, that is, one in thermal equilibrium.

The *pn* Junction in Thermal Equilibrium   Let us consider the distribution of electrons and holes in an n^+p step junction, that is, an abrupt junction that is more heavily doped on the *n* side. We choose to analyze an n^+p junction for several reasons. First the analysis is simplified in comparison to the analy-

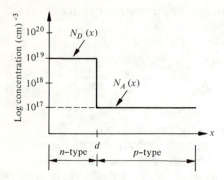

Figure II.9 An approximation of the doping profile for a *pn* junction. Such a junction, called a step junction, may be made by the processes described in Chapter 7.

sis of a general *pn* junction. Secondly, the analysis will also apply, with only a change of a few words to a p^+n junction. Thirdly, the doping distribution in "real" structures, examples of which are given in Chapter 7, often approximates an n^+p step junction. For simplicity, let us idealize the doping profile as shown in Figure II.9. One might think at first that the electron and hole concentrations corresponding to the doping profile of Figure II.9 would be those derived from Table II.1. This first estimate is shown in Figure II.10. However, such very abrupt changes of carrier concentrations as shown in Figure II.10 cannot exist. Such a distribution would imply infinite diffusion currents at the step, because of the infinite slopes of the carrier concentrations. Such impossibly large diffusion currents would quickly reduce the concentration differences by moving electrons to the right and holes to the left across the junction boundary. However this carrier motion would be eventually self-arresting because the buildup of sufficient negative charge on the

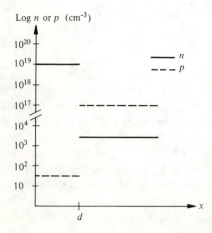

Figure II.10 First estimate of the carrier concentration profile in the junction of Figure II.9.

right and positive charge on the left would hinder further motion. This little scenario about the consequences of having an abrupt boundary between *p*- and *n*-type regions does suggest, in fact, the correct equilibrium physical situation. The electrons and holes redistribute themselves so as to produce an electric field large enough to oppose the diffusive flow, so that no further net motion of electrons and holes occurs.

Let us suppose that separate *n*-type and *p*-type semiconductors were suddenly joined to form the structure of Figure II.9. Initially, then, the carrier distributions would indeed be just as depicted in Figure II.10. It is important to realize that there is no charge or electric field associated with the distribution shown in Figure II.10, because it is made up of two electrically neutral pieces of semiconductor. Thus there are no electric forces tending to "push" the electrons and holes. However, because of the large concentration differences, holes move by diffusion to the left and electrons to the right across the boundary. This produces a new distribution, Figure II.11. The semiconductor was initally electrically neutral However, after the redistribution it is not neutral; positive charge has moved to the left and negative charge to the right. Positive charge accumulating on the left causes the potential of the left-hand side to increase relative to the right-hand side. Even with no external applied voltage, there is a voltage difference (a "built-in voltage") between the *p* side and *n* side of the junction. The potential difference implies an electric field directed from left to right. The motion of carriers then ceases, because the electric field opposes further motion. In other words, the tendency of the electrons to diffuse to the right is opposed by the action of the electric field, which pushes them to the left.

The potential distribution in the *pn* junction is shown schematically in Figure II.12. From this figure, we see that at positions between the places $x = x_n$ and $x = x_p$, the values of p and n are in transition between their values on one side of the junction and their values on the other. We refer to the region $x_n < x < x_p$ as the *transition region*. To the left of the point $x = x_n$, and

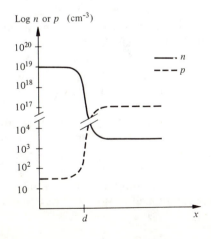

Figure II.11 The carrier distributions in the vicinity of a *pn* junction in thermal equilibrium. With the distribution of Figure II.10, the semiconductor is neutral; however, after the redistribution shown here, there is an excess positive charge on the left and excess negative charge on the right.

to the right of the point of $x = x_p$, the semiconductor is unaffected by the presence of the junction. These regions outside the transition region are called "bulk" regions, because there the carrier concentrations (when no external voltage is applied) are the same as if no junction existed. In the transition region, however, the carrier concentrations deviate from the bulk values.

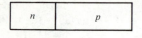

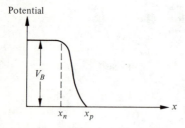

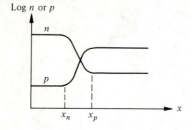

Figure II.12   The equilibrium carrier concentration and potential as a function of distance in a *pn* junction.

Comparing the initial distribution of Figure II.10 with the final distribution of Figure II.11, it can be seen that there has been a net motion of electrons to the right and holes to the left. Part of the reduction of hole concentration on the right and electrons on the left is by recombination; that is, some of the electrons that flowed into the *p* side recombined with holes, annihilating both particles. The charge that results in the potential difference shown in Figure II.12 may be found by comparing Figure II.10 with Figure II.11. In the region $x_n < x < x_p$ both the electron and hole concentrations are small compared with the concentration of dopant atoms. Therefore in the left-hand side of the transition region there is a positive charge density consisting mainly of immobile donor ions. In the right-hand side of the transition region there is a negative charge density, mainly of immobile acceptor ions. These two charge densities, positive and negative, are the source of the potential difference illustrated in Figure II.12. The region $x_n < x < x_p$ is also called the *space-charge region,* or *depletion region.* The latter term stems from the fact that the electron and hole concentrations are reduced in this region; hence it is relatively depleted of carriers.

It is possible to solve for the magnitude of the built-in voltage by using the fact that in thermal equilibrium no current, either of electrons or holes, flows. Mathematically, this may be expressed by an equation stating that the diffusion and drift currents just cancel. For example, the total current due to motion of electrons is given by Equation (II.19):

$$J_e = q\mu_e nE + qD_e \frac{dn}{dx} \tag{II.19}$$

The total electron current J_e must be zero in thermal equilibrium. By setting Equation (II.19) to zero, it is possible to derive the equilibrium electric field in terms of the slope of the electron concentration dn/dx, and further to find the magnitude of the built-in voltage. Equation (II.19) is first set to zero:

$$q\mu_e nE = -qD_e \frac{dn}{dx} \tag{II.21}$$

Using the Einstein relationship, Equation (II.17), and solving for E

$$E(x) = -\frac{kT}{q} \frac{1}{n} \frac{dn}{dx} \tag{II.22}$$

which may be written more simply as

$$E(x) = -\frac{kT}{q} \frac{d}{dx} \ln n \tag{II.23}$$

The electric field is, by definition, the negative derivative of the potential

$$E = -\frac{dV}{dx} \tag{II.24}$$

This relationship may be substituted into Equation (II.23), which then after rearrangement has the form

$$dV = \frac{kT}{q} d(\ln n) \tag{II.25}$$

Both sides of this equation are differentials; thus the difference in potential between the left-hand and right-hand sides is given by $(kT/q) \Delta(\ln n)$, where $\Delta(\ln n)$ stands for the difference in $(\ln n)$ between left-hand and right-hand sides. If the electron concentration on the n side is denoted by n_n, the electron concentration on the p side is denoted by n_p, and the potential difference is denoted by V_B (built-in voltage), then

$$\boxed{V_B = \frac{kT}{q} (\ln n_n - \ln n_p) = \frac{kT}{q} \ln \frac{n_n}{n_p}} \tag{II.26}$$

Note that the n-type side is higher in potential than the p-type side by this voltage V_B. The electrons, having a negative charge, are pushed toward the left and the holes are pushed toward the right by the potential difference. In both cases the diffusive flow is in the opposite direction. In terms of the previous discussion, the barrier voltage V_B is just the potential required so that both the total electron current and the total hole current equal zero.[5]

EXAMPLE II.3

Find the magnitude of the built-in voltage in a silicon *pn* junction that has an *n* region with doping $10^{19}/cm^3$ and a *p* region with doping $10^{16}/cm^3$.

SOLUTION

The electron concentration away from the junction in the *n* region is given by the doping. Thus $n_n = 10^{19}/cm^3$. The electron concentration in the *p* region is given by

$$n_p = \frac{n_i^2}{p} = \frac{(1.5 \times 10^{10})^2}{10^{16}} = 2.25 \times 10^4$$

Thus according to Equation (II.26)

$$V_B = 0.026 \; \ln \frac{10^{19}}{2.25 \times 10^4} = 0.88 \text{ V} \qquad \blacksquare$$

Forward Bias   We now wish to compute the particle currents in a *pn* junction with external bias applied. We consider first forward bias, in which the *p* region is made positive with respect to the *n* region, as shown in Figure II.13.

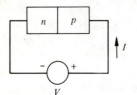

Figure II.13   A *pn* junction with bias applied. In forward bias both V and I are positive.

Of key importance in the analysis of the *pn* junction is the question of the distribution of the applied voltage across the structure. There are two homogenous bulk regions, separated by a transition region. It is a simple matter to show that the resistance of the bulk regions is typically a few ohms or less; this is deferred until Example II.5. According to the ideal diode equation, Equation (5.1), applied forward voltages of about 0.7 V typically result in currents in the milliampere range. With a bulk resistance of only a few ohms,

---

[5] The built-in voltage arises from the requirement that the current be zero under the condition of zero applied bias. Hence, if a wire is connected from the *n* region to the *p* region, no current flows. In fact a contact potential is set up at each contact such that the total potential drop going around the complete circuit is zero. A simple analogy involving gas molecules exists. In the atmosphere, a pressure difference with height is set up because the tendency of the gas molecules to escape is opposed by the force of gravity. However, this pressure difference cannot be used to produce a flow; if a connection (with a pipe) is made from the top to the bottom of the atmosphere, no flow occurs. Instead a pressure drop is set up within the pipe, just opposing the external pressure drop between its ends.

it is clear that most of the applied voltage must appear across the transition region. It is quite reasonable then to make the simplifying assumption that the entire applied voltage appears across the transition region. It is a simple matter later (if in fact it is necessary) to correct for the voltage dropped across the bulk regions, much as if external resistors were present. When the polarity of the applied voltage, as defined in Figure II.13, is positive, V is opposite in sign to V_B; hence, in moving across the transition region from p side to n side, a voltage rise of $V_B - V$ occurs. The voltage is sketched in Figure II.14. When V is positive, the potential barrier is lowered. (As shown in Example II.3, the built-in voltage is on the order of 0.9 V. We shall see that, typically, the applied forward bias is less than this value.)

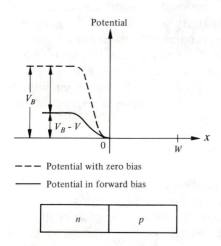

Figure II.14 The potential distribution in a *pn* junction in forward bias. A positive potential applied on the *p* side with respect to the *n* side reduces the potential barrier by the amount of the applied bias V. For convenience, the point $x = 0$ is defined to be the right-hand edge of the depletion region. The point $x = W$ is the contact to the *p* region.

Changing the voltage across the transition region of a *pn* junction has the effect of upsetting the delicate equilibrium in which drift plus diffusion currents just cancel each other. Electrons on the *n* side see a reduced field opposing their passage across the junction, as do also the holes on the *p* side. Diffusion of electrons to the right, which formerly was exactly cancelled by the action of the electric field, now is no longer fully cancelled, so a flow of electrons to the right takes place. Similarly holes are now able to move to the left. In both cases the electrostatic potential barrier opposing such motion has been lowered. If we consider an n^+p junction, then the source of electrons on the left greatly exceeds the source of holes on the right, so the electron current predominates.

Let us now compute the magnitude of the electron current. Electrons flow across the transition region and into the *p* region. This process has already been called minority carrier injection. The electrons diffuse away from the junction (there is no electric field in either of the bulk regions) and eventually recombine with holes somewhere in the *p* region. It is quite common, because of low series resistance, to make the bulk regions quite narrow; thus it is often the case that most of the injected minority carriers, here electrons, diffuse completely across the bulk region and recombine at the contact. It is the property of a "good" contact to be an efficient annihilator of minority carriers. A structure that is narrow, in the sense that the injected minority carrier can diffuse across the bulk region before it recombines, is called a short-base diode. In the short-base n^+p diode then, all of the electrons that are injected into the p region diffuse across it and recombine at the p contact. Therefore the electron current is constant throughout the p region and may be evaluated anywhere within it.

To compute the electron diffusion current it is only necessary to know the derivative of the electron concentration, that is, dn/dx. We therefore desire to determine the electron concentration as a function of position within the p region. The electron concentration may be readily evaluated at the p contact, that is, the point $x = W$ (Figure II.14). By our definition of a good contact, any injected carriers instantaneously recombine there. Hence the electron concentration is maintained at its equilibrium value at $x = W$. According to Table II.1, the equilibrium electron concentration in p-type material equals n_i^2/N_A, where N_A is the acceptor concentration. To distinguish equilibrium from nonequilibrium values, we attach a subscript zero to the equilibrium values. In this notation

$$n_W = n_{p0} = \frac{n_i^2}{N_A} \tag{II.27}$$

Here n_W stands for the electron concentration at the point $x = W$ and n_{p0} stands for the equilibrium electron concentration in the p region.

Now let us evaluate the electron concentration at the edge of the transition region (that is, at $x = 0$), which we will call n_T. To do this we make use of an equation resembling Equation (II.26), which relates the carrier concentrations on either side of the transition region to the voltage across the transition region. Strictly speaking Equation (II.26) applies only in thermal equilibrium, but it may be shown to apply also to a junction with bias applied if V_B is replaced by $V_B - V$, that is, the modified voltage across the transition region. Restating, then, Equation (II.26), we have that

$$n_T = n_n \exp\left\{-\frac{q(V_B - V)}{kT}\right\} \tag{II.28}$$

In thermal equilibrium (that is, when $V = 0$), Equation (II.28) reduces to

$$n_{T0} = n_{p0} = n_{n0} \exp\left\{\frac{qV_B}{kT}\right\} \tag{II.29}$$

To continue the analysis, we wish to eliminate n_n and n_{n0} from Equations (II.28) and (II.29). It is generally the case that n_n, the electron concentration in the n region, is unmodified by the applied voltage. It is only the minority carrier concentrations that are affected by the bias. Therefore n_n and n_{n0} are equal, and may be eliminated by combining Equations (II.28) and (II.29). The result is the desired value of n_T, the electron concentration at the edge of the transition region.

$$n_T = n_{p0} \exp \left\{ \frac{qV}{kT} \right\} \tag{II.30}$$

Again, n_{p0} is just the equilibrium electron concentration on the p side, equal to n_i^2/N_A. Equation (II.30) shows that the effect of applying a forward bias is to increase the electron concentration of the p side of the transition region in proportion to the exponential of the applied voltage.

Having found the values of the electron concentration at the contact and at the transition region edge, we can now compute the electron diffusion current in the p region. The electron diffusion current density is given by

$$J_e = qD_e \frac{dn}{dx} \tag{II.15}$$

We have only to compute dn/dx and Equation (II.15) may be evaluated. In the short-base diode no electrons recombine in the p region, so that the electron current is constant through this region. Furthermore a constant electron diffusion current requires that dn/dx be constant, from Equation (II.15). The electron concentration profile must therefore be as given in Figure II.15. The values are known at $x = 0$ and $x = W$ from Equations (II.27) and (II.30), and dn/dx must be constant in between. Clearly dn/dx is equal to $(n_W - n_T)/W$. The electrons flow to the right and the actual current flows to the left, that is,

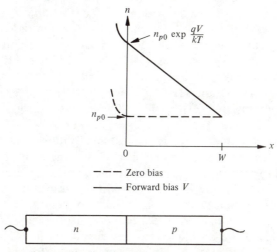

Figure II.15   The electron concentration profile in the p region in forward bias. The effect of the applied bias V is to increase n at the edge of the transition region by the factor $\exp\{qV/kT\}$.

in the negative x direction. The magnitude of the current density may now be evaluated from Equation (II.15):

$$J_e = qD_e \frac{dn}{dx} = qD_e \frac{(n_W - n_T)}{W} \qquad (II.31)$$

Substituting Equations (II.30) and (II.27) to eliminate n_W and n_T, we have

$$J_e = -\frac{qD_e n_i^2}{N_A W} \left(\exp \left\{ \frac{qV}{kT} \right\} - 1 \right) \qquad (II.32)$$

The minus sign indicates that the current flow is in the $-x$ direction, as defined in Figure II.15.

We have assumed from the outset that the hole current is small compared to the electron current because of the greater source of electrons in the n^+p junction. With little effort this assumption may now be verified. If an analysis for the hole current were made, an expression quite analogous to Equation (II.32) would be obtained, except D_e would be replaced by D_h, N_A by N_D, and W by the width of the n region. As assumed, the very large doping asymmetry ($N_D \gg N_A$) results in a much larger electron current than hole current.

Let us also reexamine our assumption that all the electron current flows through the p region by diffusion. In other words let us show that the electron diffusion current, Equation (II.32), is the total injected electron current. To prove this it is necessary to show that no sizable drift current of minority carriers occurs. Let us proceed by assuming the contrary, that is, that a significant flow of electrons in the p region can occur by drift. Then with a simple logical sequence we can show that this leads to a contradiction, thus establishing the original assertion. Assume, then, that in addition to the diffusion current density given by Equation (II.32) there is a large drift current density of electrons in the p region. Since drift current is proportional to concentration [Equations (II.11) and (II.12)] there must be an even larger hole drift current density consisting of the flow of holes toward the transition region; let us follow them into the n region. From the arguments of the previous paragraph, we know that there is only a small diffusion current of holes in the n region; therefore there must be a very large drift current density of holes in the n region. In this case there must be an even greater drift current density of electrons in the n region toward the transition region. This now gigantic electron current density, which must flow on through the p region, is evidently much larger than the electron current originally assumed, giving a contradiction. Thus we conclude that any significant flow of electrons in the p region occurs by diffusion. A similar argument holds for holes in the n region.

We have established that Equation (II.32) is the total current density in a forward-biased n^+p junction. To complete the analysis it only remains to compute the current under reverse bias.

Reverse Bias Just as in the case of forward bias, the current that flows in response to reverse bias may be evaluated by first determining the minority carrier concentrations and then calculating the diffusion currents. As in forward bias, it is assumed that the applied voltage appears across the transition region. The potential is plotted as a function of position in Figure II.16(a). The same arguments that led to Equation (II.32) may be used in reverse bias. However in this case V is negative and the electron concentration is reduced at the junction edge, $x = 0$. The increased electric field shown in Figure II.16(a) sweeps away any electrons that wander up to the junction edge, reducing the concentration. The electron concentration at the contact, $x = W$, is still the equilibrium value (n_{po}). Just as in forward bias, there is no recombination in the p region of the short-base diode, so the diffusion current, and hence the derivative of the electron concentration, are both constant through the p region. The electron concentration profile is given in Figure II.16(b). The electron diffusion current density may be evaluated from the slope of the concentration profile of Figure II.16(b), and again Equation (II.32) is obtained. Diffusion now takes electrons from the contact to the transition region (instead of the other way as in forward bias). In reverse bias the slope of the electron concentration is limited to the maximum value of (n_{po}/W), because the concentration has its equilibrium value of n_{po} at the contact, and is no less than zero at the junction, regardless of how large a reverse voltage is applied. For this reason the reverse current is limited to a small value.

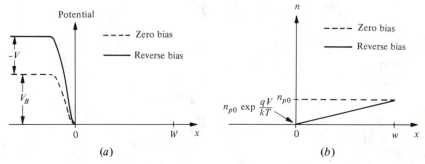

Figure II.16 The potential distribution and electron concentration profile in a reverse-biased *pn* junction. (a) The applied bias increases the voltage drop in the transition region. (b) The applied bias reduces the concentration at the edge of the transition region. For any negative voltage greater in magnitude than a few tenths of a volt, the factor $\exp\{qV/kT\}$ is essentially zero.

The hole current may be obtained in a similar fashion. However, because of the very limited supply of holes on the more heavily doped side of the n^+p junction, the hole current is negligible compared to the electron current. The total current density in forward or reverse bias is therefore given by Equation (II.32). To obtain the diode current, rather than the current density, we multiply Equation (II.32) by the diode area A. We have defined the current as positive into the *p* region (Figure II.13), whereas the current density given by Equation (II.32) flows out of the *p* region. Hence the total current flowing into the *p* region is given by

$$I = I_S \left(\exp \frac{qV}{kT} - 1 \right) \tag{II.33}$$

where

$$I_S = \frac{q D_e n_i^2 A}{W N_A} \tag{II.34}$$

To summarize the results of this section, the *I-V* characteristics of a typical silicon diode are calculated in an example. Further numerical examples are treated in the problems.

EXAMPLE II.4

Compute and sketch the room temperature *I-V* characteristics for an n^+p silicon junction with the following characteristics: $W = 10^{-3}$ cm, $N_A = 1 \times 10^{19}/$cm^3, $N_D = 2 \times 10^{16}/$cm^3, $A = 10^{-4}$ cm^2. Assume that $D_h = 8$ cm^2/sec, $D_e = 25$ cm^2/sec, $n_i = 1.5 \times 10^{10}/$cm^3.

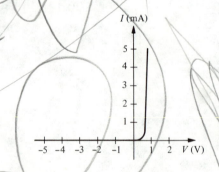

SOLUTION

From Equation (II.34)

$$I_S = \frac{q A n_i^2 D_e}{W N_D}$$

$$= \frac{1.6 \times 10^{-19} \times 10^{-4} \times (1.5)^2 \times 10^{20} \times 25}{10^{-3} \times 2 \times 10^{16}}$$

$$= 4.5 \times 10^{-15} \text{ A}$$

The *I-V* characteristics, from Equation (II.33), are given in the sketch. ∎

EXAMPLE II.5

Assume that a voltage of 0.65 V is applied to the junction of Example II.4. Find the current, and evaluate the voltage drops in the bulk *n* and *p* regions.

SOLUTION

For a forward bias of 0.65 V, the current equals $4.5 \times 10^{-15} \times \exp\{0.65 \, q/kT\}$ $= 4 \times 10^{-4}$ A. The resistance of the bulk regions is given by the resistor formula $R = \rho(L_R/A_R)$, where L_R is the length of the resistor and A_R is the area. We may evaluate the resistivity ρ from Equation (II.14). For the *n* region, $n \gg p$ and $\rho = 1/q \, \mu_e \, n = 3 \times 10^{-4}$ Ω cm. Thus the resistance of the *n* region is $3 \times 10^{-4} \times 10^{-4}/10^{-4} = 3 \times 10^{-4}$ Ω. For the *p* region, $p \gg n$, and $\rho = 1/q \, \mu_h \, p = 4 \times 10^{-1}$ Ω cm. The resistance of the *p* region is therefore $(4 \times 10^{-1} \times 10^{-3}/10^{-4}) = 4$ Ω. With a current of 4×10^{-4} A, the combined voltage drops across both bulk regions is therefore 1.6×10^{-3} V. ∎

Limitations of the Ideal Diode Equation The characteristics of a real diode differ from Equation (II.33) in a number of ways. When a very large forward bias is applied, so that large currents flow, the series resistance of the bulk regions leads to an additional voltage drop, much like that of an external series resistance. Furthermore, at very large currents, large numbers of minority carriers are injected into the bulk, leading to a modification of both the minority and majority carrier concentrations. These effects cause the current to increase rather more slowly with voltage than predicted by Equation (II.33), and thus result in a somewhat larger than predicted voltage drop.

At very small forward bias (for example, V < 0.4 V in silicon), a different effect leads to a modification of the *I-V* characteristics. Consider the n^+p junction discussed in this chapter, in which electrons are injected into the *p* side when forward bias is applied. If defects, called recombination centers, are present in the device in sufficient numbers, some of the would-be injected electrons recombine with holes even before they cross the transition region. This electron flow constitutes an extra current, called the space-charge region recombination current, which is in addition to the normal injection current. Hence at low forward bias the current is somewhat larger than predicted by Equation (II.33). However this is often of little consequence. The silicon *pn* junction, when forward biased, generally is biased in the range 0.6 to 0.8 V, in which case the space-charge recombination current is generally negligible in comparison with the ideal (diffusion) current.

In reverse bias the ideal diode equation predicts a very small current I_S, independent of reverse voltage. Generally in silicon, this so-called saturation of the reverse current is not observed. Rather a somewhat larger reverse current, which is slightly voltage dependent, is observed. This observed current is known to arise from electrons and holes that originate from (are born within) the space-charge region, and therefore is named the space-charge generation current. It is also of very little significance, simply because it is so small. Whereas the theoretical value of I_S may be 10^{-14} A, the current at 1 V reverse bias may typically be 10^{-11} to 10^{-9} A. In most applications such a small current is negligible, so it is of little consequence that it is larger than the theoretical current.

A comparison of an actual diode characteristic with the theoretical results of Equation (II.33) is given in Chapter 5, Figure 5.15. It is noted there that in the normal diode operating range, the characteristics agree well with the simple theory. Of course if the diode is to be used in the range of very low or very high current, deviations from the ideal behavior predicted by Equation (II.33) are to be expected.

Summary

- The product of electron and hole concentration is constant in semiconductors in thermal equilibrium.

- Current flow in semiconductors takes place by drift and diffusion. Drift currents flow in response to an electric field. Diffusion currents flow in the presence of a concentration gradient.

- Even with no external voltage applied, the carrier concentration gradients in a *pn* junction set up a built-in field across the junction.

- In a forward-biased *pn* junction, minority carriers cross the junction boundary. In an n^+p junction the current consists primarily of electrons injected into the *p* region. The current increases exponentially with increasing forward bias.

- In a reverse-biased junction, the current is limited to a small value, typically less than a nanoampere in silicon *pn* junctions.

References

Treatments at approximately the same level as this book:

Adler, R. B., A. C. Smith, and R. L. Longini. *Introduction to Semiconductor Physics.* New York: John Wiley & Sons, 1964.

Pederson, D. O., J. J. Studer, and J. R. Whinnery. *Introduction to Electronic Systems, Circuits and Devices.* New York: McGraw-Hill, 1966.

At a more advanced level:

Jonsher, A.K. *Principles of Semiconductor Device Operation.* New York: John Wiley & Sons, 1960.

Grove, A. S. *Physics and Technology of Semiconductor Devices.* New York: John Wiley & Sons, 1967.

Problems

II.1   A certain silicon crystal is doped with 0.1 part per million phosphorus. Find the equilibrium electron and hole concentration at room temperature.

II.2   Compute enough points to plot the electron and hole concentration in silicon at room temperature as a function of $N_D - N_A$ over the range $N_D - N_A = 10^8/cm^3$ to $10^{18}/cm^3$. (Use log-log coordinate paper.)

II.3   Compute and plot the velocity of holes in silicon as a function of the applied electric field, over the range $0 < E < 10^3$ V/cm. (It is interesting that for much higher fields, the simple mobility relationship is not obeyed. In fact for very large fields, the velocity saturates at approximately 10^7 cm/sec.)

II.4   A bar of silicon with dimensions $0.1 \times 0.1 \times 3$ cm is doped with 5×10^{17} phosphorus atoms/cm^3. (1) Compute n and p. (2) Find the resistivity. (3) Find the resistance between the ends of the bar. (4) A voltage is applied between the ends of the bar to produce a current of 1 mA. Find the average time required for an electron to drift the length of the bar.

II.5   Plot the resistivity of silicon as a function of $N_D - N_A$ over the range $N_D - N_A = 10^8/cm^3$ to $10^{18}/cm^3$. (The resistivity at doping levels in the range 10^{16} to $10^{18}/cm^3$ is somewhat lower than computed here, because the mobility is reduced.)

II.6   Calculate the diffusion current density of electrons in a region of silicon 1 μm wide in which the electron concentration at one boundary is $10^6/cm^3$ and at the other boundary is equal to $10^6 \times$ exp $\{qV/kT\}$. Evaluate for (1) $V = 0$, (2) $V = 0.2$ V, (3) $V = 0.6$ V, (4) $V = 0.8$ V. (These dimensions and numbers are representative of diffusion flow in the base of a transistor.)

II.7   An abrupt pn junction is formed in silicon between a p type region containing 10 parts per million boron, and an n-type region containing 50 parts per million phosphorus as well as 10 parts per million boron. Find the built-in voltage.

II.8   Copper has a resistivity of about 1.5×10^{-6} Ω cm, and a free carrier concentration of about 10^{23} electrons/cm^3. (1) Find the electron mobility. (2) Find the average drift velocity for a field of 1 V/m.

II.9   The p and n regions are both 5 μm wide in the junction of Problem II.7. The junction area is 10^{-3} cm^2. Compute the room temperature electron current at (1) 0 V, (2) 0.2 V, (3) 0.4 V, (4) 0.5 V, (5) 0.6 V, (6) 0.7 V.

II.10   Find the saturation current of an abrupt pn junction that is constructed from 1 Ω cm p-type silicon and 0.01 Ω cm n-type silicon, and has an n region width of 2 μm, and a p region width of 5 μm. Assume that the junction area is 10^{-3} cm^2.

II.11   Evaluate the voltage drops in the n and p regions of the junction of Problem II.9 at a forward bias of 0.60 V.

II.12   The expression for current flow in a long-base diode is similar to the short-base diode expression derived here, except that the base width W is replaced by the diffusion length of minority carriers. The diffusion length of minority carriers is the distance an average carrier wanders in its lifetime. Evaluate the room temperature saturation current in an n^+p long-base diode with the following specifications: the electron diffusion length (in the p region) equals 2×10^{-3} cm, the n side doping equals $10^{18}/cm^3$, the p side resistivity equals 1 Ω cm, the junction area equals 10^{-3} cm^2.

II.13   A graph of n_i in silicon as a function of temperature is given in Figure II.17. Find and plot the electron and hole concentration over the range of 25 to 200 °C in an *n*-type silicon sample that is doped with 5×10^{14} phosphorus atoms/cm³.

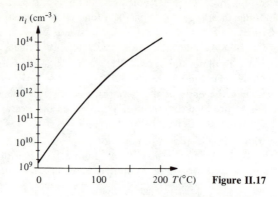

Figure II.17

II.14   Find and plot the conductivity of the sample of Problem II.13, over the same temperature range. Assume that the mobilities of the carriers vary as $(T/300)^{-5/2}$.

II.15   Using the data given in Figure II.17, plot the *I-V* characteristics of the junction of Problem II.9 at (1) 25 °C, (2) 100 °C, (3) 200 °C.

II.16   Just as an abrupt *pn* junction produces a built-in field and a built-in voltage, a gradual change of impurity concentration produces a built-in field. Equations (II.19) to (II.25) apply to any region of a semiconductor in thermal equilibrium, and offer a means of evaluating any built-in field if one exists. Suppose a doping profile of the form $N_D = 10^{19} \, e^{-X/L}$ (cm⁻³), where $L = 10^{-4}$ cm, exists in a certain region in a semiconductor. Further assume that the electron concentration equals N_D (which is approximately true). Show that there is a constant built-in field in this region, and evaluate the magnitude of the field. (This profile is representative of the base doping in *pnp* transistors.)

II.17   Equation (II.7) gives the electron concentration in a semiconductor with any doping. Derive an analogous equation for holes from Equations (II.5) and (II.6).

II.18   From Figure II.17 show that in the neighborhood of room temperature the intrinsic carrier concentration increases by a factor of about 2.5 for every 10 °C increase in temperature. How much does the current increase, at fixed forward bias of 0.65 V for every 10 °C increase in temperature? How much does the forward voltage at constant current decrease for every 10 °C increase in temperature (starting at 0.65 V)?

Index

ac, 14
Acceptors, 187
Active circuits, 100
Adder circuits, 386–390
Alpha, 234
Alternating current, 14
Ammeter, ideal, 30
Amplification, power, 514
  voltage, 111, 440
Amplifier, 110
  common base, 466
  common emitter, 438
  differential, 473
  direct coupled, 488
  input stage, 473
  multistage, 446
  operational, 112, 498
Amplitude, 132
Amplitude modulation, 212
Analog signal, 2
AND gate, 116, 214
Angular frequency, 132
Area economics, 275, 316
Asynchronous circuits, 392

Balance, 471
Bandwidth, 162, 531
Base, transistor, 229
Base overdrive, 323
Basewidth modulation, 449
BCD code, 598
Beta, 235
Bias point, 70, 426
Bias stabilization, 442–444
Biasing circuits, 428
Binary arithmetic, 597
  addition, 597
  subtraction, 600
Binary counter, 414
Binary numbers, 3, 594
Binary signal, 114
Bipolar transistor, 227
Bit, 115
Bode plot, 85, 539
Bonding electron, 180
Boolean algebra, 361
Branch, 14
Break frequency, 85
Breakdown, junction, 195, 261
Bridge circuit, 50
Buffer, 513
Built-in field, 190
Built-in voltage, 613, 616

Capacitance, 23
  effects on frequency response, 452
  impedance of, 149
  parallel connection of, 25
  series connection of, 25
  working voltage of, 25
Capacitor, bypass, 447
  coupling, 438
  filter, 208
Carrier concentration, 602, 604
Carriers, 180
Channel, MOS transistor, 551
Charge, definition, 6
Chip, 268
Circuit elements, 13, 17, 31
Circuit laws, 15
Circuit models, 105
Circuit symbols, 12, 31
Circuits, 11
  analog, 3
  combinational, 360
  continuous-state, 3
  digital, 3
  discrete-state, 3
  nonlinear, 68
  sequential, 360, 391
Clamp circuit, 212
Clock pulse, 392
Codes, 598, 600
Collector, transistor, 229
Combinational circuits, 360
  analysis, 379
  synthesis, 361
Common-base circuit, 466
Common-collector circuit, 483
Common-emitter circuit, 438
  biasing, 428, 442
  maximum voltage gain, 444
Common-mode rejection ratio, 477
Comparator, 421
Complex numbers, 135, 139
Conduction, electron and hole, 609
Conductivity, 609
Corner frequency, 85
Counters, 403
  binary, 414
  decimal, 411
  ripple, 416
Crystal growth, 276
Crystals, 180
Current, definition, 7
Current density, 8
Current divider, 45

Current flow in semiconductors, 606
Current source, 29
  construction, 479
Current-voltage characteristic (see I-V characteristic)
Current-to-voltage converter, 520
Cutoff state, 320

dc, 14
Decibel, 85
Demodulator, 212
DeMorgan's theorem, 366
Dependent sources, 101
Desensitivity, 525
Differential amplifier, 473
  bias requirements of, 474
  common mode rejection in, 477
  input resistance of, 476
Differentiator, 525
Diffusion, of electrons and holes, 191, 609
  of impurities, 282
Diffusion current, 609
Digital signal, 114
Digital systems, 359
Diode logic circuits, 214–219
Diodes, 180
  characteristics, 192–195
  small-signal behavior, 205
  (see also pn junction diode)
Diode-transistor logic (DTL), 329, 342
  fan-in capability, 338
  fan-out capability, 337
  logic levels, 331
  NAND gate, 329
  transfer characteristics, 335, 343
Direct current, 14
Direct-coupled circuits, 488
Distortion, 82, 527
Donors, 186
Dopants, 185
Drain, MOS transistor, 549
Drift, in amplifiers, 531
Drift current, 606

Early effect, 257, 449
Ebers–Moll equations, 247, 249
Einstein relation, 611
Electric field, 10
Electron, 180

Electron concentration, 184, 604
Emitter, transistor, 229
Emitter follower, 483
Encapsulation, 290
Energy, 10
Epitaxy, 277
Equivalent circuit, 54
Excitation table, 394
EXCLUSIVE OR gate, 119, 386
Extrinsic semiconductor, 186, 604

Fan-in, 338, 570
Fan-out, 123, 338, 571
Feedback, 499
Field-effect transistor, 551
Filter circuit, 85–87, 162
First-order circuit, 78
Flip-flop, 350, 391
    D, 421
    excitation table, 394
    J-K, 393
    S-R, 393
    T, 422
    truth table, 392
Flip-flop programming, 394
Forced response, 83
Forward-biased junction, 188, 616
Fourier theorem, 167
Free electron, 180
Frequency, 133
Frequency response, 451, 531
Full adder circuit, 387, 390

Gain (*see* Amplification)
Gain-bandwidth product, 532
Gate, logic, 116
    AND, 116, 214
    EXCLUSIVE OR, 119, 386
    NAND, 118, 329, 567
    NOR, 118, 567
    OR, 116, 218
Gate, MOS transistor, 550
Generation, of carriers, 184, 603
Graphical circuit analysis, 68, 199,
    256
Ground, 15

Half adder circuit, 124, 386
Header, 268
Holes, 182
h-parameters, 127, 260
Hybrid model, 127, 450

IC (*see* Integrated circuit)
Ideal circuit elements, 17
Ideal diode equation, 192, 622
Ideal op-amp technique, 503
Impedance, 148
Impurities in semiconductors, 185
Inductance, 25
    impedance of, 149
    mutual, 26
    parallel connection of, 26
    series connection of, 26
Injection current, 190, 228, 617
Input impedance, 535

Input resistance, 110, 440
Instability in amplifiers, 537
Insulators, 180
Integrated circuits, bipolar, 268
    characteristics, 471
    components, bipolar transistors,
        295, 306
    capacitors, 308
    diodes, 302
    interconnections, 303
    isolation junctions, 293
    MOS transistors, 549
    packages, 291
    resistors, 299
    special devices, 305
    cost, 275, 316
    layout, 310, 587
    MOS, 547, 584
Integrator, 523
Interstage coupling, 488
Intrinsic semiconductor, 185, 604
Inverter, 119, 323–329, 563
Inverting amplifier, 517
Inverting input, 473, 502
Ionization, 6
I-V characteristic, current source,
        32
    diode, 193
    resistor, 32
    Schottky diode, 349
    transistor (bipolar), 254
    transistor (MOS), 557
    voltage source, 32
I-V relationship, 17

J-K flip-flop, 393
Junction breakdown voltage, 195
Junction isolation, 273, 293

Karnaugh maps, 369
    4-variable maps, 376
Kirchhoff's laws, 15, 151

Large-signal diode model, 202
Level shifter, 488
Lifetime, 184
Linearity, 470, 527
Linearization, 426
Load line, 70, 199, 256
Loading effects, 107–108, 336-
        338, 507
Logarithmic amplifier, 545
Logic circuits, 116
    analysis of, 379
    synthesis of, 381
    (*see also* Gate)
Logic expression, 360
    equivalence of, 364
    mapping, 369
Logic families, 317
    DTL, 329
    MOS, 567
    TTL, 345
Logic function, 360
Logic operations, 362
Long-base diode, 625

Loop current, 38
Loop current method, 38, 42, 152
Low-pass filter, 85

Majority carriers, 187
Maps, 369
Mask making, 290
Masking, 274
Maximum power transfer, 166
Memory, electronic, 575
Metal, 180
Metallization, 288
Miller effect, 461
Minimal expressions, 367, 371
Minimization, 369
Minority carriers, 187
    collection of, 190, 228
    injection of, 190, 228, 617
    storage of, 196
Mixer, 206
MKS units, 5
Mobility, 608
Model, for circuit, 105
    hybrid-π, 450
    simplified-π, 435–436
    for transistor, 431
MOS circuits, 563
    flip-flop, 572
    inverter, 563
    memory, 575
    NAND gate, 567
    NOR gate, 567
    shift register, 580
MOS transistor, 549
    characteristics, 554, 557
    as a circuit element, 559
    circuit symbol, 551
    gate protection, 562
    as a load device, 560
    n channel, 553
    p channel, 553
    structure, 558
Multistage amplifiers, 446
Mutual inductance, 26

NAND gate, 118, 329, 567
Natural response, 75
Negative logic, 114
Newton–Raphson algorithm, 71
Node, 14
Node method, 35, 42, 152
Noise, 531
Non-inverting amplifier, 515
Nonlinear circuit element, 68
NOR gate, 118, 567
Norton equivalent, 66, 156
n^+p junction, 611
npn transistor, 231, 249, 258–260
n-type semiconductor, 186
Number conversion, binary to
        decimal, 596
    decimal to binary, 597
Number systems, 594
Numerical circuit analysis, 70
Nyquist criterion, 538

Octal numbers, 600
Octave, 97
Offset, in amplifiers, 471
Ohm's law, 18–19
Open circuit, 13
Operating point, 70, 426
Operational amplifier, 112, 498
    circuits, 508
        frequency response of, 531
        input impedance of, 535
        output impedance of, 535
    ideal model, 503
    typical characteristics, 504–505
OR gate, 116, 218
Oscillation, 537
Oscillator, 164
Output impedance, 535
Output resistance, 111, 440
Oxidation, 278

Parallel connection, 18
    of capacitances, 25
    of impedances, 150
    of inductances, 26
    of resistances, 22
Passband, 452
Passive circuit, 99
Peak detector, 212
Perfect rectifier, 199
Phase, 84, 132
Phasor, 140, 143
Photolithography, 280
Planar structure, 274
pn junction diode, 187, 611
    breakdown, 195
    characteristics, 193
    forward bias, 188, 616
    reverse bias, 189, 621
    temperature effects, 193, 221
    thermal equilibrium, 611
    time constants, 196
pnp transistor, 231, 248
Positive logic, 114
Potential, electrical, 9
Potentiometer, 530
Power, 10, 146
Power transfer theorem, 166
Principal node, 14
Propagation delay, 385
p-type semiconductor, 187
Pulsed sequential circuits, 392
Push-pull amplifier, 486

Q, 161
Quartz, 276, 278

Random access memory, 575
Recombination, 184, 603
Rectifier, 200
Rectifier circuits, 207
Reference node, 15
Relay, 319

Resistance, 18
    impedance of, 149
    input, 110, 440, 476
    output, 111, 440, 483
    parallel connection of, 22
    power dissipation in, 20
    series connection of, 22
Resistivity, 609
Resolution, 292
Resonant circuit, 160
Response, 74, 167
    forced, 83
    natural, 75
Reverse-biased junction, 189
rf circuit, 159
Rise time, 534
RMS average value, 173

Saturation, in amplifiers, 470
Saturation current, 193, 245
Saturation region, 322
Schottky diode, 309, 348
Semiconductor, 180
Semiconductor diode (see pn junc-
    tion diode)
Semiconductor technology, 275
Sequential circuit, 360, 391
    analysis, 399
    synthesis, 403
Series connection, 18
    of capacitances, 25
    of impedances, 150
    of inductances, 26
    of resistances, 22
Shift register, 415
Short circuit, 13
Short-base diode, 618
Signal, 1
    analog, 2
    digital, 2, 114
Silicon diode characteristic, 193
Silicon dioxide, 276, 278
Silicon transistor characteristic,
    258–261
Silicon wafer, 276
Silicon-gate technology, 592
Sinusoidal functions, 132
Slew rate, 533
Small signals, 432
    models, 431
    symbols, 436
Source, current, 29
    dependent, 101
    MOS transistor, 549
    voltage, 28
S-R flip-flop, 393, 573
Stability, 537
State diagram, 395
Steady-state response, 75
Sum of products expression, 367
Summing amplifier, 518
Superposition, 90

Switch, 318
Switching algebra, 361
Symbols, circuit diagram, 12
Synchronous circuits, 392
System, 4

Thermal equilibrium, 185, 611
Thévenin equivalent, 59, 156
Time constant, 80, 453
Timing diagram, 324
Transconductance, 466
Transformer, 26
Transient response, 74
Transistor, biasing, 438, 442
    breakdown, 261
    circuit symbols, 232
    construction, 231, 295
    input characteristics, 252
    large-signal equations, 249
    models, 431–434, 448–450
    modes of operation, 254–256
    output characteristics, 252
    switch, 319
Transistor operation, active mode,
    233, 239
    common-emitter connection,
        252, 438–446
    cutoff, 255
    saturation, 255
    speed limitation, 262, 458
    temperature dependence, 265
Transistor types, bipolar, 227–228
    npn, 231, 249, 258–260
    pnp, 231, 248
    field effect, 551
    MOS, 549
Transistor-transistor logic (TTL),
    345
Transition table, 406
Transition-excitation table, 407
Truth table, 116, 364
Tunnel diode, 226
Two-level logic, 383

Units, MKS, 5

V-I characteristic (see I-V charac-
    teristic)
Voltage, definition, 9
Voltage amplification, 111
Voltage divider, 43
    generalized, 153
Voltage doubler circuit, 225
Voltage follower, 511
Voltage regulator, 543
Voltage source, 28
Voltmeter, 30

Worst-case design, 327

Zener diodes, 196